OBJECT-ORIENTATION, ABSTRACTION, AND DATA STRUCTURES USING

SCALA

SECOND EDITION

CHAPMAN & HALL/CRC
TEXTBOOKS IN COMPUTING

Series Editors

John Impagliazzo
Professor Emeritus, Hofstra University

Andrew McGettrick
Department of Computer
and Information Sciences
University of Strathclyde

Aims and Scope

This series covers traditional areas of computing, as well as related technical areas, such as software engineering, artificial intelligence, computer engineering, information systems, and information technology. The series will accommodate textbooks for undergraduate and graduate students, generally adhering to worldwide curriculum standards from professional societies. The editors wish to encourage new and imaginative ideas and proposals, and are keen to help and encourage new authors. The editors welcome proposals that: provide groundbreaking and imaginative perspectives on aspects of computing; present topics in a new and exciting context; open up opportunities for emerging areas, such as multi-media, security, and mobile systems; capture new developments and applications in emerging fields of computing; and address topics that provide support for computing, such as mathematics, statistics, life and physical sciences, and business.

Published Titles

Paul Anderson, Web 2.0 and Beyond: Principles and Technologies

Henrik Bærbak Christensen, Flexible, Reliable Software: Using Patterns and Agile Development

John S. Conery, Explorations in Computing: An Introduction to Computer Science

John S. Conery, Explorations in Computing: An Introduction to Computer Science and Python Programming

Iztok Fajfar, Start Programming Using HTML, CSS, and JavaScript

Jessen Havill, Discovering Computer Science: Interdisciplinary Problems, Principles, and Python Programming

Ted Herman, A Functional Start to Computing with Python

Pascal Hitzler, Markus Krötzsch, and Sebastian Rudolph, Foundations of Semantic Web Technologies

Mark J. Johnson, A Concise Introduction to Data Structures using Java

Mark J. Johnson, A Concise Introduction to Programming in Python

Lisa C. Kaczmarczyk, Computers and Society: Computing for Good

Mark C. Lewis, Introduction to the Art of Programming Using Scala

Mark C. Lewis and Lisa L. Lacher, Introduction to Programming and Problem-Solving Using Scala, Second Edition

Published Titles Continued

Mark C. Lewis and Lisa L. Lacher, Object-Orientation, Abstraction, and Data Structures Using Scala, Second Edition

Efrem G. Mallach, Information Systems: What Every Business Student Needs to Know

Bill Manaris and Andrew R. Brown, Making Music with Computers: Creative Programming in Python

Uvais Qidwai and C.H. Chen, Digital Image Processing: An Algorithmic Approach with MATLAB®

David D. Riley and Kenny A. Hunt, Computational Thinking for the Modern Problem Solver

Henry M. Walker, The Tao of Computing, Second Edition

Aharon Yadin, Computer Systems Architecture

CHAPMAN & HALL/CRC
TEXTBOOKS IN COMPUTING

OBJECT-ORIENTATION, ABSTRACTION, AND DATA STRUCTURES USING
SCALA

SECOND EDITION

Mark C. Lewis
Lisa L. Lacher

CRC Press
Taylor & Francis Group
Boca Raton   London   New York

CRC Press is an imprint of the
Taylor & Francis Group, an **Informa** business

A CHAPMAN & HALL BOOK

CRC Press
Taylor & Francis Group
6000 Broken Sound Parkway NW, Suite 300
Boca Raton, FL 33487-2742

© 2017 by Taylor & Francis Group, LLC
CRC Press is an imprint of Taylor & Francis Group, an Informa business

No claim to original U.S. Government works

Printed on acid-free paper
Version Date: 20161109

International Standard Book Number-13: 978-1-4987-3216-1 (Paperback)

Visit the Taylor & Francis Web site at
http://www.taylorandfrancis.com

and the CRC Press Web site at
http://www.crcpress.com

Contents

List of Figures

List of Tables

Preface

Thank you for purchasing *Object-Orientation, Abstraction, and Data Structures Using Scala*. This book is intended to be used as a textbook for a second or third semester course in Computer Science. The contents of this book are an expanded second edition of the second half of *Introduction to the Art of Programming Using Scala*. The first half of that book became *Introduction to Programming and Problem Solving Using Scala*. This book assumes that the reader has previous programming experience, but not necessarily in Scala. The introductory chapter is intended to quickly bring the reader up to speed on the Scala syntax. If you already know Scala, you can skim this chapter for a refresher or skip it completely.

To the Student

The field of Computer Science has remarkable depth. It has to in order to match the capabilities and flexibility of the machines that we call "computers". Your first steps into the field should have shown you how to structure logic and break down problems, then write up your instructions for how to solve the problems in the formal syntax of some programming language. In this book, we want to build upon those capabilities. We want to give you the tools of object orientation to help you structure solutions to larger, more complex problems without experiencing mental overload in doing so. We also want to expand on your knowledge of abstraction so that you can make your code more powerful and flexible. Of course, all of this is implemented using the Scala programming language, which provides powerful constructs for expressing both object orientation and abstraction.

One of the key ways that we illustrate these concepts is through the creation of data structures. Data structures form one of the cornerstones of the field of computer science, and understanding them is essential for anyone who wants to be an effective software developer.[1] We strive to show you not only how these data structures can be written, but the strengths and weaknesses of each one, so that you can understand when you should consider choosing different ones.

This book also spends a fair bit of time looking at libraries that provide the functionality you need to do real programming. This includes things like GUIs, multithreading, and networking. The goal is to put you in a position where you can solve problems that you are interested in. These are also vitally important topics that grow in importance every day, so gaining a basic understanding of them early is something that we think is important.

There are a number of resources available to you that help support the material in this book. Much of this material is kept at the book's website, http://www. programmingusingscala.net/. We teach our own classes using a flipped format, and there

---

[1]One of the authors recently optimized a section of a professional program that was taking 90 seconds to run down to 0.3 seconds simply by using more appropriate data structures for key operations.

are video playlists for every chapter in this book available at `https://www.youtube.com/channel/UCEvjiWkK2BoIH819T-buioQ`. Even if you are not using this book as part of a course that uses a flipped format, the videos can do something significant that the book cannot. They can show you the dynamic aspect of programming. Much of the video content is live coding where we talk through our thought processes as we construct code. This is a valuable aspect of programming that prose on paper struggles to communicate. Where the videos are weak is in total depth of coverage. The book goes into much deeper detail on many topics than what is covered in the videos.

The playlists on the YouTube channel go beyond the contents of this book. They include the previous book and a number of more advanced topics. We also post more videos on a regular basis, so feel free to subscribe to see when additional material becomes available.

Lastly, we have put all of the code in this book in a GitHub repository (`https://github.com/MarkCLewis/OOAbstractDataStructScala`) that you can go through and pull down. The code is organized into separate packages for each chapter. Occasionally in the text we present changes to earlier code without showing the entire file that results. Finished files can always be found in the code repository. We have done the same for the code written in the videos. It is available at `https://github.com/MarkCLewis/OOAbstractDataStructScalaVideos`.

We hope you enjoy this step on your journey to learning computer science and developing your programming skills.

To the Instructor

Why Scala?

Probably the most distinguishing aspect of this book is the choice of Scala as the implementation language. As an instructor, you inevitably feel multiple influences when selecting the language to use for introductory courses, especially given that you do not really want to focus on language at all. You want to focus on concepts, but you know that those concepts have to be taught in some language so that students can actually get practice. So you really want a language that will let you teach the concepts that you feel are most important without getting in the way. Given this, the next obvious question is, why is Scala a good language for teaching the concepts covered in this book?

There are several pieces to this answer associated with the different groups of concepts that are covered in this book. First, Scala is an object-oriented language. Much of the discussion of the language focuses on the functional features of the language, and while those are wonderful in many ways and the expressivity they bring is used in this book on many occasions, the reality is that Scala is more object oriented than it is functional. It is more object oriented than Java, C++, Python, or most of the other languages one might consider using to teach this material, with the possible exception of Smalltalk and Ruby. All values in Scala are objects, which simplifies many details compared to languages where there are distinctions between primitives and objects.[2] There is no concept of "static", which is replaced by singleton object declarations which fit much better in the OO model. Even the functions are really just objects with an implementation of the *apply* method.

---

[2] A favorite example of this is converting between a commonly used type. In Java, why do you say i = (int)x, but i = `Integer.parseInt(s)`? Why can you not make an `ArrayList<int>` but you can make an int [] which has very different syntax? In Scala you can say `x.toInt` and `s.toInt` and you can use the same `Int` type with all of your collections which all have uniform syntax.

Scala provides many different tools for creating abstraction in code. You have the standard OO approaches through polymorphism with inheritance and type parameters. You also have the approaches found in functional languages created by the ease of passing functions around. Support for pass-by-name adds yet another tool for creating abstraction in code that is not available in many languages.

Some instructors will also be happy to know that these abstraction tools are all implemented in a statically typed manner. We consider knowledge of type systems to be an integral part of programming and CS in general. Static type systems have their own formal logic that can help to structure student thinking and there are certain topics that simply cannot be taught in dynamically typed languages.

These factors all contribute to the use of Scala for teaching data structures. The language works very well for teaching students how to construct type-safe collections of their own from the ground up.

Beyond the keywords in the title of the book, we like to introduce interesting and useful libraries, which are abundant in both Scala and the broader JVM environment. The ability to use the advanced multithreading libraries in Scala while also leaning on Java libraries for things like networking, is remarkably advantageous.

One of the key features of Scala is that it is not opinionated. It lets you do things your way. Some people lament that in professional environments. However, in the classroom, that means that you can teach what and how you want. You can use declarative constructs if you want to focus on higher level thinking or use imperative constructs when you want your students to have to deal with the details. A general idea behind Scala is that there is no silver bullet for most problems in computer science, so a language should provide you with options to do things in the way that you feel is best. We believe that this matters as much in the classroom as in other venues.

Lastly, there are the influences beyond pedagogy. There are considerations of how a language fits into the full departmental curriculum and how it impacts students outside of your curriculum. Scala works well beyond just the introductory courses. It is broadly used in web development, has a significant range of tools for parallelism, and thanks to the Spark library, it has become a dominant force in the area of data analytics. As a result, it is not hard to envision Scala being used in later courses in the curriculum. As for real-world usage, we would argue that Scala sits in the sweet spot for pedagogy. It is being broadly used outside of academia. At the time of this writing, it was number 14 on the latest RedMonk language ranking, sitting at roughly equal position on both the GitHub and StackOverflow axes of that ranking. This is high enough that you can point students, parents, and colleagues to real-world usage. However, it is far enough down that students are less likely to be able to find answers to all of your programming assignments with a simple Google search.

While there is no perfect language for anything, we do believe that for these reasons, and likely others that we have missed, Scala is a good option for introductory programming courses.

Structure and Contents

This book is pretty much what the title says it is. We strive to give students a firm foundation in the key principles of object orientation while building their ability to use abstraction in various forms, including multiple forms of polymorphism. A lot of this is communicated in the form of data structures, and we believe that this book can serve as a textbook for a basic course on data structures. We present basic order analysis and discuss the relative strengths and weaknesses of the various implementations of different Abstract Data Types along with showing implementations of those ADTs in Scala.

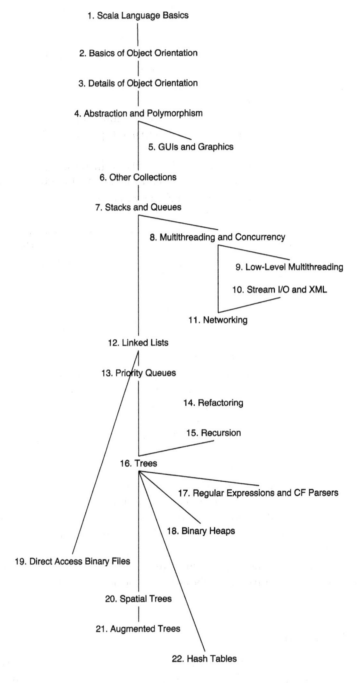

FIGURE 1: This figure shows the relationships between the chapters in this book to help you pick which ones you want to cover, and in what order.

There are also a few additional topics thrown in that we feel can go well in a CS2 course. These include GUIs, basic graphics, multithreading, and networking. We picked these because they allow students to write more "real" programs that they will hopefully find interesting. Also, in the case of multithreading and networking, these topics grow in importance every day, and having some exposure to them early is something we feel benefits students.

This book does assume that students have had at least one semester of previous programming instruction. If your students are coming to this book without having used Scala previously, the first chapter is intended to be a quick introduction to the key points of the language. We provide an appendix with a basic refresher on recursion for those students who did not cover that previously. We do not include basic sorting and searching. If your students have not previously seen that, you will want to cover it in some form at or before chapter 4.

There is a lot of material in this book. It is not expected that any course will cover all of it in one semester. If your students used *Introduction to Programming and Problem Solving Using Scala* in a previous semester, then chapters 1 and 5 can be easily skipped. Even with that, there is still more than most instructors would wish to cover in a semester. This is intentional, as we expect that you will focus on different material based on whether you are using this for a 2nd or 3rd semester course. Figure 1 shows a basic representation of the chapter dependencies in this book. A line between chapters implies that the later chapter uses material covered in the earlier one. As you can see, there are a number of chapters off to the side that can be skipped or covered in a different order without running into problems.

As mentioned in the section addressed to the student reader, there are a number of online resources associated with this book. The online videos were created by us for use in our own flipped classroom setting. If you use a blended approach in any way, we strongly encourage you to make use of these videos. Producing videos takes roughly as long as writing a book, and they allow students to see live coding, which we consider to be extremely valuable. There is also a GitHub repository with all of the code that we wrote for this book. We hope that you find these, and other resources on the course website, to be of value in your teaching.

Projects

One of the key issues with teaching object orientation is that the strengths of the OO approach really come out when programs get longer. In order to highlight this with students, you have to have them see and work on larger pieces of code. Beginning in chapter 5, we create a project that is utilized in most of the later chapters in the book to illustrate how to use the concepts that are discussed in an application that is a few thousand lines in length.

We also provide both exercises and projects at the end of each chapter. The exercises are intended to be simple, one-off tasks that students can use to test their knowledge of the contents of that chapter. The projects are continued across multiple chapters to give students a chance to build their own larger applications. Table 0.1 shows which chapters the projects appear in so that you can determine which ones are best suited to the material that you intend to cover from this book.

The table uses short names for the projects. Here are brief descriptions as well. Long descriptions can be found in the projects section of chapter 3.

- MUD - This is a multi-player text-based game. It works particularly well for highlighting the actor style of parallelism, but can be done without that.

- Spider - This is a web spider/crawler that follows links on web pages to collect information from a site.

Chapter	MUD	Spider	Game	Math	Image	Sim	Stocks	Movies	L-Systems
1. Scala Basics									
2. OO Basics									
3. OO Details	X	X	X	X	X	X	X	X	X
4. Polymorphism	X	X	X						
5. GUIs	X	X	X	X	X	X	X	X	X
6. Collections	X	X	X	X	X	X	X	X	X
7. Stacks & Queues	X	X	X	X	X	X	X	X	X
8. Multithreading	X		X	X	X	X	X	X	X
9. Low-Level Threads	X	X							
10. Streams	X	X	X	X	X	X	X	X	X
11. Networking	X	X	X	X	X	X	X	X	
12. Linked Lists	S	S	S	S	S	S	S	S	S
13. Priority Queues	X	X	X	X	X	X		X	
14. Refactor									
15. Recursion	X	X	X	X	X	X	X	X	X
16. Trees	X	X	X	X	X	X	X	X	X
17. Regex/Parsers	X	X	X	X	X	X	X	X	X
18. BinaryHeaps	S	S	S	S	S	S		S	
19. Binary Files	X	X	X	X	X	X			
20. Spatial Trees	X	X	X	X	X	X			X
21. Augmented Trees	X	X	X	X	X	X			X
22. Hash Tables	S	S	S	S	S	S	S	S	S

TABLE 0.1: This table shows which projects appear as options in each of the different chapters. The 'X' represents a unique project description. An 'S' indicates that there is a standard description for students to use their own implementation of a data structure in their project.

- Game - This option has students create a graphical game. There are technically two paths to this, one for a standalone game and one for a networked game.

- Math - This is short for "Mathematics Workbench". It has students write something like a baby Mathematica® worksheet program.

- Sim - This is short for "Simulation Workbench" in which students write a variety of types of simulations and integrate them into a single, larger project. The simulations include N-body physics simulations as well as discrete event simulations.

- Stocks - This is a networked, stock portfolio management system that involves pulling stock data from the web and helps users manage their stock portfolio.

- Movies - This project involves keeping an inventory of movies with the eventual goal of setting up a little business that rents/streams movies to users. Users can track rental requests, favorites, and movies already rented.

- L-Systems - In this project, students play with the formal grammars known as L-systems that can have interesting turtle graphic interpretations that produce fractals and structures that resemble biological forms.

Chapter 1

Scala Language Basics

This book assumes that you have previous programming experience, though perhaps not with Scala. This chapter is intended to get you up to speed on the basic syntax and semantics of the Scala language. If you have previous experience with Scala, this can be a refresher. If your previous experience is in some other language, this chapter will help introduce you to the key concepts that you need to use in later chapters.

Programming languages are generally classified by different PARADIGMs. A paradigm is an approach that a language takes to organizing code and solving problems. The four main paradigms of programming are IMPERATIVE, FUNCTIONAL, OBJECT-ORIENTED, and LOGIC. Scala is generally described as both functional and object-oriented, but it is not strictly functional, as it allows the imperative style as well. This combining of paradigms is becoming more common, and even older languages, like Java and Python, have gained features that also allow them to use aspects of all three of these paradigms. In Scala, one often prefers more functional approaches, while languages like Java and Python are naturally imperative with functional features.

Through the course of this book, and hopefully later in your education, the meanings of these terms and the features of the various paradigms will become clearer. The discussion in this book will generally explicitly say whether code is written in a way that is more functional or more imperative, and everything in Scala is fundamentally object-oriented.

1.1   First Application

In this book, we are writing our programs as applications in Scala. If you happened to use *Introduction to Programming and Problem Solving Using Scala* [9] previously, most of that book used the REPL and the scripting environment. Applications were only introduced at the very end of that book. Here they will be the standard, though simple expressions can be entered and tested using the REPL as well. The appendices, which are available on the book website[1] introduce some options for how you might want to work with Scala. Pick the tool that you like best. The rest of this book is generally tool agnostic and code is presented independent of the way in which you will compile and run it.

Applications in Scala are created by putting a `main` method in an `object` declaration. The following code shows the traditional "Hello World" program.

```scala
object HelloWorld {
  def main(args:Array[String]):Unit = {
    println("Hello world!")
  }
}
```

Depending on your background, different things here might look unfamiliar, so we will go through the elements one at a time and compare them to what you might have seen previously. For comparison, and to aid the following discussion, here are similar programs written in Java and Python.

```java
public class HelloWorld {
  public static void main(String[] args) {
    System.out.println("Hello world!")
  }
}
```

```python
def main():
    print("Hello world!")


main()
```

This Python code is more complex than is really required. The definition of a function is included to make it more similar to the others for comparison purposes.

The main code of the Scala application is inside of an `object` declaration. We will discuss the meaning of the `object` declaration more in chapter 3 where we will also contrast it to the more commonly seen `class` declaration. For now, we will just say that this creates what is called a SINGLETON OBJECT. An object is something that combines both data and

---

[1]http://www.programmingusingscala.net

functionality. This declaration begins with the keyword `object` followed by the name we want to give our singleton object, then there is a block of code enclosed in curly braces. It is customary to use the CAMEL CASE naming scheme in Scala, where each word begins with a capital letter. The names for `objects` also start with a capital letter.

All declarations in Scala begin with a keyword. The `object` keyword declares singleton objects. The only thing we have put inside the body of our `object` is a METHOD called `main`. Methods are declared with the `def` keyword. That is followed by the name of the method, then the PARAMETER LIST with types in parentheses, the result type after a colon, then an equal sign and the body of the method. The names of methods should also use camel case naming, but they should begin with a lowercase letter.

The body of the `main` method is identified by curly braces. Inside is a single line that calls the `println` method and passes it the `String` that we want to print.

If you are coming to Scala from Java, a lot of this will look familiar. You are used to seeing `class` instead of `object`, and while you would not use `def`, you would have `public static` on your `main` method. In Java, you put the types before the things they modify, like the method result and the type of the argument, `args`. In Scala, types come after things such as methods, variables, etc., when they are specified, and they are separated by a colon. If you see a colon in Scala, the thing to the right of it will be a type that is attached to the thing to the left of it. You also notice that the type is `Array[String]` in Scala instead of the `String[]` that is used in Java. Unlike Java, there is also no semicolon at the end of the line calling `println`. Like Java, Scala does use semicolons to separate statements, the difference is that Scala will infer semicolons at the end of lines where they make sense so that you do not have to type them. You can still put semicolons at the end of lines, or use them to separate multiple statements on a single line, but such style is typically frowned on in Scala.

If you are coming from a Python background, there are good odds that you have just been writing functions with `def` and not putting them inside of `classes`. In Python, there are no curly braces, and indentation is meaningful to the program. In Scala, it is the curly braces that are meaningful and indentation, which you really should continue to use, is for the purposes of human readability.

Unlike Java, Python is a scripting language, and so simple programs like this one are often just written as executable lines of code with no declarations at all.[2] There are good odds that your introductory course using Python showed you how to declare functions using `def`, but it likely did not show you `class` declarations, or other aspects of object-oriented programming. Also, while there is some meaning to the concept of "main" in Python, it is only by convention that programmers write a `main` function. The language does not require it. When writing a Scala application, there has to be a `main` method that takes a single argument of type `Array[String]` that is declared inside of an `object`.[3]

The biggest difference between Python and Scala is the fact that types are specified at all. Python is a dynamically typed language. The system it uses for checking types is commonly called "duck typing".[4] Languages that use duck typing check that values have the right operations at runtime. If they do not, an error is produced. Scala, on the other hand, is a statically typed language. This means that, in general, all types are checked

---

[2]Scala scripts are the same way. Writing "Hello World" as a Scala script requires just the single line `println("Hello world!")`. The additional lines are provided here because we are focusing on applications, not scripts.

[3]In chapter 4 we will see that you can write applications in a manner where you do not explicitly create the `main` method.

[4]This term is based on the idea that if it walks like a duck and quacks like a duck, then it must be a duck. So if a value supports the operations that are needed by a piece of code, then clearly it was of the right type.

for correctness by the compiler before the program is run. In this chapter, you will see that in many places, Scala will figure out types for you. However, there are some places, like the parameters to functions, where they have to be specified by the programmer. This might seem tedious, but it allows the compiler to catch a lot of errors earlier, which is very beneficial for large code bases or projects that are being worked on by many people.[5]

1.2   Comments

Comments are an essential part of any real-world programming project. Good code is written in a way so that other people can read it. There are many things that you can do to help make code readable, including using names that are meaningful. Still, there are times when you really need to explain what is going on in plain English or to provide additional information related to the code. This is where comments come in.

Scala style comments are just like Java style comments. You can make a single line comment with //. Anything that you put after the // on that line will be part of a comment. Multiline comments begin with a /* and end with a */.[6] These comments can include almost any text you want, including newlines. Just note that a */ ends the comment.

Scala also has an automatic documentation generating tool called ScalaDoc that builds web-based documentation from code. This tool looks for multiline comments that start with /**. The extra * does not matter to the compiler, as the /* starts a comment, but it tells the documentation generator that the comment holds information that is intended to go in the documentation. For an example of the type of page generated by ScalaDoc, see figure 1.1 later in this chapter, which shows the official Scala library documentation that is generated with ScalaDoc.

The following code shows what our first application might look like if we added some comments.

```scala
/** Our first application. */
object HelloWorld {
  /** The entry method that is run for this application.
   *
   * @param main the arguments from the command line
   */
  def main(args:Array[String]):Unit = {
    println("Hello world!") // This call to println sends text to standard output.
  }
}
```

ScalaDoc comments go directly before the declarations that they provide information for. You can see from the comment on main, that there are some special elements that begin with @ that provide additional information. In addition to @param, you will see the use of @return and @constructor later in the book. ScalaDoc is heavily modeled off of JavaDoc, so if you have experience with using JavaDoc comments, the same attributes will generally work with ScalaDoc as well.

The comment on println in this example would generally be considered superfluous.

---

[5]One could argue that the lack of compile time type and syntax checking is a big part of the reason why large projects written for Python 2 have been slow to migrate to Python 3.

[6]Note that multiline comments do not have to span multiple lines. They can begin and end on the same line.

It is included primarily to demonstrate single-line comments, but including comments like this, which only duplicate information that is clearly visible in the code, is a bad idea. Be careful about comments in your code. They need to be meaningful and provide additional information to the reader. Over-commenting clutters code and can lead to the code and the comments getting out of sync. Stale comments that say something that is no longer true about the code are worse than no comments at all.

Single-line comments in Python start with the hash character (#). A comment may appear at the start of a line or following a whitespace after a line of code. Multi-line comments start with triple double quotes (""") or triple single quotes (''') before the comment and then end with matching triple double quotes or triple single quotes at the end of the comment.

1.3   Variables and Types

In your previous programming experience, you inevitably used variables to give names to the values that were used in your programs. In Scala, there are two different keywords that are used for declaring variables: `val` and `var`. The following two lines show simple examples of these two types of declarations.

```
val name = "Pat Smith"
var age = 18
```

Here again, there are differences between Scala and other languages you might have learned. In Python, you do not declare variables, you simply assign values to names and start using them. This difference exists in part because Python is dynamically typed, so the names do not have types associated with them. A name can hold a numeric value on one line, then be assigned a string value on the next, and a list value on a third. Doing so is probably a bad idea that leads to code that is hard to work with, but Python does nothing to prevent you from doing it.

If your background is in Java, you are used to declaring variables, but you are used to putting types in the place where Scala puts `val` or `var`. You probably immediately noticed that this code does not specify types. That is because Scala does local type inference. That means that Scala will try to figure out the type of things based on information on the current line. As long as it can, you do not have to specify the type yourself. You can if you wish though. The previous two lines could have been written, using explicit types, in the following way.

```
val name:String = "Pat Smith"
var age:Int = 18
```

Consistent with what we saw for functions previously, the types are given after the variable names and separated from the names by a colon. Note that both types start with a capital letter. In Java, primitive types begin with lowercase letters and object types begin with capital letters. In Scala, there are no primitive types, and all types begin with capital letters.[7]

---

[7]In Scala, the types that were primitives in Java are generally called value types, and the types that came from `Object` are called reference types. The distinction between these is discussed more in chapter 4.

Another difference between Scala and Java is that variable declarations in Scala have to be given an initial value.[8] Inside of methods you cannot have declarations like these where you would give them values later.

```
val name:String
var age:Int
```

This makes the language safer, as uninitialized variables are common sources of errors in languages that allow uninitialized variables.

An obvious question at this point is, what is the difference between a `val` declaration and a `var` declaration. The answer is that you can change what value is referenced by a `var` declaration, but you cannot for a `val` declaration. In Java terms, the `val` declaration is `final`, and the `var` declaration is not. So you cannot do later assignments to a `val` declaration.

```
name = "Jane Doe" // This is an error: cannot assign to a val
age = 19          // This compiles just fine
```

A general rule of thumb in Scala is that you should use `val` declarations as much as possible, and only go to `var` declarations when needed. This is part of the functional paradigm. Odds are good that your previous exposure to programming has largely used the imperative paradigm. In this paradigm, re-assignment to variables is common. The values of variables are called the STATE of the program, and assignment changes state. The functional paradigm typically avoids changing or mutating the state. This actually makes it easier to think about the logic of the program when the value of a variable, once created, does not change.

One last thing to note deals with variable naming conventions. In Scala, variable names should start with lowercase letters, just like method names. If the name was made up with multiple words, it would use camel case with the first word in the variable name starting with a lower case letter and each subsequent word capitalized. The names were also picked to have meaning. Hopefully this is something that you learned previously: that you should choose meaningful variable names. Using names that are not meaningful makes programs much harder to work with, especially for programs that are used over a long period of time and have multiple programmers editing them.

You can get started in Scala with only a few different types. The following list gives the basics in alphabetical order.

- `Boolean` - This type is used for decision logic. It has only two values, `true` and `false`, that can be used as literals in your programs.

- `Char` - This type is used to represent single character values. You can make literal characters by putting the character you want in single quotes. For example, 'a' or '0'.

- `Double` - This type represents general numbers that might not be integers. They are stored using the IEEE double precision floating point standard. It is worth noting that these are not equivalent to the real numbers from mathematics, which have infinite precision. Numbers that include decimal points, like 3.14, or that use scientific notation, like 1e100, have this type.

- `Int` - This type is used to represent integer values. They are stored as 32-bit, signed binary numbers. If you enter a number that does not have a decimal point or scientific notation, it will have this type.

---

[8]In chapter 4, we will see that members of `classes` and `traits` can be declared without giving them values, but in Scala that means they are abstract, so the semantics are different from Java.

- `String` - This type is used for representing sequences of characters used for text. You can create a `String` literal in your program by putting text between double quotes, like `"Hello"`.

- Tuples - These are ordered groups of types. You can make them in your code by putting comma-separated values inside of parentheses. So `(1, 3.14, "Hello")` is a tuple with the type `(Int, Double, String)`.

- `Unit` - This is a type that represents a value that carries no information. There is only one instance of `Unit`, and it is written as `()`. You can get the values out of a tuple with methods called `_1`, `_2`, etc. or using patterns, which will be discussed later.

The fact that every value in Scala is an object means that you can call methods on all values, regardless of their type. To convert between types, there are methods like `toInt` or `toString` that implement meaningful conversions. So if you had a variable called `num` with the value `"42"`, you could use `num.toInt` to get the value as an integer. For the numeric types, you can also call `MinValue` and `MaxValue` to get the range of allowed values.

This is by no means a complete list of types. Such a list cannot be made, as there are effectively an infinite number of types. A lot of this book will focus on how we create our own types in Scala. This is just the minimum set needed to get you up and running in Scala. We will see more later on.

There are a few other details of the `Char` and `String` types that need to be considered here as well. The normal `Char` and `String` literals cannot include line feeds, and there are challenges dealing with some other characters as well. For that reason, there are escape characters that likely mirror those that you have seen previously involving a backslash and various characters. Scala also support RAW STRINGS, which are `String` literals that can span multiple lines and do not treat the backslash as an escape character. The raw strings in Scala begin and end with three double quotes. You can put anything inside of them except for a sequence of three double quotes.

`Strings` in Scala support concatenation using the plus sign. So you if have the variables `name` and `age` defined above, you can do something like this.

```scala
val message = name + " is " + age + "."
```

This will build a longer `String` and give it the name `message`. This syntax is rather long, and can get unwieldy if you need to append many variables into a formatted `String`. For that reason, Scala also supports STRING INTERPOLATION. To make an interpolated `String`, simply put the letter `s` in front of the double quotes. That allows you to insert variables with a dollar sign. The previous declaration could be written in the following way using an interpolated `String`.

```scala
val message = s"$name is $age."
```

If you need a more complex expression in the interpolation, you can put it in curly braces after the dollar sign. For example, if you wanted to do arithmetic on a value and interpolate the result you could do something like this.

```scala
s"$age times two is ${2*age}."
```

The curly braces are required after the second dollar sign.[9]

---

[9]There is another form of interpolation where you prefix the double quote with an `f` and then use C-style formatting in the `String`. We will not use that type of string interpolation in this book.

1.4   Statements and Expressions

As you are already aware, functions/methods are built out of STATEMENTS. A statement in a programming language is much like a sentence in natural language. It is something that stands on its own. You might have spent less time concerned with the EXPRESSIONS that typically make up statements. An expression is a piece of code that has a value and a type. It evaluates to something. This is different from a statement, which just does something. Variable declarations are statements that associate a name with something, but they do not provide any value as a result.

The functional aspects of Scala mean that most things that are not declarations are expressions. This is a significant difference from Java and Python that will be explored more in section 1.5. You can probably start writing expressions in Scala that work just fine based on what you know from your previous studies because so many things are common across programming languages. However, there are differences that are worth noting.

First off, things that look like operators in Scala are really just methods. For example, when you enter `4+5`, Scala actually sees `4.+(5)`. So `+` is the name of a method defined in the type `Int`. This works with methods that have alphabetic names as well as with those that have symbolic names, other than the fact that spaces are required when you use alphabetic names. As an example, the `Int` type also has a method called `min`. So it is valid to call `4.min(5)`, but you can also use the much easier to type `4 min 5` to get the same result.

Precedence rules in Scala are based on the first character in the operator/method, and it follows the same ordering that is used in other programming languages. Most operators are left associative. In math, this means that operators of the same precedence are done from left to right. So $2 - 3 + 4$ is $(2 - 3) + 4$, not $2 - (3 + 4)$. Operators/methods that end with a colon in Scala are right associative. Not only are they evaluated from the right to the left, the method is called on the object that is to the right of the operator. We will see an example of this in section 1.7.2.

If you come from the Java world, you will find that all of the operators that you are used to from that language work in the same way in Scala with one significant exception, the `==` operator. In Java, `==` is always an identity check. That is why you normally had to use the `equals` method when comparing `Strings`. In Scala, the `==` method calls `equals`, so it will behave the way you want for everything built into the standard libraries. If you really want to check for identity, use the `eq` method/operator. In Python, you use the `==` the same as you would in Scala; however, if you want to check for identity, you would use the "is" operator.

Scala also defines `*` on the `String` type that takes an `Int`. This allows you to do the following.

```
"hi"*5     // this is "hihihihihi"
```

This is especially handy when you need to repeat a string a variable number of times, often to produce some type of formatting. This is also something that Python users will find familiar.

However, there are a few more differences between Scala and Python. First off, in Scala there is no `**` operator provided for doing exponentiation. You can do exponentiation with the `math.pow(x:Double, y:Double):Double` method from the library. Similarly, there is no `//` operator to do floor division. If you divide two `Ints`, the result will be a truncated `Int`. If you divide two `Doubles` or a `Double` and an `Int`, the result will be a `Double` that is as close as possible to the result of that division. The following four expressions illustrate this.

```
9 / 5      // this is 1 in Scala and Python 2.x, but 1.8 in Python 3.x
9.0 / 5    // this is 1.8 in all
9 / 5.0    // this is 1.8 in all
9.0 / 5.0  // this is 1.8 in all
```

If you do an operation between two arguments of different types, the less general type is elevated to the more general type, and the operation is done between those values.

There are also differences between Scala and Python in the area of logical operators. While Python uses **and**, **or**, and **not**, as the logical operators for Boolean values, Scala uses the C-style operators of **&&**, **||**, and **!** respectively. The **&&** and **||** are short circuit operators, so if the value can be determined by the first argument, the second argument is not evaluated. If you happened to learn Python 2.x, you might also have seen **<>** used to represent not equal. Scala, like Python 3.x, will only accept **!=** for not equal.

1.4.1   Lambda Expressions/Closures

A LAMBDA EXPRESSION is a language construct that allows you to write a function as an expression. This is a functional programming construct that has found its way into pretty much every major programming language because it is so useful. Both Java and Python have lambda expressions, though it is possible that you have not encountered them because they are often considered to be more advanced constructs in those languages. In Scala, they are common place, and you really want to know how to create them to create more expressive code.

The main syntax for a lambda expression in Scala uses the **=>** symbol, which is often read as "rocket". To the left of the rocket you put the argument list, parentheses are needed if there is more than one argument or to specify a type, and to the right you put the expression for the body of the function. Here is an example where we make a function with a lambda expression, give it a name, and use it.

```
val square = (x:Double) => x*x
println(square(3))    // prints out 9.0
```

We declare **square** to be the name of the function created by our lambda expression. Note that you call the lambda expression just like you would a function, by putting the argument(s) in parentheses after the name. The syntax shown here specifies the type of the argument for the lambda expression, and Scala figures out the result type. As a general rule, we do not have to specify things that Scala can figure out, and most of the time, we will leave that off. Most uses of lambda expressions are in situations where the type of the lambda expression is completely known.

We can define the same function using a different approach where we put the type on the name and the lambda expression itself does not have any types.

```
val square:(Double)=>Double = x => x*x
```

Here we explicitly give the type of the lambda expression. These types include a rocket, just as the lambda expressions do. So **(Double)=>Double** means a function that takes a **Double**, represented by the type in parentheses to the left of the rocket, and produces a **Double**, shown by the type to the right of the rocket. The parentheses could include multiple types separated by commas if the function has multiple parameters. They can also be left off completely if there is only a single parameter. So in this case, the type could have been written as **Double=>Double**. Because the type has been specified early, the lambda expression does not require a type specification on the parameter. This is more similar to what you will do most of the time when you write lambda expressions.

There is also a shorter syntax that works for creating lambda expressions that works for simple expressions where any parameters appear only once each and in order. In these situations, you can use underscores as place holders for the parameters, and you do not use the rocket at all. The `square` function does not fit that rule because the parameter x is used two times. Instead, we can make a function called `twice` that simply doubles the value of the input. Here it is defined using the rocket syntax.

```
val twice:Double=>Double = x => x*2
```

Using the underscore syntax, it can be written in the following way.

```
val twice:Double=>Double = _*2
```

The rules for exactly when you can use the underscore syntax are somewhat complex. A practical approach is that if you are using the underscore syntax and you get odd error messages, you should switch to the rocket syntax.

One last thing to note about lambda expressions is that they should typically be short. The ones shown here have a single, simple expression as the body of the lambda. This is the ideal usage. You can use curly braces to make more complex expressions, but if your lambda expression is more than a few lines long, it probably should not be written as a lambda expression. You should likely pull it out and make a full method with an informative name instead. These can be passed into places that call for functions just as easily as lambda expressions can.

1.5   Control Structures

Every language includes a number of different elements that control the flow of execution through a program. Most of these constructs are very similar across most languages. For example, you inevitably worked with an `if` construct previously. This section will cover the various control structures that exist in Scala, then compare and contrast them with what you might have seen previously.

1.5.1   Conditionals

The most basic control structures are conditionals. Every language needs some way of saying "if". In most languages, that is exactly how you say it. Scala is not an exception. The syntax for an `if` looks like `if (`*conditional*`)` *trueCode* `else` *falseCode*. The *conditional* can be any expression with a `Boolean` type. The *trueCode* is what will be evaluated if the condition is `true` and the *falseCode* is what will be evaluated if the condition is false. The `else` and the *falseCode* are both optional, but you will probably include them in Scala more than you might in other languages. If *trueCode* or *falseCode* need to include multiple statements, curly braces are used to form a block of code.

That description probably sounds very much like what you have seen previously. Where the `if` in Scala differs from that seen in most non-functional languages is that it can be used as either a statement or as an expression. In both Java and Python, the `if` is only a statement. It is not an expression because it does not evaluate to a value. In Java and Python, when you need to make a conditional expression, you do so with the ternary operator, `?:`. Scala has no need for the ternary operator because `if` works just fine as an expression.

To make this clear, consider the simple example of determining the price of a ticket at a movie theater. Assume we have a variable called `age` and that the price of a ticket is $8 for kids under 12 and seniors over 65, and $12 for everyone else. We can write this in an imperative way using a `var` declaration and the `if` as a statement as follows.

```
var price = 12
if (age < 12 || age > 65) {
  price = 8
}
```

While this is perfectly valid Scala code, this is not how you would normally write this in Scala. As was mentioned earlier, we would prefer to use a `val` declaration when possible. In this situation, that is easy to do if you use the `if` as an expression instead of a statement. In that situation, the same code looks like the following.

```
val price = if (age < 12 || age > 65) 8 else 12
```

Here the `if` produces a value, and that value is given the name `price`. Doing it this way means that we cannot accidentally mess up the value of `price` later by doing some other assignment to it.

It is worth noting that when you use the `if` as an expression, you almost always want the blocks of code for the true and false conditions to have the same result type. It really does not make much sense for the true option to produce an `Int` and the false option to produce a `String`. Such a line of code might be valid Scala code that compiles, but the value that you get from it will have an unusual type that you cannot do much with.[10]

Scala allows you to nest any language construct inside of any other language construct. One implication of this is that you can put an `if` inside of another `if`. There is no special syntax for doing this. Normally, when you nest things, the nested code should be indented inside of curly braces. One common exception to this rule is when you have `if` statements or expressions nested in the `else` part of another `if`. In that situation, it is commonly accepted to format your code in the following way.

```
if (cond1) {
  ...
} else if (cond2) {
  ...
} else if (cond3) {
  ...
} else {
  ...
}
```

For those coming from Python, you should know that there is no `pass` statement in Scala. It would be rather unusual to include a condition that you do not want to do anything for in Scala, but if you do, you can leave the curly braces empty. Note that doing so only really makes sense when you are using the `if` as a statement and not an expression. If you use it as an expression, that means that you are expecting it to result in a value, and the empty code will produce the type `Unit`, which generally will not be helpful.

Scala includes a second conditional statement called `match`. This is a multiway conditional that is decided on a single value. For those with a Java background, it might sound like a `switch` statement, but it is a far more powerful construct than the `switch`. The general syntax for `match` is as follows.

---

[10]The details of the type that is produced will be discussed in chapter 4.

```
expr match {
case pattern1 => expr1
case pattern2 if cond => expr2
...
}
```

Each `case` has a PATTERN and an optional IF GUARD. An IF GUARD can be used to filter the results even further. Patterns play a role in a number of different constructs in Scala. They can be as simple as a value or a name, but they can be more complex. For example, you can make patterns that match tuples. The following example shows a `match` that could be used as part of the classic "Fizzbuzz" problem. In this problem, you are supposed to print all the numbers from 1 to 100 except that you substitute "fizz" for values that are divisible by 3, "buzz" for values that are divisible by 5, and "fizzbuzz" for values are that divisible by both. This code assumes that the value in question has the name i.

```
(i % 3, i % 5) match {
case (0, 0) => "fizzbuzz"
case (0, _) => "fizz"
case (_, 0) => "buzz"
case _ => i.toString
}
```

The way a `match` is evaluated is that we go from the first case down, and use the first one that matches. There are several things to note in this example. The underscore is used as a wildcard. It matches anything, without giving it a name. You could replace the underscores with variable names, and the values of that part of the expression would be bound to that name. You would do that if you wanted to use that value in the expression for that `case`. Order clearly matters as the last case matches everything.

Like the `if`, the `match` in Scala is an expression. It produces a value given by the evaluation of the expression for that `case`. This is being used in the Fizzbuzz example above. Note that the code does not print or change anything, it simply produces different values based on the value of i. As with the `if`, this is generally only meaningful if every `case` results in the same type.

The expression in each `case` can be arbitrary code, and it can span multiple lines. You do not need curly braces for this, as Scala knows that the current `case` ends when it sees the next occurrence of the `case` keyword. If you are used to the `switch` statement in Java, you know that it requires the user to put `break` statements at the end of `cases` to prevent the execution from falling through. That is not the case with `match` in Scala. Indeed, there is no `break` in Scala, and control only goes to one `case` on an execution of a `match`. It is also worth noting that your `cases` need to cover all possibilities. An error will occur if you execute a `match` and none of the `cases` actually match the expression.

1.5.2 Loops

Another class of language constructs are loops, which give you the ability to have code execute multiple times. The most general loop is the `while` loop, which exists in both Java and Python. The syntax in Scala is identical to that in Java. It looks like the following.

```
while (condition) body
```

The *body* of the loop is generally longer, so most of the time you will see it put in curly braces though they are not technically required if there is only a single statement. The while loop is a pre-check loop that checks if the condition is true before executing the body. Scala, like Java, but unlike Python, also includes a post-check do-while loop. The syntax of this loop is as follows.

do *body* while(*condition*)

The only difference between the two is that the *body* of the do-while loop will always happen at least once, because the condition is not checked until after it has happened. The *body* of the while loop might not happen at all. Both the while and do-while loops should only be used as statements. If you use them as an expression, the result is always of type Unit.

The more interesting loop in Scala, and most other languages, is the for loop. In its simplest usage, the for loop in Scala is like the for-each loop in Java or the for loop in Python. That is to say that it runs through the elements of a collection. A basic usage of the for loop in Scala might look like the following.

```
for (i <- 1 to 10) {
  println(i)
}
```

Looking at this code, it should be pretty clear that it will print the numbers 1 to 10 on separate lines. A more general view of this basic syntax is as follows.

for (*pattern* <- *collection*) *body*

The *pattern* can be as simple as a variable name, as seen in the example above, or something more complex, like tuples. If the *pattern* does not match an element of the collection, it is simply skipped over, and no error occurs. We will see more about the collections in section 1.7.2. The 1 to 10 from the example above produces a type called a Range and it is actually calling the method to on the Int 1, and passing it an argument of 10. There is also an until method that makes the upper bound exclusive. The arrow between the *pattern* and the *collection* is often read as "in". The combination of *pattern* <- *collection* is called a generator.

The for in Scala can do a lot more than just this simple iteration through a single collection. The example above uses the for loop as a statement that does things, but does not produce a value. By including the yield keyword before the body, we can turn the for into an expression.

for (*pattern* <- *collection*) yield *body*

This expression will produce a new collection with the values that the *body* evaluates to on each time through the loop. So the following will declare squares to be a collection with the values 1, 4, 9, ..., 100.

```
val squares = for (i <- 1 to 10) yield i*i
```

You can also specify multiple generators in a for loop that are separated by semicolons. You can also specify if guards and variable declarations. These are also separated by semicolons. Instead of using semicolons, you can replace the parentheses with curly braces, and put newlines between each of the different elements in the for. The following example illustrates multiple generators, a variable definition, and an if guard to create a variable called `evenProducts` that contains the even elements from the products of the numbers between 1 and 10.

```scala
val evenProducts = for (i <- 1 to 10; j <- 1 to 10; product = i*j;
    if product % 2 == 0) yield product
```

This formatting with multiple things on the same line is often challenging to read. You can put newlines after the semicolons, but Scala also allows `for` loops that use curly braces in place of the parentheses. When this syntax is used, semicolons are put on new lines, just like for normal Scala code. Using this, the previous `for` loop looks like the following.

```scala
val evenProducts = for {
  i <- 1 to 10
  j <- 1 to 10
  product = i*j
  if product % 2 == 0
} yield product
```

This style of formatting with the curly braces using multiple lines is considered proper Scala style when there are multiple generators, `if` guards, or variable declarations.

It was mentioned in section 1.5.1 that there is no `break` in Scala. This also applies to loops. It simply is not a keyword in the language. Similarly, the `continue` statement that you might have used in loops in Java or Python does not exist in Scala.

1.5.3   Error Handling

Exception handling is a common way of dealing with errors in many languages. Java has a `try/catch/finally` statement and Python has a `try/except/finally` statement for doing this. Scala also includes a `try/catch/finally`, but the syntax is a bit different from that of Java. The general syntax is like the following.

```scala
try
  expression
catch {
  case pattern => expression
  case pattern => expression
  ...
}
finally expression
```

The `catch` or the `finally` can be left off. The `cases` in the `catch` should have an appropriate *pattern* for the type of error they are set up to handle. The following example assumes we have a variable called `str` that holds a `String` that we would like to convert to an `Int` using `toInt`. If the value of `str` is not a valid integer value, this will throw a `NumberFormatException`. The one `case` uses a pattern that matches on the type so that if this happens, a value of 0 is produced. The `e` that you see in this code is just a variable name, that will contain a reference to an object of type `NumberformatException` if `str` is not a valid integer value.

```scala
val num = try {
  str.toInt
} catch {
  case e:NumberFormatException => 0
```

}

One thing that this should make clear is that the `try/catch` in Scala is an expression. For that reason, you generally want the result types of the `catch`es to match that of the expression in the `try`. Note that if you have a Python background, you might think of `str` as a library function. There is no standard function in Scala called `str`. It is just a variable name we are using here that indicates a variable with the type `String`.

As with Java and Python, the `finally`, if provided, holds code that should always happen, whether an exception is thrown or not.

1.6   Declarations and Scope

In section 1.1 we said that declarations in Scala all begin with keywords. The simple first example used the `object` and `def` keywords to declare a singleton object and a method inside of that `object`. In section 1.3 we saw how you can use `val` and `var` declarations for variables. There are a few other declarations that you can make in Scala, and some rules associated with them that are significant for you to understand.

The basic syntax for an `object` declaration looks like this.

```
object ObjectName {
  object code and declarations
}
```

The code that you put in an `object` will be executed when that `object` is first used elsewhere. Most of the time, you will not put statements directly in an `object`, instead you will put declarations, particularly method and variable declarations. The full details of how and why you do that is covered in chapter 3.

The general syntax for a method declaration is as follows.

```
def methodName(argumentList) : ResultType = expression
```

The colon and the *ResultType* are optional for non-recursive methods, but it is recommended that you include them for any method where the return type is not immediately obvious. This helps other programmers read your code, and also allows Scala to provide more meaningful error messages if you do something wrong. To illustrate this, consider a method that calculates the square. One could write it like this.

```
def square(x:Double) = x*x
```

That is actually a reasonable thing to do for this method because it is very short, and it is very clear that because `x` is a `Double`, the method will result in a `Double`. However, now consider that something is going wrong in the program, and you want to print out the value of `x` every time that the `square` is called. That means we have to insert a `println` statement into the method, which will also require the addition of curly braces. One might accidentally write the following code.

```
def square(x:Double) = {
  x*x
  println(s"Squaring $x") // Makes the result type Unit, which is not what we want.
```

```
}
```

This code still compiles, but it does not do what you want, and will likely produce errors elsewhere in your code. Why is that? It is because the result type of a block of code is the type of the last expression in it. In this case, `println` is the last statement, which produces a type of `Unit`. When we leave off the `ResultType`, we allow Scala to infer it, and it is possible for us to put something in the code that does not match what we want. If we had specified the result type, as in the following code, this would produce a syntax error on the square method, indicating to us that the `println` really needs to come before the `x*x` so that we are producing the desired value.

```
def square(x:Double):Double = {
  x*x
  println(s"Squaring $x") // This produces an error at this point because println
    produces Unit, not Double.
}
```

Both Java and Python have **return** statements. Indeed, those languages require that they be used any time a method/function is supposed to return a value. Scala also has a **return** statement, but its use is strongly discouraged. As you have seen, methods in Scala have an expression for the body, and blocks of code in Scala are just multi-statement expressions that result in whatever is produced by the last statement inside of them. There are going to be some times when it is really helpful to produce a result in the middle of a method and stop the method execution at that point. In those situations, using a **return** might be acceptable to simplify the logic, but it is not the style that you should go to regularly. Note that this is a very functional aspect of Scala. Functional languages are more about evaluating expressions than manipulating state, and the approach of just giving back the last expression fits with that style.

The other declarations that we have not discussed yet are **class**, **trait**, and **type**. The **class** declaration is covered in detail in chapter 3 and the **trait** declaration is covered in chapter 4. The basic syntax for a **class** is as follows.

```
class ClassName(argumentList) {
  class code and declarations
}
```

The *argumentList* and the parentheses around it are optional. Other elements of the **class** declaration area are introduced later as they are needed. The **trait** declaration syntax is very similar to that for **class**, except that **traits** cannot take arguments.[11]

The **type** declaration is used to give an alternate name to a type in Scala. For example, you might have a program where you use 3-tuples of Doubles to represent points. Instead of typing in (Double, Double, Double) in lots of places in your program, it would be superior to have an alternate name like Point3D. That could be accomplished with the following declaration.[12]

```
type Point3D = (Double, Double, Double)
```

There is more that can be done with **type** declarations, but those uses require the concepts of abstract types and declarations that will be covered in chapter 4.

---

[11]This is the case as of Scala 2.12. One of the changes being considered for the language is the ability to include arguments for **traits**.

[12]There are better ways to define a Point3D that are more type safe and provide greater functionality. However, this example shows a simple use of a **type** declaration.

Now that we have seen at least the syntax for the various declarations in Scala, one of the things you need to understand in relation to them is SCOPE. When something is declared, it can only be used through some parts of the program, outside of that, it does not exist. The general scoping rules in Scala are basically the same as they are in Java, but if you come from a Python background, things will be a fair bit different.

Scope is determined by the curly braces in Scala, which define blocks of code. Inside of code blocks in methods, declarations exist from the point of the declaration until the closing curly brace that it was nested in. So, things that you declare in methods and other executable blocks of code, can only be used after the point of declaration. In the bodies of `object`, `class`, and `trait` declarations, the scope is anywhere inside of the curly braces. That means that you can use the things that you declare before the point of declaration. The following simple application shows some of these scoping rules.

Listing 1.1: ScopeApp.scala

```scala
/** This application demonstrates scope in Scala programs. Comments are
  * included to show when certain declarations can not be used.
  */
object ScopeApp {
  /** Entry point for the application.
    *
    * @param args command line arguments
    */
  def main(args:Array[String]):Unit = {
    // Can't use factValue up here before it is declared.
    println("Printing some factorials.")
    val factValue = 5

    // Note that the methods are being called here above their declarations.
    println(imperativeFactorial(factValue))
    println(recursiveFactorial(factValue))
  }
  // Can't use factValue here because the scope has closed.

  /** Calculate factorial in an imperative manner.
    *
    * @param n the value to calculate the factorial of
    */
  def imperativeFactorial(n:Int):Int = {
    var prod = 1
    for (i <- 2 to n) {
      prod *= i
    }
    // Can't use i outside of the for loop.
    prod
  }
  // Can't use n or prod here because the scope has closed.

  /** Calculate factorial with a recursive method.
    *
    * @param n the value to calculate the factorial of
    */
  def recursiveFactorial(n:Int):Int = if (n < 2) 1 else n*recursiveFactorial(n-1)
}
```

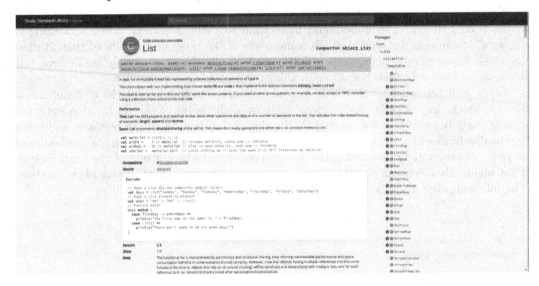

FIGURE 1.1: The Scala API looks like this when viewing scala.collection.immutable.List. The right column shows the various `object`s, `class`es, and `trait`s. Selecting one shows you the visible members inside of it. This is generated using ScalaDoc on the library code.

It is also significant to note that only `object`, `class`, and `trait` can be declared at the top level in Scala applications; these do not have to be inside something else. All other statements, including declarations, have to appear inside of one of those three.[13] This means that you cannot put `val` or `def` declarations outside of the `object` in the above example.

1.7   Essential Libraries

A lot of the power of a programming language comes from the libraries that have been written for it. This is a big part of the popularity of languages like Java and Python, and it is a strong point for Scala as well. The primary implementation of Scala, which is what we use in this book, compiles to and runs on the Java Virtual Machine (JVM). This gives you full access to all of the Java libraries. Scala also has its own set of significant libraries. In this section, we are going to introduce some of the key libraries that you need to know. More libraries will be introduced as needed later in the book.

We are only going to dedicate space to covering key elements of the libraries in this book. That means that you need to know how to find out more about what is in the standard libraries and the methods that you can call on the standard libraries. You do this using the API (short for Application Program Interface). You can find the standard API for Scala at `http://www.scala-lang.org`. Figure 1.1 shows a screenshot of the Scala 2.12 API with one of the types that we will discuss in section 1.7.2 selected.

The right column shows the top level `object`s, `class`es, and `trait`s that are part of the standard library. The little circles have the letters 'o', 'c', or 't' to indicate which one of those it is. You will notice that in many cases there is both an 'o' and a 'c' or a 't'. This

---

[13]Scala also has a scripting environment that does not have this restriction. In scripts, anything can go at the top level. This helps with writing short programs.

indicates that there is a companion object, a subject that will be discussed in chapter 3. Whatever is selected on the right will have its information displayed on the left. It begins with general documentation, then shows the list of members that you can use if you scroll down.

You will also notice on the right side that there names that do not have the circles next to them that are indented. These are for PACKAGES. These should be familiar to people coming from Java, as Scala uses the same style of packages as Java does. If you are coming from Python, the packages are similar to modules in that they organize code and help to prevent names from clashing. All the code in the standard libraries for Scala is under the `scala` package and its sub-packages.

As a general rule, you should not try to memorize APIs. You will remember the things that you use frequently and look up everything else. The API is extremely large, so what really matters most is that you learn how to access and use it. You will probably want to keep a tab open with the API in it anytime you sit down for a significant coding session.

1.7.1 Standard Input and `import` Statements

Our initial programs are all going to be text based, at least until we get to chapter 5 where we learn about Graphical User Interfaces. As such, being able to take input from the user is going to be a significant capability. The Scala libraries provide a number of methods with names like **readInt**, **readDouble**, and **readLine** that you can use to do simple reading of values from standard input. These are in the **StdIn** object in the **scala.io** package. The full name for this **object** then is **scala.io.StdIn**, and the full names of the methods are things like **scala.io.StdIn.readInt**.

For obvious reasons, we do not want to have to type in those full names every time that we read something from the user. Scala has **import** statements that allow us to use shorter names. All that an **import** statement does is it tells the Scala compiler that it should go looking in certain places for names that it does not find locally.

Scala has a lot more options for **import** than Java does. The basic syntax is to follow the keyword **import** with the name of the thing you want to bring into the current scope. Scala allows you to import any defined name, and the **import** statements follow standard scoping rules, so you can use **import**s that only bring in names for a small section of the code, but normally they appear at the top of files and their scope is the whole file.

If you just wanted to use the **readInt** method in your code, you might use the following.

```
import scala.io.StdIn.readInt
```

A lot of the time, you want to bring in all the names defined in some package or object. You can do this by putting an underscore as the last part of what is being imported.[14] So if you are going to use several of the **read** methods, you might use this.

```
import scala.io.StdIn._
```

By default, every Scala program effectively includes the following three **import** statements. You never need to manually include these imports within your application.

```
import java.lang._
import scala._
import Predef._    // This is short for scala.Predef._
```

---

[14]If you are familiar with Java, you will note that the underscore in Scala is playing the role that the * played in Java. Note that the Scala **import** goes further than the Java one because you can import more than just packages and static methods.

The first one is also imported by default in Java. The third one shows that imports are additive. Because the second line imports everything inside of the `scala` package, we do not need to specify that to import the `Predef object` in that package. It is customary to leave off the `scala` package from the beginning of `import` statements. The earlier example for using multiple `read` methods is often written in the following way.

```
import io.StdIn._
```

There are a few more features of the Scala `import` statement. Instead of an underscore, which pulls in everything, you can also specify names to pull in explicitly by putting them in curly braces. So if you only want `readInt` and `readLine`, you might do the following.

```
import io.StdIn.{readInt, readLine}
```

Scala also allows you to rename or hide certain names in an `import`. This is helpful when there are conflicts between names in different packages. For example, the `java.util` package includes the names `List`, `Map`, and `Set`, which conflict with similarly named items in the Scala collections. So using `import java.util._` would generally be a bad idea. Say that we wanted to use a number of Java collections, but we do not want to hide the Scala `List` type, and we are not going to use the Java `Map` or `Set` types at all. We might use the following `import` statement.

```
import java.util.{List => JavaList, Map => _, Set => _, _}
```

This renames `java.util.List` to `JavaList`, so that we can refer to it without hiding the `List` type in Scala, which will be discussed in the next section. Renaming to an underscore hides the name so it is not imported. This is what we are doing for `Map` and `Set`. The underscore at the end brings in all the other names from `java.util` without modification.

Now that you know about `import` statements, we can return to using those `read` methods to get information from standard input. The following little application shows the use of `readLine` and `readInt`.

Listing 1.2: StandardInput.scala

```
import io.StdIn._

/** Simple example showing standard input. */
object StandardInput {
  /** Main method to run application.
    *
    * @param args the command line arguments
    */
  def main(args:Array[String]):Unit = {
    println("What is your name?")
    val name = readLine()
    println("How old are you?")
    val age = readInt()
    println(s"$name is $age years old.")
  }
}
```

Note that the parentheses after the calls to `readLine` and `readInt` are not required, but because they cause side effects, it is recommended that you include them. In general, empty argument lists can be left off in Scala for declaring or calling methods. The compiler will require that calls match local declarations in this regard.

It is left as an exercise for the reader to go look in the API and see the full list of **read** methods in **StdIn**. One important thing to note about these methods is that they all read a full line of input. So, for example, you cannot use **readInt** and have the user put two or more numbers on a single line. If you want to allow that type of input, you will use **readLine** and parse the input yourself. The next section covers methods that make this type of thing easy.

1.7.2 Basic Collections

Probably the most significant library elements when it comes to language usability are those for collections. In Scala, all collections are part of the standard library. In Java, the array types, signified by [] on the type, are language-level constructs that are not part of the library. Python goes even further, supporting a variety of collections as built-in types such as **list**, **set**, and **dict**. There are pros and cons to having collections as part of the language instead of the libraries. Having them built into the language allows them to use special syntax, for example, Python uses brackets for **lists** and braces for **dicts**. Java, on the other hand, only has special syntax for arrays, and use of other collections, like **ArrayList** or **HashMap**, requires the more verbose object-oriented syntax. The downside of built-in collections is that languages that rely on them typically do not do a good job of handling collections that are not built in. The syntax of such collections is distinctly different from the built in collections, and typically much less elegant. That is part of the challenge for Java collections other than arrays, and applies to user-created collections in Python as well.

Scala has a very flexible and expressive syntax that allows code written in libraries to look and act much as if it were part of the language itself. This means that all the collections can be put in the libraries, and users can add their own collections that can function just like the standard ones. A reasonable portion of this book is dedicated to writing data structures that we will put in the form of Scala collections.

For now, we are going to focus on the two most basic collections in Scala, the **Array**, and the **List**. We will also look at the **Range** collection type. All three of these store ordered sequences of values of a specified type. We saw the use of **Array[String]** as the type of the parameter to **main** in our previous examples. The **Array** is a fixed-length, mutable sequence. The **List** is an immutable sequence that has efficient prepending; note this is in contrast to arrays which cannot be prepended or appended efficiently. The **Range** represents a fixed set of elements, as was seen in the simple examples of the **for** loop.

1.7.2.1 Array and List

Arrays and Lists can be created in a number of different ways. The simplest approach looks like the following.

```
val nums = Array(6, 8, 3, 4, 1, 7) // Makes an Array[Int]
val words = List("Scala", "is", "fun", "to", "code", "in") // Makes a List[String]
```

This treats the **Array** and **List** objects as functions and passes them arguments that are supposed to go into a new collection of that type. Scala infers the type of the contents from the arguments that are passed in.

Once you have the collection, you can access the members by index using parentheses.

```
println(nums(2)) // prints 3
println(words(0)) // prints Scala
```

This is different from Java and Python, where indexing arrays/lists is done with square brackets. As was mentioned earlier, the `Array` type is mutable, so you can alter elements by assigning values to them using this same kind of syntax. However, this will not work with a `List`, because they are immutable.

```scala
nums(2) = 99            // After this nums is Array(6, 8, 99, 4, 1, 7)
words(0) = "Won't work" // Error: cannot assign to a List. Message says no update.
```

Coming from an imperative background, the fact that you cannot mutate the `List` via assignment might seem a bit odd, and you might think that there is not much point to a collection that you cannot mutate. We'll come back to this shortly, as immutable collections are a cornerstone of functional programming, and are extremely helpful for some types of tasks.

1.7.2.2   Creating Collections Using `fill`

The syntax shown above for making `Array`s and `List`s is fine for short sequences, but it does not work very well for large collections. The methods that we will typically use for large `Array`s and `List`s are `fill` and `tabulate`. These are methods that exist on the `Array` and `List` objects. We'll start with a few example invocations of `fill`.

```scala
val zeros = Array.fill(20)(9)           // An Array with 20 9s in it.
val emptyStrings = Array.fill(10)("")   // An Array with 10 empty Strings.
val randoms = List.fill(100)(math.random) // A List with 100 random Doubles
    between 0 and 1
```

These calls show two significant elements of Scala syntax that we will introduce briefly here, and discuss in full in section 1.8. These features are CURRYING and PASS-BY-NAME. All three use currying, which is a common feature of functional programming languages. Currying is where arguments to a function are passed as separate argument lists instead of a single, longer argument list. To use currying in this way, there is not anything complex about it. You just need to know that the `fill` method takes a first argument list with one `Int` that specifies how many elements will go in the `Array` or `List`, then a second argument that gives the code used to get the values.

Note that the way we describe the second argument list is intentional. This is due to the second language feature that is used by `fill`. The second argument is passed-by-name. The normal semantics that are used in Java, Python, and by default in Scala are called PASS-BY-VALUE. With this semantic, arguments are evaluated before the function/method is called, and the value it evaluates to is passed in. Using pass-by-name, the code itself is passed in, and it is re-evaluated every time it is used. This does not matter for the first two uses of `fill`, but it is important for the third. Using pass-by-value semantics, that last call would have produced a `List` with 100 occurrences of the same random number. The fact that pass-by-name semantics are used means that we get 100 different random numbers. This concept can take a bit to get your head around, but it is a very powerful feature of Scala that we will see a number of times through this book. For that reason, we cover it in more detail in section 1.8.5.

1.7.2.3   Creating Collections Using `tabulate`

The `fill` method is useful for many situations where we want to create large `Array`s or `List`s, but it has a significant limitation; the code used to initialize the values does not have any information about the index that the value is going into. If you need that information, you should use the `tabulate` method. Consider a simple example where we want to make an `Array[Int]` that stores the values from 1 to 100. This can be done with `fill` if we write extra code using a `var` declaration, but it is done very nicely with the following use of `tabulate`.

```
val count = Array.tabulate(100)(i => i+1)
```

As you can see, calls to `tabulate` are very similar to those to `fill`. It is curried with two argument lists. The first is the length of the collection to create. The second is a function that takes a single `Int` that is the index it is making the value for. Here we use a lambda expression with the rocket syntax to specify that function. The indices of the collection are 0, 1, 2, ... 99. These are passed into the lambda expression as the values of i. This lambda expression produces a value one larger than the input, so we get back 1, 2, 3, ... 100. Note that as was mentioned in the section on lambda expressions, no types are specified. That is because the `tabulate` method itself tells Scala that it takes a function with a single parameter that is an `Int`. Because this is a very simple lambda expression, we would often use the underscore notation instead, as follows.

```
val count = Array.tabulate(100)(_+1)
```

The example of the Fizzbuzz `match` expression also works nicely with `tabulate`. The way it was written earlier, it was an expression that assumed we had an `Int` named i. We can now edit that expression slightly, and use it as the body of the function argument for `tabulate` to get a `List` with the different values for Fizzbuzz. That would look like the following.

```
val fizzBuzz = List.tabulate(100)(i => ((i+1) % 3, (i+1) % 5) match {
case (0, 0) => "fizzbuzz"
case (0, _) => "fizz"
case (_, 0) => "buzz"
case _ => (i+1).toString
})
```

If you do not like retyping the expression i+1 so often, you can introduce a new variable in the lambda expression, but this requires also adding another set of curly braces, because the body of the lambda expression is no longer just a single `match` expression, it also includes the variable declaration.

```
val fizzBuzz = List.tabulate(100)(index => {
  val i = index+1
  (i % 3, i % 5) match {
  case (0, 0) => "fizzbuzz"
  case (0, _) => "fizz"
  case (_, 0) => "buzz"
  case _ => i.toString
}})
```

1.7.2.4 Creating Arrays Using new

We can also make large `Arrays` in Scala using the `new` syntax, which should be familiar to those coming from Java. This syntax is not commonly used in Scala, except perhaps in libraries. As it happens, the data structures that we will write are library-style code, so we need to introduce this.

```
val emptyNums = new Array[Double](100) // Makes an Array with 100 values of 0.0
```

```
val emptyStrings = new Array[String](100) // Makes an Array with 100 values of
    null:String
```

The second line demonstrates why this approach is not used much in Scala. The uninitialized values for `String` and many other types in Scala, as in Java, is `null`. This is the name for a reference that does not refer to anything, and using them causes `NullPointerExceptions`. Anyone with a Java background will be very familiar with these types of errors as they are extremely common in Java. They are very uncommon in Scala, as the style of Scala rarely leads to `null` values in normal code. We will generally avoid anything that would produce a `null` value, but for some things it will be unavoidable, and we simply have to be careful.

1.7.2.5   Creating Lists Using Cons

There is another way to built `Lists` as well using the `::` operator, which is generally read as "cons". We already saw that the `List` type is immutable. So once you make a `List`, there is nothing you can do to change it. You cannot alter the values stored in it, nor can you add additional elements. However, it is efficient to make a new `List` by prepending an element with cons. Here is a simple example that uses cons to build a list.

```
1 :: 2 :: 3 :: Nil  // same as List(1,2,3)
```

The value `Nil` at the end is the name given to the empty `List`. It was briefly mentioned in section 1.4 that when you use the operator notation for symbolic methods that end with a colon, they are right associative instead of left associative. To highlight this, we can rewrite that expression with explicit parentheses to show how things are grouped, or using the dot notation to treat `::` as a method call.

```
1 :: (2 :: (3 :: Nil))
Nil.::(3).::(2).::(1)
```

The `::` method is a method of `List`, not a method of `Int`, so it only makes sense if things are grouped this way.

The power of the cons operator comes from the ability it gives us to efficiently build collections, especially when we do not know how many elements are supposed to go into the collection. Consider the following example where we made a `List`, then make another one that has one additional element.

```
val noOne = 2 :: 3 :: 4 :: Nil
val oneToFour = 1 :: noOne
```

This is efficient for the `List` type, but it would not be for an `Array`.

What about building a `List` from user input?[15] One way to do this is to use an imperative style with a `var` declaration for the `List`. The following method takes the number of elements to read in and uses a `for` loop to read in that many values.

```
def readList(n:Int):List[Int] = {
  var build = List[Int]()
  for (i <- 1 to n) {
    build ::= readInt() // equivalent to build = readInt() :: build
  }
  build.reverse
}
```

---

[15]We already know how to do this with `List.fill(n)(readInt())`, but it makes a good demonstration of using cons as well.

There are two things to note about the declaration of **build**. First, it is a **var**. This is essential because consing onto a **List** does not modify the original **List**, it creates a new one, and we have to do an assignment to that **var** to remember the new one. Second, instead of using **Nil**, we use **List[Int]()**. This is because the type of **Nil** is actually **List[Nothing]**, so the line that conses the new values on would not compile if we had used **Nil**. Also note that in the last line of the method, we have to call **reverse**. This is because the cons does a prepend, so **build** comes out in the opposite order of what the users input. It is possible to append to a **List**, but it is not efficient, and you should not use it.[16]

In practice, code that uses the **List** type is generally written in a functional rather than imperative way. Instead of using loops and **vars**, we typically use recursion. Here is another version that provides the same functionality using recursion instead.[17]

```scala
def readList(n:Int):List[Int] = {
  if(n < 1) Nil
  else readInt() :: readList(n-1)
}
```

The **readList** methods that we just showed both take the number of elements as an input. The fact that they know the size up front means that they could just as easily be written for **Arrays** or using **fill**. The power of **::** with a **List** is clearer when we do not have a fixed length. For example, consider the following method that reads values and adds them to the list until a negative number is encountered, at which point it stops reading.

```scala
def readPositive():List[Int] = {
  val i = readInt()
  if(i < 0) Nil
  else i :: readPositive()
}
```

1.7.2.6 Range

The section on the **for** loop also briefly introduced another collection type, the **Range**. In that section, we saw that expressions like **1 to 10** produce a **Range** and that this is actually calling the method **to** on the **Int**, 1, and passing it 10. The **to** method is inclusive on the upper end. Because indexes in programming languages, including Scala, often start at 0 instead of 1, it is often helpful to have a **Range** that is exclusive on the high end. Imagine we have an **Array** called **arr**, and we want to run through the valid indices on it. You could use the expression **0 to arr.length-1**, but you have to remember that -1 at the end or you will go out of bounds. For this reason, there is an **until** method that makes a **Range** that is exclusive on the high end, so it is more common to use **0 until arr.length**.[18]

If you want the values in the **Range** to count by something other than 1, you can use the **by** method. The expression **1 to 10 by 2** produces a **Range** with the values 1, 3, 5, 7, and 9. You can also use **by** to count backwards. There are no elements in **10 to 1**, but **10 to 1 by -1** does what you would expect.

All of these examples have used **Ints** for the bounds. You can use other numeric types. For example, **'a' to 'z'** does exactly what one would think it does. However, if you use

---

[16]If you are interested in seeing how you append to a **List** or add elements to either end of an **Array**, you can go look that up in the API.

[17]Technically, this version will not work as well if n gets bigger than about 10,000. For that we would need to write a slightly different version that is tail recursive.

[18]It is worth noting that sequences also have a method called **indices** that produces a **Range** with the valid indices for that collection.

`Doubles` for the bounds, you have to specify a step using `by`. It makes sense that integral types count by 1 by default. Such a default is less applicable for non-integral values. Instead, you can to do things like `0.0 until 10.0 by 0.1`.

1.7.2.7 Methods

We now know how to make `Arrays`, `Lists`, and `Ranges`, but we have not seen many of the methods that can be called on them. It is those methods that really make those collections powerful. Readers are strongly encouraged to go look in the API for a full list of the methods. We want to point out some of the key ones here, especially the higher-order methods. These are methods that take functions as arguments and/or produce functions as results. These methods are very powerful and useful, but there are good odds that you have not seen their equivalent in other languages.[19]

Some of the methods on these collections have names that make their functionality rather obvious. These include things like `contains`, `endsWith`, `indexOf`, `isEmpty`, `lastIndexOf`, `length`, `nonEmpty`, `size`, and `startsWith`.[20] We'll look at other significant methods, grouping them based on functionality. One key thing to notice is that because these methods all work on `Lists` and `Ranges`, as well as `Arrays`, they do not mutate their collections. Instead, the ones that return collections return new ones and leave the old ones unaltered. For each method, we give a short example. Many of the examples assume one of these two declarations:

```scala
val nums = Array(6, 8, 3, 4, 1, 7)
val words = List("Scala", "is", "fun", "to", "code", "in")
```

Also note that the type `Seq[A]` is used for a general sequence of the type that our collection contains and `Seq[B]` is a collection that might have a different element type.

- **Get/Remove Values**

    - `drop(n:Int)` - Gives back a collection with the first n elements removed.

      `nums.drop(3)`   `// Array(4, 1, 7)`

    - `dropRight(n:Int)` - Gives back a collection with the last n elements removed.

      `nums.dropRight(2)` `// Array(6, 8, 3, 4)`

    - `head` - Gives back the first element of the collection.

      `nums.head`   `// 6`

    - `last` - Gives back the last element of the collection.

      `nums.last`   `// 7`

    - `patch(from:Int, that:Seq[A], replaced:Int)` - Produces a new collection where **replaced** elements beginning at index **from** have been replaced by the contents of **that**. This can be used to effectively insert into or remove from a collection

---

[19]Both Python and Java, starting with Java 8, include higher-order methods/functions in their libraries, but they are not often taught in introductory courses.

[20]Both `length` and `size` return integer lengths of the collections.

```
nums.patch(3, Nil, 3)        // Array(6, 8, 3)
nums.patch(3, List(99, 98), 1) // Array(6, 8, 3, 99, 98, 1, 7)
```

- slice(from:Int, until:Int) - Produces a new collection that has the elements
  with indices in the range [*from, until*). Note that the higher end is exclusive.

```
nums.slice(2, 5)  // Array(3, 4, 1)
```

- splitAt(index:Int) - Produces a tuple where the first element is a collection
  of the elements before the provided index and the second element is the elements
  at and after that index.

```
words.splitAt(3)  // (List("Scala", "is", "fun"),List("to", "code",
    "in"))
```

- take(n:Int) - Gives back a new collection with the first n elements of the current
  collection.

```
nums.take(4)       // Array(6, 8, 3, 4)
```

- takeRight(n:Int) - Gives back a new collection with the last n elements of the
  current collection.

```
nums.takeRight(4) // Array(3, 4, 1, 7)
```

- **Set-like Methods** - These are actually multiset methods as they do not require
  uniqueness.

  - diff(that:Seq[A]) - Gives back a new collection that is the multiset difference
    of the current collection and **that**. For every element which appears in **that**, the
    first matching occurrence is removed from the current collection.

```
nums.diff(List(4, 5, 6, 7)) // Array(8, 3, 1)
```

  - distinct - Produces a new collection where each element from the current col-
    lection appears only once. The order is based on the order of the first occurrence
    of each element in the current collection.

```
List(1, 2, 3, 1, 6, 5, 2, 1).distinct // List(1, 2, 3, 6, 5)
```

  - intersect(that:Seq[A]) - Gives back a new collection that has the elements
    which occur in both the original collection and **that** in the order they appear in
    the original collection.

```
nums.intersect(List(4, 5, 6, 7)) // Array(6, 4, 7)
```

  - union(that:Seq[A]) - Produces a new collection with the elements of **that** ap-
    pended to the end of the current collection. Use this in conjunction with **distinct**
    if you do not want the duplicates.

```
nums.intersect(List(4, 5, 6, 7)) // Array(6, 8, 3, 4, 1, 7, 4, 5, 6, 7)
```

- **Various Other Methods** - These are useful, but do not fit into the other categories.

— `max` - Only works for collections of numeric values and produces the maximum element.

```
nums.max          // 8
```

— `min` - Only works for collections of numeric values and produces the minimum element.

```
nums.min          // 1
```

— `mkString` - Builds a `String` by concatenating the `String` representations of the contents together. There are three versions of this. The first version takes no arguments. The second takes a single `String` that is inserted between the elements. The third version takes three `String` arguments. The first is put at the beginning, the third is put at the end, and the second is put between.

```
nums.mkString                  // "683417"
nums.mkString(", ")            // "6, 8, 3, 4, 1, 7"
nums.mkString("(", ", ", ")")  // "(6, 8, 3, 4, 1, 7)"
```

— `product` - Only works for collections of numeric types and produces the product of the elements.

```
nums.product      // 4032
```

— `sorted` - Produces a new collection with the elements sorted in their natural order. This only works for types that have been written with a natural ordering. Note that in the natural ordering of `Strings`, capital letters come before lowercase letters.

```
nums.sorted       // Array(1, 3, 4, 6, 7, 8)
words.sorted      // List("Scala", "code", "fun", "in", "is", "to")
```

— `sum` - Only works for collections of numeric types and produces the sum of the elements.

```
nums.sum          // 29
```

— `zip(that:Seq[B])` - Produces a new collection of tuples where the first element of each tuple comes from the first collection, and the second element comes from that. The result has a length equal to the shorter of the current collection and that.

```
nums.zip(words)   // Array((6,"Scala"), (8,"is"), (3,"fun"), (4,"to"),
    (1,"code"), (7,"in"))
```

— `zipWithIndex` - Gives back a new collection of tuples where the first elements are the members of this collection and the second elements are the indices of the first elements.

```
words.zipWithIndex // List(("Scala",0), ("is",1), ("fun",2), ("to",3),
    ("code",4), ("in",5))
```

We saw earlier that you can efficiently prepend elements to a `List` with `::`. You can also efficiently pull off the first element of a `List`. The **head** method gives you back the first element and the **tail** method gives you back the rest of the `List` after the first element.[21] This allows you to process `Lists` one element at a time efficiently, and it is typically done with recursive methods. To demonstrate this, we can write methods that calculate length and the sum of `List[Int]`.

```scala
def listLength(lst:List[Int]):Int =
    if(lst.isEmpty) 0 else 1 + listLength(lst.tail)

def listSum(lst:List[Int]):Int =
    if(lst.isEmpty) 0 else lst.head + listSum(lst.tail)
```

These methods can also be written in a different way. Both **Arrays** and **Lists** can be used with patterns. In both cases you can do things like the following.

```scala
val Array(n1, n2, n3, _, _, _) = nums // Declares n1 = 6, n2 = 8, n3 = 3
val List(w1, _, _, _, w2, _) = words // Declares w1 = "Scala", w2 = "code"
```

This approach has limited use though, because the pattern only matches collections of the correct length. When working with `Lists`, you can use `::` in a pattern to separate the first element from the other elements.

```scala
val w1::rest = words  // Declares w1 = "Scala", rest = List("is", "fun", "to",
    "code", "in")
```

These patterns can also use `Nil` to match the empty list. This gives us another way to write `listLength` and `listSum` using **match**.

```scala
def listLength(lst:List[Int]):Int = lst match {
  case Nil => 0
  case h::t => 1 + listLength(t)
}

def listSum(lst:List[Int]):Int = lst match {
  case Nil => 0
  case h::t => h + listSum(t)
}
```

This use of patterns for getting the parts of `Lists` is actually the more common style in Scala as opposed to using the **head** and **tail** methods.

We have already seen that Scala has conversion methods like **toInt**, **toDouble**, and **toString** that can be used to convert basic types. For that reason, it probably should not come as a surprise that the collections have methods like **toList** and **toArray**. If you look in the API, you will see that there are quite a few other options for doing conversions between collections. This is sufficient for now though, and it allows you to change one to the other when needed. For example, you might need to build a collection when you do not know the length, but then mutate it later. You could use a `List` to efficiently build the original collection, then call **toArray** on that `List` to get an `Array` that you can mutate as needed.

1.7.2.8 Higher-Order Methods

These are all methods that take a function as an argument. That gives them a great deal of power that you do not get from the standard methods. Odds are that you have not

---

[21]You can call the **tail** method on **Arrays** as well, but it is very inefficient.

used these types of methods before, so we will discuss the three you will likely use the most, then give a quick list of the others.

The simplest higher-order method to understand is `foreach`. It takes a single argument of a function, and calls that function with each element of the collection. The `foreach` method results in `Unit`, so it is not calculating anything for us, the function that is passed in needs to have side effects to make it useful. Printing is a common use case for `foreach`.

```scala
nums.foreach(println) // Print all the values in nums, one per line.
words.foreach(println) // Print all the values in words, one per line.
```

In this usage, we have simply passed in `println` as a function. That works just fine, but if we wanted to print twice the values of each of the numbers, we would likely use an explicit lambda expression.

```scala
nums.foreach(n => println(2*n)) // Print twice the values in nums.
```

The fact that `foreach` does not produce a value limits its usefulness. The higher-order methods you will probably use the most do give back values. They are `map` and `filter`. The `map` method takes a function that has a single input of the type of our collection, and a result type of whatever you want. This function is applied to every element of the collection and you get back a new collection with all the values that function produces. Here are two ways to get back twice the values in `nums` using `map`.

```scala
nums.map(n => n*2)    // Array(12, 16, 6, 8, 2, 14)
nums.map(_*2)         // Array(12, 16, 6, 8, 2, 14)
```

This example used a function that took an `Int` for input and resulted in an `Int`. Those types do not have to match though. Here we use `map` to calculate the lengths of all the values in `words`.

```scala
words.map(_.length)   // List(5, 2, 3, 2, 4, 2)
```

The function passed here takes a `String` as input and results in an `Int`. Note that any function that is passed to `map` on `words` has to take a `String` as the input type. However, the result type can be whatever you want to calculate.

In section 1.7.1, we mentioned that the `read` methods only read a single value per line and that we would describe how to get around that later. The simplest approach, if the types on the line are uniform, is to use `map`. There is a method on `String` called `split` that breaks the `String` into an `Array[String]` based on a delimiter that you pass in. Consider the following usage.

```scala
"1 2 3 4 5".split(" ") // Array("1", "2", "3", "4", "5")
```

Here we start with a single long `String` that has multiple integers separated by spaces. Calling `split` with a space as the delimiter gives us an `Array[String]` that has those values split apart. They are still all of type `String` though, so we need to convert each element over to an `Int`. This is where `map` comes into play. Converting values is a special form of applying a function to them, and `map` is intended to apply a function to all the elements of a collection and collect the results. We can get the desired `Array[Int]` in the following way.

```scala
"1 2 3 4 5".split(" ").map(_.toInt) // Array(1, 2, 3, 4, 5)
```

If we were reading input from a user and we wanted them to enter multiple numbers on a single line separated by spaces, we might use code like this.

```scala
val values = readLine().split(" ").map(_.toDouble)
```

This would make **values** as an **Array[Double]** with whatever numbers the user typed in.

The third commonly used higher-order method that we will discuss is **filter**. As the name implies, this method lets some things go through and stops others. It is used to select elements from a collection that satisfy a certain PREDICATE. A predicate is a function that takes an argument and results in a **Boolean** that tells us if the value satisfies the predicate. Here are three different example uses of **filter**, all of which use the underscore shorthand for the lambda expression that is the predicate.

```scala
nums.filter(_ < 5)       // Array(3, 4, 1)
nums.filter(_ % 2 == 0)  // Array(6, 8, 4)
words.filter(_.length > 2) // List("Scala", "fun", "code")
```

You can use **map** and **filter**, in conjunction with some of the other methods listed above, to complete a lot of different tasks when processing collections. To illustrate this, imagine that you have a **List** of grades that are stored in tuples with a **String** for the type of grade and an **Int** with the grade value. We want to write a method that will find the average of the grades that have a particular type. We can write that in Scala with the following code.

```scala
def gradeAverage(grades:List[(String, Int)], gradeType:String):Double = {
  val matches = grades.filter(_._1 == gradeType)
  matches.map(_._2).sum.toDouble/matches.length
}
```

Remember, the contents of tuples can be accessed with methods like _1. These are not zero referenced, so _1 is the first element in the tuple. The lambda expression in the filter could be written using rocket notation as **t => t._1 == gradeType** if you find that to be easier to understand. A similar conversion could be done for the lambda expression passed to **map**.

Hopefully that helps you to understand the basics of higher-order methods and how they can be used. The following list shows some of the other higher-order methods that are available on the Scala sequences. We will continue to use **nums** and **words**, so here are their definitions to refresh your memory.

```scala
val nums = Array(6, 8, 3, 4, 1, 7)
val words = List("Scala", "is", "fun", "to", "CODE", "in")
```

- **Predicate Methods** - These methods all take a predicate as the main argument. The **filter** method would be part of this list.

  - **count(p: A => Boolean)** - Gives back the number of elements that satisfy the predicate.

    ```scala
    nums.count(_ < 5)            // 3
    ```

  - **dropWhile(p: A => Boolean)** - Produces a new collection where all members at the front the satisfy the predicate have been removed.

    ```scala
    nums.dropWhile(_ % 2 == 0)   // Array(3, 4, 1, 7)
    ```

  - **exists(p: A => Boolean)** - Tells you where or not there is an element in the sequence that satisfies the predicate.

```
nums.exists(_ > 8)          // false
words.exists(_ == "code")   // true
```

- filterNot(p: A => Boolean) - Does the opposite of filter. It gives back a new collection with the elements that do not satisfy the predicate.

```
nums.filterNot(_ < 5)         // Array(6, 8, 7)
nums.filterNot(_ % 2 == 0)    // Array(3, 1, 7)
words.filterNot(_.length > 2) // List("is", "to", "in")
```

- find(p: A => Boolean) - Gives back an Option[A] with the first value that satisfied the predicate. See section 1.7.3 for details of the Option type.

```
nums.find(_ % 2 == 1)     // Some(3)
words.find(_.length < 2)  // None
```

- forall(p: A => Boolean) - The counterpart to exists. This tells you if the predicate is true for all values in the sequence.

```
nums.forall(_ < 9)         // true
words.forall(_ == "code")  // false
```

- indexWhere(p: A => Boolean) - Gives back the index of the first element that satisfies the predicate.

```
nums.indexWhere(_ < 5)          // 2
words.indexWhere(_.contains('o')) // 3
```

- lastIndexWhere(p: A => Boolean) - Gives back the index of the last element that satisfies the predicate.

```
nums.lastIndexWhere(_ < 5)          // 4
words.lastIndexWhere(_.contains('o')) // 4
```

- partition(p: A => Boolean) - Produces a tuple with two elements. The first is a sequence of the elements that satisfy the predicate and the second is a sequence of those that do not. So s.partition(p) is a shorter and more efficient way of saying (s.filter(p), s.filterNot(p)).

```
nums.partition(_ < 5)        // (Array(3, 4, 1), Array(6, 8, 7))
words.partition(_.length > 2) // (List("Scala", "fun", "code"),
    List("is", "to", "in"))
```

- prefixLength(p: A => Boolean) - Tells how many elements at the beginning of the sequence satisfy the predicate.

```
nums.prefixLength(_ > 5)          // 2
words.prefixLength(!_.contains('o')) // 3
```

- takeWhile(p: A => Boolean) - Produces a new sequence with the elements from the beginning of this collection that satisfy the predicate.

```
nums.takeWhile(_ % 2 == 0)   // Array(6, 8)
```

- **Folds** - These methods process through the elements of a collection moving in a particular direction. The ones that go from left to right are listed here, but there are others that go from right to left. These methods are a bit more complex and can take a while to understand. Once you understand them, you can replace most code that would use a loop and a **var** with one of them.

  - `foldLeft(z: B)(op: (B, A) B)` - This curried method produces an element of type B, which is the type of the first argument, the first argument to the operation, and the output of the operation. It works by going from left to right through the collection, and for each element it passes the accumulated value and the next collection element into the **op** function. The result is used as the accumulated value for the next call. **z** is the first value for the accumulator, and the last one is the result of the function.

    ```
    nums.foldLeft(0)(_ + _)       // 29
    nums.foldLeft(1)(_ * _)       // 4032
    words.foldLeft(0)(_ + _.length) // 18, which is the sum of all the
        lengths of the strings
    ```

  - `reduceLeft(op: (A, A) A)` - This method works like `foldLeft`, but it is a little more restricted. There is no initial accumulator, as the first call to **op** is passed the first and second elements. The output of that is then passed in with the third element and so on. The value after the last element is the result of the reduce.[22]

    ```
    nums.reduce(_ + _)       // 29
    nums.reduce(_ * _)       // 4032
    ```

  - `scanLeft` - Does what a `foldLeft` does, but returns a sequence with all the intermediate results, not just the last one.

    ```
    nums.scanLeft(0)(_ + _)       // Array(0, 6, 14, 17, 21, 22, 29)
    nums.scanLeft(1)(_ * _)       // Array(1, 6, 48, 144, 576, 576, 4032)
    words.scanLeft(0)(_ + _.length) // List(0, 5, 7, 10, 12, 16, 18)
    ```

- **Other Methods**

  - `flatMap(f: A => Seq[B])` - This works in the same way as **map**, but is useful when the function produces sequences and you do not want a `List[List[B]]` or `Array[Array[A]]`. It flattens the result to give you just a basic sequence. The example here shows a comparison with **map**.

    ```
    Array(2, 1, 3).map(1 to _)    // Array(Range(1, 2), Range(1), Range(1, 2,
        3))
    Array(2, 1, 3).flatMap(1 to _) // Array(1, 2, 1, 1, 2, 3)
    ```

  - `maxBy(f: A => B)` - Gives back the maximum element based on comparison of the B values produced by passing the elements of this collection into **f**.

    ```
    words.maxBy(s => s(0))       // "to", this is looking at the first letter
        of each word in the sequence and finding the max
    ```

---

[22]Technically the accumulator can be some other type B that is a supertype of A, but we have not yet covered what that means.

- `minBy(f: A => B)` - Gives back the minimum element based on comparison of the B values produced by passing the elements of this collection into `f`.

  ```
  words.minBy(s => s(0))      // "Scala", this is looking at the first
       letter of each word in the sequence and finding the min
  ```

- `sortBy(f: A => B)` - Produces a new collection with the elements sorted based on the B values produces be passing the elements of this collection to `f`. The following example sorts the **Strings** based on the second character.

  ```
  words.sortBy(s => s(1))      // List("Scala", "in", "to", "code", "is",
       "fun")
  ```

- `sortWith(lt: (A, A) => Boolean` - Produces a new collection with the elements sorted based on the comparison function that is passed in. The `lt` function, short for "less than", should return true if the first argument should come before the second argument. The following example uses a greater-than comparison to sort in reverse order.

  ```
  words.sortWith((s1, s2) => s1 > s2) // List("to", "is", "in", "fun",
       "code", "Scala")
  ```

If you have some experience with Python, you might be wondering about list comprehensions. They do not exist in Scala, though the `for` loop is actually referred to as a `for` comprehension. Indeed, the `for` loops in Scala all get converted to calls to various higher-order methods, including `foreach`, `filter`, `map`, and `flatMap`. You can get the functionality that you might be used to from Python list comprehensions from these other constructs in Scala.

1.7.3   The `Option` Type

Another significant type in the Scala libraries is the `Option` type. This type should be used whenever you may or may not have a value. `Options` come in two different flavors[23] called **Some** and **None**. The **Some** type holds a single value. The **None** type is used to represent the absence of a value.

If you are accustomed to Java, you know that most libraries have methods that return a `null` when nothing is found. This is part of why `NullPointerExceptions` are so common in Java. Starting with Java 8, the `Optional` type was introduced. If you have used this, you have some idea of how Scala's `Option` works. If you have a Python background, you know that you can return **None** from functions when there is not a valid value to give back.

The last section mentioned that the `find` method has a result type of `Option`. That makes it a good place to start the discussion. Consider these two examples that were shown above.

```
nums.find(_ % 2 == 1)      // Some(3)
words.find(_.length < 2)   // None
```

The first one is looking for an odd number and finds that the first one is the value 3, so it gives back **Some(3)**.

This is simple enough, but it leads to the question of how does one do things with an instance of `Option`. In particular, how does one get the value when something is found so

---

[23]We will learn to call these subtypes in chapter 4.

that further operations can be performed with it? The `Some[Int]` is not an `Int` and you cannot do math with it; for that you need to pull out the value. There is a method called `get` on the `Option` type that provides the value if you have an instance of `Some`, but it throws an exception if you have a `None`, so you have to be careful what you call it on, and most of the time you simply will not call it.

The general approach for dealing with an `Option` is with a `match` using pattern matching for the `cases`. The following example shows how this can work.

```
nums.find(_ % 2 == 1) match {
case Some(n) => println("Found "+n)
case None => println("No odd numbers")
}
```

Note that `Some` as a pattern needs a single argument that will be bound to the value that is found if one is found. The `None` is a pattern that stands alone as there is no value. Also, this shows why it is uncommon to use the `get` method. Pattern matching provides an alternative approach for extracting the value in a `Some`.

You can use `match` in pretty much any situation because it works as both a statement and an expression. However, for a lot of uses, it is a rather verbose option. A lot of times when you want a value, you have a default that you want to use in the situation where you get a `None`. In this situation, you can use `getOrElse`. This method takes a single argument that is the default value that you want to use. So if you know all the values in `nums` are positive, and you want a -1 to indicate nothing was found, you might do the following.

```
val oddPositive = nums.find(_ % 2 == 1).getOrElse(-1)
```

The `Option` type also has most of the other methods that we just described for the collections such as `foreach`, `map`, `filter`, and `flatMap`. You can almost think of an `Option` as being like a little collection that holds one item or zero items, but not more.

1.7.4 Text Files

Another basic capability that you need to know early on is how to read from and write to text files. The proper way to read from a text file in Scala is using `scala.io.Source`. You typically construct an instance of this by calling `fromFile` on the `Source` object. Each instance of `Source` is an `Iterator[Char]`. Often you want to deal with whole lines instead of individual characters. To do that, you can call the `getLines` method, which gives you back an `Iterator[String]`.

The difference between an `Iterator` and either an `Array` or a `List` is that the contents of the `Iterator` are consumed when they are used, so you can only go through the contents of an `Iterator` once. This is appropriate when working with files, as it reduces the amount of memory that you have to use. In addition to supporting most of the methods that we saw for `Array` and `List`, the `Iterator` has the methods `hasNext` and `next`, which tell you if there is another element, and then gives you the next element. There are two main ways that you can go through the contents of an `Iterator`. One is using the higher-order methods like `map` and `filter`. Another is using a `while` loop that goes as long as `hasNext` returns true and calling `next` in the body of the loop to pull out data.

If every line of a file is the same, it is easiest to use something like `map`. Consider a file that stores numbers separated by commas.

```
val source = io.Source.fromFile("matrix.txt")
val matrix = source.getLines.map(line => line.split(",").map(_.toDouble)).toArray
source.close
```

The first line creates an instance of `scala.io.Source` by calling `fromFile` and gives it the variable name `source`. The second line of code has several commands chained together, so let's break them apart. `source.getLines` gives you back an `Iterator[String]`. Then a `map` method is called with a function that converts each line to an `Array[Double]`. This is done by calling `split` on each line (which gives back an `Array[String]`) and then mapping the `Strings` to `Doubles` using a call to `map`. Lastly, there is a call to `toArray` at the end of the line that creates `matrix`. This call is critical. First off, there is the fact that `Iterators` are consumed when you go through them, so without the call to `toArray`, you could only go through the values in `matrix` once, which is not ideal for many algorithms. In addition, when you close the `Source`, you can no longer read from the `Iterator`. Attempts to do so will throw an exception. As such, you typically call something like `toArray` or `toList` when you have the data in the form that you want so that it will all be loaded into memory to use.

Now imagine that the text file was grades for a class and the format had the student's name on one line, followed by three lines with their quiz, assignment, and test grades respectively, each separated by commas. Using `map` on the lines is problematic for this situation because not every line contains the same data.[24] Instead, we will use a `while` loop in conjunction with the `hasNext` and `next` methods.

```scala
val source = io.Source.fromFile("grades.txt")
val lines = source.getLines
var students = List[(String, Array[Int], Array[Int], Array[Int])]()
while(lines.hasNext) {
    val name = lines.next
    val quizzes = lines.next.split(",").map(_.toInt)
    val assignments = lines.next.split(",").map(_.toInt)
    val tests = lines.next.split(",").map(_.toInt)
    students ::= (name, quizzes, assignments, tests)
}
students = students.reverse
source.close
```

The variable `students` is declared to be an empty `List` of the four tuple (`String`, `Array[Int]`, `Array[Int]`, `Array[Int]`). Each time through the loop, another tuple is consed to the `List` after the values have been read from the file with four different calls to `next`. The line with `students = students.reverse` at the end is just there to put the items in the order that they appear in the file. Note that you could also write this using a recursive function without needing the `var`. If you make the first line of the file store how many students are in the file, you could do this with `List.fill` as well.

If you have familiarity with Java, note that you can also use `java.util.Scanner` just fine in Scala. However, it is not a Scala collection, so you will be lacking methods like `map` and `filter`, which can produce much more succinct and often easier to both read and code.

A general rule for the Scala libraries is that they include code that improves on what appears in the Java libraries or provides functionality that the Java libraries do not have. Scala does not include anything new or different if the Java equivalent is superior or sufficient. Writing to file happens to be an instance of the latter situation. Writing to a text file should be done with `java.io.PrintWriter`. You can make a new instance of `PrintWriter` with `new` and passing to it the name of the file you want to write to. Once you have an instance, you can call the `print` and `println` methods to write to it. As with all files,

---

[24]There is a method called **grouped** on collections that we did not introduce that could put the lines into groups of four, and you could use `map` on that grouping, but we would rather demonstrate the use of a `while` loop here.

you should remember to close it when you are done. So code writing to a file might look something like the following.

```
val pw = new java.io.PrintWriter("NewFile.txt")
...
pw.println(...)
...
pw.close()
```

1.8 Other Language Features

There are a few other features of the Scala language that we will introduce quickly because they are unlike what you are used to, and they are going to come up later on. Some were mentioned earlier in the chapter, but were not fully explained at that time.

1.8.1 Unfinished Code

When you are writing a large piece of software, you often approach things in a top-down manner. It is helpful to be able to test the parts of code you have written before you actually complete the full implementation. You cannot run your code to see if it works, though, until it actually compiles. In Java, a common way to deal with this is to put in **return** statements with simple default values like 0 or **null**. In Python you can use **pass** in places of unwritten code.

To help with this situation, the creators of Scala added the symbol ??? to the standard library that you should use in place of unimplemented code. So if you get to the point where you need to include a complex method to get your code to compile, but you do not want to spend the time writing it yet, you could include the following implementation.

```
def complexMethod(info:SomeType):ResultType = ???
```

The ??? has the type **Nothing**. As we will be discussing in chapter 4, this makes it safe to use as the value for any result type. If you actually call this method, the ??? will throw a **NotImplementedError**. The advantage of this over some default code that has the correct type is that it is easier to search for in the code and cannot be ignored when it is executed because it produces a run-time error.

1.8.2 Named Arguments

We give names to method parameters, and use those names in writing the methods. However, those names typically are not used in any way at the point of the call to the method. There are situations where it can be nice to be able to refer to those names. Consider a method where you pass in multiple values with the same type. Perhaps a method where we are passing in student grades for quizzes, assignments, and tests. Each one is a List[Int] as in the following code.

```
def gradeWork(quizzes:List[Int], assignments:List[Int], tests:List[Int]):Unit = ???
```

As a user of this method, it would be easy to forget the order that those arguments should be given. Also readers of the code might have a hard time telling what a line of code like the following means.

```
gradeWork(List(86,94), List(45, 99), List(100))
```

This can be fixed by using named arguments. Instead of just giving the argument values in order, you can give the name of the parameter for that argument, followed by an equal sign and the value. When you do this, the order of the arguments does not have to match the order of the parameters on the method. Consider the following call, which does the same thing as the one above.

```
gradeWork(tests = List(100), quizzes = List(86,94), assignments = List(45, 99))
```

Here the tests are passed in first instead of last. Anyone reading this code knows immediately what each of the three arguments means.

You can also mix regular and named arguments. The regular arguments have to come first. Every argument after the first named argument must also be named.

1.8.3   Default Parameter Values

Scala also allows you to provide default values for parameters. When calling a method that has default values, you do not have to give arguments for things with defaults. Consider the following definition that provides defaults of Nil for the **quizzes** and **tests**.

```
def gradeWork(quizzes:List[Int] = Nil, assignments:List[Int], tests:List[Int] =
  Nil):Unit = ???
```

This method can be called passing only the value of **assignments** using a named argument like the following.

```
gradeWork(assignments = List(45, 99))
```

For this call, both **tests** and **quizzes** will be empty lists with no grades.

Any parameter can have a default, but it typically makes sense for default parameters to be at the end of the list. This makes it easier to use non-named arguments. The following example illustrates that you cannot leave off the **quizzes** argument unless the call uses a named argument, because **assignments** is required, and **quizzes** has to appear before it.

```
gradeWork(Nil, List(45, 99))
```

1.8.4   Curried Functions/Methods

Section 1.7.2.1 introduced the **fill** and **tabulate** methods for making **Arrays** and **Lists**. Both of these methods used a format we referred to as currying. Recall that when currying, arguments are passed in separate argument lists instead of making a single longer argument list. Curried functions are a rich topic in the area of functional programming. The basic idea is that you can call such functions without providing all the argument lists, and the result is a function.

The syntax for creating curried methods in Scala is actually very simple. You simply put multiple argument lists after the name of the method. A simple, though not very useful example, is a curried method to add two numbers.

```
def add(x:Int)(y:Int):Int = x+y
```

This is how `fill` and `tabulate` are declared, and we will use this style for some examples later in this book as well. This method can be called in the way one might expect.

```
add(3)(4)      // 7
```

If we choose to call this method with only one argument, we have to be explicit and tell Scala that we are doing this on purpose, and not just forgetting the second argument. We do this by putting an underscore after the last argument list. Here is an example of doing so.

```
val add3 = add(3)_
nums.map(add3) // Array(9, 11, 6, 7, 4, 10)
```

The name `add3` is bound to the function that we get from calling `add(3)_`. The second line shows that this is a function when it uses it as the argument to `map` to add three to every element of `nums`.

1.8.5 Pass-by-Name

Section 1.7.2.1 also mentioned that the second argument to the `fill` method used pass-by-name semantics. The normal way that arguments are evaluated is called pass-by-value. Using pass-by-value semantics, complex expressions are evaluated first, then the value is passed into the function.[25] With pass-by-name, the code for an expression is bundled up into something called a THUNK that can be executed later. The thunk is executed every time that the name is used inside of the method.

The syntax for declaring an argument to be passed in by name is to put a rocket before the type without any argument list. For example, `=> Int` is a pass-by-name argument that produces an `Int` when evaluated.

To help make this clearer, it is useful to see an example. Consider a function that takes a single `Int` by value and makes a tuple with that value in it three times.

```
def makeInt3Tuple(a:Int):(Int, Int, Int) = (a, a, a)
```

If we call this passing in the value 5, we get the value that one would expect.

```
makeInt3Tuple(5)      // (5, 5, 5)
```

Now we want to define the same style of function, but with the value passed in by name. Note that all that is really changed here is the addition of `=>` in front of the `Int` on the type.

```
def makeInt3TupleByName(a: => Int):(Int, Int, Int) = (a, a, a)
```

Nothing interesting happens if we call this with 5 either.

```
makeInt3Tuple(5)      // (5, 5, 5)
```

In order to get something interesting to happen, we have to pass in an argument that has a side effect. The simplest way to do that is to pass a block of code that prints something then gives back a value.

---

[25]In Scala one should actually picture it as being references to the value objects that are passed into functions, but that distinction is not significant here.

```
makeInt3Tuple({println("eval"); 5})     // (5, 5, 5) and prints "eval" once
```

The syntax here is using the curly braces to make a compound expression with multiple statements. It is still just an expression though, so we can pass it as an argument.

When you call this using the pass-by-value semantics, the word "eval" will only pass once because the expression is evaluated only once, then the value is passed to the function. However, if we call this using the by-name version, the word "eval" will print three times, one for each time the argument, a, is invoked in the method.

```
makeInt3TupleByName({println("eval"); 5})     // (5, 5, 5) and prints "eval" three
    times
```

The results of all of those calls were still rather boring. A more interesting example would be to have an expression that evaluates to a different value each time it is executed. One way to get that is to use a method that generates random numbers, but that makes it challenging to see what is really going on. Instead, we can declare a var that is mutated in the expression and is part of the result of the expression. In this example, we increment the var by one and give it back.

```
var i = 0
makeInt3Tuple({i += 1; i})          // (1, 1, 1)
```

The result using the pass-by-value semantics should not be too surprising. The expression is evaluated once before the call, so i is incremented to 1, and that value is passed into the function. The call with pass-by-name semantics really illustrates what is different about that approach.

```
var i = 0
makeInt3TupleByName({i += 1; i})     // (1, 2, 3)
```

Here, every time that the argument is used in the method, the value of i is incremented. So the first value in the tuple is 1, the second is 2, and third is 3. Unlike all the other calls, this one creates a tuple with different values.

Passing blocks of code is not uncommon in Scala, especially when using pass-by-name. For that reason, the Scala syntax allows you to leave off the parentheses and only have curly braces for any argument list with a single argument.[26] So that last example could be rewritten in the following way.

```
var i = 0
makeInt3TupleByName {i += 1; i}     // (1, 2, 3)
```

We'll finish off this section with an example that shows the power of combining pass-by-name semantics and curried methods. Consider the following simple while loop.

```
var i = 0
while(i < 10) {
  println(i)
  i += 1
}
```

What we are going to do is write a method called myWhile that we can use in exactly the same way. To do this, we are going to think of the while loop as a method that takes two separate argument lists. The first is the condition, which evaluates to a Boolean, and the

---

[26]Leaving off the parentheses is the preferred way of implementing this in Scala.

second is the body, which evaluates to Unit. Inside the method, we need to check if the condition is true. If it is, we do the body and then start from the beginning again. In code, this looks like the following.

```scala
def myWhile(cond: => Boolean)(body: => Unit):Unit = {
  if(cond) {
    body
    myWhile(cond)(body)
  }
}
```

After you have this, you can replace the while in any loop you write with myWhile and it will work in the same way.

1.9 The Read, Evaluate, Print Loop (REPL)

The code segments shown in this chapter have been shown in the way that they might appear in your application code with the results of significant expressions in comments after those lines. Most of the code that you write for this book will be written as applications, so this is the style that we want to focus on. However, Scala provides another environment that is remarkably helpful for testing small pieces of code called the REPL, short for "Read, Evaluate, Print Loop". The name basically says what the REPL does. It reads a statement from the users, executes it, and if it has a value, prints that value. It then waits for the next statement from the user.

You can get a command line REPL by simply running the scala command without arguments. You can also get a REPL in various environments like Eclipse and IntelliJ. Under Eclipse you do this by right-clicking on the project and selecting Scala ¿ Create Scala interpreter from the context menu. This REPL will be in the scope of the project and have access to all the code that you put in the project.

Here are a few lines from the previous section being executed in a REPL. Notice that after each input, a response is provided to show you information about what was just entered.

```scala
scala> def makeInt3TupleByName(a: => Int):(Int, Int, Int) = (a, a, a)
makeInt3TupleByName: (a: => Int)(Int, Int, Int)

scala> makeInt3TupleByName({println("eval"); 5})
eval
eval
eval
res0: (Int, Int, Int) = (5,5,5)

scala> var i = 0
i: Int = 0

scala> makeInt3TupleByName({i += 1; i})
res1: (Int, Int, Int) = (1,2,3)

scala> i
res2: Int = 3
```

You might notice that the two basic expressions are followed by lines that have **res0**, **res1**, and **res2**. These are names given to the values as if there had been **val** declarations. This allows you to refer back to things that were calculated earlier in your REPL session. You should use the REPL when you need to test a single expression or some small piece of code. Simple examples later in this book will also utilize the REPL format.

This concludes our quick introduction to the basics of the Scala language. This should provide you with the basic information that you need to continue on with more complex topics.

1.10 Putting It Together

To illustrate how you might write a simple application, we will write a small shipping calculator application. Our company charges the following rates:

- If the package weighs 2 pounds or less, the flat rate is $2.50

- If the package weighs over 2 pounds but not more than 6 pounds, the flat rate is $4.00

- If the package weighs over 6 pounds but not more than 10 pounds, the flat rate is $5.00

- If the package weighs over 10 pounds, the flat rate is $6.75

We want to write an application that asks the user how many packages they want to ship, then to enter the weight of each package, and finally display the total shipping cost for those packages. The code below shows a more imperative approach to solving this problem.

```scala
import io.StdIn._

object ShippingCalc {
  def main(args: Array[String]): Unit = {
    println("How many packages would you like to ship?") // This call to println
      sends text to standard output
    val numPackages = try {
      readInt()
    } catch {
      case _: NumberFormatException => 0 // If the user does not enter an integer
        value then the value will default to zero
    }
    var totalShippingCost = 0.0 // Be careful to initialize with a double value
    for (n <- 1 to numPackages) {
      println("What is the weight of package "+n+"?")
      totalShippingCost += calculateShippingCharge(readDouble())
    }
    println("Your total shipping charge is $"+totalShippingCost)
  }

  def calculateShippingCharge(weight: Double): Double = {
    if (weight <= 2) 2.5
    else if (weight <= 6) 4.0
    else if (weight <= 10) 5.0
    else 6.75
```

```
24    } // end of calculateShippingCharge
25  }
```

Although this code basically solves the problem, we should be striving to use a more functional approach. The code below is very similar to the first set of code, but is more functional which is the preferred style in Scala.

```
1   import io.StdIn._
2
3   object ShippingCalcS {
4     def main(args: Array[String]): Unit = {
5       println("How many packages would you like to ship?") // This call to println
              sends text to standard output
6       val numPackages = try {
7         readInt()
8       } catch {
9         case _: NumberFormatException => 0 // If the user does not enter an integer
              value then the value will default to zero
10      }
11      val shippingCharges = for (n <- 1 to numPackages) yield {
12        println("What is the weight of package "+n+"?")
13        calculateShippingCharge(readDouble())
14      }
15      println("Your total shipping charge is $"+shippingCharges.sum)
16    }
17
18    def calculateShippingCharge(weight: Double): Double = {
19      if (weight <= 2) 2.5
20      else if (weight <= 6) 4.0
21      else if (weight <= 10) 5.0
22      else 6.75
23    } // end of calculateShippingCharge
24  }
```

As you can see, all mutable values have been removed. Each iteration of the for loop in line 12 yields a value which ultimately results in **shippingCharges** containing a collection of **Double** values. Thus, we can use the **sum** method in line 16 to calculate the sum for the display.

1.11   End of Chapter Material

1.11.1   Summary of Concepts

- Scala is a general purpose programming language that supports object-oriented programming well as both functional and imperative approaches to writing software.

- Scala applications are written as singleton **objects** that define a method called **main** which takes a single argument of type **Array[String]**.

- Methods are declared with the keyword **def** followed by the name of the method, an argument list in parentheses, a return type, an equals sign, and the body of the method.

- You can use `//` to mark a single-line comment, or `/*` followed by `*/` for a multiline comment. You can also use the `scaladoc` tool to convert commented code directly into API documentation. Comments for this should begin with `/**` and go directly before the thing they describe.

- You can declare variables using the keywords `val` and `var`. A `var` declaration can be reassigned to reference a different value.

- Lambda expressions in Scala are also known as Anonymous Functions or Closures and can be written as literals.

  - The rocket notation has an argument list and body separated by `=>`.
  - A shorter notation uses underscores for arguments. This only works for certain functions.

- The most basic conditional is `if`.

  - The syntax is `if`(*condition*) *trueExpression* `else` *falseExpression*.
  - The *condition* needs to be an expression with the `Boolean` type.
  - In Scala it can be used as an expression with a value or as a statement. As an expression, you need to have an `else` clause.
  - Curly braces can define blocks of code that function as a large expression. The value of the expression is the value of the last statement in the block.

- There is another type of conditional construct in Scala called a `match`.

  - A `match` can include one or more different `case`s.
  - The first `case` that matches the value being matched will be executed. If the `match` is used as an expression, the value of the code in the `case` will be the value of the `match`.
  - The `case`s are actually patterns. This gives them the ability to match structures in the data and pull out values.
    * Tuples can be used as a pattern.
    * Lowercase names are treated as `val` variable declarations and bound to that part of the pattern.
    * You can use `_` as a wildcard to match any value that you do not need to give a name to in a pattern.
    * After the pattern in a `case` you can put an `if` guard to further restrict that `case`.

- Scala has several types of loops that can be used: the `for` loop, the `while` loop, and the `do-while` loop.

- Scala's `for` loop is a for-each loop that iterates through each member of a collection. It has many options that give it a lot of flexibility and power.

  - A generator in a `for` loop has a pattern followed by a `<-` followed by a collection that is iterated through. The `<-` symbol should be read as "in".
  - A `Range` type can specify ranges of numeric values. The methods `to` and `until` can produce `Range`s on numeric types. The method `by` can adjust stepping. Floating point `Range`s require a stepping.

– The `yield` keyword can be put before the body of a `for` loop to cause it to produce a value so that it is an expression. When you have a `for` loop yield a value, it produces a collection similar to the one the generator is iterating over with the values that are produced by the expression in the body of the loop.

– The left side of a generator in a `for` loop is a pattern. This can allow you to pull values out of the elements of the collection, such as parts of a tuple. In addition, any elements of the collection that do not match the pattern are skipped over.

– `if` guards can be placed in `for` loops. This is particularly helpful when using `yield` and the values that fail the conditional check will not produce an output in the result.

• Exceptions are a way of dealing with unexpected conditions or other errors in code. When something goes wrong, an exception can be `thrown`. It is then up to the code that called the function where the error occurred to figure out what to do about it.

– Exception handling is done through the `try-catch` expression. Code that might fail is put in a `try` block. The `catch` partial function has cases for the different exceptions the coder knows how to handle at that point.

– When an exception is thrown, it immediately begins to pop up the call stack until it comes to a `catch` that can handle it. If none exists, it will go all the way to the top of the call stack and crash that thread.

• Scope is determined by the curly braces in Scala.

• You can declare a new type using the keyword `type` followed by the name you want to use for that type. Unlike `val`, `var`, and `def`, it is the general style for type names to begin with capital letters.

• Only `object`, `class`, and `trait` can be declared at the top-level in Scala applications; these do not have to be inside something else. All other statements, including declarations, have to appear inside of one of those three.

• The Scala libraries contain packages which are groups of code that have common functionality.

• Names that include full package specifications can be very long. `import` statements can be used to allow the programmer to use shorter names.

• You can use an underscore as the last element of an `import` statement to pull in everything in the specified scope.

• Scala also allows you to rename or hide certain names in an `import`. This is helpful when there are conflicts between names in different packages.

• Instead of an underscore, which pulls in everything, you can also specify names to pull in explicitly by putting them in curly braces.

• `Arrays` and `Lists` are collections that allow us to store multiple values under a single name.

• Both can be created in a number of ways.

– `Array(e1,e2,e3,...)` or `List(e1,e2,e3,...)` to make short collections where you provide all the values.

- `new Array[Type](num)` will make an `Array` of the specified type with the specified number of elements. All elements will have a default value. This is the least common usage in Scala.
- Use the `::` operator with `Lists` to prepend elements to the front and make a new `List`. Use `Nil` to represent an empty `List`.
- `Array.fill(num)(byNameExpression)` will make an `Array` of the specified size with values coming from repeated execution of the specified expression. This will also work for `Lists`.
- `Array.tabulate(num)(functionOnInt)` will make an `Array` with `num` elements with values that come from evaluating the function and passing it the index. Often function literals are used here. This will also work with `List`.

- `Arrays` have fixed size and are mutable. You can access and mutate them using an index in parentheses: `arr(i)`.

- `Lists` are completely immutable, but you can efficiently add to the front to make new `Lists`. They can be accessed with an index, but using `head` and `tail` is typically more efficient.

- The Scala collections, including `Array` and `List`, have lots of methods you can call on to do things.

  - There are quite a few basic methods that take normal values for arguments and give you back values, parts of collections, and locations where things occur.
  - The real power of collections is found in the higher-order methods. The primary ones you will use are `map`, `filter`, and `foreach`.

- You can use an `Option` type when you are not certain if you will have a value or not. You can think of an `Option` as being like a list that holds one item or zero items.

- Files are important to applications as values stored in memory are lost when a program terminates. Files allow data to persist between runs.

- One way to read from files is using `scala.io.Source`.

  - A call to `Source.fromFile` will return an instance of `BufferedSource` that pulls data from the file.
  - `Source` is an `Iterator[Char]`. The `Iterator` part implies that it is a collection that is consumed as values are pulled off. It gives individual characters.
  - The `getLines` method returns an `Iterator[String]` with elements that are full lines.

- The `split` method on `String` is useful for breaking up lines into their constituent parts. It takes a delimiter as a `String` and returns an `Array[String]` of all the parts of the `String` that are separated by that delimiter.

- For some applications it is easier to read data with a `java.util.Scanner`. This does not provide a Scala-style collection, but it has methods for checking if certain types of values are present and reading them in a way that is more independent of line breaks.

- You can use the `java.io.PrintWriter` to write text data out to a file.

- Scala added the symbol ??? to the standard library that you can use as a placeholder for unimplemented code. The ??? has the type `Nothing`. Although your code will compile if you call a method that contains ???, it will crash with a `NotImplmentedError` if it is called.

- Scala allows named parameters. Instead of just giving the argument values in order, you can give the name of the parameter you want to count that argument for, followed by an equal sign and the value. If used, the order of the arguments does not have to match the order of the parameters on the method.

- You can also mix regular and named arguments. The regular arguments have to come first. Every argument after the first named argument must also be named.

- Scala allows you to provide default values for parameters. When calling a method that has default values, you do not have to give arguments for things with defaults.

- A curried function is a function that takes one set of arguments and returns a function that takes another set of arguments. When currying, arguments are passed in separate argument lists instead of making a single longer argument list.

- By default, arguments in Scala are passed by value. The value of the reference is passed into the function. This automatically makes an alias of the object. If you want to protect a mutable object from unintended changes when it is passed to a function, you need to make a defensive copy.

- Scala also allows pass-by-name parameters. Instead of passing a value, these pass a chunk of code called a thunk that is evaluated every time the parameter is used.

- Scala provides a Read, Evaluate, Print Loop (REPL) environment that is helpful for testing small pieces of code. The REPL reads a statement from the users, executes it, and if it has a value, prints that value. It then waits for the next statement from the user. You can get a command line REPL by simply running the `scala` command with arguments.

1.11.2 Exercises

1. Look in the Scala API and see the full list of `read` methods in `scala.io.StdIn`. What method is used to read in a single character?

2. API scavenger hunt. Find the following methods in the Scala API. To answer each question, write out the method name and the description listed in the API.

   (a) What method can be used to capitalize the first character of a string?

   (b) Given a string that contains someone's first and last name, what method can be used to find the index of the space that would be present between the two names?

   (c) What method can be used to determine if a `BigInt` is possibly a prime number?

   (d) What method can be used to convert a `BigDecimal` to a `Double`?

   (e) What method can be used to copy an `Iterable` to an `Array`?

3. The `reverse` method can be called on `String`. Use this to write an application where the user inputs a word and you tell them whether or not it is a palindrome.

4. Write an application that asks the user to enter a number of seconds. You should display the appropriate number of days, hours, minutes, and seconds in that many seconds.

5. The last example in the chapter showed two ways of calculating shipping costs using a `for` loop. Now re-write those examples in as many ways that you can think of using higher-order methods on a `Range`.

6. Find how many even numbers are in a list of integers in the following ways.

   (a) Using a `while` loop.

   (b) Using a `for` loop.

   (c) Using the `count` higher-order method.

   (d) Using `filter` higher-order method.

   (e) Using `map` higher-order method.

   (f) Using recursion.[27]

7. Write a currency converter application that reads in an amount of money in U.S. dollars and converts it to the following currencies: Bitcoin, British Pound, Canadian Dollar, Chinese Yuan, Euro, Indian Rupee, Mexican Peso, and the Swiss Franc. You can find the conversion rates on the Internet. You should write separate functions for each currency conversion.

8. Write a set of functions to do the following operations on 2-tuples of `Int` as if they were the numerator and denominator of rational numbers.

   (a) Addition

   (b) Subtraction

   (c) Multiplication

   (d) Division

9. Think of as many ways as you can to make an `Array[Int]` that has the values 1–10 in it. Write them all to test that they work. You should be able to do this at least five different ways.

10. Think of as many ways as you can to make an `List[Int]` that has the values 1–10 in it. Write them all to test that they work. You should be able to do this at least five different ways.

11. Write several versions of code that will take an `Array[Int]` and return the number of even values in the `Array`. Each one will use a different technique. To test this on a larger array, you can make one using `Array.fill(100)(util.Random.nextInt(100))`.

    (a) Use a recursive function.[27]

    (b) Use the `count` higher-order method.

    (c) Use the `filter` higher-order method in conjunction with a regular method.

    (d) Use the `map` higher-order method in conjunction with a regular method.

---

[27]If you are not familiar with recursion, you can refer to an introduction in the appendices on the book website at `http://www.programmingusingscala.net`.

12. Make a text file that contains a list of the 50 states and the state's abbreviation (you can find many sample files on the Internet). Each line should contain a state name and abbreviation which are separated by some symbol such as a comma. Write an application that reads in this information and then creates a new file that only contains the abbreviations. For an added challenge, you can practice using different methods that will convert the abbreviation to all lower case, all upper case, or mixed case.

Chapter 2

Basics of Object-Orientation and Software Development

The real focus of this book is object-orientation. We start that journey in this chapter by talking about the nature of object-orientation and how we break down problems in an object-oriented way. We will continue the discussion in chapter 3 where we cover more details of how object-orientation is done in Scala specifically.

Scala is a completely object-oriented language. Chapter 1 spends a lot of time discussing the various methods that can be called on different types of objects. That chapter did not discuss anything about how to actually write object-oriented code. There are good odds that your introduction to programming did not cover this much either. That is generally intentional. Introductions to programming often focus on building up your logic skills and teaching you to break up problems into different functional units. Hopefully at this point your problem-solving skills have reached a sufficient level that we can take the next step to have you doing full object-oriented decomposition of problems.

2.1   The Meaning of Object-Orientation

The basic idea of an object is that it is something that contains both data and the functionality that operates on that data. It should be a reasonably self-sufficient package that handles particular responsibilities in a program. There are many other features that come along with object-orientation in most languages, but not all languages do things the same way. The only consistent feature is the grouping of data and functionality that is often called ENCAPSULATION. The power of this comes from the ability to pass objects into other code and have them know how to do the things that they are supposed to do. Having the methods attached to the object allows different objects to potentially do things in different ways. How we create this effect in Scala will be discussed more in chapter 4.

Chapter 1 showed how you can group data with tuples in Scala. The tuples do not have functionality to go with the data, and the data elements do not get meaningful names associated with them. They also are not unique types with meaning. A tuple of (`String`, `Int`) could be the combination of a student name and grade, a city and the distance to it, a product name and inventory number, or many other things. Scala cannot tell which one you were meaning because to Scala, all (`String`, `Int`) tuples are the same.

Scala is a CLASS-BASED object-oriented programming language.[1] That means that the way that we create object-oriented programs is to write `class`es in our programs and those `class`es define types and are used to make objects. Inside the `class`es we put declarations of the various members that we want objects, also called instances, created from that class to have. These members can be any valid Scala declaration, including other `class`es. For now we will primarily focus on the member data, often called properties, and the member functions, generally called methods.

One of the main goals of object-orientation is to allow you to think about programs at a higher level and not worry about details so much. You were able to do this to some extent without object-orientation because when you call a function, you do not have to think about the details of how it does what it does, you just need to know that it will do what it is supposed to do. For example, you have no knowledge of exactly what goes on when you print something or read input. Those things might seem simple, but there are actually many complex layers between the function/method that you call and the addition of text to a display, or getting keystrokes from the keyboard into a variable. You do not have to care about all those details, you can simply call `println` or `readLine`.

In an object-oriented program, you wrap responsibilities up in objects that can have multiple methods that are each supposed to do something. It might also have data associated with it, but the key is that when you are passed an object, it knows how to do a set of tasks and it makes sure that all of the different tasks work together properly. Often, when you use an object, you do not really know what data is in it or how the various methods work. Those details are hidden away, which means that you generally do not have to think about them or want to think about them when writing code that uses the object.

2.2 What Are `class`es?

We've talked a fair bit about `class`es, but what exactly are they other than some construct that we can declare in our code. Common analogies for `class`es and their relationship to instance objects are a cookie cutter and the cookies or a house blueprint and the house. You have one cookie cutter, but you can use it to create many different cookies. Every cookie made with a particular cookie cutter will be the same in certain key ways. In the same way, a blueprint can be used to make many different houses, but they will all come out as close copies of one another. The `class`es that you write in your code serve as the blueprints for making instance objects.

The declarations in a `class` tell us what members go into the instance objects created

---

[1]Both Java and Python are also class based. Most object-orientated programming languages are. The most notable exception is JavaScript, which allows you to create objects and add things to them without having classes. JavaScript 6, more formally known as ECMAScript 6, adds class syntax to the language, but it is only syntactic sugar on top of the prototype-based object-orientation that the language has provided since its creation.

from that class. The members can be any type of declaration in Scala. We will typically focus on member data, declared with `val` or `var`, and member functions, declared with `def`.[2] The member data is sometimes referred to as properties and the member functions are almost always referred to as methods. The syntax for a `class` is as follows.

```scala
class TypeName(arguments) {
  // Methods and Member Data
}
```

The curly braces on a `class` declaration, like all curly braces in Scala, define a new scope. This scope is associated with the `class` and anything you want to put in the class goes inside of them. You can also put normal code inside of the curly braces. That code will be executed every time an object is instantiated with `new`. Like code in a function, code put in the `class` is executed from top to bottom. It is important to note that while you can use things that are declared in the body of `class`es before their declaration, you need to be careful of this when it comes to code and member data. Member data is created and initialized in order from top to bottom along with code execution. So having code that is directly in the body of the `class` use data before it is defined typically causes problems. This is not an issue for data that is used inside of methods that are called after the construction of the object. To avoid the errors this can cause, simply put all member data at the top of a `class`.

The combination of the arguments and the code in the `class` that isn't in methods specifies how objects will be built and it commonly referred to as the CONSTRUCTOR. In other languages, constructors are written explicitly as separate methods. In this book, we will often use the term primarily to refer to the arguments that are passed in at the time of instance constructions.

In chapter 1 we used the example of student grades a few times. We will continue with that example here to demonstrate the creation and usage of a `class`. We will start with a very basic implementation. This class takes the relevant data as input, and has methods that calculate averages.

```scala
class Student(firstName: String, lastName: String, quizzes: List[Int],
    assignments: List[Int], tests: List[Int]) {
  def quizAverage: Double = if (quizzes.isEmpty) 0 else quizzes.sum.toDouble /
    quizzes.length

  def assignmentAverage: Double = if (assignments.isEmpty) 0 else
    assignments.sum.toDouble / assignments.length

  def testAverage: Double = if (tests.isEmpty) 0 else tests.sum.toDouble /
    tests.length

  def grade: Double = quizAverage*0.2 + assignmentAverage*0.5 + testAverage*0.3
}
```

The way it is written has some problems if we try to use it. The calculations of averages are just fine, but we cannot actually refer to the values passed in like `firstName` or `lastName` in outside code. Consider the following application in an `object` called `GradeBook` that makes a student, then passes it to a method that should print it.

```scala
object GradeBook {
  def main(args: Array[String]): Unit = {
```

---

[2]In later chapters we will make significant use of nesting `class`, `trait`, and `object` declarations as well.

```scala
    val student = new Student("Jane", "Doe", List(97, 80), List(100), List(89))
    printGrade(student)
  }

  def printGrade(s:Student):Unit = {
    println(s.firstName+" "+s.lastName) // Does not compile. firstName and lastName
        are not members.
    println(s.grade)
  }
}
```

As you can see from the comments, the first line in `printGrade` has an error. This is because arguments to `classes` are not members of the class by default. Member data in `classes` requires a `val` or `var` declaration. We can turn them into members by putting `val` or `var` in front of the argument name.

In this case, we probably want the names and the grades to all be members of `Student`. Whether we make them all `vals` or have some `vars` depends on whether we want the instances of our `class` to be immutable.[3] Either way, we are probably going to want to have methods that "add" grades to a student. The quotes are used because if `Student` is an immutable type, then we cannot really change the grades or an instance of it. Instead, the methods that support that functionality would create a new instance of `Student` with the additional grades. If we make it mutable, then we would use `var` declarations for the grade lists and prepend new grades to them.

We are going to choose to make the `Student` mutable, but it is worth seeing what a method like `addQuiz` might look like for an immutable `Student`.

```scala
// This would go inside of an immutable version of Student.
def addQuiz(newQuiz:Int):Student = new Student(firstName, lastName,
    newQuiz::quizzes, assignments, tests)
```

The key thing to note with this code is that the current value of `quizzes` is not modified. The `::` operation does not alter a `List`, it makes a new one. So this code makes a new student and passes it that new `List`. This is not as inefficient as it might sound. The fact that `String` and `List` are immutable means that most of the data used in the two instances of `Student` are shared. None of the real data is copied. Sharing data like that would not be safe if the data were mutable, as would be the case if we used an `Array` instead of a `List`.

The following code shows a more complete implementation of a mutable version of `Student`. The arguments to the `class` have been put on separate lines. This style is common when the argument list gets longer.

```scala
class Student(
    val firstName: String,
    val lastName: String,
    private var quizzes: List[Int] = Nil,
    private var assignments: List[Int] = Nil,
    private var tests: List[Int] = Nil) {
  def quizAverage: Double = if (quizzes.isEmpty) 0 else quizzes.sum.toDouble /
      quizzes.length

```

---

[3]People often refer to `classes` being immutable, but in reality, this generally means that the instance objects created from the `class` are immutable. For most of this book, we will use this shortened approach, but it is important for you to know what is really meant.

```scala
 9    def assignmentAverage: Double = if (assignments.isEmpty) 0 else
         assignments.sum.toDouble / assignments.length
10
11    def testAverage: Double = if (tests.isEmpty) 0 else tests.sum.toDouble /
         tests.length
12
13    def grade:Double = quizAverage*0.2 + assignmentAverage*0.5 + testAverage*0.3
14
15    def addQuiz(newQuiz: Int): Boolean = {
16      if (newQuiz >= -20 && newQuiz <= 120) {
17        quizzes ::= newQuiz
18        true
19      } else false
20    }
21
22    def addAssignment(newAssignment: Int): Boolean = {
23      if (newAssignment >= -20 && newAssignment <= 120) {
24        assignments ::= newAssignment
25        true
26      } else false
27    }
28
29    def addTest(newTest: Int): Boolean = {
30      if (newTest >= -20 && newTest <= 120) {
31        tests ::= newTest
32        true
33      } else false
34    }
35  }
```

With this implementation, the code in the `GradeBook` object above compiles just fine because `firstName` and `lastName` are now `vals`. We have also added three methods that allow us to add new values to the three different grade types. These methods have safety checks to make sure that the grades are in an acceptable range between -20 and 120.[4] This range would not work for every grading scheme, but the purpose is to illustrate that we can do bounds checking. The methods return a `Boolean` that tells whether or not the grade was actually added.

Note that the three members that store the grades are preceded by `private` as well as `var`. The keyword `private` is a visibility modifier. It tells Scala what parts of code are allowed to access things inside of a `class`, `object`, or `trait` declaration. Things that are `private` can only be accessed inside of the current declaration. If there is no visibility modifier, as is the case with `firstName`, `lastName`, and all the methods, then we say that the member is public.[5] Public members can be accessed outside of the `class` using the dot notation as is done in the `printGrade` method earlier. With the grade lists set as private, we have made those bounds checks more meaningful. If the lists were public, any code could add a grade outside the range with code like `student.tests ::= -100`. The fact that `tests` is private means that outside code cannot access it directly and add grades outside the allowed range.

---

[4]Yes, there are teachers who have grading schemes that allow negative grades.

[5]Scala does not have a keyword for `public`. It is simply the default visibility if nothing else is specified.

Argument or Internal Member?

Member data can be declared either in the arguments to a `class` or in the body of the `class`. How should you choose which one to use? The answer to this question is based on whether the outside code that calls `new` and creates an instance of the `class` should be able to provide different values for that member data. If the answer is "yes", then that member should probably be part of the argument list. If not, then it should be declared in the body of the `class`. To think of it the other way around, if the data member will always start with a particular default value, there is no reason to make the user pass in that default value, and the member should be declared inside of the body of the `class`.

In the case of **Student**, we could have considered moving the grade declarations into the body of the class and setting them all to `Nil` to start with in this mutable implementation. However, if we ever ran into the situation where we wanted to create an instance of **Student** with multiple grades, that would have made it much harder to create. Making the grade list arguments with default values is ideal if many usages start with no grades because those cases do not have to pass in extra arguments, but the exceptions that want to start with grades can easily do so.

The ability to make things private is a significant aspect of most object-oriented languages. It allows programmers to have the language enforce safety. If you have a data member that should not be changed in arbitrary ways, or a method that should not be called by outside code, you make it private, and outside code cannot accidentally do something wrong. If it tries to, it will produce a syntax error. This also helps enforce "separation of interface and implementation". This is a very critical aspect of object-oriented programming that we will address more in coming chapters. The idea is that outside code should only depend on the behavior of the public methods and the presence of the public data. These public members are collectively referred to as the PUBLIC INTERFACE or just the interface of the `class`.[6] The outside code should not depend on how the methods are implemented or on anything that is internal to the `class`. Doing so breaks encapsulation and makes code more brittle. The idea is that you should be able to freely change the inner details of a `class`, as long as the public interface stays the same.[7]

---

[6]Those with Java background might be familiar with the `interface` construct in that language. While the idea is roughly the same, we are using the term interface here in a more general, language-independent manner.

[7]Adding methods to the public interface is also safe. Where things get challenging is when you change or remove elements of the public interface. How big a problem this is depends on how much other code uses this `class`. Just imagine if for some reason they decided to remove the `length` method from `String`. Many millions of lines of code would be broken by such a change.

private or **Not?**

private or **Not?**

For every member declaration that you put in your **class**, you have to make the decision about whether it should be visible to outside code. There is no hard set rule for this, but we can make a few suggestions. Any data that is mutable should probably be **private**. That means all **vars** as well as mutable constructs like **Arrays**, even as **val** declarations, should generally be private. Why? Imagine you are working on a big project that has one million lines of code with a team of developers. Somehow a data member gets messed up and has a value that it should not. Would you rather look for all assignments to that field across the full million lines, or just have to look in that one **class**, because only the code in that **class** can directly alter that value? The other key question that you should ask for all declarations is whether it should be accessed/visible outside of the **class**. For example, if you have a method that is only used to help out with some other method, it should probably be **private**. Later on we will see situations where we have a **class** declared inside of another **class** or **object** that is a detail of the implementation, and not something we want outside code to know about. That inner **class** will be **private**, as will be any methods that have that type as an argument or a result because outside code does not know about that type and could not use those methods.

One might argue that the last implementation was too restrictive in regard to things being **private**. We do not want any random part of the code to be able to alter the grades, but that does not mean that nothing else should even be able to see the grades. In that implementation, outside code has no access to the individual grades, only the averages. To allow outside code to see the grades without being able to change them, we can add methods that provide access to them. In Java, it is standard to give such methods names that start with "get". In Scala, the methods have the natural names and the member data is given the same name prefixed with an underscore. Here is what that looks like in the code.

```scala
class Student(
    val firstName: String,
    val lastName: String,
    private var _quizzes: List[Int] = Nil,
    private var _assignments: List[Int] = Nil,
    private var _tests: List[Int] = Nil) {

  // Check that grades are in the proper range.
  require(_quizzes.forall(q => q >= -20 && q <= 120), "Quiz grades must be in the
      range of [-20, 120].")
  require(_assignments.forall(a => a >= -20 && a <= 120), "Assignment grades must
      be in the range of [-20, 120].")
  require(_tests.forall(t => t >= -20 && t <= 120), "Test grades must be in the
      range of [-20, 120].")

  def quizAverage: Double = if (_quizzes.isEmpty) 0 else _quizzes.sum.toDouble /
      _quizzes.length

  def assignmentAverage: Double = if (_assignments.isEmpty) 0 else
      _assignments.sum.toDouble / _assignments.length
```

```scala
def testAverage: Double = if (_tests.isEmpty) 0 else _tests.sum.toDouble /
    _tests.length

def grade:Double = quizAverage*0.2 + assignmentAverage*0.5 + testAverage*0.3

def addQuiz(newQuiz: Int): Boolean = {
  if (newQuiz >= -20 && newQuiz <= 120) {
    _quizzes ::= newQuiz
    true
  } else false
}

def addAssignment(newAssignment: Int): Boolean = {
  if (newAssignment >= -20 && newAssignment <= 120) {
    _assignments ::= newAssignment
    true
  } else false
}

def addTest(newTest: Int): Boolean = {
  if (newTest >= -20 && newTest <= 120) {
    _tests ::= newTest
    true
  } else false
}

def quizzes = _quizzes

def assignments = _assignments

def tests = _tests
}
```

Now outside code could do something like `student.tests` to get the current `List` of test grades. Because the `List` type is immutable, they cannot change it or do anything to it. This gives us a good way of viewing values without allowing outside code to change them in ways that should not be allowed.

Note that this approach would need to be modified if the grades were stored in `Arrays` instead of `Lists` because `Arrays` are mutable. If you ever have a method that gives back one of your private mutable values, the outside code that gets it now has the ability to change those values. To prevent this, when `Arrays` are passed around and you do not want things to be changed, you have to make DEFENSIVE COPIES. This leads to the topic of whether it is better to use mutable or immutable data for things. Many people think of mutable data as being more efficient and easier to work with, but if you actually take care to do things like make defensive copies, or if your program is multithreaded,[8] the immutable options can be superior in both performance and ease of programming.

One other thing was added to this code—three `require` statements at the top of the `class`. These calls are intended to close a weakness that we had previously where grades outside of the accepted range could get into an instance at creation. Recall that code written in the body of a `class` is executed when a instance is created using `new`. The `require` function is part of the standard library, and it does what the name implies. It has to have one `Boolean` argument that is evaluated and must be true. If it is not, an exception is

---

[8]Full details on multithreading are covered in chapters 8 and 9.

thrown. A second argument with the message to print if it fails is optional, but strongly recommended.

2.3 Software Development Stages

The software development process can be broken up into a number of different types of activities. There are roughly five stages involved in the production of any large software system.[9] There are many different approaches to organizing these five stages, but the general nature of them is always the same, and they all have to be involved. The five stages are as follows.

Analysis This is where things begin. During analysis, the developers figure out what it is that the program is supposed to do. This is a problem definition phase and is often called the "what" phase. The more detailed the analysis, the easier the other stages will be. In practice, it is not possible to get an initial analysis that covers everything, and inevitably, things change over time on real products. For this reason, while all software must start with analysis, it is not something you finish and leave; you generally have to come back to it. Analysis is very independent of programming and only looks at what you want to do, not how to do it.

Design After you have sufficient analysis and understand the problem you are trying to solve, the development turns to figuring out how to produce the desired result in code. This is the design phase and is often called the "how" phase. Once again, a more complete design will make life easier later on, but realize it is basically impossible to make a perfect and complete design initially, so this is something that will also be revisited. The design phase discusses code, but does not actually involve writing code.

Implementation The middle of the five stages is where you actually write the code. This book, and much of your introduction to computer science, focuses on developing your ability to write code. However, this is only one phase in the life cycle of software development, and when things are done well, it can actually be a small phase relative to others. This is the focus early on because you cannot really do any of the other phases, with the possible exception of analysis, unless you have a strong grasp of how the implementation will work.

Testing and Debugging At this point you should have already learned that just writing a program is not the biggest hurdle you have to clear. Finding the errors in the program can take far longer. Testing and debugging is the process of finding and fixing errors. It is an ongoing process in software development and it often takes significantly longer than the implementation.

Maintenance After the software is deemed complete and sufficiently free of errors/bugs, it is released. This ushers in the maintenance phase. Maintenance is something that is hard to simulate in a classroom environment, but is critical for most professional software developers. People who pay money for a piece of software often want to know that someone will be there to correct errors that are found later on. Also, customers will often think of additional features they would like to see implemented in the

---

[9]Depending on what source you look at, requirements gathering and deployment could be added, taking the total number of stages higher.

software. Maintenance is the general process of dealing with issues in the software after it has been released.

You will come back to these different steps in many later CS courses. The field of software engineering is all about how to do these things in a way that produces the best results with the least amount of effort.

2.4 Analysis

So what exactly is entailed in the analysis phase of software design? What do you have to do in order to be able to say that you know what problem you are solving? One key way to approach this question is to think about how the software is going to be used. Picture a user running the software. What different types of users are there? What are the different ways they can interact with the system? By answering these questions, you can come to truly understand the main functionality of the software.

There are many different approaches to doing software analysis. Simply writing down prose that describes the system is an option, but not a very good one because it is dense and requires anyone viewing the document to do a significant amount of reading to get the needed information from it. For this reason, there are diagramatic approaches to analysis that can communicate key ideas quickly. The Unified Modeling Language (UML) is a communications tool which defines standards for 14 different diagram types that are significant for object-oriented programming. These UML diagrams can be used as an effective communication tool so everyone understands how the software system is put together. Each diagram can be backed up by prose that provides details which are not captured in the diagram. One diagram commonly used diagram during the analysis phase is the use case diagram. The use case diagram is one that happens to be focused on what the software does and how the users interact with the system.

A use case diagram includes various actors and the use cases they participate in. Actors often represent humans in different roles, but they can also represent software from outside of the current system that communicates with that system asking it to do things. Figure 2.1 shows a very simple use case diagram with a single actor and a few basic use cases. As you can see, actors are represented with stick figures.

Figure 2.1 has a large box in it that represents the system being analyzed. This is the standard notation for use case diagrams. Note that the actor is outside of the system. This makes sense, unless you happen to be putting people into your software. This is also why other pieces of software are represented as actors. They are not part of what you are focusing on. They are external and send in commands to make things happen.

Inside the system there are a few bubbles with text in them that represent use cases. There are lines from the actor to the "Login" and "Logout" use cases to indicate that these are actions that the "User" can initiate. The bubbles represent things that can be done in the software and the lines indicate who can do it. If your system had both regular users and administrators as two types of actors, the administrator would have lines to use cases that were not available to the normal user. Actions like deleting a user account should not be part of the options given to normal users.

There are two other use case bubbles shown with the text "Enter Credentials" and "Re-validate". These both have annotated arrows connecting them to the use case for "Login". The arrow pointing from "Login" to "Enter Credentials" is annotated with the text "<<include>>". This indicates that when the "Login" use case is invoked, it can include

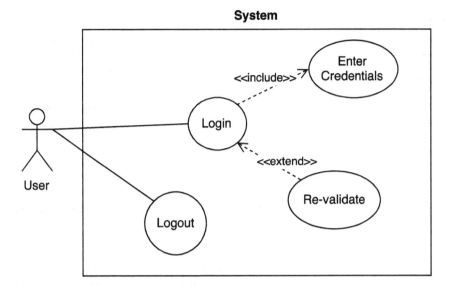

FIGURE 2.1: This is a simple example of a UML use case diagram. Actors are drawn as stick figures. The system being analyzed is a box. The ways in which the actors use the system are drawn as circles.

having the user enter his/her credentials. The arrow pointing from "Re-validate" to "Login" is annotated with "<<extends>>". This implies that the source of the arrow is a subtype of its destination. So having the user re-validate is shown to be a special case of logging in.

Use case diagrams might seem a little silly, but they can force you to really think about all the different actions that a user should be able to do in your application and then to see how those different actions fit together. That can be very useful as you enter the design step of building your software.

2.5 Design

The analysis phase should not involve thinking about code at all. It is all about functionality and interaction with users. The design phase is when you begin thinking about how the problem will be solved using code. When you are working in the object-oriented paradigm, the design process involves figuring out the types of objects that you need in the program and laying out `classes`, `objects`, and `traits` for them. We focus mostly on the `classes` at this point.

Hopefully you now have some idea of what `classes` are, what things should potentially become `classes`, and how to declare a `class` in your code based on the discussion in section 2.2. Real programs do not have just one `class`; they have many, each of which should probably go in a different file.[10] Thinking about a large program with a lot of `classes` in terms of code does not work very well, but programmers often do need to think about

---

[10]Unlike Java, Scala does not enforce that you can only have a single public `class` in a file. There is no concept of public or non-public `classes` in Scala. However, for code organization purposes, you only put multiple `classes` in a file if they are extremely closely related.

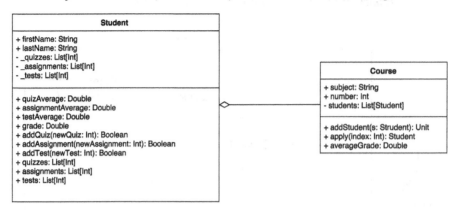

FIGURE 2.2: This shows the UML class diagram for the `Student` class from section 2.2 along with a `class` for a course that contains multiple students.

programs with many `class`es, and they also need to communicate their thoughts to other programmers. To do the higher-level work of designing an application and breaking it into appropriate classes, it is helpful to work with sketches instead of actual code.

All the UML diagrams are intended to be "whiteboard compatible". This basically means that they are easy to draw out. You can use drawing software to create UML diagrams; however, it is more common for programmers to just draw them on a whiteboard while thinking through a problem and trying to communicate ideas with one another. Another common UML diagram that you are likely to use is the class diagram. Class diagrams are nothing more than boxes with 3 sections and text. Each box is used to represent a `class`. When drawn out all the way, the box is broken into three sections. The top section has the name of the `class`, the middle section shows member data/properties, and the bottom section shows methods. Each member is generally preceded with a symbol that shows its visibility. A "+" means that something is public. A "-" means that it is private. There is also a "#" that indicates that it is protected, a visibility that will be discussed in chapter 4. Figure 2.2 shows what our `Student class` and a second class for a course would look like when drawn this way.

When there are multiple `class`es, lines can be used to illustrate relationships between them. This figure shows a connection between the `Course` and `Student class`es. There are multiple types of arrows the are used in class diagrams. We'll see more of them over time. This particular style of arrow represents aggregation. It implies that a `Course` includes multiple `Student`s.

These relationship lines are very important for helping to understand the structure of your software. If you have a box that has a large number of lines pointing to it, that immediately tells you that `class` is critical and that changes to it are more likely to break the software than changes to `class`es that are not very connected. What you really do not want to have happen to your software is to see a diagram with lines going all over the place so that everything is highly connected.[11] If you see that, you know that the software shown was constructed in a way that will be very hard to maintain.

Like the use case diagram, the class diagram does not show you everything about a system. However, it does communicate quite a bit of information in a format that people can read quickly and easily. The information that it imparts is particularly significant for helping them understand how the system works. The most obvious piece of information is

---

[11]The proper term for connections between elements of code is coupling, and you generally prefer weak coupling (fewer connections) over strong coupling.

the classes/types that are used in the system. It also gives you a quick reference to the capabilities of the different elements through the methods and member data.

Depending on what you are doing and what you are trying to communicate, some of the sections in a class might be left out of a class diagram. A lot of the time, you do not really care about all the methods, and often you do not even care about the details of the member data. In those situations, you are just drawing boxes with class names and linking them together to indicate appropriate relationships. That is not a problem. That is another key to UML diagrams—they do not have to be exhaustive. You use them to communicate ideas and can easily leave out details that go beyond the level you are currently considering. This is an advantage over doing design work in code, as compilers are not very forgiving about leaving out most of the details.

There is a lot more to UML class diagrams, including not only other types of arrows to represent different relationships, but also boxes for packages and annotations on the relationships between types. These different aspects will be introduced as needed in later chapters.

2.6 Bank Example

It is worth taking some time to run through an example problem that most readers should be familiar with, and doing a little analysis and design. We will then also present a bit of the implementation to help you see how everything fits together.

At this point, most of the problems that you are coding are given to you by an instructor or they come out of a book like this one. That means that you are typically told what you should do, and often given some direction on how to do it. Up to this point, the problems have probably also generally been quite small. For that reason, when you get an assignment or project, you might normally just sit down and start typing. That is to say that you might just jump to the implementation phase. If that describes how you work, then it is a bad habit that you should try to break. Especially as problems get bigger, it becomes more important to have a good idea about how you are going to approach them before you start writing any code. Failure to spend time thinking about a problem early on typically leads to much more time being spent correcting problems later when you realize that the approach you have taken has problems. Spending some time to draw out your classes and think about what members they have can save you significant time in the long run.

The problem that we want to look at is the example of a bank. There are lots of things that could be modeled in a bank. One might first think about customers and accounts, as that is the side of banking that most of us interact with, but they are also businesses with branches and employees. A single program should likely not be used to model both sides of that. So we will choose to model the customer interaction side instead of the internal business workings.

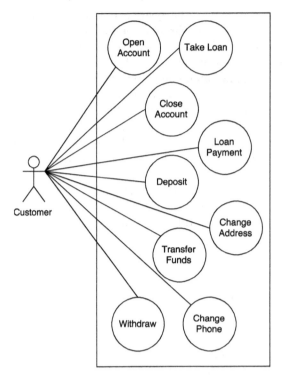

FIGURE 2.3: This figure shows a use case diagram for a potential bank application that models the customer-facing aspects of a bank.

2.6.1   Bank Analysis

We can start off by doing some analysis of the problem. We could write a long paragraph telling everything that we want customers to be able to do, but the use case diagram in figure 2.3 should get the idea across well enough.

2.6.2   Bank Design

Based on the potential actions shown in this use case diagram, we can immediately see some options for different classes that we could write. Options include **Account**, **Loan**, **Customer**, **Address**, and **PhoneNumber**. We probably want a **class** that represents a full bank to keep track of different objects. Figure 2.4 shows a class diagram with a potential design for this program laying out those six classes. The figure also includes one other element that is drawn as a rounded rectangle. These are used in UML object diagrams. In most object-oriented languages, including Java and Python, these constructs would not mix in the same diagram. However, because Scala has **object** declarations for singleton objects, it makes sense to include them alongside the **class**es in a class diagram. The **object** is included here to remind you that a full application has to have at least one **object** declaration that will hold the **main** method of the program.

Most of the methods and member data should be fairly well explained just by their names. If this were a formal design, we would want additional prose to make certain it was clear what each method was supposed to do. One thing that might stand out is that there are two methods in the **Bank** with the name **findCustomer**. This is because there are two different ways that we might want to find a customer, either by their name or by their ID.

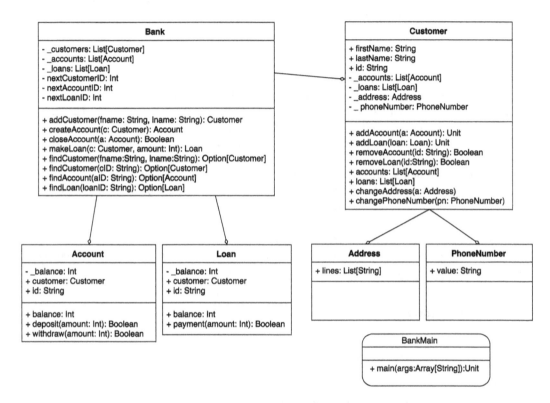

FIGURE 2.4: This figure shows a class diagram for a potential bank application that models the customer-facing aspects of a bank.

When two methods have the same name, we say that the method is overloaded. Overloaded methods have to have different arguments so that the compiler knows which one you are referring to. In this case, one version takes a single `String` argument, while the other takes two `String` arguments. The return types could also be different, but that is not required, and it is not sufficient to do the overloading. If you have two methods that only vary in their return type, you will get a syntax error.

Another aspect of this design that might jump out to readers is that all the "find" methods in `Bank` produce an `Option` type. That is because there is a chance that no matches will be found. It is proper Scala style to use `Option` in this situation instead of doing something like returning `null`. The use of `Option` is safer, and less error prone, as it allows the compiler to check to make sure that the programmer is doing something about the possibility that the value is not present.

2.6.3 Bank Implementation

Using this design, we can build an actual implementation. We begin with the `Bank` class itself. Note that each of the methods is preceded by a Scaladoc-style comment that describes the method, the arguments, and the return value. The first line in the file is a `package` declaration. This will be discussed in more detail in section 2.7.

The `Bank` takes three arguments for existing customers, accounts, and loans. Making these into arguments is required if we later wish to have this information restored from a file. All three arguments provide default values, so that they do not have to be specified when they are not needed. Note that they are also declared to be `private var`. They will change as new ones are created and removed, but when they are changed, certain things must be done to keep the implementation consistent.

There are also three `private var` declarations that are used to prevent us from having the same IDs for customers, accounts, and loans. Their initial values are calculated based on the lists that are passed in using a `foldLeft` method that will run through and find the largest ID value currently in use.

```
 1  package oobasics.bank
 2
 3  /**
 4   * Represents a bank with customers, accounts, and loans.
 5   */
 6  class Bank(
 7      private var _customers: List[Customer] = Nil,
 8      private var _accounts: List[Account] = Nil,
 9      private var _loans: List[Loan] = Nil) {
10
11    private var nextCustomerID: Int = _customers.foldLeft(-7)(_ min _.id.toInt) + 7
12    private var nextAccountID: Int = _accounts.foldLeft(-13)(_ min _.id.toInt) + 13
13    private var nextLoanID: Int = _loans.foldLeft(-17)(_ min _.id.toInt) + 17
14
15    /**
16     * Add a customer to this bank. Creates a new customer with the specified name.
17     * @param fname the first name of the new customer.
18     * @param lname the last name of the new customer.
19     * @return the new customer that was created and added.
20     */
21    def addCustomer(fname: String, lname: String, address: Address, phoneNumber:
            PhoneNumber): Customer = {
22      val id = nextCustomerID.toString
```

```scala
23      val c = new Customer(fname, lname, "0" * (8 - id.length) + id, address,
            phoneNumber)
24      _customers ::= c
25      nextCustomerID += 7
26      c
27    }
28
29    /**
30     * Create a new account for a specified customer. The customer must be a member
31     * of this bank.
32     * @param c the customer to add an account to.
33     * @return the account that was created.
34     */
35    def createAccount(c: Customer): Account = {
36      require(_customers.contains(c), "The customer must be a member of this bank.")
37      val id = nextAccountID.toString
38      val acc = new Account(0, c, "0" * (8 - id.length) + id)
39      _accounts ::= acc
40      nextAccountID += 13
41      acc
42    }
43
44    /**
45     * Close a specified account and remove it from the associated customer.
46     * @param a the account to close.
47     * @return tells if the account was found and closed.
48     */
49    def closeAccount(a: Account): Boolean = {
50      val index = _accounts.indexOf(a)
51      if (index < 0) false else {
52        a.customer.removeAccount(a.id)
53        _accounts = _accounts.patch(index, Nil, 1)
54        true
55      }
56    }
57
58    /**
59     * Make a loan to a specified customer for a specified amount. The customer
60     * must be a member of this bank.
61     * @param c the customer to add the loan to.
62     * @param amount the initial amount of the loan.
63     * @return the loan that was created.
64     */
65    def makeLoan(c: Customer, amount: Int): Loan = {
66      require(_customers.contains(c), "The customer must be a member of this bank.")
67      val id = nextLoanID.toString
68      val loan = new Loan(amount, c, "0" * (8 - id.length) + id)
69      _loans ::= loan
70      nextLoanID += 17
71      loan
72    }
73
74    /**
75     * Find a customer by their name.
76     * @param fname the first name of the customer you are looking for.
```

```
77      * @param lname the last name of the customer you are looking for.
78      * @return the customer if one is found.
79      */
80     def findCustomer(fname: String, lname: String): Option[Customer] = {
81       _customers.find(c => c.firstName == fname && c.lastName == lname)
82     }
83
84     /**
85      * Find a customer by their ID.
86      * @param cID the ID of the customer being searched for.
87      * @return the customer if one is found.
88      */
89     def findCustomer(cID: String): Option[Customer] = _customers.find(_.id == cID)
90
91     /**
92      * Find an account by ID.
93      * @param aID the ID of the account being searched for.
94      * @return the account if one is found.
95      */
96     def findAccount(aID: String): Option[Account] = _accounts.find(_.id == aID)
97
98     /**
99      * Find a loan by ID.
100      * @param loanID the ID of the loan being searched for.
101      * @return the loan if one is found.
102      */
103     def findLoan(loanID: String): Option[Loan] = _loans.find(_.id == loanID)
104   }
```

Most of the methods in this code should be reasonably self-explanatory. One thing that might seem odd is the third argument in lines like this.

```
val acc = new Account(0, c, "0" * (8 - id.length) + id)
```

The "0" * (8 - id.length) pads out the ID string so that it is always 8 digits in length with leading zeros if needed. There are several other things to note. First, is the use of patch when an account is closed. Here patch returns an updated loans list after it has replaced "1" element starting at index "index" with the value of "Nil". This effectively removes the loan. The next thing to note is the use of find in all of the methods that begin with "find". find returns the first element of a collection that matches the predicate function shown in parenthesis. The last thing to note is the use of require at the beginning of a number of methods to specify a constraint that has to be met.

Next up, we look at the implementation of Customer. Again, there are a number of values passed in at construction. The ones that can be mutated are made private and are given names that begin with an underscore so that nicely named accessor methods can also be provided. There are two members for the lists of Accounts and Loans that are declared in the class. They were put there based on the idea that when a Customer is created, it never has any of these, so there is no point in passing them in.

```
1  package oobasics.bank
2
3  /**
4   * Represents a customer at the bank.
5   */
6  class Customer(
```

```scala
   val firstName: String,
   val lastName: String,
   val id: String,
   private var _address: Address,
   private var _phoneNumber: PhoneNumber) {

 private var _accounts = List[Account]()
 private var _loans = List[Loan]()

 /**
  * Adds an account to this customer. The account must be associated with this
  * customer. Does nothing if this is already a current account.
  * @param a the account to add.
  */
 def addAccount(a: Account): Unit = {
   require(a.customer == this, "Account being added to wrong customer.")
   if (!_accounts.contains(a)) _accounts ::= a
 }

 /**
  * Adds a loan to this customer. The loan must be associated with this customer.
  * Does nothing is this is already a current loan.
  * @param loan the loan to add to this customer.
  */
 def addLoan(loan: Loan): Unit = {
   require(loan.customer == this, "Loan being added to wrong customer.")
   if (!_loans.contains(loan)) _loans ::= loan
 }

 /**
  * Remove the account with the specified ID from this customer.
  * @param id the account ID to remove.
  * @return tells if the account was there to remove.
  */
 def removeAccount(id: String): Boolean = {
   val index = _accounts.indexWhere(_.id == id)
   if (index < 0) false else {
     _accounts.patch(index, Nil, 1)
     true
   }
 }

 /**
  * Remove the loan with the specified ID from this customer.
  * @param id the loan ID to remove.
  * @return tells if the loan was there to remove.
  */
 def removeLoan(id: String): Boolean = {
   val index = _loans.indexWhere(_.id == id)
   if (index < 0) false else {
     _loans.patch(index, Nil, 1)
     true
   }
 }
```

```
62    /**
63     * Provides public access to the accounts.
64     * @return the list of accounts.
65     */
66    def accounts: List[Account] = _accounts
67
68    /**
69     * Provides public access to the loans.
70     * @return the list of loans.
71     */
72    def loans: List[Loan] = _loans
73
74    /**
75     * Change the value of this customer's address.
76     * @param newAddress the new address of this customer.
77     */
78    def changeAddress(newAddress: Address): Unit = {
79      _address = newAddress
80    }
81
82    /**
83     * Change the phone number of this customer.
84     * @param newPhoneNumber the new phone number of this customer.
85     */
86    def changePhoneNumber(newPhoneNumber: PhoneNumber): Unit = {
87      _phoneNumber = newPhoneNumber
88    }
89  }
```

Each of the methods includes a Scaladoc comment, and the implementations do not contain anything that should stand out as unusual after seeing what was done in Bank.

Up next is the Account class. The primary functionality of this class is keeping track of a balance. There are methods for doing deposits and withdrawals that have checks to make sure that the amount is valid. Also note that the balance is stored as an Int. This is because the arithmetic on a Double is inexact. For example, $1.0 - 0.9 - 0.1$ is not zero. That's a problem for money, so we use integer values to represent the amount of money in cents instead of a floating point value in a Double.

```
1   package oobasics.bank
2
3   /**
4    * Represents an account with the bank. Potentially a checking or savings account.
5    */
6   class Account(
7       private var _balance: Int,
8       val customer: Customer,
9       val id: String) {
10
11    customer.addAccount(this) // Make sure the customer knows about this account.
12
13    /**
14     * Provides public access to the current balance for this account.
15     * @return the value of the balance.
16     */
17    def balance = _balance
```

```
18
19   /**
20    * Makes a deposit to the account. Deposit values must be positive.
21    * @param amount the amount of money to deposit.
22    * @return tells is the deposit occurred successfully.
23    */
24   def deposit(amount: Int): Boolean = {
25     if (amount > 0) {
26       _balance += amount
27       true
28     } else false
29   }
30
31   /**
32    * Makes a withdraw from the account. The amount must be positive and less
33    * than the current balance.
34    * @param amount the amount to withdraw.
35    * @return tells if the withdraw occurred successfully.
36    */
37   def withdraw(amount: Int): Boolean = {
38     if (amount > 0 && amount <= _balance) {
39       _balance -= amount
40       true
41     } else false
42   }
43 }
```

One thing that is lacking from this implementation that any actual bank would have to include is logging transaction histories. Transactions that change account values would have to be recorded in some format that can be audited later on.

Simpler still is our implementation of a `Loan class`. Again, the primary information is a balance and the functionality includes making payments. Both the `Loan` and the `Account` could be extended with functionality for calculating interest, but that is not done here.

```
1   package oobasics.bank
2
3   /**
4    * Represents a loan that the bank has made to a customer.
5    */
6   class Loan(
7       private var _balance: Int,
8       val customer: Customer,
9       val id: String) {
10
11    customer.addLoan(this) // Add this loan to the customer.
12
13    /**
14     * Provides public access to the balance left on this loan.
15     * @return the value of the balance.
16     */
17    def balance = _balance
18
19    /**
20     * Make a payment on this loan. Only works for amounts greater than zero and
21     * less than or equal to the current balance.
```

```
22    * @param amount the amount of the current payment.
23    * @return tells if the payment went through successfully.
24    */
25   def payment(amount: Int): Boolean = {
26     if (amount > 0 && amount < _balance) {
27       _balance -= amount
28       true
29     } else false
30   }
31 }
```

This example implementation finishes off with the **Address** and **PhoneNumber** classes. We have left these mostly empty. As written, they store simple values. They are written as **class**es though, because proper implementations would need to validate the values being used. Instead, these **class**es include a comment that starts with TODO. This is a good thing to do in your code because IDEs will make a list of these comments so that you can easily find them.

```
1  package oobasics.bank
2
3  class Address(val lines: List[String]) {
4    // TODO - Code to verify that this is a valid address.
5  }
```

```
1  package oobasics.bank
2
3  class PhoneNumber(val value: String) {
4    // TODO - Code that will verify this is a valid phone number.
5  }
```

When you consider each of these **class** implementations, take particular note of how each one is specifically responsible for handling its own data and functionality. Customers keep track of their accounts, but they do not directly alter any of the information in an account, that can only be done through the methods in the **Account class**. This is how it should be. When you are building object-oriented programs, always keep this type of separation in mind. Try to make the "units" of your program as self-contained as possible.

2.6.4 Making an Application

In order to complete this example, we need to use these **class**es in a sample application. At this point, we only have the ability to do text interfaces, GUIs will be introduced in chapter 5, so we will make an example that provides a text menu and allows the user to create and manipulate the different elements in the bank in appropriate ways. This application is written in the **object** named **BankMain** below.

The **main** method in this application starts by making a new instance of **Bank**, an integer called **option** to track the user's menu selection, and three different **Option** types to keep track of the current customer, account, and loan. The values of customer, account, and loan are initially set to **None** to indicate that there are no current customers, accounts, or loans. This is followed by a **while** loop that goes until the user selects to quit. It prints the relevant information, then reads the user input and executes a large **match** statement to perform the proper operation based on the user selection.

```
1  package oobasics.bank
```

```scala
import io.StdIn._

/**
 * Primary text-based interface for our bank.
 */
object BankMain {
  def main(args: Array[String]): Unit = {
    val bank = new Bank
    var option = 0
    var customer: Option[Customer] = None
    var account: Option[Account] = None
    var loan: Option[Loan] = None

    while (option != 99) {
      println(menu)
      println("Selected Customer: "+customer.map(c => c.firstName+"
        "+c.lastName).getOrElse("None"))
      println("Selected Account: "+account.map(a => a.id+"
        ("+a.customer.firstName+" "+a.customer.lastName).getOrElse("None"))
      println("Selected Loan: "+loan.map(l => l.id+" ("+l.customer.firstName+"
        "+l.customer.lastName).getOrElse("None"))
      option = readInt()
      option match {
        case 1 => customer = Some(createCustomer(bank))
        case 2 => customer = selectCustomer(bank)
        case 3 => account = customer.map(c => createAccount(bank, c))
        case 4 => account.foreach(a => closeAccount(bank, a))
        case 5 => loan = customer.map(c => makeLoan(bank, c))
        case 6 => account = selectAccount(bank)
        case 7 => loan = selectLoan(bank)
        case 8 => account.foreach(a => deposit(bank, a))
        case 9 => account.foreach(a => withdraw(bank, a))
        case 10 => account.foreach(a => checkAccountBalance(a))
        case 11 => loan.foreach(l => payLoan(bank, l))
        case 12 => loan.foreach(l => checkLoanBalance(l))
        case 13 => customer.foreach(c => changeAddress(c))
        case 14 => customer.foreach(c => changePhone(c))
        case 99 =>
        case _ => println("That is not a valid option. Please select again.")
      }
    }
    println("Goodbye.")
  }

  private def createCustomer(bank: Bank): Customer = {
    println("What is the customer's first name?")
    val firstName = readLine()
    println("What is the customer's last name?")
    val lastName = readLine()
    println("What is the customer's address? (End your input with a blank line.)")
    val address = readAddress()
    println("What is the customer's phone number?")
    val phoneNumber = new PhoneNumber(readLine())
    bank.addCustomer(firstName, lastName, address, phoneNumber)
```

```scala
54    }
55
56    private def selectCustomer(bank: Bank): Option[Customer] = {
57      println("Do you want to find a customer by name or id? (name/id)")
58      var style = readLine()
59      while (style != "name" && style != "id") {
60        println("Invalid response. Do you want to find a customer by name or id?
              (name/id)")
61        style = readLine()
62      }
63      if (style == "name") {
64        println("Enter the first and last name of the customer separated by a space.")
65        val names = readLine().trim.split(" +")
66        bank.findCustomer(names(0), names(1))
67      } else {
68        println("Enter the customer ID you are looking for.")
69        val id = readLine().trim
70        bank.findCustomer(id)
71      }
72    }
73
74    private def createAccount(bank: Bank, customer: Customer): Account = {
75      bank.createAccount(customer)
76    }
77
78    private def closeAccount(bank: Bank, account: Account): Unit = {
79      bank.closeAccount(account)
80    }
81
82    private def makeLoan(bank: Bank, customer: Customer): Loan = {
83      println("How much is the loan for?")
84      val amount = readInt
85      bank.makeLoan(customer, amount)
86    }
87
88    private def selectAccount(bank: Bank): Option[Account] = {
89      println("What is the ID of the account?")
90      bank.findAccount(readLine())
91    }
92
93    private def selectLoan(bank: Bank): Option[Loan] = {
94      println("What is the ID of the loan?")
95      bank.findLoan(readLine())
96    }
97
98    private def deposit(bank: Bank, account: Account): Unit = {
99      println("How much do you want to deposit?")
100     val worked = account.deposit(readInt())
101     if (worked) println("The deposit was successful.")
102     else println("The deposit failed.")
103   }
104
105   private def withdraw(bank: Bank, account: Account): Unit = {
106     println("How much do you want to withdraw?")
107     val worked = account.withdraw(readInt())
```

```
108      if (worked) println("The withdraw was successful.")
109      else println("The withdraw failed.")
110    }
111
112    private def checkAccountBalance(account: Account): Unit = {
113      println(account.balance)
114    }
115
116    private def payLoan(bank: Bank, loan: Loan): Unit = {
117      println("How much do you want to pay on the loan?")
118      val worked = loan.payment(readInt())
119      if (worked) println("The loan payment was successful.")
120      else println("The loan payment failed.")
121    }
122
123    private def checkLoanBalance(loan: Loan): Unit = {
124      println(loan.balance)
125    }
126
127    private def changeAddress(customer: Customer): Unit = {
128      println("Enter a new address for "+customer.firstName+" "+customer.lastName+".")
129      println("You can have multiple lines. Enter a blank line when done.")
130      val address = readAddress()
131      customer.changeAddress(address)
132    }
133
134    private def changePhone(customer: Customer): Unit = {
135      println("Enter a new phone number for "+customer.firstName+"
             "+customer.lastName+".")
136      val phoneNumber = new PhoneNumber(readLine())
137      customer.changePhoneNumber(phoneNumber)
138    }
139
140    private def readAddress(): Address = {
141      def helper(): List[String] = {
142        val input = readLine()
143        if (input.isEmpty()) Nil
144        else input :: helper()
145      }
146      new Address(helper())
147    }
148
149    private val menu = """Select one of the following options.
150  1. Create Customer
151  2. Select Customer
152  3. Create Account
153  4. Close Account
154  5. Make Loan
155  6. Select Account
156  7. Select Loan
157  8. Deposit to Account
158  9. Withdraw from Account
159  10. Check Account Balance
160  11. Make Payment on Loan
161  12. Check Loan Balance
```

```
162   13. Change Address
163   14. Change Phone
164   99. Quit"""
165   }
```

Each of the options is implemented in its own method. The Scaladoc comments have been left off of this code. This was done in part to keep the length of the code more reasonable. The other reason is that different parts of code have different commenting requirements. In this case, the `class`es in the previous section are library code. Libraries need to be well documented, especially those for things like the language API, because they are going to be used by multiple programmers, who often do not have access to the code or should not be spending time looking at the code to determine how things are done. Our main `object` here does not contain anything that is used outside of itself. This is reflected by the fact that the methods, other than `main`, are all `private`. This makes documentation comments unneeded, though explanation comments could be beneficial for more complex sections of code.

Note that we also could have written this by putting the code from the various `private` methods into the `match` to make one really long `main` method. That approach would require comments to be understandable. Breaking the work up into multiple methods makes the code easier to understand and maintain. The method names act like little comments that provide information to the reader. Having separate methods also potentially makes the code easier to test and debug.

Because many of you may still be very new to Scala, let's take a closer look at some of the lines of code in the `match`. In line 23, we see that `createCustomer` is called and we pass our `bank` object as an argument. This function gathers all the necessary customer data from the user and if it can add the new customer to the bank, it returns the customer and we set the current `customer` to the `Some` of that customer, which is the customer that was just created.

Before we look at the other `case`s, we need to keep in mind that `customer`, `account`, and `loan` are `Option` types that will contain either `None` or `Some` and if it is `Some`, then they will contain the current customer, account, or loan value respectively.

Now let's look at how `map` works on an `Option` type. In line 25 we see `map` called on `customer`. If `customer` is `None`, then nothing will happen; however, if `customer` is `Some` then `map` will apply the `createAccount` method to the current customer value, which will subsequently add that customer to the `bank` and return that customer's account to `account`. You can see a similar usage of `map` in line 27.

Next we will look at how `foreach` is working on `Option` type. `foreach` applies a function to every element in a collection. In line 26 we see `foreach` called on `account`. If `account` is `None`, then nothing will happen; however, if `account` is `Some` then `foreach` will apply the `closeAccount` method to the current account value which will subsequently remove that account from the `bank`. You can see a similar usage of `foreach` in lines 30 through 36.

The key aspect of this code that readers really want to take in is how the `class`es from the previous section are used. When designing your `class`es, you need to keep in mind how various things will be used together. Where and how will objects be instantiated? How are they organized? You will get better at designing with experience.

2.7 Making Packages

You might have noticed that all of the files in the implementation in the previous section begin with the line `package oobasics.bank`. This is a `package` declaration. Large software projects that have a lot of code in them need to be organized. This is done by adding packages. In Java, packages are always represented as directories and creating packages in IDEs will generally mirror this arrangement. Scala does not require that source files be organized into a directory structure that mirrors the packages, but doing so is recommended to make it easier to find files. Packages can be nested just as directories can be nested. When directories are nested, their names are separated by forward or backslashes, depending on your operating system. Package names are separated by dots, just like those used between object and method names.

In a source file, you specify a package using the `package` keyword. If everything in a file is in a single package, you can place a line at the top of the file that specifies the `package`. You can also follow a package declaration with curly braces and the code inside that block will be in the specified package. In this book we will use the first approach based on the idea that packages should be reasonably large groupings of code and single files should not get too long.

Note that in IDEs, you need to be careful that the `object` and `package` names in files match those of files and directories if you want to run `main` method as applications in the IDE. If the file has a `package` declaration that does not match the directory location, or the name of the file does not match the name of the `object`, the IDE will likely tell you that it cannot find the `object` when you try to run it.

2.8 End of Chapter Material

2.8.1 Summary of Concepts

- The basic idea of an object is that it is something that contains both data and the functionality that operates on that data.

- `class` serve as the blueprints for making instance objects. The declarations in a `class` tell us what members go into the instance objects created from that `class`.

- Arguments to `class`es are not members of the `class` by default. Member data in `class`es requires a `val` or `var` declaration. We can turn them into members by putting `val` or `var` in front of the argument name.

- Member data can be declared either in the arguments to a `class` or in the body of the `class`. If the data member will always start with a particular default value, there is no reason to make the user pass in that default value, and the member should be declared inside of the body of the `class`. Otherwise that member should probably be part of the argument list.

- The keyword `private` is a visibility modifier. It tells Scala what parts of code are allowed to access things inside of a `class`, `object`, or `trait` declaration. Things that are `private` can only be accessed inside of the current declaration.

- Scala does not have a keyword for `public`. It is simply the default visibility if nothing else is specified.

- Public members can be accessed outside of the `class` using the dot notation.

- The `require` function is part of the standard library, and can be used to perform a check during runtime. It is intended to be used as a precondition of a method. It has to have one `Boolean` argument that is evaluated and must be true. If it is not, an exception is thrown. A second argument with the message to print if it fails is optional, but strongly recommended. There is a related method called `assert` that should be used to generally state when some condition should be true.

- The **software life cycle** is the different steps taken in creating a piece of software.

  - The process of figuring out what the program is actually supposed to do is called **analysis**.

  - Laying out how code is going to be written is called **design**.

  - **Implementation** is the actual act of writing the software.

  - After code has been written, you have to **test** and **debug** to correct any errors that have been put into the code.

  - After software is released, the continued work on it for enhancements and further bug fixes is called **maintenance**.

- **UML** stands for Unified Modeling Language and it is a standard for a number of diagram types that are used for the software development process. This chapter introduced two of them.

  - The **Use Case Diagram** is a style of diagram used during analysis to help consider the ways in which users and other outside entities interact with the software.

  - A **Class Diagram** is a diagram that shows a rough sketch of major code groupings like `class`es, `object`s, and `trait`s as well as relationships between them.

2.8.2   Exercises

1. Describe the different phases of the software life cycle.

2. Imagine that you are writing software that has to model a car. Draw a UML diagram for how you might break down that problem.

3. Imagine that you are writing software for a retail clothing store that specializes in clothing for nurses. The store sells a variety of jackets, tops, and pants. Draw appropriate use case and class diagrams. One class you might consider making is a ClothingItem class that manages clothing inventory details. Make sure you include a CashRegister class that allows for the purchase and return of inventory.

4. You are asked to design a simple multi-player trivia game. Assume that each question will have multiple answers associated with it, but that only one answer is correct. The player that answers the most questions correctly wins. Draw appropriate use case and class diagrams to help with the analysis and design of this problem.

5. Write a class named Employee that holds information about an employee such as name, ID number, job title, department, email, and phone number. Once you have written the class, write a program that creates 3 employees.

6. Write a class named Movies that holds information about a home movie collection. It should hold information such as title, year, description, genre, and rating. Once you have written the class, write a program that creates 5 movies.

7. Annmarie Eslinger runs a community adult education program. The community has fall, spring, and summer sessions. Annmarie needs to have a system that will track the adult ed classes as well as the students who register for each class. A class is typically only offered once a week for an 8-week session. Each class will have a maximum number of people who are allowed to register for that class. Some classes may have an additional fee associated with them for things such as supplies, court rental fees, certification costs, etc. Draw appropriate use case and class diagrams to help with the analysis and design of this problem.

8. Speedy Automotive is a local automotive shop that specializes in paint and body repair, but also performs services such as oil changes. When a customer brings a car to the shop, the shop manager gets the customer's personal information. Then they record information about the car such as make, model, and year. The manager will also want to provide quotes for the services or repairs requested by the customer. These quotes should show estimated labor charges, estimated parts charges, estimated disposal charges, sales tax, as well as the total estimated charges. Draw appropriate use case and class diagrams to help with the analysis and design of this problem.

9. Zoomafoo day care is a local animal day care business that provides both day care and shortterm boarding services for cats and dogs. Their staff includes both cat and dog specialists. Animal clients can be there either half days or full days. Their owners can request medication administrations, extra walk sessions, extra play time sessions, or specialized feeding. Each of these additional services carries an additional charge. Zoomafoo also needs to be able to record and bill for the required services. Draw appropriate use case and class diagrams to help with the analysis and design of this problem.

10. Beautiful Lawns is a local lawn maintenance company. They provide complete lawn care services for their customers which include mowing, blowing, edging, weed control, fertilization, insect control, over seeding, and core aeration. Customers can call Beautiful Lawns and request a quote. A customer service representative needs to gather general customer information as well as what services the customer wants, how often they want the services performed, and the size of the yard. Draw appropriate use case and class diagrams to help with the analysis and design of this problem.

2.8.3 Projects

There are no specific projects for this chapter. Projects related to this chapter will be available at the end of the next chapter as we wanted to make sure you had a firm understanding of object-orientation before you started them.

Chapter 3

Details of Object-Orientation in Scala

Chapter 2 introduced the key concepts of object-orientation and how you should think about problems in an object-oriented way. In this chapter, we will explore some of the details of Scala as a language and how object-orientation is handled.

3.1   The 2D Vector classes

To help motivate and illustrate many of the examples in this chapter, we are going to create a number of different implementations of 2D vectors. In case you do not recall from your previous math and physics course (or the movie *Despicable Me*®), a vector is a quantity that has a magnitude and a direction. They can be stored in a number of different ways, including polar coordinates and Cartesian coordinates. We will use the Cartesian coordinates. This means that each instance stores values for x and y components. We will also provide basic operations for doing things like adding two vectors. Figure 3.1 illustrates the addition of two vectors to produce a third vector.

You might wonder why we would want a class for this. After all, chapter 2 described how we want to focus on the things or responsibilities in a problem when we design our classes. There are lots of applications in areas like math and physics that use the concept

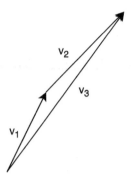

FIGURE 3.1: This figure shows three vectors, where $v_3 = v_2 + v_1$.

of a vector, but they can also be useful for anything that has locations and movements. That could be the sprites (small images) for characters in a graphical game, or perhaps entities in some type of simulation. There are several projects that appear at the end of chapters in this book that could benefit from having a vector in 2D or 3D, so it is an example that many of you might find useful to include in the code you write while going through this book.

Our `classes` will store the x and y components and provide basic operations on them. In order to demonstrate different approaches, we are going to have two basic lines of these types, one for immutable vectors, and one for mutable vectors. Our first versions of these are shown here, and they primarily use elements that were discussed in chapter 2.

```
/**
 * A basic 2D immutable vector.
 */
class Vect2D_a(val x: Double, val y: Double) {
  def plus(v: Vect2D_a) = new Vect2D_a(x + v.x, y + v.y)
  def minus(v: Vect2D_a) = new Vect2D_a(x - v.x, y - v.y)
  def scale(c: Double) = new Vect2D_a(x * c, y * c)
  def magnitude = math.sqrt(x * x + y * y)
}
```

As you can see, methods called `plus`, `minus`, `scale`, and `magnitude` are provided. The `plus`, `minus`, and `scale` methods all produce new instances of the `class` without modifying the original in any way. Because this `class` is immutable, a fact that should be apparent from the fact that the only data values are `val Doubles`, it has to be done this way as we cannot modify the instances once they are created.

Given what we currently know, a good approach to testing our `class` is to create an object with the same name as the class and give it a `main` method so it can be run as an application. The only reason for naming the object the same as the class is to keep things nicely organized.[1]

```
object Vect2D_a {
  def main(args: Array[String]): Unit = {
    val v1 = new Vect2D_a(1, 2)
    val v2 = new Vect2D_a(2, 2)
```

---

[1]An `object` that is declared in the same file as a `class` and which has the same name is called a companion object. These will be discussed in detail in section 3.3.1.

```
5      val v3 = v1.plus(v2)
6      println(v3.magnitude)
7    }
8  }
```

This code simply creates two instances of Vect2D_a and then uses the **plus** method to add them. It finishes off by printing the magnitude of the resulting vector, which should be 5.0 as the components of **v3** have lengths of 3 and 4.

One could also choose to make a mutable version of the 2D vector. An implementation of that might look like the following.

```
1  /**
2   * A basics 2D mutable vector.
3   */
4  class MutableVect2D_a(private var _x: Double, private var _y: Double) {
5    def x = _x
6    def y = _y
7    def setX(newX: Double): Unit = _x = newX
8    def setY(newY: Double): Unit = _y = newY
9    def plus(mv: MutableVect2D_a): MutableVect2D_a = {
10     _x += mv.x
11     _y += mv.y
12     this
13   }
14   def minus(mv: MutableVect2D_a): MutableVect2D_a = {
15     _x -= mv.x
16     _y -= mv.y
17     this
18   }
19   def scale(c: Double): MutableVect2D_a = {
20     _x *= c
21     _y *= c
22     this
23   }
24   def magnitude = math.sqrt(x * x + y * y)
25 }
```

Note that in this implementation, the names **x** and **y** are given to methods that provide access to the **private vars** named _x and _y. There are also methods that start with the word **set** to assign new values to these members. The use of "setter" methods is standard in Java and other languages, but we will see shortly that Scala provides an alternate approach that you should use when you want outside code to be able to set member variables.

A short section of code that uses this **class** is shown here. It is very much like that shown for Vect2D_a, except that we do not introduce a variable for v3. Instead, the **plus** method here mutates the value of v1 and we print the magnitude of that instance at the end.

```
1  object MutableVect2D_a {
2    def main(args:Array[String]):Unit = {
3      val v1 = new MutableVect2D_a(1, 2)
4      val v2 = new MutableVect2D_a(2, 2)
5      v1.plus(v2)
6      println(v1.magnitude)
7    }
8  }
```

3.1.1 `this` Keyword

The `plus`, `minus`, and `scale` methods in this implementation mutate the values of `_x` and `_y`, then give back an instance of `MutableVect2D_a`. The reason that they give back an instance is that this allows you to chain together calls. For example, you could do `v1.plus(v2).minus(v3).scale(5)`. This would add v2 to v1, then subtract v3 from v1, and finally scale the value of v1 up by a factor of 5. In order to make this work, we have to have a way to refer to the current object that a method is called on inside of that method.

Referring to the current object is not something we have to do all that often. By default, when you write code in a `class` that calls a method or uses a member of that `class`, the call is made on the current object or the value from the current object is used. All calls are implicitly made on the current instance. That is exactly what you want most of the time and given the scoping of variables, it feels very natural. This implicit specification of an object prevents you from having to specify one. That is fine in general, but for situations like our mutable `plus` method, we need to be able to put a name to the current object. This type of situation also arises when you need to call a function/method that needs an instance of the object as an argument and you want to use the current instance. To refer to the current instance in a method, you use the `this` keyword. That is the reason that the word `this` appears on the last line of the `plus`, `minus`, and `scale` methods.

Scala is implicitly adding `this.` in front of your calls to methods and when you access local members. For example, the line

```
_x += mv.x
```

is really short for

```
this._x += mv.x
```

Scala simply added the `this` for you. You can add in references to `this` explicitly in your code when they are not required, but only do so if you feel that they are truly helpful to the reader. Otherwise, they make the code longer without any real benefit. Normal Scala style would have you only use the `this` keyword when it is actually required.

3.2   Special Methods and Symbolic Names

There are a few method names that are special when you put them into `class`es in Scala. In a way, these special methods are rules that you have to learn as special cases. Such things often weigh upon a language. In this case there are not many of them, and the power they provide is quite remarkable.

3.2.1   Scala Naming Rules/Symbolic Names

To this point, all of the names we have used for variables, methods, `class`es, etc., in Scala have used names that are allowed in most all programming languages. These names begin with an underscore or a letter, followed by any number of letters, numbers, or underscores. This is what is allowed in Java and Python. Scala also allows operator symbols to be included in names with certain restrictions. The operator symbols are +, -, *, /, %, |, &, ^, ~, !, <, >, =, ?, $, \ and :. Names can be made that include only symbol characters, or have a standard name followed by an underscore, followed by symbol characters.

This means that you can write a method named +. Indeed, the `Int` and `Double` types do just that so that we can add them. The ability to use those methods as operators is a general capability of the Scala language. You can use any method call that takes a single argument using operator syntax. That means you can leave out the dot and the parentheses and just put the operator between the two operands. What makes this more interesting in Scala is that methods with symbolic names are easily distinguished from other tokens. So when the Scala compiler sees v1+v2, it knows that the + is a separate token from the v1 and v2 without having spaces between them. On the other hand, `a min b` needs the spaces to make things clear, but it demonstrates that a method called `min` can be used as an operator just as easily as a method called +.

We can see how this is useful by making a second version of our vector `class` that has symbolic operator method names.

```scala
class Vect2D_b(val x: Double, val y: Double) {
  def +(v: Vect2D_b) = new Vect2D_b(x + v.x, y + v.y)
  def -(v: Vect2D_b) = new Vect2D_b(x - v.x, y - v.y)
  def *(c: Double) = new Vect2D_b(x * c, y * c)
  def /(c: Double) = new Vect2D_b(x / c, y / c)
  def magnitude = math.sqrt(x * x + y * y)
}
```

The real advantage of this comes out in the usage of the `class`. With this version, we can say v1 + v2 instead of v1.plus(v2).

```scala
object Vect2D_b {
  def main(args: Array[String]): Unit = {
    val v1 = new Vect2D_b(1, 2)
    val v2 = new Vect2D_b(2, 2)
    val v3 = v1 + v2
    val v4 = v3 * 3
    println(v3.magnitude + " " + v4.magnitude)
  }
}
```

Recall that v1 + v2 is shorthand for v1.+(v2). Similarly, v3 * 3 is short for v3.*(3). So the method is being called on the first value, with the second value as the argument to the method. What we have written here does not allow us to write 3 * v3 because that would try to use a method called * in the `Int` or `Double` type. It is possible to write code in Scala that allows this, but it requires the use of a feature called implicit conversions, which we do not introduce here but are covered in the appendices on the book's website.

In addition to names like + and *, you can also make names like ** and *+++. As with so many things in life, just because you can use symbolic operator names for methods does not mean that you should. They have their place. Part of the reason that we picked the example of a 2D vector was to demonstrate symbolic methods. They should only be used in situations where their meaning is clear. Putting mathematical operations on mathematical constructs is a great usage of symbolic method. The Scala collections also make use of symbolic methods, for example, ++ is generally used to concatenate collections. However, if you abuse symbolic methods, it can lead to code that is very difficult to read. Changing from text names that people can read to arbitrary symbol sequences will make it nearly impossible for you, or others, to understand what is going on.

As an example of this, we might want to add methods for dot products to our 2D vector class. If you do not recall what those are, that is not a problem here. What matters is that in mathematics, these are commonly represented as symbols, $a\dot{b}$ for a dot product and $a \times b$

for a cross product. Those are not valid operator symbols, indeed, you do not find keys for them on most keyboards. It might be tempting to make up your own combination of symbols for them such as *+ for dot product and *+* for cross product. Even if you could rationalize these choices, it would be a bad idea, as other people reading your code might not find your choices to be meaningful. It is better to stick with the method names dot and cross. Those can be used just as well with the infix notation, and all readers will know what the code is doing. When added to our earlier code, they might look like the following.

```scala
def dot(v: Vect2D_b) = x * v.x + y * v.y
def cross(v: Vect2D_b) = x * v.y - y * v.x
```

In the case of the mutable 2D vector, the method that had been called plus should not be called +. This is because it does not just add vectors, it modifies the contents of one of the vectors. This type of modification is typically done with assignment. So instead of calling the method +, it is better to call it +=. Here is a version of the mutable 2D vector class with appropriate symbolic operator method names.

```scala
class MutableVect2D_b(private var _x: Double, private var _y: Double) {
  def x = _x
  def y = _y
  def setX(newX: Double): Unit = _x = newX
  def setY(newY: Double): Unit = _y = newY
  def +=(mv: MutableVect2D_b): MutableVect2D_b = {
    _x += mv.x
    _y += mv.y
    this
  }
  def -=(mv: MutableVect2D_b): MutableVect2D_b = {
    _x -= mv.x
    _y -= mv.y
    this
  }
  def *=(c: Double): MutableVect2D_b = {
    _x *= c
    _y *= c
    this
  }
  def /=(c: Double): MutableVect2D_b = {
    _x /= c
    _y /= c
    this
  }
  def magnitude = math.sqrt(x * x + y * y)
}
```

This class defines +=, -=, *=, and /= for performing our standard operations in a manner that mutates the original value. Using of this class might look like the following.

```scala
object MutableVect2D_b {
  def main(args:Array[String]):Unit = {
    val v1 = new MutableVect2D_b(1, 2)
    val v2 = new MutableVect2D_b(2, 2)
    v1 += v2
    println(v1.magnitude)
  }
```

₈ `}`

As before, we have the mutating methods return the current object so that they could be done in sequence in a manner like `v1 += v2 -= v3 *= 5`. This type of thing is commonly done in many libraries, including the Scala collections. However, one could make a strong argument that it is not highly readable in this situation. If you do not find this style to be readable, you can easily make the assignment methods produce `Unit` instead of the current object, and then this type of code will not compile. Instead, all the operation will have to happen on separate lines.

Parsing Assignment Operators

We first saw operators like `+=` when working with types like `Int` and `Double`. For example, you might see a line like `i += 1` inside of a `while` loop to increment some integer variable. This usage is short for `i = i + 1` as there is no `+=` operator on the `Int` type. The `Int` type is immutable, so a `+=` operator does not make sense. That means that `+=` can either be a call to a method named `+=`, or it can be a call to the `+` method with an assignment to a `var`.

In general, when Scala sees a `+=` in an expression, or any other symbolic operator followed by an equal sign, it first checks if the object to the left of the operator has a method with that name. If it does, that method will be called. If such a method does not exist, Scala will try to interpret it as an assignment to a `var` with the method having the name that comes before the equal sign. The following code illustrates this difference with our vector types.

```scala
object AssignmentOps {
  def main(args: Array[String]): Unit = {
    var v1 = new Vect2D_b(1, 2)
    val v2 = new Vect2D_b(2, 2)
    v1 += v2   // v1 = v1.+(v2)

    val mv1 = new MutableVect2D_b(1, 2)
    val mv2 = new MutableVect2D_b(2, 2)
    mv1 += mv2 // mv1.+=(mv2)
  }
}
```

As the comments indicate, the first usage of `+=` performs the method called `+` and then does an assignment to a `var` instead of calling a method called `+=`.

Understanding this is not just helpful for writing and using code for mutable types, it also can be useful for understanding error messages. For example, if you were to accidentally declare `v1` to be a `val` instead of a `var` in this example, the error message you would get says, "`value += is not a member of oodetails.Vect2D_b`". The solution to this problem is not to introduce a `+=` method into the type, but instead to make the declaration use a `var`. Keep this in mind. Error messages often point toward a possible solution, but since the compiler does not know what you were trying to do, that solution might not be the correct one for your situation.

Precedence (highest at the top)
other special characters
* / %
+ −
:
= !
< >
&
^
all letters
assignment operators

TABLE 3.1: This table shows the precedence rules for operators in Scala. The precedence is based on the first character in the operator.

3.2.1.1   Precedence and Associativity

When you call a method using the normal method notation, the order in which things happen is obvious. This is because the argument to the method is found inside of parentheses and the parentheses determine the order. When you use operator notation, there have to be PRECEDENCE rules. You already know that 2+3*4 is 14, not 20. That is because multiplication has higher precedence than addition. This is true in Scala as it is in math. However, the + and * are simply names for methods in `Int` in this expression. So how does Scala decide that the * should happen before the +? When methods are used in operator notation, Scala uses the first character in the name to determine precedence. Table 3.1 lists the different levels of precedence that are used.

In the case of + and *, the first character happens to be the only character. The table tells us that * has higher precedence than + so it is done first. If you put methods called *+* and +*+ in a class, the *+* would have higher precedence by virtue of starting with *. If you also put a method named ^*^, it would have lower precedence than either of the other two because of the low precedence of ^. These types of precedence rules exist in most other programming languages as well. The only difference with Scala is that you have the capability of defining your own methods that can be used as operators and have the appropriate precedence.

What if two operators are are the same level of precedence? Which comes first? For example, is 5-3+2 equal to 4 or 0? If the − happens first you get 4. If the + happens first you get 0. The order in which operators of the same precedence occur depends on ASSOCIATIVITY. You are used to operators being LEFT-ASSOCIATIVE. This means that the furthest left operator of a given precedence is done first. Such is the case of + and − and hence the value of the expression 5-3+2 is 4. However, there are some operations that should be RIGHT-ASSOCIATIVE. An example of this from standard math is exponentiation. For example $3^{3^3} = 3^{(3^3)} \neq (3^3)^3$. The numbers are grouped from right to left here, not the normal left to right of other operations.

An operator in Scala is right-associative if it ends with a :. In addition to being right-associative, it is called on the object to the right side of it and the value on the left side is the argument to the method. We have actually seen this behavior before with the :: operator.

```
3 :: 4 :: lst
```

This expression only makes sense if the :: operator is right-associative. The :: operator is not defined on Int, it is defined on List. So Scala sees this as the following.

```
3 :: (4 :: lst)
```

This happens to be the same as this.

```
lst.::(4).::(3)
```

There are quite a few other methods in the API that work in this way. We will see some more in chapter 6.

3.2.2 Unary Operators

Not all operators take two arguments. Operators that take only a single argument are called UNARY OPERATORS. Consider the - to make a number negative. With a value like -5 you could say it is just part of the number. However, with -(a+7) the role of the - is clearly that of an operation that is performed on the result of a+7. The +, !, and ~ methods are other examples of prefix unary operators. The last two perform Boolean and bit-wise negation respectively.[2]

As it happens, negation is an operation that is defined on 2D vectors as well. Given a vector (x, y), the negation of that vector should be $(-x, -y)$. You can write special methods in a class that are interpreted as prefix unary operators. To do so, simply name the method unary_*op*, where *op* is replaced by +, -, !, or ~. So if we want to be able to do negation on our most recent immutable 2D vector, we could add the following method.

```
def unary_-() = new Vect2D_b(-x, -y)
```

Once that is added to the class, you can do things like -(v1+v2).

This operation makes perfect sense on our immutable 2D vector where all the operations make new instances of the class. It is not as clear that it makes sense in the mutable version, where the methods tend to mutate existing instances. Just as we did not want a method named + to mutate the instance it was called on, most programmers would be very surprised if having an expression that included -mv1 were to change the value of mv1 to point in the opposite direction. On the other hand, including one method that produces a new instance amid all the other methods that do mutation would likely confuse people using the library. For these reasons, it is probably best to not include unary_- in the mutable version.

3.2.3 Property Assignment Methods

Another form of method that Scala interprets in a special way is property assignment methods. These are only significant in classes that allow mutation. In the MutableVect2D_b class, we had included methods called setX and setY. This style is common in Java and Python, but it is not the preferred method in Scala. Using the "setter" methods, one would write a line like the following.

```
v2.setX(3)
```

This works, and is reasonably readable, but the creators of Scala felt that it would be much nicer to be able to do that with a line like this.

---

[2]Bit-wise negation operates on the individual bits in an integer value, turning all zeros to ones and all ones to zeros. This operator is not used frequently, but comes in very handy when doing bit manipulation operations.

```
v2.x = 3
```

After all, this is how you would set the value of x if it were a public **var** declaration. Indeed, there were several design choices in Scala that were made to make it so that one could not distinguish the use of member data and methods in code outside of a **class**. Part of this is allowing methods to be called without parentheses, as we see in our 2D vector example with **def** x and **def** y. Having property assignment methods completes the ability to interchange member data with methods.

If you put a method in a **class** that has the form **def** *name_*=(value:Type), it allows you to write code that does assignment to *name*. So we want to replace setX and setY in our mutable 2D vector with the following.

```
def x_=(newX: Double): Unit = _x = newX
def y_=(newY: Double): Unit = _y = newY
```

These methods simply set the values of _x and _y. It turns out that this is what would be produced by the compiler if we simply had a public **var** in the **class**. For this reason, it is actually acceptable in Scala to write your code with a public **var** when there are no restrictions on the values that a property of the **class** can take on. This is not generally acceptable in Java. The reason is that if you later decide that you need restrictions, putting those in place in Java would require rewriting code in any place that used the **class**. In Scala, it only requires adding accessor and assignment methods for that property and changing the declaration to use a different name with a **private var**.

To help you understand this, consider the bank account example from chapter 2. We do not want to have a public **var** for the balance, because that would allow any outside code to arbitrarily set the balance. That means that code could easily set a balance to be negative, and when the value is set, that change is not automatically logged. In practice, we really need all alterations for the balance to either be deposits or withdrawals. That does not mean that we cannot give programmers using our account the ability to act like they are setting the balance with an assignment.

To help you understand this, let us assume that you did want to be able to enter something like myAccount.balance = 700. We could have this ability with a public **var**, but we decided that was too little control. The functionality could be retained by adding the following method.

```
def balance_=(newBalance:Int) = bal = newBalance
```

Thanks to the way Scala handles assignment operators, this will let both of the following lines work.

```
myAccount.balance = 700
myAccount.balance += 40
```

So doing += will work like a deposit and -= will work like a withdrawl. The downside is that neither returns a **Boolean** to let you know if it worked. As written, neither checks the value being deposited or withdrawn either.

This particular method leaves things too open. We might as well have the **var** because we are doing exactly what the **var** would do. However, the advantage of an assignment method is that it can be more complex and have more logic behind it. For example, the balance_= method could be altered to the following.

```
def balance_=(newBalance: Int): Unit = {
  if (newBalance > _balance) deposit(newBalance - _balance)
  else if (newBalance < _balance) withdraw(_balance - newBalance)
```

```
}
```

This version will reuse the code from **withdraw** or **deposit** including any error checking, logging, etc. Using this version, the act of trying to assign a negative value will not alter the balance, and ideally a message to that effect would be logged.

Note that you only include an assignment method for properties that are mutable so there are many situations where they are left off. If you are programming in a more functional style, you will not have these assignment methods. It is only when objects specifically need to be able to mutate that these are helpful. The real advantage in that case is that the code can look like a normal assignment, but it will call the method that can do various types of error checking.

3.2.4   The apply Method

Another method that Scala treats in a unique way is the **apply** method. This is the method that allows objects in Scala to be treated as functions. You write **apply** just like any other method that you might put in a **class**. You can call it just like any other method as well. However, Scala will allow you to call the **apply** method without using a dot or the name **apply**. When you remove those, it makes it look like you are treating the object as a function. Indeed, this is how all functions work in Scala. As was said very early on, everything in Scala is an object. That includes functions. A function is just an object of a type that has an **apply** method that takes the appropriate arguments.

One use of **apply** is how you have been indexing into all of the collection types. Consider the following code that you could write in the REPL.

```
scala> val arr = Array(5, 7, 4, 6, 3, 2, 8)
arr: Array[Int] = Array(5, 7, 4, 6, 3, 2, 8)

scala> arr(3)
res0: Int = 6

scala> arr.apply(3)
res1: Int = 6
```

The call to **arr(3)** is actually just a shortcut for **arr.apply(3)**. Scala is doing nothing more here than assuming the presence of the call to **apply**.

We can illustrate the creation of an **apply** method in the 2D vector. In this construction, we assume that the x component has an index of 0 and the y component has an index of 1. All other indices are invalid.

```
def apply(index:Int): Double = index match {
  case 0 => x
  case 1 => y
  case _ => throw new IndexOutOfBoundsException(s"2D vector indexed with $index.")
}
```

This might not seem like a very useful thing to do, but when done in conjunction with some of the techniques that we will learn about in later chapters, it can greatly broaden the capabilities of our **class** to be used in manners similar to other collections. As it is, it provides the ability to access the different components in loops.

As with the other special methods, you should not just add an **apply** method to everything because you can, even if it seems to make sense. Consider the bank example. The **Bank class** has multiple accounts in it, and you might be tempted to make an apply method that works similarly to `findCustomer`, looking at a customer by name or ID. The problem is that this `apply` method would be very ambiguous. The name of the `apply` method is fixed, so you cannot easily indicate if you should be passing in a name or an ID, or if it is intended to be used to look up customers, accounts, or loans.

Another common usage of `apply` will be discussed in section 3.3.1. This one will be very generally applicable to your `classes`.

When Parentheses Matter

We have seen that methods that do not take arguments can be written and called without parentheses. As a rule of thumb, methods that have side effects, especially mutation, should always be written and called with parentheses. Methods that do not have side effects, which could almost be replaced by a `val`, should be written and called without the parentheses.

Things change when you call a function, an object with an `apply` method. The parentheses are what tells Scala that you want to invoke the function, even if the function does not take any arguments. To see an example of this, consider the following REPL session. We begin by defining a function, `f`, that does not take any arguments and calls `math.random`. When you create a function in Scala with the `=>` notation or the underscore notation, Scala makes a new object that does whatever you specified as the body of the function in an appropriate `apply` method. This is followed by using `f` with and without parentheses.

```
scala> val f = () => math.random
f: () => Double = $$Lambda$1247/3927958430411c6d44

scala> f()
res0: Double = 0.7870833012654959

scala> f
res1: () => Double = $$Lambda$1247/3927958430411c6d44
```

The call to `f()` is what you want to do most of the time. If you do not put the parentheses, you get back the object for the function itself. There are situations where this is useful, such as passing the function to some other function or method, but you need to be aware that the parentheses are required in order to invoke the `apply` method.

It is interesting to note that if you explicitly call `apply`, because it is a method, the parentheses are not required.

```
scala> f.apply
res5: Double = 0.721683560738538
```

3.2.5   The update Method

The companion method to `apply` is `update`. This method gets called when you do an assignment into an expression that would otherwise be a call to `apply`. Consider the example array named `arr` declared in the last section. You can alter a value in the array with an expression like `arr(3) = 99`. The `arr(3)` part of this would normally be an invocation of `apply`, but that will not work in this case because you cannot do an assignment into the result of a function. To make this possible, Scala expands that expression out to `arr.update(3, 99)`.

One of the primary differences between mutable and immutable collections is simply that mutable collections define an **update** method, and immutable ones do not. This is why, if you ever tried to assign a value to an index in a `List`, the error message that you got said `error: value update is not a member of List[Int]`.

Clearly we do not want to put an **update** method in our immutable 2D vector, but it makes perfect sense to put one in our mutable 2D vector. Such a method might look like the following.

```scala
def update(index:Int, value:Double): Unit = index match {
  case 0 => x = value
  case 1 => y = value
  case _ => throw new IndexOutOfBoundsException(s"2D vector indexed with $index.")
}
```

After you have included this in the **class**, it is possible to write lines of code like `v2(0) = 99`, which changes the x value to 99.

Advanced Visibility Options

Scala has more flexibility in visibility modifiers than what has been described so far. Either `private` or `protected` can be followed by square brackets and the name of an enclosing scope of a `package`, `class`, `trait`, or `object` type. (The `trait` construct will be introduced in chapter 4.) This will make that member publicly visible for everything at that scope and closer to the declaration. The primary benefit of this is that you can give special access to closely related bits of code without making a member public to the whole world.

For example, while we have not shown it, all the code for this chapter is in a package called `oodetails`. Imagine we had some method named `somewhatPrivateMethod` in a **class** that should not really be called by any random code, but which was useful for many of the **classes** and/or **objects** in our package. We could prefix that method with `private[oodetails]` to provide exactly this access. To code outside the package, it will appear as if it is private. However, code inside the package will see it as if it were public.[3]

The square brackets can also include the keyword `this` to make something less visible than the standard `private`. Declaring a member to be `private[this]` means that is it `private` not only to other pieces of code, but also to any other instances of the enclosing **class**. Normally if a method gets hold of a reference to another object of the same type, it can access `private` members. This will not be the case if the member is `private[this]`.

A good example of this usage would be our mutable 2D vector. Using the plain `private` visibility, we could accidentally have one 2D vector modify the values in another one by directly altering `_x` and `_y`. It would be an even more important addition to the `Account` and `Loan` **classes** in our bank example. As we wrote them earlier with plain `private`, one instance of `Account` could change the balance in another instance of `Account` without going through the `deposit` or `withdraw` methods. By changing the `_balance` to be `private[this]`, we could have the language prevent that type of behavior.

3.3   object **Declarations**

Earlier we saw `object` declarations as they are used to create applications. Once again, the `object` declaration creates a single object, commonly called a singleton object, in the current scope with the members and methods that you put in it. Unlike a `class`, you cannot pass an `object` any arguments. This is because you do not make the `object` with **new**. The declaration of the `object` is what creates the object.

Just like with a `class`, you can put methods and other members inside of an `object` declaration, and you can provide them with appropriate visibility modifiers. Any code that you put in the body of the `object` will be executed when it is created. If it is at the top level in a file, not contained in anything else, that will happen the first time it is referred to. If it is nested in some other scope, then the `object` will be created and its interior code will be executed when that code is executed.

Even though an `object` declaration does not create a new type, we use a capital letter as the first letter in the name to distinguish it from normal instantiated objects. To use the members or methods of an `object`, use the `object` name and call it just like you would for an instance of a class. `object` declarations can be used in any scope where you want an object, but you only want one object of that type. At the top level they are typically used as a place to organize functions, but they have far more versatility than just providing a place to collect methods.

3.3.1   Companion Objects

The most common use of `object`s is as COMPANION OBJECTS. A companion `object` is an `object` that has the same name as a `class` and appears in the same file as the `class`. The companion `object` has access to **private** members of the class it is a companion with. Similarly, the `class` can see **private** elements in the `object`.

You might have wondered why it is that when we are building objects in Scala, sometimes we use **new** and sometimes we do not. For example, you can make a List with code like `List(1, 2, 3, 4)`. The reality is that making a new object always invokes **new**. When you do not type it, it means that you are calling other code that does. When you use the name of the type without **new** to build an object, you are calling the **apply** method on the companion `object` and that **apply** method is calling **new**. So `List(1, 2, 3, 4)` is really short for `List.apply(1, 2, 3, 4)`.

It is very common in the Scala libraries to include a companion object with various **apply** methods for whatever forms you want users of your `class` to be able to use to create them. This is particularly helpful when the construction of the instance object requires significant work processing the data that the user provides into the form that the `class` needs for construction. The bank account could be an example of this. Though we used a very simple method to generate new account numbers without using companion objects, it is very likely that banks would want to use more complex algorithms, and it would make sense to have code in the `Account` companion object do the appropriate calculations. Below, we can see an example of a class with a companion object with various **apply** methods used for constructing a student. The full code for the `Student` `class` was on page 57.

```scala
class Student(
    val firstName: String,
    val lastName: String,
    private var _quizzes: List[Int] = Nil,
```

```scala
    private var _assignments: List[Int] = Nil,
    private var _tests: List[Int] = Nil) {
  ...
}


object Student {
  // This apply method takes the same arguments as the class and just passes them
       through
  def apply(firstName: String, lastName: String, quizzes: List[Int] = Nil,
            assignments: List[Int] = Nil, tests: List[Int] = Nil): Student = {
    new Student(firstName, lastName, quizzes, assignments, tests)
  }
  // This apply method reads in student information from a file
  def apply(filename: String): Student = {
    val source = io.Source.fromFile(filename)
    val lines = source.getLines()
    val firstName = lines.next
    val lastName = lines.next
    val quizzes = lines.next.split(" ").map(_.toInt).toList
    val assignments = lines.next.split(" ").map(_.toInt).toList
    val tests = lines.next.split(" ").map(_.toInt).toList
    source.close
    new Student(firstName, lastName, quizzes, assignments, tests)
  }
}
```

The first `apply` method simply takes values that are passed through to the constructor for the `class`. It might be invoked with `Student("Jane", "Doe")`. The second `apply` method takes the name of a file and builds a new instance of `Student` based on the contents of that file. That second version could be invoked with `Student("StudentData.txt")`.

Companion objects are also places where you put methods that are generally helpful for dealing with the responsibilities of a `class`, possibly including other options for building them. Examples of these types of methods are `fill` and `tabulate`, found in the `Array` and `List` companion objects.

---

private Constructors and Multiple Constructors

Consider the example mentioned above where bank accounts use a complex algorithm to generate IDs, and those algorithms are implemented in the `apply` methods of the companion object. In this situation, you would probably want to force outside code to use these `apply` methods instead of directly calling `new` on the `Account`. You can do this by making the CONSTRUCTOR for the `class` `private`.

In languages like Java and Python, constructors are written much like separate methods, so it is immediately obvious how to make them private. In Scala, the arguments to the `class` play the role of the primary constructor. In order to make this `private`, simply put the `private` keyword in front of the argument list outside of the parenthesis. The following shows what that might look like for the bank account.

```scala
class Account private(
    private var _balance: Int,
    val customer: Customer,
    val id: String) {
```

Scala also provides a syntax for creating multiple constructors. That is to say that you can have different sets of arguments that are passed in to **new** at object instantiation. The syntax is to define methods with the name **this**. Such methods must be set equal to a call to a constructor of the form **this(...)**, where the parentheses have other arguments. For example, if bank accounts were commonly created with a balance of zero, we might include the following secondary constructor.

```scala
def this(c:Customer, id:String) = this(0, c, id)
```

Note that all constructors must eventually wind up calling the primary constructor, which is the argument list for the **class**.

This syntax for multiple constructors is presented for completeness. If there are different sets of arguments that can be used for instantiating objects, it is more common to have a companion object with multiple **apply** methods than to have multiple constructors.

3.4   Final Versions of 2D Vectors

We can take everything that we have introduced in this chapter, and put it all together in final versions of our mutable and immutable 2D vectors. We begin with the immutable version, which is simply called **Vect2D** to indicate that it is the final version. You should go through it carefully to see how the elements of each of the previous sections fits together in the completed code.

```scala
package oodetails

class Vect2D private(val x: Double, val y: Double) {
  def +(v: Vect2D) = Vect2D(x + v.x, y + v.y)
  def -(v: Vect2D) = Vect2D(x - v.x, y - v.y)
  def *(c: Double) = Vect2D(x * c, y * c)
  def /(c: Double) = Vect2D(x / c, y / c)
  def magnitude = math.sqrt(x * x + y * y)
  def dot(v: Vect2D) = x * v.x + y * v.y
  def cross(v: Vect2D) = x * v.y - y * v.x
  def unary_-() = Vect2D(-x, -y)
  def apply(index:Int): Double = index match {
    case 0 => x
    case 1 => y
    case _ => throw new IndexOutOfBoundsException(s"2D vector indexed with $index.")
  }
}

object Vect2D {
  def main(args: Array[String]): Unit = {
    val v1 = Vect2D(1, 2)
    val v2 = Vect2D(2, 2)
    val v3 = -(v1 + v2)
    val v4 = v3 * 3
```

```
25    println(v3.magnitude+" "+v4.magnitude)
26   }
27
28   def apply(x:Double, y:Double) = new Vect2D(x, y)
29 }
```

The companion object has both a **main** that provides simple tests of some of the features, as well as an **apply** for creating new instances of **Vect2D**. There is nothing that prevents a companion object from also being an application. This can be helpful for simple testing. We will learn how to do proper testing in chapter 7. Note that all the places where we create new instances of **Vect2D** have had the **new** removed, with the exception of the one in the companion object's **apply** method.

Up next is the mutable version. This code is a bit longer, even though we left out methods like **dot** and **cross**. This is primarily because the modification of the current instance requires two separate statements, plus we return the current instance so that calls can be chained. This class also includes an **update** method and property assignment methods, which did not make sense in the immutable version.

```
1  package oodetails
2
3  class MutableVect2D private(private[this] var _x: Double, private[this] var _y:
       Double) {
4    def x = _x
5    def y = _y
6    def x_=(newX: Double): Unit = _x = newX
7    def y_=(newY: Double): Unit = _y = newY
8    def +=(mv: MutableVect2D): MutableVect2D = {
9      _x += mv.x
10     _y += mv.y
11     this
12   }
13   def -=(mv: MutableVect2D): MutableVect2D = {
14     _x -= mv.x
15     _y -= mv.y
16     this
17   }
18   def *=(c: Double): MutableVect2D = {
19     _x *= c
20     _y *= c
21     this
22   }
23   def /=(c: Double): MutableVect2D = {
24     _x /= c
25     _y /= c
26     this
27   }
28   def magnitude = math.sqrt(x * x + y * y)
29   def apply(index:Int): Double = index match {
30     case 0 => x
31     case 1 => y
32     case _ => throw new IndexOutOfBoundsException(s"2D vector indexed with $index.")
33   }
34   def update(index:Int, value:Double): Unit = index match {
35     case 0 => x = value
36     case 1 => y = value
```

```
37      case _ => throw new IndexOutOfBoundsException(s"2D vector indexed with $index.")
38    }
39  }
40
41  object MutableVect2D {
42    def main(args:Array[String]):Unit = {
43      val v1 = MutableVect2D(1, 2)
44      val v2 = MutableVect2D(2, 2)
45      v1 += v2
46      println(v1.magnitude)
47      v2.x = 3
48      v2(0) = 99
49    }
50
51    def apply(x:Double, y:Double) = new MutableVect2D(x, y)
52  }
```

The key thing to note is that unless you have a good reason to do otherwise, you should probably tend to make immutable **class**es in Scala. There are definitely times when making things mutable can be beneficial, but often, those benefits will not be as strong when you consider the complexities that are added to reasoning about and debugging the code. The immutable code can also have benefits when we begin doing multithreading, and when we have to do defensive copying to prevent other code from modifying our mutable instances.

3.5   case classes

One of the most useful elements of the Scala language is **case classes**. A **case class** is just like any normal **class**, but with a number of features added in without you having to write them. **case class**es are mostly used for simple immutable collections of data. It would actually be reasonable to make our **Vect2D** into a **case class**, but then we would not get to demonstrate several of the features in it as they would be done automatically.

By adding the word **case** in front of your declaration, Scala adds the following features for us. Some of these are things that we have discussed already, some we will discuss in coming chapters, and some are only briefly mentioned in the on-line appendices.[4]

- All arguments are **val** members by default, so you do not need to type that.

- Implementations of **hashCode**, **equals**, and **toString** are all provided. These will be discussed in chapter 4.

- A copy method with named parameters is provided, so that you can easily make near copies that differ in key fields.

- Causes the **class** to extend[5] **scala.Product**, giving it a few extra methods such as **productArity**.

- Causes the **class** to extend **Serializable**. We will discuss what this means in chapter 10.

---

[4]The appendices are available at http://www.programmingusingscala.net.
[5]Inheritance, which uses the keyword **extends**, is discussed in chapter 4.

- A companion object with the following methods.

  - An `apply` method for creation, so you never use **new** with a **case class**.
  - An `unapply` method that provides pattern matching. The `unapply` method is a more advanced feature and is only discussed in the on-line appendices.[6] This section looks at the pattern matching, which is the real benefit to users.
  - There are also methods called `curried` and `tupled`. Both of these create instances like `apply`, but the first takes the values one at a time in a curried manner and the other takes all the values in a single tuple argument.

The most important aspects listed here are the `copy` method and the ability to do pattern matching. To help illustrate these features, we will use the following **case class** for our examples.

```
case class Student(name: String, assignments: List[Int], tests: List[Int],
    quizzes: List[Int]) {
  ...
  def calcAverage:Double = ???
}
```

This **case class** represents a student in some program, and stores their names and grades. The grades are broken into three different groups. We have left the body of the **case class** mostly empty, as those details do not alter the examples. Indeed, it is fairly common for **case class**es to not have bodies. They simply store their data and have the methods that are implicitly provided to them.

As **case class**es are used to represent immutable values, one cannot assign to their member data. Instead, one needs to make a modified copy of the instance. This is such a common thing to do, that the **case class** has a method to help you do it. The `copy` method has as many parameters as the **case class**, but they are all named parameters that default to the current values. This means that you can actually call the `copy` method with no arguments and get an exact copy, though doing so is pointless.[7] The more meaningful usage of `copy` comes when you pass in new values for some subset of the arguments. Consider this expression, which adds a new test grade called `testGrade` to a student called `s` and gives it the name `s2`.

```
val s2 = s.copy(tests = testGrade :: s.tests)
```

The first thing in parentheses, `tests`, is an argument name. It tells Scala that what follows the equals sign should be used as the value of the parameter with that name. All the other parameters take their default values.

This might seem really inefficient, but it is not, because the fields are not all copied. Instead, both `s` and `s2` will share the same `name`, `assignments`, and `quizzes`. Only `tests` is not shared, and because of the way that the `List` and cons work, even most of the `tests` `List` is memory that the two share.[8]

For many Scala developers, the most significant thing that you get when you add the `case` keyword to a `class` is pattern matching. We have seen some uses of pattern matching before with tuples and `List`s. This is a common feature in functional programming languages. We will do a lot of pattern matching in chapter 8 when we talk about actors and the Akka library.

---

[6]The appendices are available at `http://www.programmingusingscala.net`.

[7]There really is never a reason to make an exact copy of an immutable object. The original cannot be changed, so you should pass around the original instead of making copies.

[8]The details of how cons works are part of the discussion in chapter 12.

In general, patterns can appear in Scala in places where you would create new variables, excluding method arguments. This means that we can use patterns with `val`, `var`, `for` loops, or `match` expressions. Consider that we have a second `case class` for an instructor in our system, something like the following.

```scala
case class Instructor(name: String, students:List[Student])
```

Somewhere in the program we get a variable called `participant` that we know refers to either a `Student` or an `Instructor`, but we do not know which.[9] If it is a `Student`, we want to get back the average of their assignment grades. If it is an `Instructor`, we want to get the average of the assignment averages of all of that instructor's students. That could be accomplished with the following code.

```scala
val assignmentAve = participant match {
  case Student(n, a, t, q) =>
    if (a.length == 0) 0.0 else a.sum.toDouble / a.length
  case Instructor(n, ss) =>
    val averages = for (Student(sn, a, t, q) <- ss) yield {
      if (a.length == 0) 0.0 else a.sum.toDouble / a.length
    }
    if (averages.length == 0) 0.0 else averages.sum.toDouble / averages.length
}
```

In this code we see `case`s with patterns for both `Student` and `Instructor`. The pattern has the name of the `case class` with variable names for each of the arguments. The arguments are actually their own patterns. A name that starts with a lowercase letter just happens to be a pattern that matches anything and binds the value of whatever it matches to that name. This example uses short names. You could certainly choose to use longer names if you wished. A general rule of thumb is that the larger the scope of a variable, the longer and more descriptive the name should be. Many of the variables created in these patterns have a scope of only one line, so single-letter names are not completely unacceptable here as they would be for variables with a larger scope.

You see that in the `case` for `Instructor`, there is another use of a pattern for `Student` inside of a `for` loop. Recall that one of the great strengths of patterns in `for` loops is that if a value does not match the pattern, the loop simply skips to the next value. In this situation, all elements in the `List[Student]` have to match the pattern, but that does not always have to be the case.

Every so often, you might find that you create a `case class` that does not need any arguments at all. In this situation, you should use a `case object`. These behave in many ways like a `case class`, but they do not take any arguments, as is standard for singleton objects.

Advanced Pattern Matching

Pattern matching is a powerful feature with a lot of options. We want to run through some of them here. First, in any situation, you can do a match on types. Assume that the `Student` and `Instructor` classes were declared as plain `class`es instead of `case` classes, with all of their fields declared as `val`s. That type of declaration would not

---

[9]Given what we know now, this is an unlikely situation as the type of `participant` would be something like `AnyRef`, which is not particularly useful. In chapter 4, we will see how such a situation might arise in a more meaningful way.

automatically get pattern matching capabilities, but we could still match on types as is done here.

```scala
val assignmentAve2 = participant match {
  case stu:Student =>
    if (stu.assignments.length == 0) 0.0 else stu.assignments.sum.toDouble /
        stu.assignments.length
  case ins:Instructor =>
    val averages = for (stu:Student <- ins.students) yield {
      if (stu.assignments.length == 0) 0.0 else stu.assignments.sum.toDouble /
          stu.assignments.length
    }
    if (averages.length == 0) 0.0 else averages.sum.toDouble / averages.length
}
```

Putting a colon in a pattern works just like putting a pattern elsewhere in Scala; it specifies a type. In this case, it specifies a constraint so that the pattern is only matched if the value has that type.

In general, patterns are built out of other patterns. This allows you to nest patterns. Given the `case class` for `Instructor` above, we could have a pattern like the following.

```scala
Instructor(iname, Student(sname, a, firstTest :: otherTests, q) :: Nil)
```

This pattern will match an instructor with a single student, and that student must have at least one test grade. Values will be bound to the names `iname`, `sname`, `a`, `firstTest`, `otherTests`, and `q`. This is actually nesting patterns four layers deep. At the top is the pattern for the `Instructor`. The second argument to that uses a pattern for a `List`, whose first element is a pattern on `Student`. Finally, the third argument for the `Student` pattern is another `List` pattern.

Finally, there is the issue of naming things. In our earlier examples, we used variable names for every field, even the ones that we never used. If you are not going to use some part of a pattern, there really is not a reason to come up with a name for it. In those situations, you should use the `_` pattern. Like a variable name, it matches anything, but you cannot refer back to it. In the examples above, we only cared about the assignment grades, in that situation, we should use a pattern like `Student(_, a, _, _)` so that we do not have to come up with names for the things that we are not going to use.

It is common to have the last `case` in a `match` expression be `case _`. This provides a default for the expression as it matches anything, without caring what it matches or giving it a name.

Sometimes you actually want a name for a pattern that you are also going to break apart. You can do this by putting a name followed by an `@` in front of a pattern. Consider the complex pattern for an `Instructor` from above. If you wanted names for the original `Instructor` and `Student` objects matched in that pattern, you could modify it in the following way.

```scala
instr@Instructor(iname, stu@Student(sname, a, firstTest :: otherTests, q) ::
    Nil)
```

This creates additional names called `instr` and `stu` that are bound to those parts of the pattern.

3.6   Encapsulation/Separating Interface from Implementation

Now that we have had a little time to play with object-orientation, and we have looked at many of the details of writing object-oriented code in Scala, it is time to take a little step back, and discuss the bigger picture. One of the biggest challenges in software development is controlling complexity. Object-orientation helps in certain ways with organizing code in a manner that can make the act of coding less complex. We want to take a bit of time to look at how it can do that.

The binding of data and functions together is called ENCAPSULATION. It might not seem like much, but it can have a profound impact on the way that software is developed. Having functionality bound into them makes objects far more self-sufficient. It can also free the programmer from having the think about many types of details. As software grows and becomes more complex, the ability to ignore various details becomes more important. When you are writing code that is dealing with an object, you do not want to have to think about how every method does exactly what it does. You just want to know that the object will handle its responsibilities in an appropriate manner.

One of the goals we would like to strive for in our object-oriented programs is to have SEPARATION OF INTERFACE AND IMPLEMENTATION. The interface of an object, or some other module of code, is the set of constructs that you can interact with directly. For objects that are instances of a `class`, that would be the public members and methods. The implementation of that interface is what makes the interface do what it is supposed to do. You want these things separate for a few reasons.

One reason is so that modifications to the details of the implementation do not break other code. If the interface and implementation are truly separated, then the code that makes calls to the interface does not depend at all on the details of the implementation. All that it depends on is that the elements of the interface adhere to a particular CONTRACT. That contract tells you what information is needed, what information you will get back, and what side effects, if any, will happen in between. It should not tell you how things are done in between. It should be possible to completely redo the details of how things happen, as long as the new approach does not violate the contract, and still have any code using the implementation work just fine.

Another reason for trying to keep interface and implementation separated is that the interface is typically much simpler than the implementation, and it benefits coders to not have to keep too many details in their head at once. Software has the ability to be arbitrarily complex. This is both a blessing and a curse. The flexibility that this provides is part of why software can do so many different things. When the complexity of the software goes beyond what the programmer can deal with, it is essential that some parts of the complexity can be safely ignored.

The reality is that you have been taking advantage of this separation in everything you have done in your previous programming experience, and you will continue to do so as long as you program computers. When you call `println`, you do not really care about all the details that go on between your call and having characters appear on a display. Thankfully you do not have to think about them either. All those details could change and as long as `println` keeps its contract of sending your information to the standard output you will not care that it has been changed. This is not just true of `println`, but of most of the Scala library code that we have been, and will be, using.

How do we separate interface and implementation in our own code? This is where the `private` visibility really comes into play. We have previously discussed the use of `private` from the standpoint of not wanting to allow other code to mess with things that it should not. Equally important is the idea that other code should not depend on the private details in order to work. Methods that outside code does not need to call should not be available to it. Generally speaking, outside code should not know the details of how you are storing your data either. For example, outside code should not care if we use `Int` or something else to store the balance of our bank accounts, as long as everything works as expected. Starting in chapter 12, we will be making `private class`es that are implementation details that no other code should really know about.

Consider our 2D vector examples. Early on in the process, we made the decision to store the data as Cartesian coordinates. It is also possible to store the information in polar coordinates. In that case, we store a length, r, and an orientation angle, θ, which we would write in the code as `theta`. The two representations are related by $x = r * cos\theta$ and $y = r * sin\theta$. The operations that we wrote are a bit easier to implement in Cartesian coordinates, but perhaps over time we find that we are doing more things that would be more efficient using the polar representation. At that time, we might decide to change the internal representation from storing `x` and `y` to `r` and `theta`. Thanks to the design of Scala, and the ability to change between `var`s and a combination of `def`s and property assignment methods, we can make this change, even if we had started with `x` and `y` being public `var`s. However, in Java, and most other object-oriented languages, starting with public member data would preclude this change. To enable this change, you have to start by making all the data `private` so that code that uses it does not depend on it. To state it differently, if the original design does not separate the interface from the implementation, you will be stuck with that implementation, even if another one makes more sense.

The takeaway message is that you should only make things public when the code that uses your `class`es needs them to operate. By doing this, you can produce a stronger separation between interface and implementation, which will make your code more robust and resistant to change.

3.7 Revisiting the API

As you work through this book, you will need to have a certain reliance on the Scala API. In the next few chapters we will be uncovering many elements of Scala that will help you to understand different parts of the API and you will be expected to go into the API to get information on what you can do with the libraries beyond what is presented in the book.

We have already learned enough to help a bit with understanding what some things mean in the API. When you open up the API and click on "scala" under "root" on the right side you see something that looks like 3.2. There are three different declaration styles in Scala that are indicated with the small circles next to the names in this list.

The circles next to the names contain one of three letters. The meaning of these letters is as follows.

- c - For a `class`.

- o - For an `object`.

- t - For a `trait`, which we will learn about in chapter 4. For now you can think of a `trait` as being like a `class`.

When you click on one of these, you are shown the methods and members that are defined

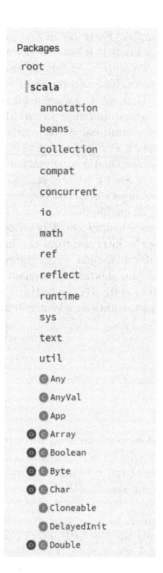

FIGURE 3.2: This is part of the right column that you see when you enter the API. The little circles with the letters c, o, and t indicate whether it is a `class`, `object`, or `trait`.

in them, whether it be a `class`, an `object`, or a `trait`. The API itself is built by running a program called `scaladoc` on scala code to generate HTML descriptions of what is in the code.

When there is an "o" next to a "c" or a "t", it is a companion object. If you want to know if you can build instances without the direct use of `new`, look there and see if there is an `apply` method defined. If there is, it generally implies that the `apply` method is the preferred way to make instances of that type.

3.8   `import` Options

The usage of `import` that we have seen so far (with all the `import` statements at the top of the file) with a name that possibly ends in an underscore is the most basic usage in Scala. It mirrors the way `import` has to be used in Java. There are a lot of additional options in Scala that we have not really considered. Here is a list of some details of `import` starting with the general usage. Each one includes situations where you might want to use it in different ways.

- So far, we have been putting `import` statements at the top of the file. This is not required as they can go anywhere. `import` follows normal scoping rules. So if you put the `import` at the top of the file, then the names brought in by it are visible through the whole file. However, if you put it inside of curly braces, the names that are `import`ed will only be visible through the end of the current scope.

- As we have seen before, you can use an underscore as a wildcard to represent everything in a certain scope.

- If you do not want to `import` everything, but you do want to `import` several things, you can put multiple names inside of curly braces. For example, `import scala.io.StdIn.{readLine, readInt}` will bring in the `readLine` and `readInt` methods, but not others.

- Names of `package`s and `class`es in Scala are truly nested. So if you have `import`ed all the contents of some `package`, and want to then `import` a sub-`package`, you can leave off the first part of the name for the sub-`package`. For example, after you do `import scala.io._` you can do `import StdIn._` instead of `import scala.io.StdIn._`[10] For that matter, because `scala._` is `import`ed by default, you can leave the `scala.` off of everything.

- So far we have primarily used `import` to bring in the contents of packages. It can be used to bring a lot more into the current scope. You can import package names or the members of objects. Things like the members of objects should generally be done in limited scopes.

- You can give a different name to things that you `import` using a rocket, `=>` inside of the curly braces. Put the name the thing normally goes by on the left side and the name you want to use on the right side. For example, if you need to use `java.util.List`,

---

[10]Some caution is required when doing this. Some class names occur in multiple packages and some package names occur in multiple places as well. Using full names will always bring in what you want. Relative `import`s can interfere with one another in odd ways. This can be especially problematic when copying imports or small sections of code from one file to another.

you probably do not want it hiding the `List` in Scala. To get around this, you could do `import java.util.{List => JUList}` so that the name `JUList` refers to the Java version. You can also hide a name by putting an underscore to the right of the rocket. So if you want to `import` all of the `java.util` package except the `List`, you could do this: `import java.util.{List => _, _}`.

- Due to the nesting of packages, there are times when a particular name could refer to more than one package. To remove ambiguity, the name `_root_` can be used to refer to the base below all packages. So `import _root_.scala.io.Source` can be used to refer to the `Source` type even if there are conflicts with the names `io` or `Source`.

3.9   End of Chapter Material

3.9.1   Summary of Concepts

- Objects are constructs that combine data and the functionality that operates on that data.

- The standard way of making objects in Scala is to define a `class` and instantiate objects from it. You can think of a `class` as a blueprint for making objects.

  - A `def` declaration in a `class` defines a method. A `val` or `var` declaration defines member data.

  - `class`es can take parameters. By default they are not members. Adding `val` or `var` in front of the name makes it member data.

  - By default, constructs declared inside of a `class` are public and can be seen by any code. A big part of the power of object-orientation comes from the ability to hide things so that other code cannot get to it. The `private` modifier makes it so no code outside of the `class` can see the declaration. There is also a `protected` modifier that makes declarations visible only to subtypes.

  - A big part of the benefit of making things `private` is that it can be used to hide implementation details. This facilitates the separation of interface and implementation.

  - To enhance the expressivity of the language, Scala allows a broader range of names than most languages and interprets some method names in special ways.

    * In addition to the "standard" names that start with characters or underscores and then have characters, numbers, or underscores, Scala allows two other types of names. One of those is names that include only operator characters. The other is a "normal" name followed by an underscore and one or more operator symbols. Operator syntax, dropping the dot and the parentheses, works for any method with a single argument. Operator precedence is determined by the first character. Operators ending in : are right associative.

    * Unary prefix operators can be declared with a method name of `unary_` followed by symbols including `+`, `-`, `~`, and `!`.

    * Scala does not have a strong syntactic distinction between methods and members. You can mimic a property of the class by writing a method with no arguments or parentheses along with a special method for assignment. That method has the name of the property followed by `_=`.

  * Function call syntax is expanded out to calls to the `apply` method.
  * Indexed assignment is expanded out to a call to `update`.
  - You can have multiple methods with the same name as long as they have different arguments. This is called overloading.

- Scala also allows `object` declarations which create singleton objects. These cannot take arguments and are not instantiated with `new`. A single instance just exists for using.

  - Scripts are good for small programs, but larger programs are written and run as applications. The entry point to an application is defined by a `main` method in an `object`. Top-level declarations are split to different files. Code is compiled with `scalac` and run using `scala`.

  - An `object` with the same name as a `class` can be declared in the same file and it becomes a companion object. Companions have access to `private` declarations.

- The `case` keyword on a `case class` effectively adds a bit of code to a `class`. It makes all parameters into `val` declarations. It also adds a `copy` method and code for pattern matching.

- You can do a lot of different things with `import` statements. They can go anywhere in the code, rename things, or hide them.

3.9.2 Exercises

1. If you previously wrote scripts in Scala, you can turn them into an application by embedding everything into an `object` in the `main` method. The results of this should compile with `scalac` and the run with `scala`.[11]

   The result of this simple conversion is not generally ideal. Any method or `case class` declarations should generally be pulled out of `main`. The methods likely should go inside of the `object`, but outside of `main`. If those methods are not generally useful, they should be made `private`. The `case class`es could go inside of the `object` where they might or might not be `private`, depending on how generally useful they are. However, if they truly stand on their own and have meaning, they should go outside the `object` and into a separate file bearing their name.

   The results of this modification might not compile. That will depend on the quality of the original code and whether you used variables that were declared outside of methods in the methods. If you use variables defined outside of the methods in their body, you have one of two choices. Generally your first choice should be to pass those variables in by adding extra parameters to the method. If a variable really deserves to be a data member/property of the `object`, it can be moved up a level so it too is in the `object`, but not in `main`.[12]

   Your goal for this exercise is to run through this process on a number of scripts that you wrote earlier. When you do this, it is recommended that you make a subdirectory for each script you are converting, then copy files into there. This way you not only

---

[11]Remember that when you run an application with `scala` you give only the name of the `object`, not the name of the file. Most of the time the file name should start with the `object` name so you are just leaving off the `.scala`.

[12]Note that if a variable is set by user input, it almost certainly needs to stay in `main`. Having singleton `object`s that request users to enter input will lead to very odd and unexpected behaviors in larger programs.

preserve the original script, you make it easier to compile that single application. Doing this on a number of scripts will really help to build your feel for how this new approach differs from what we had been doing previously.

2. Every semester you build a schedule of the courses that you will take in the following semester. On the book's website, you will find a file that contains a listing of courses. You can either use this file or create your own. Ultimately, you want help the user decide which courses to take by providing them with options. The user will have to select how many hours they will want to take, the courses they are interested in taking as well as how much they want to take those courses. The user will have to tell your program how many courses should be taken the next semester and a rank of how much they "want to take" a course. The program will eventually print out all the schedule combinations with the courses that make the cut.

    Make this into an object-oriented application. To do this, you want to build a `Course` `class`. One of the methods should probably be one that takes another `Course` and determines if the two overlap or not. There might also be methods designed to do specific types of `copy` operations that you find yourself doing frequently, such as making a copy of a course at a different time or with a different professor.

    The fact that the choice of professor often impacts how much you want to take a course, you could split this off and make a `class` called `Professor` that has the name of the `Professor` and a favorability rating. The `Course` could then hold a reference to `Professor` and have a rating that is independent of the professor. A method can be written to give you the combined favorability. This way you can easily copy course options with different professor selections and have that change automatically taken into account. As an added challenge, you can also build a mutable `Schedule class` that keeps track of what the user wants in a schedule along with what has been met so far from the courses that have been put inside of the schedule.

    As a hint, you can find ideal schedules using recursion, if you are familiar with that approach. Another approach would be using the combinatorial methods, `permutations` and `combinations`, of `Array` or `List`.

3. Design a program that will allow a user to keep track of recipes and items they have in their pantry or refrigerator. You will create an object-oriented application and possibly make mutable `class`es which will have methods. Besides having `class`es for `Pantry`, `Recipe`, and `Item`, you might want to have a `class` that keeps track of all the known `Item` types that have been used in the program. That `class` should have methods that facilitate looking up items by names or other attributes so that users do not wind up creating duplicates of `Item` objects.

4. Make a music library object-oriented application to organize your music. You should keep track of data such as name of the song, the artist, the album, the year released, and anything else you find significant. The program should be able to display the information for all the songs as well as allow the user to narrow down the display by a specified artist, album, or genre. They should also be able to add and delete information.

    You can also make types for play lists or other groupings of information that would benefit from encapsulation. Put the main script into an `object` with a `main` and you have an application.

5. For any of the things you code as applications, you can put in proper `scaladoc` comments and then run `scaladoc` on your source to see the HTML files that are generated.

6. convert an example of a **class** that has public **vars** to a **class** that has private **vars** with accessor and assignment methods.

3.9.3 Projects

There are several different projects listed here that are developed through the rest of this book. Information on these and how different projects in each chapter are related to different end projects as well as some sample implementations can be found at the book's website. The projects in this chapter have you getting started by laying out some initial foundation and gaining some experience with what you are doing. Note that code you write this early is not likely to survive all the way to the end of the book. You are going to learn better ways of doing things and when you do, you will alter the code that you have written here.

1. This project will grow throughout the book. Build a simple text role-playing game (RPG). This is a kind of game where the player explores a world by moving between rooms and spaces in the game using commands like "west" or "open the box". Often there are battles to wage or puzzles to solve. These games often blend gaming with storytelling. Some text adventure game examples include *Adventureland*, *Zork*, *Rogue*, and *World War II: Heroes of Valor*. Modern adventure games tend to be graphical, like *Portal* and *Life Is Strange*, however the thread of this project is not graphical.

   This game will eventually become a multi-user game, sometimes called a MUD, once you complete the networking chapter. If you want to experience what a MUD is like to help you understand where this is going, you can visit SlothMUD by running **telnet slothmud.org 6101** on the command line. For this project, you should make a number of **class**es to represent things like characters, rooms, and items in the game. Have it so that they can be read in from file and written out to file. Create an **object** with a **main** method that you can use to run the game and have a single player walk through the world. In your game, a player can walk between rooms.

   Your program will read in from a map file that you will write by hand and let the user run around in that map by using commands like "north" to move from one room to another. The map file will have a fairly simple format right now and you will create your own map file using a text editor. The format of the map file should start with a line telling the number of rooms and then have something like the following. You can change this if you want to use a slightly different format:
   ```
   room_name
   long line of room description
   number_of_links
   direction1
   direction2
   destination2
   ...
   ```
   This is repeated over and over. (The number of rooms at the top is helpful for putting things in an **Array**.) There is a link on the book website to a sample map file, but you do not have to stick exactly to that format if you do not want to. You might deviate if you are thinking about other options you will add in later.

   The interface for your program is simple. When you run the program, it should read in the map file and keep all the map information stored in an **Array[Room]** where **Room** is a **Class** you have made to keep the significant information for each room. You may also want to include a class to represent an **Exit** from a room.

Most of the original text adventure games had some type of puzzle solving where the items could interact with one another or with rooms in certain ways to make things happen. You must add a "help" command which prints out the objective of the game and how to play. This should read and print the content of a file so you do not have to type in the full help in print statements. For example, you should tell the player how to play the game, start the player off in a room, describe the room to the player if they enter it, accept player commands such as "south" or "quit", etc.

A full MUD implementation will also include computer-controlled characters. At this point only include one type of character. Make it so that character moves randomly around the map as the player executes commands. If the player enters the same room as that character, the description should show them much as it would items in the room. What this is building towards is a minimal RPG with weapons/combat, spell casting, or something else that can be executed with a time delay.

2. Another option for a project is a web spider[13] for data collection and processing. This option can be customized to fit lots of fields. The basic idea of a web spider is that it loads pages and follows links to get to other pages. Search engines like Google® use spiders to find data and index it so that people doing searches can find it. This project will be much more focused so that you do not require an entire farm of servers to make it run.

The idea here is that you are spidering one site or a short list of sites for particular types of files. You might be looking for certain data sets, images, box scores, or anything else that you can find a way to recognize. You might want to have several possibilities. The program will compile that information and then provide useful ways for the user to view it.

For this project option, you can think about what data you want to collect and set up some `class`es to store that information. Pull down one or two files manually (consider using `wget` in a Linux environment) and do a minimal display of the information from those files.

3. Soon you will be able to use graphics and thus be able to create a graphical game. This project helps you set the stage for your game. Eventually, you will be able to include networking so that more than one person can play. The possibilities for your games are extremely broad and range from puzzle games like *Tetris* where you have `class`es for `Board` and `Piece` or a role-playing game where you have `Units` and `Cities` or `Characters` and `Items`. Decide what `class`es you need for your game and implement them for this project.

4. For the more numerically inclined you might consider the option of a simulation workbench. This project will have you write a program that can perform a number of different styles of numerical simulations. By the end of the semester, this project will evolve to the point where you can distribute work across multiple machines and view what a simulation is doing. You will also implement some spatial data structures to make integration of long-range forces more efficient. You might even be able to do things like simulate cloth using a particle mesh technique.

For now, make an application that does Newtonian gravity for a number of particles. Break this problem up into different `class`es for the particles, the system of particles, and the integrator. Your program should also have the ability to save to file and load from file.

---

[13]These are also called web crawlers.

To help you get started, we will describe Newtonian gravity and the first numerical integrator that you will use. Newtonian gravity is characterized by the equation $F_G = -\frac{Gm_1m_2}{d^2}$, where F_G is the force due to gravity, G is the universal gravitational constant, m_1 and m_2 are the masses of the two bodies, and d is the distance between them. In order to use this equation, we are going to make a few changes to it. First, we are going to set $G = 1$. In the standard units of meters, kilograms, and seconds (mks), the value of this constant is 6.67×10^{-11}. However, we are not constrained to work in mks units, and it just makes things easier to work in units where $G = 1$. Second, this is a scalar equation. It is great for discussion in a physics class, but we need things to move around in a 3D space, which means that we are going to need a vector version. This gives the magnitude of the force. The reality is that it is directed along the line between the two bodies and d is really a vector quantity that we should write as $\vec{d}$. So if we say that $\vec{d}$ points from the first body to the second body, then $\vec{F}_G = \frac{m_1m_2}{|\vec{d}^2|}\frac{\vec{d}}{|\vec{d}|} = \frac{m_1m_2}{|\vec{d}^3|}\vec{d}$. Given what we have learned, you can write a nice `class` for representing the 3D vectors.

Each particle in the simulation will have a position and a velocity, which are both vectors, as well as a mass. You could add a radius if you want, but we are not going to be using one yet. To advance the system, we have to figure out how to change the position and the velocity over time. In reality these are differential equations that would be integrated with infinite accuracy, but that is not numerically feasible. The approach that we are going to take to this is to use a simple kick-step approach. First, we calculate forces and give the velocities a kick, then we add the velocities into the positions to get new positions. We do this repeatedly using a small time step, which we will call Δt here, but which you will likely call `dt` in your code. You might start off trying a value of $\Delta t \approx 0.01$.

To glue these pieces together we have to remember that $\vec{F} = m\vec{a}$, where a is the acceleration, which is the time derivative of velocity. After you have used the formula for $\vec{F}_G$ to calculate the total force on a particle from all of the other particles, you can then set $\vec{v} = \vec{v} + \vec{a}\Delta t$. You do that for all the particles to do the "kick" part of the integration. Once you have all updated velocities, you run back through and set $\vec{x} = \vec{x} + \vec{v}\Delta t$, where $\vec{x}$ is the position of a particle.

At this point, all you can really do is read initial conditions in from a file and write positions and velocities out to a file. Then you can read those positions and velocities into some other program and plot them. As a quick sanity check you can test your program with two particles. The first one is at $(0, 0, 0)$ with a velocity of $(0, 0, 0)$ and a mass of 1.0. The second one is at $(1, 0, 0)$ with a velocity of $(0, 1, 0)$ and a mass of 10^{-20}, which you will write as 1e-20 for Scala. When you integrate this system, if you have done things correctly, the second body should orbit around the first one roughly once every 2π time units.

5. This project involves you pulling data for a number of different stocks into a program and processing them with the eventual goal of building an application that could be used to help users manage their stock portfolio. Yahoo has stock downloads of CSVs that you can pull down to use which can be found at urlfinance.yahoo.com. Once you are at the site, enter the stock you would like to search for in the search field and you should see the main display page for that stock. Then select "Historical Prices", which is currently located on the left of the page. This will bring up the first page of historical prices for that stock. At the bottom of this list on the first page, you will find an option that will allow you to download this to a spreadsheet. From that point,

you should be able to export it to a CSV file so that it can be read into your program. You should consider downloading several stocks for your portfolio.

Besides thinking of what **class**es you will need to store this information, you should consider what activities you will want to perform. For instance, users will want to track the stock's symbol, company name, which stocks the user owns, how much they own, when they purchased them, how much they purchased them for, current value, sell date, sell price, etc. You will probably want to provide users with a daily view of their portfolio, the monthly averages of their portfolio or of selected stocks, and the min, max, and mean values. They will want to know their total assets, daily stock price fluctuations, and the percent of change for each stock held. Eventually you will want to track trends and create graphs that depict the distribution of stock, stock price fluctuation, and comparisons of initial and final investment values. You should also add the ability for the user to add or remove stocks from their portfolio.

You should also implement a minimum investment amount so that your portfolio system is only managing more significant portfolios. For instance, it is not uncommon for management companies to expect a minimum investment of $10,000. It is recommended that users should keep a balanced portfolio, thus it should contain several stocks. As a rule of thumb, no one stock should have a portfolio value of more than one third of the total portfolio value. Your application should help the user maintain this balance by not allowing the user to purchase too much of one stock and advising them to sell stock in order to maintain balance.

6. This project involves keeping an inventory of movies with the eventual goal of setting up a little business that rents/streams movies to users. A relatively complete list of all the Region 1 DVDs that exist can be found at `http://www.hometheaterinfo.com/dvdlist.htm`. You can also find movie information at sites like `http://www.omdbapi.com/` and `http://www.imdb.com/interfaces`. You may find it interesting to play with these data sets, but for the purposes of this project, we want you to consider what **class**es you will need in order to store this information and what methods you will need to initially implement.

   To begin with, you should consider allowing the user to search for movies by name, list all the movies in a specific genre, find out which titles are produced by a specific company, etc. You should also allow users to add their own titles. These titles would be movies that they do not see on the list, but would like you to consider acquiring.

   You might consider adding a **class** for actors that keeps track of what movies they have been in so that users could find movies based on a favorite actor or actress. As you will plan on allowing the user to rent these movies, you will have to keep track of how many copies of each movie you have available, how many are currently being rented, dates of rental, etc.

   The users will also want to browse your selection and create lists of movies they will want to watch. They should also be allowed to prioritize the order of that list. Also allow the users to remove and change movies in their watch request list. They also might like to keep a separate list of their favorite movies and rank those as well. You can decide if you want to limit that list to something like their top 10 or top 25. You might also consider helping the user keep track of what movies the user has already rented. You might find it useful to keep track of each movie's overall rental history for your own purposes so that you can analyze rental trends. Be creative and set up the business as you think would be most successful.

7. One of the early educational programming languages, called Logo, made graphics easy

to use by implementing a turtle graphics system (`http:///en.wikipedia.org/wiki/Turtle_graphics`). The idea of turtle graphics is that you have a cursor, typically called the turtle, that has a position, an orientation, and pen settings. The turtle can turn and move. When it moves, it can either draw or not.

A simple way to encode instructions for a turtle is with a `String`. Although we cannot graph a turtle movement yet, we can create the `String` that will represent the movement. Different characters tell the turtle to do different things. A 'F' tells the turtle to move forward while drawing. A lowercase 'f' tells it to move forward without drawing. The '+' and '-' characters will it to turn to the left and right, respectively. Other characters can be added to give the system more power. The amount that the turtle moves or turns for each character is considered to be a fixed parameter. Using this system, one could draw a square by setting the angle to 90 degrees and using the string "F+F+F+F". Two squares that are separated could be made with "F+F+F+FfF+F+F+F".

You can use turtle graphics to draw fractal shapes generated with L-systems. You can find a full description of L-systems in "The Algorithmic Beauty of Plants", which can be found online at `http://algorithmicbotany.org/papers/#abop`. L-systems are formal grammars that we will use to generate strings that have turtle representations. An L-system is defined by an initial `String` and a set of productions. Each production maps a character to a `String`. So, the production `F -> F-F++F-F` will cause any F in a `String` to be replaced by F-F++F-F. The way L-systems work is that all productions are applied at the same time to all characters. Characters that do not have productions just stay the same.

So, with this example you might start with `F`. After one iteration you would have `F-F++F-F`. However, this will grow quickly. After the second iteration you would have `F-F++F-F-F-F++F-F++F-F++F-F-F-F++F-F`. The next iteration will be about five times longer than that.

The productions for an L-system can be implemented as a `List[(Char,String)]`. You can use the `find` method on the List and combine that with `flatMap` to run through generations of your `String`. Later in the text, we will be able to convert these strings into graphics using the turtles.

There are a number of features of the L-systems implementation that call for object-orientation. Starting at the top, an L-system should be represented by a `class` that includes multiple `Productions` as well as an initial value. The `Production` type should be a `class` that includes a `Char` for the character that is being changed and a `String` for what it should result in.

One of the advantages of encapsulating a production in an L-system is that you can make probabilistic productions. These start with a single character, but can result in various different `Strings` depending on a random value. Whether a `Production` is deterministic or not does not really matter to the whole system as all the system does is ask for what a given character will be converted to.

Using a `String` to represent the state of the L-system at any given time works fine, but there is some benefit to creating your own `class` that is wrapped around a `String`. This `class` might also include information on what generation you are at in applying the productions. The advantage of this is that the type makes it clear this is not any random `String`, instead it is the state of an L-system.

Write an application that can read an initial string and a set of productions from a file, then has the ability to advance the string through multiple generations of productions.

Additional exercises and projects, along with data files, are available on the book web site.

Chapter 4

Abstraction and Polymorphism

Understanding the concept of abstraction is critical for improving your capabilities as a programmer. The ability to express our thoughts in more abstract ways is a significant part of what has allowed modern software systems to scale to the size they are today. Abstraction comes in many forms, and you have dealt a little with abstraction already. At the simplest level, writing a function or a method that takes arguments is a form of abstraction. The one function can work for as many values as you want. Consider the basic example of a function that squares numbers. You can write the code 3 * 3 or 5 * 5, but a function like the following is an abstract representation of that procedure for the Double type.

```scala
def square(x: Double): Double = x * x
```

A more interesting form of abstraction that we already know enough to create is writing methods that take functions as arguments. A simple example of that would be a method that reads values from input and combines them in various ways as specified by the calling code. To understand this, consider the following two methods.

```scala
def sumValues(n: Int): Double = {
```

```scala
  if (n < 1) 0.0 else readDouble() + sumValues(n - 1)
}

def multiplyValues(n: Int): Double = {
  if (n < 1) 1.0 else readDouble() * multiplyValues(n - 1)
}
```

Both methods read a specified quantity of numbers.[1] The first method takes the sum of the numbers. The second one takes the product. We'd like to abstract over the operation and have one method that can do either of these, or more.

If you look at the two methods, there are two ways in which they differ. They return different values for the base case, and they use different operations for combining the values. To make a more powerful version, we need to abstract those aspects that vary by passing in arguments to specify those aspects of the calculations. The following code does exactly that.

```scala
def combineValues(n: Int, base: Double, op: (Double, Double) => Double): Double =
    {
  if (n < 1) base else op(readDouble(), combineValues(n - 1, base, op))
}
```

One argument has been added to specify the value returned in the base case. Another argument has been added to specify the operation. This is the more interesting one because it is a function that takes two `Doubles` and produces a `Double`. To get the behavior that we had from `sumValues` and `multiplyValues` respectively, we would use the following calls.

```scala
combineValues(5, 0, _ + _)
combineValues(5, 1, _ * _)
```

The real beauty of abstraction is that your abstract version of code is typically more powerful beyond the original goals. For example, we can use `combineValues` to find minimum and maximum as well. The following invocations could be used to produce those results.

```scala
combineValues(5, Double.MaxValue, _ min _)
combineValues(5, Double.MinValue, _ max _)
```

Indeed, `combineValues` will allow us to combine user input values for any associative operators. The only real limitation is that they have to work with the type `Double`.

The above examples should have made it clear that abstraction helps us to write less code. Without the `combineValues` method, we could wind up copying and pasting `sumValues` many times, making minor alterations for each of the different operations that we wanted to support. Abstraction allows us to write a single version that is far more capable.

The primary shortcoming of `combineValues` is that it only works with the `Double` type. The primary goal of this chapter is to show you how you can abstract over types in Scala. This is a key aspect of many object-oriented languages, and it is essential for writing highly flexible, powerful code.

---

[1]These can be rewritten with `while` or `for` loops. The code is a bit longer, so we stick with basic recursive implementations. If you are not comfortable with recursion, you are encouraged to read the on-line appendix at http://www.programmingusingscala.net.

4.1   Polymorphism

One of the most powerful ways to add abstraction into code is through POLYMORPHISM. From the Greek roots, the word literally means "many shapes". In the context of programming though it means many types. So far the code that we have written has been monomorphic. This means that our code only worked with one type. A good example of when this is a limitation comes in the form of sorting and searching algorithms.

Sorts are extremely common algorithms in computing. There are numerous situations where it is advantageous to have data in a particular order. The purpose of a sort is to arrange a data set in a particular order. We will consider comparison sorts, where we have the ability to ask if one element should come before another, and we use that information to put an entire collection in proper order according to that comparison. There are many ways of doing these sorts. We don't really care about the sorting itself, only the manner in which we can abstract it. For that reason, we will use a bubble sort. If you have previously covered sorts in your studies, you know that this isn't a very efficient sort, but it is simple to understand and easy to write.[2]

Our initial bubble sort is a monomorphic sort, that sorts an `Array[Double]` in place. The term "in place" means that a sort puts the result in the same memory that the values started in. Typically they only use a small, fixed amount of memory beyond the memory for the input array.[3]

```scala
def bubbleSort(arr: Array[Double]): Unit = {
  for (i <- 0 until arr.length - 1; j <- 0 until arr.length - i - 1) {
    if (arr(j + 1) < arr(j)) {
      val tmp = arr(j)
      arr(j) = arr(j + 1)
      arr(j + 1) = tmp
    }
  }
}
```

This method takes an `Array[Double]` as input and moves the items in it around to sort them, so the result is `Unit`. Basically, the method is called for side effects, and doesn't produce a result value. The sort begins with a for loop that has two generators for variables i and j. Inside, it checks if two adjacent elements, those at indexes j and j+1, are out of order. If they are, lines 4–6 swap those elements. You should trace through those three lines to make certain that you understand how the swap works, because it is a fairly common thing to see in imperative code.

What we are most interested in here is that this sort works with the type `Double` and nothing else. If you want to sort `Int`s instead, you have to copy the method and change `Double` to `Int`. If you wanted to sort some type of `class` that you had written, you would have to make another copy, and then change `Double` to the name of your `class`, then alter the code or the comparison. The key point is that this sort only works with the one type and nothing else. Getting sorts for multiple types requires a lot of code duplication.

This type of duplication is wasteful and it leaves us with overly long code that is prone

---

[2]If you haven't studied sorts previously, you don't really need detailed knowledge of them at this point, but you should strongly consider learning about them before you get to chapter 15, as that chapter will spend time on how to efficiently implement some more advanced sorting algorithms.

[3]The formal way to describe the memory overhead here is $O(1)$. Chapter 7 includes a more complete description of this notation and the general topic of asymptotic orders.

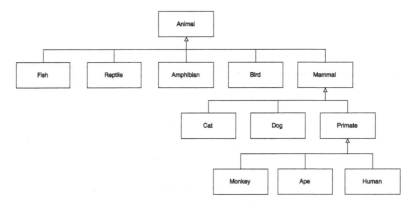

FIGURE 4.1: This figure shows a limited set of options for the subtyping relationships of animals as a UML diagram.

to errors and hard to work with. Plus, one would expect that with so few changes between the versions, it should be possible to make one version that handles them all. To put it a different way, we should be able to abstract out the differences so that we have one version that will work in different situations. In this case, part of that abstraction is abstracting over types. To do that we need polymorphism.

There are multiple different styles of polymorphism, but we can group them into two broad categories: UNIVERSAL and ad-hoc. Universal polymorphism implies that code can work with an infinite number of types. By contrast, ad-hoc polymorphism only works with a finite number of types. We will primarily consider universal polymorphism in two different forms, INCLUSION POLYMORPHISM and PARAMETRIC POLYMORPHISM.

4.2   Inclusion Polymorphism (Inheritance and Subtyping)

Inclusion polymorphism is a form of universal polymorphism that we get from SUBTYP-ING. That is to say, when all elements of one type are also part of another, more general, type. This type of relationship is produced by a programming language feature called inheritance. We often say that inheritance models an "is-a" relationship. There are lots of examples of this in the real world. A car is a vehicle, as is a bus. A cat is a mammal, as are dogs and humans. This second example shows another feature of the is-a relationship—it can have multiple levels. We could consider the broad category of animals. There are several subtypes of animals including birds, reptiles, amphibians, and mammals. We have already said that cats, dogs, and humans are subtypes of mammals. We could come up with similar lists in other areas too. We might also decide that we want to break things down further. We might want a type for primates, in which case, humans would move down to be a subtype of primate.

Figure 4.1 shows what this might look like as a UML diagram. You can see that there is a different type of arrowhead used to indicate the subtype relationship we get from inheritance. It has an open triangular head, and it points from the subtype to the supertype.

If we tried to model this diagram in code, each box would be a **class** or a **trait**,[4] and each subtype would specify in the code that it inherits from its supertype. When one class,

---

[4]Traits are discussed later in this chapter.

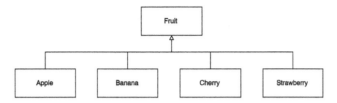

FIGURE 4.2: This is a simple class diagram showing subtypes of fruit.

say `Primate`, inherits from another class, say `Mammal`, it implies two things. One of those things is that `Primate` is a subtype of `Mammal`. The other is that `Primate` gets all the data and methods that were part of `Mammal`. The latter we will call CODE-REUSE. It helps us to not duplicate code, and might seem like the more important of the two aspects, however, the real power of polymorphism is the result of the subtyping.

So what does it mean for `Primate` to be a subtype of `Mammal`, or more generally, for a type B to be a subtype of type A? We can say that B is a subtype of A if any situation where an object of type A is needed, we can give it an object of type B and it will work.[5]

To put this into a more concrete form, consider the following example. You have a recipe that calls for three different types of fruit. The recipe gives you instructions for how to prepare the fruit and include it with the other ingredients. In this case, `Fruit` is the supertype and we have a description of how to do something with instances of that type. There are many different subtypes of `Fruit` that we could choose to use, including `Apple`, `Cherry`, `Banana`, and `Strawberry`. This particular example is illustrated using a simple UML diagram in Figure 4.2. Here we see the type `Fruit` at the top with generalization arrows connecting it to the various subtypes. The direction of the arrows in UML is significant because while subtypes have to know about their supertype, a supertype generally does not, and should not, know about subtypes.

Why Supertypes Do Not Know About Subtypes.

The reason that a supertype should not know about subtypes is that such knowledge limits the universality of the polymorphism. When the supertype is created, there might only be a few subtypes. In our fruit example, the number of subtypes is four. For the polymorphism to be universal, it should be possible to create a new subtype at a later date, without altering the supertype, and have any code that already uses the supertype work well with the new subtype.

In code, this means that you can write a function like `makeBreakfastShake` that works with type `Fruit` and it should work with instances of `Apple`, `Banana`, `Cherry`, or `Strawberry`. Consider this code that might represent that function.

```
def makeBreakfastShake(fruit: Fruit): Unit = {
  if (!fruit.canEatSkin) {
    fruit.peel()
  }
  blender += fruit
  blender += juice
  if(fruit.fractionalLiquidContent < 0.3) blender += juice // add extra juice
```

---

[5]This is formally referred to as the Liskov substitution principle.

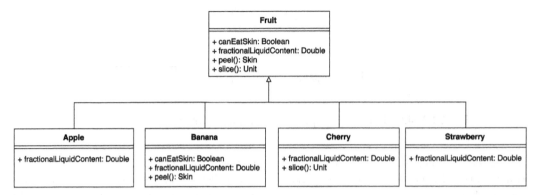

FIGURE 4.3: This class diagram for fruit adds some methods that we might want and how they might be overridden in the subclasses.

```
    blender += ice
    blender.blend
}
```

The idea is that we could call this function using code like this.

```
makeBreakfastShake(new Banana)
makeBreakfastShake(new Strawberry)
```

We want this to work because **Banana** and **Strawberry** are both subtypes of **Fruit**. The question is, what is required for this to happen? For the type **Fruit** to work with **makeBreakfastShake** it needs to have a method called **canEatSkin** that takes no arguments and returns a **Boolean**, a method called **peel** that also takes no arguments, and a method called **fractionalLiquidContent** that takes no arguments and produces a **Double**. For objects of type **Banana** to work in place of **Fruit**, they have to have those same methods.

As you can see, it turns out that the code reuse part of inheritance is not just there to reduce code duplication, it helps with subtyping. A subtype is guaranteed to have all the methods and data members of the supertype that might be called by outside code in part because it can inherit them from the supertype.

Of course, you might not always want those methods to do the same things. In this case, objects of the **Banana** type should return **false** for **canEatSkin** while objects of the **Strawberry** type should return **true**. Each type of fruit will likely also return a different value for **fractionalLiquidContent**.[6] Changing the implementation of a method in the subtype is called OVERRIDING. In Scala, when a subtype overrides a method, the keyword **override** must come before the **def** keyword for the method declaration. Figure 4.3 shows a UML class diagram using the more complete format for the classes so that you can see the methods that might be in them.

The **Fruit** type in this diagram has four different methods that pertain to all different subtypes of **Fruit**. Each of these can be implemented in some default way. The four subtypes only list the methods that they override. The other methods are present in their default implementation as a result of inheritance.

---

[6]Those with previous experience in object orientation might be tempted to have **canEatSkin** and **fractionalLiquidContent** be data members of type **Boolean** and **Double** based on the idea that it is a property of the fruit. If you did that, you would want to make that property **private**, which means you would have to add accessor methods, perhaps called **getCanEatSkin** and **getFractionalJiuceContent**, as well. Simply adding a method and overriding it in the subtypes is a simpler approach.

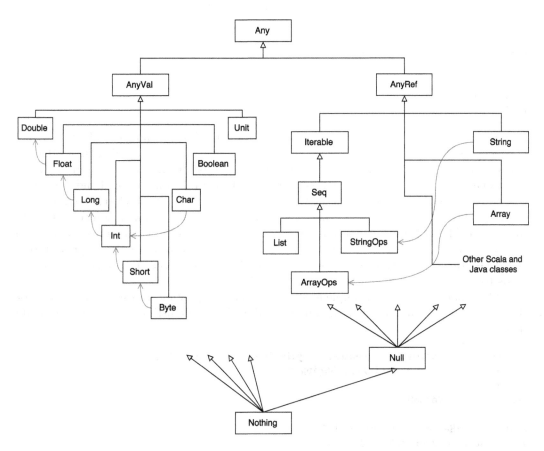

FIGURE 4.4: Diagram of general subtype relationships in Scala. This figure has been adapted from a similar figure in *Programming in Scala* by Odersky, Spoon, and Venners [10]. The thick lines with full arrowheads indicate a subtyping relationship. The thin lines with an open arrowhead indicate an implicit conversion.

4.2.1   Inheritance in the Scala Type System

All types in the Scala language are part of an inheritance hierarchy. Figure 4.4 shows a UML representation for a number of different types. Every class that you create automatically inherits from a type called **AnyRef**, which inherits from **Any**. You can look in the API to see the types **Any**, **AnyRef**, and **AnyVal** right at the top of the list of classes. If you look in **AnyRef** you will see that there are a number of different methods that will be in any **class** you create. We can demonstrate this by making a little **class** in the REPL and calling one of those methods.

```scala
scala> class Vect3D(val x:Double,val y:Double,val z:Double)
defined class Vect3D

scala> val jHat = new Vect3D(0,1,0)
jHat: Vect3D = Vect3D@24753433

scala> jHat.toString
res0: java.lang.String = Vect3D@24753433
```

Here we have a **class** for an immutable 3-D vector type that has no methods, just public members for three components. The second line creates an instance on this **class**, which is followed by a notification that the variable **jHat** is of type **Vect3D** and has a given value. The value is a bit odd, having the name of the class followed by an **@** symbol and a number. The third line shows a call to one of the methods inherited from **AnyRef**, the **toString** method. This call makes it clear where the text representation of our instance of **Vect3D** came from. The **toString** method is actually defined on the type **Any** and, as such, is safe to call on any object in Scala. For that reason, it is used extensively to get a representation of objects. Here we see that it is called by the REPL any time that the value of an object needs to be printed.

Unfortunately, the default implementation of **toString** is not all that informative. For that reason, it is something that you often override. This code listing shows how we can do that.

```scala
scala> class Vect3D(val x:Double,val y:Double,val z:Double) {
     | override def toString(): String = s"Vect3d($x, $y, $z)"
     | }
defined class Vect3D

scala> val jHat = new Vect3D(0,1,0)
jHat: Vect3D = Vect3d(0.0, 1.0, 0.0)

scala> jHat.toString
res1: String = Vect3d(0.0, 1.0, 0.0)
```

The inclusion of **override** with a **def** that has the same signature[7] as the inherited **toString** gives us a different implementation. You can see that this is now used in the REPL when the value of the object is printed as well as when the method is called directly.

4.2.2   Inheritance in Scala Code

The previous example showed an inherited method and how we can override it, but it does not actually demonstrate how to get inheritance in our code. The example uses the

---

[7]The signature of a method is the list of arguments expected by that method, and their types.

inheritance that we get by default. If we want to inherit from something other than `AnyRef`, we use the `extends` keyword. This keyword is placed after the arguments to the class and before the open curly brace that starts the body of the class.

```
class Name(args) extends Supertype(args) { body }
```

If you do not put in the `extends` keyword you get `extends AnyRef` by default.

To illustrate this, let us look at a different example that is a classic case of inheritance. We define a supertype called `Shape` that has some methods for things that we expect all shapes should be able to do.

```
class Shape {
  def area: Double = 0.0
  def perimeter: Double = 0.0
  def draw(gc: GraphicsContext): Unit = {}
}
```

There are many different types of shape that we might want to have as subtypes of this class. Here are two possible examples.

```
class Rectangle(val width: Double, val height: Double) extends Shape {
  override def area: Double = width * height
  override def perimeter: Double = 2.0 * (width + height)
  override def draw(gc: GraphicsContext): Unit = {
    gc.fillRect(0, 0, width, height)
  }
}
```

```
class Circle(val radius: Double) extends Shape {
  override def area: Double = math.Pi * radius * radius
  override def perimeter: Double = 2.0 * math.Pi * radius
  override def draw(gc: GraphicsContext): Unit = {
    gc.fillOval(0, 0, radius, radius)
  }
}
```

Each of these classes takes some arguments and extends `Shape`, then overrides all of the methods of Shape.[8]

We can now write code that takes an instance of `Shape` and passes it an instance of either `Rectangle` or `Circle` and it will work. The reason it will work is that inheritance guarantees us that any method we could call on an instance of `Shape` will be defined in the subtypes. Consider the following function and two calls to it.

```
def areaPerimeterRatio(s: Shape): Double = {
  s.area / s.perimeter
}
```

```
val circleAPR = areaPerimeterRatio(new Circle(5))
val rectAPR = areaPerimeterRatio(new Rectangle(4, 5))
```

This is the heart of polymorphism—one piece of code that can work on multiple types. This is universal polymorphism because at any point in the future we could create a new subtype of `Shape` and it would work with this code as well.

---

[8]The details of the `draw` methods will be discussed in chapter 5.

As was mentioned earlier, inheritance should be used to represent an "*is-a*" relationship. In the examples we have seen so far, it seems quite natural to say that an apple *is-a* fruit or that a rectangle *is-a* shape. If you cannot say that, then inheritance probably is not the construct you want to use. Instead, you should use composition. This is where you put an instance of the class(type) as a data member of some other **class** instead of inheriting from it. To get the functionality, you make calls on that member. Composition is used to model a "*has-a*" relationship. For example, a car "has an" engine. We would create a separate class for an engine which would be included in the car class. We will see many examples of this type of construction later in the book. It should be emphasized that you do not want to abuse inheritance. The code reuse aspect might seem great, but the subtyping should only be used when it makes sense.

There are times when you will have an *is-a* relationship and subtyping makes sense, but not all of the methods you would get through code-reuse fit. In those situations you should also refrain from using inheritance. An example of this is a square. Clearly, a square *is-a* shape. In addition, a square *is-a* rectangle. That relationship can be expressed with the following one-line class.

```scala
class Square(length: Double) extends Rectangle(length, length)
```

Here the class **Square** takes one argument for the **length** of an edge. It extends **Rectangle**, and passes that **length** as both the **height** and the **width** arguments. As this example shows, when the supertype takes arguments, they are passed in an argument list that follows the name of the supertype after extends.

Given what we have written for **Rectangle**, this implementation of **Square** is perfectly acceptable and calls to all the methods of an instance of **Square** will work as desired. That is only because our **Rectangle** type happens to be immutable. Consider this implementation of **MutableRectangle** with an associated **MutableSquare**.

```scala
class MutableRectangle(var width: Double, var height: Double) extends Shape {
  override def area: Double = width * height
  override def perimeter: Double = 2.0 * (width + height)
  override def draw(gc: GraphicsContext): Unit = {
    gc.fillRect(0.0, 0.0, width, height)
  }
}
```

```scala
class MutableSquare(length: Double) extends MutableRectangle(length, length)
```

Everything here compiles fine, and it might even seem to run fine. However, there is something truly broken with this implementation. To see this, consider the following code.

```scala
def adjustRectangle(r: MutableRectangle): Unit = {
  r.width = 20
}
```

```scala
val square = new MutableSquare(5)
adjustRectangle(square) // Warning: after this call we don't actually have a
    square.
```

The call to **adjustRectangle** does something that should not be possible. It makes a "square" that is 20 by 5. This function should work with all types of rectangles (even rectangles whose sides are all the same length and known as squares). That certainly does not seem to fit the normal definition of a square. This call is legal because we chose to make **MutableSquare** be a subtype of **MutableRectangle**.

The problem is that when we changed `Rectangle` to `MutableRectangle` we effectively introduced two new methods: `def width_=(w:Double)` and `def height_=(h:Double)`. These methods are not suitable for the `MutableSquare`, but by virtue of inheritance, our `MutableSquare` type gets them.

It is tempting to try to fix this by explicitly overriding the methods that set the `width` and `height` fields as is done in this code.

```scala
class MutableSquare(length: Double) extends MutableRectangle(length, length) {
  override def width_=(w: Double): Unit = { // ERROR: Can't override mutable
      variables.
    width = w
    height = w
  }
  override def height_=(h: Double): Unit = { // ERROR: Can't override mutable
      variables.
    height = h
    width = h
  }
}
```

This does not compile because you are not allowed to **override** methods associated with `var` declarations. This is a rule of Scala because allowing this type of thing would cause problems for code that tried to do it. Even if it were allowed, if we explicitly added those methods to `MutableRectangle` so they could be overridden, or if we used different methods to access and set the variables, it would not be a good idea. To understand why, consider what a programmer is expecting to happen when a line like `r.width = 20` is executed. Clearly the programmer who writes this line is expecting it to change the value stored in the `width` field. If the `height` field changes as well, that would be unexpected behavior and might well violate assumptions made elsewhere in the function. That could lead to bugs that are remarkably hard to find because the code would work fine when used with an instance of `Rectangle`, and reasoning about the program will not likely help the programmer find the error.

So be careful with how you use inheritance. It should only be used when you want both the subtyping *and* the code-reuse. You might also note from examples like this how many things are made more complex by having mutable **class**es. Immutability makes life simpler in many different ways.

Arguments to the Supertype

When you pass arguments to the supertype constructor, they are generally either constants, or they are expressions involving values that were passed into the constructor for the subtype. The use of `length` for the `Square` type is an example of the latter. Note that these should almost always have different names, and the ones for the subtype should probably not be `val`s or `var`s because these values do not need to be kept in the subtype because they are already being kept in the supertype.

To understand this, imagine that instead of calling the argument to `Square` `length`, we had instead called it `width`. This would potentially cause a problem for any code that we defined inside of `Square`. When you use `width` in the `Square` class, you would be using the argument called `width`, not the `val` called `width` in the supertype. This isn't likely to cause errors with a `val` in the immutable versions, but for `MutableRectangle` and `MutableSquare`, this would cause subtle bugs when the `var` in the supertype is changed, but the code in the subtype gets the original value from the argument that is hiding the var.

In the situation where an argument to the subtype is passed straight through as an argument to the supertype, you probably don't want to refer to the argument in the subtype class, even if it uses a different name and is neither a val or a var. So in our Square, we really shouldn't use length inside of the class, even if it isn't hiding anything from Rectangle.

4.2.3   private Visibility and Inheritance

One thing that is lacking in our Shape example is a color for the shapes. Since there is a draw method, it would make sense for them to have a color to be drawn in, and because all shapes would have that color value, it would make sense for it to be part of the Shape class. Here are implementations of Shape, Rectangle, and Circle where the color field is a private, mutable value in Shape.

```scala
class Shape(private var color: Color) {
  def area: Double = 0.0
  def circumference: Double = 0.0
  def draw(gc: GraphicsContext): Unit = {}
}

class Rectangle(val width:Double,val height:Double,c:Color) extends Shape(c) {
  override def area:Double = width*height
  override def circumference:Double = 2.0*(width+height)
  override def draw(gc: GraphicsContext): Unit = {
    gc.fill = color // ERROR: Can't get to the private color data member.
    gc.fillRect(0.0,0.0,width,height)
  }
}

class Circle(val radius:Double,c:Color) extends Shape(c) {
  override def area:Double = math.Pi*radius*radius
  override def circumference:Double = 2.0*math.Pi*radius
  override def draw(gc: GraphicsContext): Unit = {
    gc.fill = color // ERROR: Can't get to the private color data member.
    gc.fillOval(0.0,0.0,2.0*radius,2.0*radius)
  }
}
```

The reason it is private is because it is mutable and we do not want to allow any piece of code that gets hold of a Shape object to be able to change that value. As the comments indicate, these classes will not compile. While subtypes get copies of everything in the supertype, they do not have direct access to private elements. So using this approach, the subtypes cannot get to color unless we add a method for them to access it.

The behavior where subtypes can't access private members really is what we desire. When something is made private, that means that other parts of code outside of that class should not be able to access it. Anyone can make a subtype of a class. If that gave them access to private data, then private would not really be all that safe. Granted, if color were declared as a val instead of a var, we would not need to hide it because the Color class we will use is immutable, so other code could not change it, even if it could access it.

4.2.4 protected Visibility

Of course, there are times, like our colored shape example, when you do want to have data or methods that are only accessible to the subtypes. This is the reason for the protected visibility. When a method or data member is modified with the protected keyword, it is visible in that class and its subclasses, but not to any other code. Using this, we could modify the Shape class to the following.

```scala
class Shape(protected var color: Color) {
  def area: Double = 0.0
  def circumference: Double = 0.0
  def draw(gc: GraphicsContext): Unit = {}
}
```

With color set to be protected, the previous code for Rectangle and Circle will work.

The protected visibility is something that is not used very often. If a value really needs to be hidden it should be private. Making it protected is a signal to programmers that it is needed by subclasses, but that they need to know what they are doing if they are going to use or alter it.

4.2.5 Calling Methods on the Supertype

Sometimes when you override a method in a subtype, you still want the code from the supertype to be run. There are actually situations where you are expected to call the method in the supertype before doing additional work. To support this type of behavior, you can use the super keyword like an object that references the part of the current object that is the supertype.

This can be used to solve the coloring problem in a way that keeps the color private. Consider the following implementation of Shape.

```scala
class Shape(private var color: Color) {
  def area: Double = 0.0
  def perimeter: Double = 0.0
  def draw(gc: GraphicsContext): Unit = {
    gc.fill = color
  }
}
```

This has a private member color so it is not visible to the subtypes and is well encapsulated so we do not have to worry about other code changing it. Unlike the last version with a private color, this one has some code in the draw method that sets the fill on the GraphicsContext object to color. Having this in the code allows us to do the following in Rectangle.

```scala
class Rectangle(val width: Double, val height: Double, c: Color) extends Shape(c) {
  override def area: Double = width * height
  override def perimeter: Double = 2.0 * (width + height)
  override def draw(g: GraphicsContext): Unit = {
    super.draw(g)
    g.fillRect(0.0, 0.0, width, height)
  }
}
```

By making a call to draw on super, this version can set the color when draw is called without actually having access to the value in the supertype. The same type of thing could be done in `Circle`.

One advantage to this approach is that as long as all the subtypes of `Shape` follow the pattern of making a call to `super.draw` before they draw their own geometry, it would be possible to add other settings[9] to the `Shape` type without altering the `draw` methods in the subtypes. In large libraries, this same behavior can be problematic as it forces the supertype to stick with a certain behavior because all the subtypes, including those written by other authors, are expecting it to be maintained.

4.2.6 Abstract Classes

The versions of `Shape` that we have written so far all have some aspects that should feel less than ideal to you. In particular, providing default implementations for `area` and `perimeter` that do things like produce 0.0, or writing an empty set of curly braces for `draw` might seem less than ideal. Maybe it doesn't bother you to write a little code like that which doesn't really do anything, but there is a bigger problem with doing this which can be summed up in the following line of code.

```
val s = new Shape(Color.red)
```

This line of code compiles and runs just fine. The question is, what does it really mean? If you were to call `area` or `perimeter` on the object that `s` references, you would get back 0.0. If you were to call `draw`, nothing would be drawn.

The problem is that while `Shape` is certainly a valid type, it is not a complete specification of an instance object. This is because while we certainly feel that all shapes should have areas and perimeters, we cannot really define them in the completely general sense. We need to be able to say that a type has a method or some piece of member data, but not give it a definition. We also want `Shape` to only represent a supertype, but not implement that supertype; objects should only be instantiated for the subtypes. This can be done with the following code.

```
abstract class Shape(private var color: Color) {
  def area: Double
  def perimeter: Double
  def draw(gc: GraphicsContext): Unit = {
    gc.fill = color
  }
}
```

Two things have been changed here. We have taken away the equal sign and what follows it for `area` and `perimeter`. We have also added the keyword `abstract` to the `class` declaration.

Methods and member data that are not given a value in a `class` declaration are considered to be ABSTRACT. We do this when some member has no reasonable concrete definition. If anything in a `class` is abstract, then the `class` itself must be labeled as `abstract`. If you leave that off, you will get a syntax error. When a type inherits from an `abstract class`, it must either provide an implementation for the abstract members and methods or that subtype must also be labeled `abstract`.

Given this revised `Shape class`, the `Rectangle` and `Circle classes` shown above will work just fine and we will no longer have the conceptual difficulties produced by instantiating the `Shape class` directly. In addition, the `override` keyword is not required when you

---

[9]In chapter 5 we will see that there are other draw setting such as strokes or transformations.

implement a method that is abstract in the supertype.[10] There are two reasons for this. First, on the semantic side, you are not actually overriding an implementation in the supertype. Second, and more importantly, if the method is abstract and you do not implement it, that will generate a syntax error unless the class is labeled as `abstract`.

Why Require `override`?

You might wonder why the `override` keyword is required when you are overriding a method that is implemented in a supertype. You might also wonder why this restriction goes away when the method in the supertype is abstract. The explanation for this is the belief that typos and similar simple errors should produce syntax errors as often as possible.

To illustrate this, imagine a method with the signature `def setPrice(p:Int): Unit` that occurs in some supertype called `InventoryItem`. Perhaps a programmer adds a new subtype called `OnlineItem` which extends `InventoryItem` when the company decides to begin doing sales online. In the subtype, the programmer accidentally makes a method `def setPrice(p:Double): Unit`. If the original method were abstract, or the version in `OnlineItem` were marked with `override`, this would be an error as the argument types do not match. The mismatched argument type means that the subtype simply has two versions of `setPrice`. One of them takes an `Int` and the other takes a `Double`. That is most certainly not what the developer desired, so it is best if the compiler provides a syntax error in that situation.

What if you had implemented the method in the supertype and simply left off the `override` keyword in the subtype? Unfortunately, that would actually compile and run, but it would inevitably give interesting behavior that would take a while to debug, because it would likely take time to figure out that the method in the subtype wasn't being called. For this reason, you should only put implementations of methods in supertypes if they will be used in most subtypes. If something is going to be overridden frequently, you should strongly consider making it abstract.

4.2.7 `traits`

The `class` and `object` constructs are not the only ways to create new types in Scala. A third option, which is significant to us in this chapter, is the `trait`. A `trait` is very similar to an `abstract class` in many ways. A trait encapsulates data member and method definitions, which can then be reused by putting the traits into classes. A class can include any number of traits. The two primary differences between `traits` and `abstract classes` are that `traits` cannot take arguments[11] and you can inherit from more than one `trait`.

It was not specifically said above, but you might have noticed there was no description of how to list multiple `classes` after `extends`. This was not an oversight; you are not allowed to list multiple `classes` there. Scala only supports single inheritance of `classes`. On the other hand, you are allowed to follow the `class` with multiple `traits` or not extend a `class` and just extend one or more `traits`. The various types you inherit from are separated by the `with` keyword.

To understand why you would want to do this, consider the following example. Imagine

---

[10]Even though it isn't required, many developers prefer a style where `override` is always included to prevent errors when implementations are added or removed from supertypes.

[11]There are plans to allow `traits` to take arguments in a future version of the Scala language.

you are writing software to simulate people moving through a building to help with designing the space. You have a type called `Person` that represents everyone that can be in the simulation. The company you are working with knows there are certain categories of people who behave in significantly different ways that the simulation must handle. For example, they need types for `Parent`, `Child`, and `GeneralAdult`. The need for facilities in the building means they also need types for `Male` and `Female`. This last part creates a problem for single inheritance because you want to be able to be able to represent a type `Father` by inheriting from `Parent` and `Male`. The subtype relationship makes sense there, but if both of those type are written as `classes`, you will not be able to use them both. In class inheritance, each class can only inherit from just one superclass.

There are two ways that you could approach this problem with traits. A somewhat standard inheritance scheme might use the following structure.

```
trait Person { ... }
trait Parent extends Person { ... }
trait Male extends Person { ... }
class Father extends Parent with Male { ... }
```

Here both `Parent` and `Male` are subtypes of `Person` and `Father` inherits from both. In this case `Person` has to be a `trait`. This is because a `trait` can only inherit from other `traits`, not `classes`. Below that, either `Parent` or `Male` could have been a `class`, but not both because `Father` could not inherit from both of them. Without a pressing reason to make one of these a `class`, the choice of using `trait` for all three provides consistency.

An alternate approach to constructing these types could be to use the `traits` as mix-in types. Here is what the code could look like.

```
class Person { ... }
trait Parent { ... }
trait Male { ... }
class Father extends Person with Parent with Male { ... }
```

In this construction both the `Parent` and `Male` types are `traits` that do not inherit from the `Person`. Instead, they are mixed in with the `Person` to create the `Father`. This approach is perhaps a bit more advanced and if the `Parent` and `Male` types involve code that requires them to know they will be used with a `Person`, you will need to use self-types which are discussed in the on-line appendices.[12] What matters now is that you are aware that this option exists so that as we build different inheritance hierarchies, you will understand what is going on.

---

trait or abstract class?

A standard question programmers deal with in writing Scala is whether to choose an `abstract class` or a `trait` when coding an abstract type that will be used with inheritance. The general rule of thumb here is to prefer a `trait` because it allows the flexibility of multiple inheritance. While it cannot take arguments, any values that you would want to provide as arguments can be put into the `trait` and left undefined so that the `class` that implements them in the end will have to provide values for those.

We have mentioned a few times now that `traits` do not take arguments. So how do you

---

[12]The appendices are available at http://www.programmingusingscala.net.

get values into a **trait** at construction? One way to do this is to declare abstract values in the **trait**, then give them a definition in the subtype. This works because Scala allows things other than methods to be abstract, so things like **val** and **var** declarations can be left abstract in the **trait**, and be implemented in the subtype. To see how this works, we will flesh out one of the earlier examples with **Person** and **Mother**.

```scala
trait Person {
  val name: String
}

trait Parent extends Person {
  def children: List[Person]
}

trait Female extends Person

class Mother(val name: String) extends Parent with Female {
  override def children: List[Person] = ???
}
```

In this code, **Person** declares an abstract **val** called **name** and **Parent** declares an abstract **def** called **children**. Note that neither **Parent** nor **Male** have to define **name**. While they both inherit from **Person**, they are both abstract by default, as are all **traits**, so they can have abstract members. However, **Mother** is not abstract, so it has to include a proper definition of both **name** and **children**. The definition of **children** is done with an overridden method that we haven't given an implementation for. The definition of **name** is done in the argument list for **Mother**, where we say that the argument **name** is a **val**.[13]

4.2.8 final

When something is **abstract**, it basically means that it has to be implemented in a subtype. There are times when it is also useful to be able to say that something cannot be overridden or changed in a subtype. This requirement can be enforced with the **final** keyword. You can use **final** to modify member data or methods that you do not want to be changed in subtypes. It can also be used to modify a whole **class** when you do not want to allow there to be any subtypes of that **class**.

You might wonder why you would want to make something **final**. One answer to this question is that you want to preserve some type of behavior in the current implementation that should not be altered in subtypes. It can also be a way of telling other code that uses a **class** that there is no possibility of getting different behavior from an instance object of that type than what they would get from that specific type.

The most common example of a place where you should use **final** to preserve a behavior in a type is when a **class** defines a type that is immutable. The strengths of immutable types, and how they can be used to simplify code, is something that has been addressed quite a bit already in this book. Without inheritance, if you wrote a type to be immutable, you knew that any instance of that type could be passed to other code without fear of it being changed.[14] With inheritance, that does not automatically hold true unless you make

---

[13]There are some complexities associated with using values that are overridden in the subtype in code executed during the construction of a **trait**. Details of this are presented in the on-line appendices which are available at http://www.programmingusingscala.net. To avoid this, we recommend that you avoid putting "constructor" code in your **traits** that uses abstract values.

[14]We will also see in chapter 8 that such objects can be used safely across multiple threads.

the `class` final. If the `class` is not marked as `final`, then subtypes could add mutable data and/or override methods such that the new implementations include some reference to mutable state and this could have the potential of breaking things that were using these subtypes with the assumption that they would be like the supertype.

To help you understand this, consider the following example `class` which contains a `String` member and a method for comparing single characters in `Strings`.

```scala
class CharInStrComp(val str: String) {
  def positionCompare(str2: String, index: Int): Int = {
    if (index >= str.length) {
      if (index >= str2.length) 0 else -1
    } else if (index >= str2.length) 1 else {
      str(index).compareTo(str2(index))
    }
  }
}
```

This class is immutable. So if you have an object created from this class, it can be passed around freely. With inheritance though, it is be possible to make a mutable subtype like the following.

```scala
class CntCharInStrComp(s: String) extends CharInStrComp(s) {
  var cnt = 0
  override def positionCompare(str2: String, index: Int): Int = {
    cnt += 1
    super.positionCompare(str2, index)
  }
}
```

If you make an instance of `CntCharInStrComp`, it can be used in any place that wants a `CharInStrComp`. The only problem is that if this is done in a place where the code relies on objects being immutable, that code can now break because it is being passed as an object that is mutable. To prevent this from happening, the original version should be made `final`. You do that by simply adding the keyword `final` in front of the `class` declaration as is shown here.

```scala
final class CharInStrComp(val str: String) {
  def positionCompare(str2: String, index: Int): Int = {
    if (index >= str.length) {
      if (index >= str2.length) 0 else -1
    } else if (index >= str2.length) 1 else {
      str(index).compareTo(str2(index))
    }
  }
}
```

Now any attempt to make a subtype, whether mutable or not, will produce an error.

The `final` keyword is often underused in programming. However, it is good to get into the habit of making things `final` when there are not supposed to be supertypes or when something should not be overridden. Not only is this a safe practice to get into, code that is labeled as `final` can sometimes be compiled to faster implementations as well.

4.2.9 Method Resolution

When you are using inheritance, it is possible for you to create types for which there are multiple different method implementations. You should understand how Scala will determine which of the method implementations will be used when the method is called on an object.

If the method is defined in the `class` or `object` the instance object was created from, that is the version that will be used. If it was not defined there, a version in one of the supertypes will be used. It will look at the supertypes beginning with the last one in the list. If it is not in the last type, it will try anything that type inherits from before going to the earlier supertype elements of the list. This conversion from the full inheritance structure to a list of types to check through is called LINEARIZATION. One caveat is that if the type appears multiple times, the last one by order of appearance in the resolution list is used.

Earlier in the chapter, we listed two approaches to defining the type `Father`. Here is the code we wrote earlier.

```scala
trait Person { ... }
trait Parent extends Person { ... }
trait Male extends Person { ... }
class Father extends Parent with Male { ... }


class Person { ... }
trait Parent { ... }
trait Male { ... }
class Father extends Person with Parent with Male { ... }
```

Let us consider the two approaches to defining the type `Father` given above. The linearization for the two approaches is actually the same. In the first case, if you list all the types from the end of the list back, ignoring repeats, you get `Father`, `Male`, `Person`, `Parent`, and `Person`. The first instance of `Person` is removed leaving us with `Father`, `Male`, `Parent`, and `Person`. When you have repeats, the rule is that the last occurrence is kept. In the second approach, starting from the end goes directly to `Father`, `Male`, `Parent`, and `Person`[15]

Why Linearize?

Being able to inherit from many things is called MULTIPLE INHERITANCE and it is only allowed in some languages, C++ being the most notable case. However, it leads to many complications that many newer languages have chosen to avoid. In the case of Scala, you only get single inheritance from `class`es and multiple inheritance is only allowed with `trait`s.

The reason you are not allowed to do multiple inheritance with `class`es is that the semantics of inheriting from a `class` is to get a full copy of everything in that `class` in the subclass. With that in mind, consider the problems that are created by the UML diagram shown in figure 4.5. This situation is called the DIAMOND PROBLEM.

The C++ language follows the rule of full inclusion of superclasses in subclasses, which illustrates how bad this problem can be. Using this rule, types B and C get full copies of A in them. Then D has a full copy of B and C, meaning it has two full copies of A, a situation that is clearly problematic. (C++ includes a construct called virtual

---

[15]In both of these cases, `AnyRef` and `Any` appear at the end of the list as every `class` implicitly inherits from `AnyRef`, which inherits from `Any`.

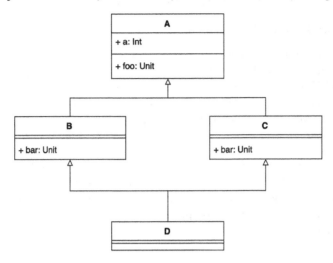

FIGURE 4.5: This UML class diagram shows what is commonly called the diamond problem. The challenge with this construct is that it can lead to significant ambiguity. Different languages have different ways of dealing with this.

---

inheritance that can take this down to one copy.) Imagine you make an instance of type D and try to use the value a, or call the methods **foo** or **bar**. Which one would it use? What code would be executed? D inherits all of those members from two places, and, in the case of **bar**, the code for the two can be different.

In C++, the programmer has to specify which of the supertypes to use in this situation. In Scala (and Perl and Python), the supertypes are "linearized" so that there is a specific order in which they are checked to resolve a method call. That removes any ambiguity. Other languages, such as Java, make it impossible to inherit multiple implementations or data values so this type of problem cannot arise, but the programmer is limited from doing certain things.

4.2.10   Inheriting from Function Types

We saw back in chapter 3 that when you treat an object like a function and pass it value arguments, Scala expands that to a call to the **apply** method on that object. When you enter the code o(i), Scala sees o.apply(i). As a result, you can easily make types that you treat like functions. In order for the type to really be a function type that you can use in places that expect functions, like calls to **map**, **filter**, **foreach**, etc., you need to have your type actually be a subtype of the function type.

To illustrate this, consider the simple example of a custom type that is a function which takes an integer and results in twice its value. We could write such a type like this.

```
class G { def apply(i: Int) = 2 * i }
```

If we instantiate an object of this type, we can use it like a function.

```
scala> val g = new G
g: G = G@28f2e328
```

```
scala> g(5)
res0: Int = 10
```

However, if we try to use this same object in a call to map, we get an error.

```
scala> List(1, 2, 3) map g
<console>:10: error: type mismatch;
 found   : G
 required: (Int) => ?
       List(1,2,3) map g
                       ^
```

As you can see from the error message, this does not work because the map method is expecting an argument that is a function of the form (Int) => ?. That is something that takes an Int argument and outputs anything. We can make a type that fits this, but we need to use inheritance to do so.

```
scala> class F extends ((Int) => Int) { def apply(i: Int) = 2 * i }
defined class F

scala> val f = new F
f: F = <function1>

scala> List(1, 2, 3) map f
res1: List[Int] = List(2, 4, 6)
```

The only difference between type F and type G is that F explicitly extends (Int)=>Int. You might note that not only does this make the call to map work, it also shows up in the creation of the object f, which displays that a function that takes one argument was created (F = <function1>).

Scala Applications Revisited

If you have ever selected the option in Eclipse to make a Scala application, you have inevitably noticed that it does not create an object with a main method. Instead, it makes an object that extends a type called App. The App type provides a main method so your code simply goes into the body of the object. You can use the variable args to get hold of any command line arguments. Here is a sample that does that.

```
object MyApp extends App {
  println(args mkString (" "))
}
```

Now that you know about inheritance, you can choose to use this method of making applications if you prefer.

4.2.11 Enumerations

A common situation when programming is the need to represent a value that can come from a small set of possibilities. A common example of this is the colors of a street light. A standard street light can be green, red, or yellow. No other values are allowed. If your program were to include a street light construct, it would be ideal if the compiler could provide errors when you accidentally used something other than one of those values. Other common examples are days of the month or months of the year. All of these are examples of what many programming languages call enumerations.

It is tempting to store these values using existing types like Int or String. You could define constants so that you do not have "magic numbers" in your code. For example, you might want to write the following.

```scala
object StreetLightColor {
  val Red = 0
  val Green = 1
  val Yellow = 2
}
```

This would allow you to use names like StreetLightColor.Red in your code. The problem with this approach is that you could accidentally do many things that shouldn't be allowed. You could set a light to the value 99, because that is a valid Int. You could add together or subtract values, which are not really meaningful operations. Basically, this approach isn't type safe.

For this reason, many languages provide a special keyword and syntax for creating enumerations. In Scala, the standard library provides a trait called Enumeration that you can inherit from to get this type of functionality. You could use the following declaration for the street light example.

```scala
object StreetLightColor extends Enumeration {
  val Red, Green, Yellow = Value
}
```

Enumeration has a type member Value representing the individual elements of the enumeration. This still allows you to use names like StreetLightColor.Red, but now the type is StreetLightColor.Value, which can only refer to the values Red, Green, and Yellow. Any other value would be a type error.

Given this declaration, you could write a simple StreetLight implementation with the following code.

```scala
class StreetLight(private var _color: StreetLightColor.Value) {
  def color = _color

  import StreetLightColor._

  def cycle: Unit = _color match {
    case Red => _color = Green
    case Green => _color = Yellow
    case Yellow => _color = Red
  }
}
```

To instantiate an object of type StreetLight you could code the following.

```scala
val light = new StreetLight(StreetLightColor.Red)
```

Note the use of an import to make it possible to use short names for the colors in the]class.

4.3 Parametric Polymorphism

The inclusion polymorphism we get from inheritance and subtyping is not the only form of universal polymorphism. There is a second type called parametric polymorphism. This type of polymorphism is something that we have been using since chapter 1, because you use it every time you deal with a List or an Array. Parametric polymorphism is what we get when we write code that has type parameters.

You have become familiar with using parentheses to pass values to methods when you call them or to classes when you construct a new instance. In the same way, you can use square brackets to pass type parameters. Consider these two val declarations.

```scala
scala> val arr1 = Array(1,2,3)
arr1: Array[Int] = Array(1, 2, 3)

scala> val arr2 = Array[Any](1,2,3)
arr2: Array[Any] = Array(1, 2, 3)
```

They both declare arrays that store the values 1, 2, and 3. Those values are all Ints so if we let Scala infer a type, as in the first case, we get Array[Int]. In the second case we explicitly pass in a parameter type of Any and we get back an Array[Any]. The syntax here is to pass any type arguments in square brackets before the parentheses that contain the value arguments. You might feel like there is not any difference between these two lists, but the type parameter changes what you can do with them.

```scala
scala> arr1.sum
res0: Int = 6

scala> arr2.sum
<console>:9: error: could not find implicit value for parameter num: Numeric[Any]
       arr2.sum
            ^
```

You are not able to do sums with type Any. You might question why you cannot sum the elements of arr2, but consider what happens here.

```scala
scala> arr2(1) = true

scala> arr2(2) = List('a','b','c')

scala> arr2
res1: Array[Any] = Array(1, true, List(a, b, c))
```

Type Any really does mean anything in Scala. Hopefully it is clear why trying to take the sum of this array should fail.

4.3.1 Parametric Types

What about using type parameters in our code to make it more flexible? You use a type parameter when you have code that you want to have work with a very broad set of types that possibly have nothing in common with one another. The most common example of this is collections—things like Lists and Arrays that should be able to hold anything you want, but where you do care about what they are holding because you need to know the type of each element when you take them out. This is a usage that we will see a lot in chapters 6, 7, 12, 13, 16, 18, and 20. For now, we can provide some simple examples to give you a feel for this use of parametric polymorphism.

We will use some code designed to help run a theme park to help illustrate this. As you write code for running the different aspects of the park, you realize that a lot of different things are recorded by time of day. So you might store various things by the hour of day, such as sales numbers, passenger counts, resource usage, or employees on duty. Without parametric polymorphism, you wind up duplicating a lot of code because while the code for the times aspect is the same, the types can be very different. Here is a first draft implementation of a class you might use to deal with this.

```scala
package polymorphism.themepark

/**
 * This is a Time of Day Values collection to help reduce code duplication
 * when dealing with values that are associated with the time of day.
 *
 * @tparam A the type of data being stored.
 */
class ToDValues[A] {
  private val values: Array[Option[A]] = Array.fill(24)(None: Option[A])

  /**
   * This allows you to get a value for a particular hour. If there isn't
   * a value, it will throw an exception.
   *
   * @param hour the hour of the day to get. Should be between 0 and 23 inclusive.
   * @return the value stored for that hour.
   */
  def apply(hour: Int): A = values(hour).get

  /**
   * This allows you to get a value for a particular hour. If there isn't
   * a value, it will return None.
   *
   * @param hour the hour of the day to get. Should be between 0 and 23 inclusive.
   * @return an Option of the value stored for that hour.
   */
  def get(hour: Int): Option[A] = values(hour)

  /**
   * Allows you to set the value in a particular hour.
   *
   * @param hour the hour of the day. Should be between 0 and 23 inclusive.
   * @param v the new value to set.
   */
  def update(hour: Int, v: A) = values(hour) = Some(v)

  /**
   * Allows you to set the value in a particular hour using a String for time.
   *
   * @param hour the hour of the day. Should be between 0 and 23 inclusive.
   * @param v the new value to set.
   */
  def update(time: String, v: A) = {
```

```
    val hour = hourFromTime(time)
    values(hour) = Some(v)
}


/**
 * This method clears the value at a particular time.
 *
 * @param hour the hour to clear.
 */
def clear(hour: Int): Unit = { values(hour) = None }

/**
 * This method clears the value at a particular time.
 *
 * @param hour the hour to clear.
 */
def clear(time: String): Unit = {
  val hour = hourFromTime(time)
  values(hour) = None
}


/**
 * Allows you to combine two sets of data using a specified function.
 *
 * @param o the other set of data.
 * @param f The function to apply to the two data types.
 */
def combine(o: ToDValues[A], f: (Option[A], Option[A]) => Option[A]):
    ToDValues[A] = {
  val ret = new ToDValues[A]
  for (((v, i) <- (values, o.values).zipped.map((v1, v2) => f(v1,
      v2)).zipWithIndex) {
    ret.values(i) = v
  }
  ret
}

override def toString(): String = "ToD :\n"+
  (for ((o, i) <- values.zipWithIndex) yield i+" : "+o).mkString("\n")

private def hourFromTime(time: String): Int = {
  time.substring(0, time.indexOf(':')).toInt +
    (if (time.endsWith("PM") && !time.startsWith("12")) 12 else 0)
}
}
```

This **class** is parametric. The name, ToDValues is followed by a type argument [A]. The square brackets distinguish this from a value argument. It is customary to use single, uppercase letters, beginning with "A" for the names of type parameters. This code could be used like this.

```
val riders1 = new ToDValues[Int]
val riders2 = new ToDValues[Int]
val worker1 = new ToDValues[String]
val worker2 = new ToDValues[String]
```

```scala
riders1(12) = 5        // same as riders1.update(12, 5)
riders1("8:24AM") = 10 // same as riders1.update("8:24AM", 10)
riders1(14) = 7
riders2("2:13PM") = 8

worker1(12) = "Kyle" // same as worker1.update(12, "Kyle")

val totalRiders = riders1.combine(riders2, (o1, o2) => (o1, o2) match {
  case (None, None) => None
  case (Some(a), None) => Some(a)
  case (None, Some(b)) => Some(b)
  case (Some(a), Some(b)) => Some(a + b)
})

println(riders1)
println(totalRiders)
```

This code creates four different instances of the class, two that store `Int`s type and two that store `String`s. It then adds some values into three of them and uses combine to add up the riders.

Having parametric polymorphism means that we can write one class and have it work with `Int`, `String`, or any other type we desire. This class still has some significant limitations. For one thing, it is hard to set up. If you had the data you want in some other sequence and you wanted to get it into here, you could not do so without writing code to manually copy it element by element. The combine method is also a bit limited because we can only combine data of the same type and get back that type. We cannot, for example, combine workers with riders. That type of functionality could be useful, but it requires we put type parameters on methods.

4.3.2 Parametric Functions and Methods

The standard approach in Scala for letting users create instance objects of a type in different ways is to create a companion object with versions of `apply` that take the different arguments you want. We then make it so that the class itself takes the proper data as an argument. Here is an example of what that might look like.

```scala
object ToDValues {
  // First attempt that lacks type parameter and does not compile.
  def apply(): ToDValues = new ToDValues(Array.fill(24)(None))
}
```

As the comment implies, this does not work. The reason is that `ToDValues` is not a valid type by itself. It has to have a type parameter to be fully specified. In order to get a type so that we can use it there, we need to give the method a type parameter. The syntax for this is straightforward. We simply put type parameters in square brackets between the name and the value parameters. Doing this, we can make two `apply` methods that look like the following.

```scala
object ToDValues {
  def apply[A]() = new ToDValues[A](Array.fill(24)(None))

  def apply[A](a: A*) = {
    val d = a.map(Option(_)).toArray
```

```
  new ToDValues[A](if (d.length < 24) d.padTo(24, None) else if (d.length > 24)
      d.take(24) else d)
  }
}
```

The second apply method allows us to have code that constructs a `ToDValues` object in the following ways.

```
  val riders1 = ToDValues[Int]()
  val riders2 = ToDValues(0,0,0,6,7,3,9)
```

What should jump out at you here is that the second usage does not specify the type parameter. This is because when type parameters are used with methods they are generally inferred. As long as the type appears in the arguments to the method, you do not have to tell Scala what to use; it will figure it out. The line that creates `riders1` requires that we tell it `[Int]` because there are no arguments, but the second one can figure out that it is working with the `Int` type.

This code works because the class has been set up to take an argument. That means the declaration would have changed from what was shown above to something like this.

```
class ToDValues[A](private val values:Array[Option[A]]) {
```

There is one very big problem with this, it is possible to make a `ToDValues` object with an array that does not have 24 elements. This is clearly bad behavior and we need to prevent it. One way to prevent it is to make it so that the **class** has a private constructor. This means that only code associated with the class or its companion object can make a call to **new** for this object. Syntactically this is done by putting the **private** keyword before the value arguments like this.

```
class ToDValues[A] private (private val values:Array[Option[A]]) {
```

Now the only way that outside code can instantiate a `ToDValues` object is through the `apply` methods of the companion object, and it is easy for us to make sure those always create instances using arrays with 24 elements.

The other addition we would like to make is to enhance the **combine** method so that it can work with different types. Ideally, we'd like to have the ability to make it so that the input types could be two distinct types and the output type is a third type. We can accomplish this by introducing two type parameters on the method like this.

```
  def combine[B,C](o:ToDValues[B])(f:(Option[A],Option[B])=>Option[C]) :
      ToDValues[C] = {
  new ToDValues((values, o.values).zipped.map((v1, v2) => f(v1, v2)))
  }
```

Now we could do things like combine workers with riders to keep track of correlations that might be used for bonuses.

If you look closely, you will see that something other than the addition of the type parameters changed here. Instead of a single value parameter list with two parameters, there are two lists with one parameter each. You should recall this is a technique called currying. This is not required, but it makes using the method easier. Here is an example usage.

```
  val totalRiders = riders1.combine(riders2)((o1, o2) => (o1, o2) match {
    case (None, None) => None
    case (Some(a), None) => Some(a)
```

```
    case (None, Some(b)) => Some(b)
    case (Some(a), Some(b)) => Some(a+b)
  })
```

If we do not curry this method, the local type inference in Scala will not automatically figure out the type of o2 and we would have to specify it in the function declaration with (o1,o2:Int).

4.3.2.1 Parametric Sorts

At the beginning of the chapter, we used the task of sorting as a motivating example for sorts. We saw that we had to copy our code and make changes when we wanted to work with a different type. Now you can see that we would like to have a method like sort[A] that could sort any type we want. A simple attempt at doing this will show us a detail of type parameters that we have not discussed that makes this more challenging.

```scala
1  def bubbleSort[A](arr: Array[A]): Unit = {
2    for (i <- 0 until arr.length - 1; j <- 0 until arr.length - i - 1) {
3      if (arr(j + 1) < arr(j)) { // Error: < not a member of type parameter A
4        val tmp = arr(j)
5        arr(j) = arr(j + 1)
6        arr(j + 1) = tmp
7      }
8    }
9  }
```

The comment on line 3 describes the error that we get if we try to compile this code. What does this error message mean? When we say that code must work with a type parameter, such as A, Scala has to make sure that it would work for any allowable A. Since there are no restrictions on this, that means that it must comply with the requirements of type Any, and the Any type does not define a method called <.

One way to get around this is to put bounds on what we allow for type A. We'll see how to do that in the next section. For now we will look at an alternate approach. Instead of assuming the type will work with a particular method, like <, we will pass in an extra argument that is a function that plays the role of <. We'll call this function lt, short for less than.

```scala
def bubbleSort[A](a: Array[A])(lt: (A, A) => Boolean) {
  for (i <- 0 until a.length; j <- 0 until a.length - 1 - i) {
    if (lt(a(j + 1), a(j))) {
      val tmp = a(j)
      a(j) = a(j + 1)
      a(j + 1) = tmp
    }
  }
}
```

This now provides us with a sort that can work for any type that we want, as long as we can define a function that represents less than for that type. As with the combine method, this works best if we curry the call to take two separate argument lists. If you load that into the REPL we can see it work like this.

```scala
scala> val rints = Array.fill(10)(util.Random.nextInt(30))
rints: Array[Int] = Array(29, 19, 15, 6, 14, 25, 1, 18, 10, 22)
```

```
scala> Sorts.bubbleSort(rints)(_<_)

scala> rints
res0: Array[Int] = Array(1, 6, 10, 14, 15, 18, 19, 22, 25, 29)

scala> val rdoubs = Array.fill(10)(math.random)
dints: Array[Double] = Array(0.10388467739932095, 0.21220415385228875,
    0.8450116758102296, 0.5919780357660742, 0.9652457489710996,
    0.9401962629233398, 0.08314463374943748, 0.1502193866199757,
    0.7017577117339538, 0.9599077921736453)

scala> Sorts.bubbleSort(rdoubs)(_<_)

scala> rdoubs
res4: Array[Double] = Array(0.08314463374943748, 0.10388467739932095,
    0.1502193866199757, 0.21220415385228875, 0.5919780357660742,
    0.7017577117339538, 0.8450116758102296, 0.9401962629233398,
    0.9599077921736453, 0.9652457489710996)
```

Here the same method has been used to sort both Ints and Doubles.

Comparators

You will often see this type of sorting done using a function that returns an Int instead of a Boolean. When you need to know if two values are possibly equal in addition to being less than or greater than, this approach is beneficial. If the values are equal, the function will return 0. If the first one is less, it will return a negative value. If the first one is greater it will return a positive value. We will see this type of usage of comparison functions in later chapters.

The strength of the approach of using a second function argument for comparison is that it makes the sort more flexible. For example, in either of the above examples we could have sorted the numbers from greatest to least instead of least to greatest by simply passing in _>_ instead of _<_. In addition, it truly places no limitations on the type A. This allows you to sort things according to any sort order you want. For example, if you had a type for students you might want to sort by last name in some contexts or by grade in others. This approach would allow you to do that because the sort order is not attached to the object type. The downside of this approach is that the person calling the sort method has to write a comparison function each time they want to use it. In the examples above, that was not a big deal because we had simple comparisons. However, there are times when that will not be the case.

4.3.3 Type Bounds

As we just saw, a normal type parameter, like [A], is treated like Any when we use objects of that type in code. While this is a very common usage, there are certainly times when you want to limit what types can be used in a type parameter so that you can safely call methods other than the small set that are available with Any. The way you do this is to put bounds on the type.

The most common form of this is putting an upper bound on a type. This is done with the <: symbolic keyword. To understand why you might want to do this, let's consider an example using code that we had before with fruit. The function, `makeBreakfastShake` that we wrote at the beginning of the chapter only used a single piece of fruit. This is not how it normally works. You would want to be able to include multiple pieces of fruit, perhaps of different types. A common example would be a shake with both bananas and strawberries. To do that, you might consider having code like this where you pass in an `Array[Fruit]`.

```scala
def makeBreakfastShake(fruits: Array[Fruit]) {
  for (fruit <- fruits) {
    if (!fruit.canEatSkin) {
      fruit.peel
    }
    blender += fruit
  }
  blender += juice
  blender += ice
  blender.blend
}
```

This code probably looks fine to you, and in many situations it will work. However, it is not perfect. Here is an example of where it runs into problems.

```scala
scala> val berries = Array(new Strawberry, new Strawberry)
berries: Array[Strawberry] = Array(Strawberry@51d36f77, Strawberry@103b1799)

scala> makeBreakfastShake(berries)
<console>:16: error: type mismatch;
 found   : Array[Strawberry]
 required: Array[Fruit]
Note: Strawberry <: Fruit, but class Array is invariant in type T.
You may wish to investigate a wildcard type such as '_ <: Fruit'. (SLS 3.2.10)
       makeBreakfastShake(berries)
                          ^
```

The variable `berries` here has the type `Array[Strawberry]`. The way `makeBreakfastShake` is written here, the argument passed in has to actually be an `Array[Fruit]`. The type `Array[Strawberry]` is not a subtype of `Array[Fruit]`. To understand why this is the case, simply imagine what would happen if such a call were allowed in this code.

```scala
def substituteCherries(bowl: Array[Fruit]) {
  if (!bowl.isEmpty) bowl(0) = new Cherry // if the bowl is not empty substitute
    the first piece                       // with a Cherry
}
```

This is fine if you call it with an `Array[Fruit]`, but if you were allowed to call this function with an `Array[Strawberry]`, you would wind up with a `Cherry` in an array that is not allowed to hold one.

The solution to this problem is implied in the error message above. The `makeBreakfastShake` function needs to say that it can take an array of any type that is a subtype of `Fruit`. This is done by specifying a type bound using `<:` as seen here.

```scala
def makeBreakfastShake[A <: Fruit](fruits: Array[A]) {
  for (fruit <- fruits) {
    if (!fruit.canEatSkin) {
      fruit.peel
    }
```

```
    blender += fruit
  }
  blender += juice
  blender += ice
  blender.blend
}
```

Using the `<:` symbol lets you put a constraint on the type parameter `A`. In this way, we can say that it is safe to assume that objects of type `A` are something more specific than `Any`. This is required in this example because `Any` does not have methods for `canEatSkin` or `peel`.

You can also specify a lower bound on a type using `>:`. We will see uses of this type constraint in chapter 12 when we are building immutable data types.

Covariant and Contravariant

It is worth noting that the problem we had using an `Array[Strawberry]` would not have occurred had we used a `List` instead of an `Array`. The version of the code shown here would work for a list of any type that is a subtype of `Fruit` without the type bound.

```
def makeBreakfastShake(fruits: List[Fruit]) {
  for (fruit <- fruits) {
    if (!fruit.canEatSkin) {
      fruit.peel
    }
    blender += fruit
  }
  blender += juice
  blender += ice
  blender.blend
}
```

This works because `List` is a covariant type, which means that the type `List[Strawberry]` is a subtype of `List[Fruit]`, something that was not the case for `Array`.

If you look in the API at `scala.collection.immutable.List`, you will see that it has a type parameter of `+A`. This means that it is covariant and that subtype relationships on the whole type match the type parameters. The details of covariant, contravariant, and invariant are more advanced and appear in the on-line appendices.[16] A simple explanation of why this works is that because the `List` is immutable, so our example of substituting a `Cherry`, which caused problems for the `Array`, is not an allowed operation on a `List`.

4.3.3.1 Type Bounds and Sorting

One of the uses of traits is as mixins that add specific functionality to types. An example of this is the trait `Ordered[A]`. This trait has one abstract method, `compare(that:A):Int`. This method does a comparison to type `A` and returns a negative value if **this** comes before **that** in the ordering, a positive value if **this** comes after **that** in the ordering, and zero if they are equal. It has concrete methods to define the normal comparison operators based on the results of **compare**. Although **that** is just a variable name and you could have given it a different name, it is standard practice to name this variable **that**.

Any object that is a subtype of **Ordered** should work well in a sorting algorithm. That fact that it is a subtype of **Ordered** means that it has a natural ordering that we want to use for the sort. This natural ordering means that we do not have to pass in a comparison function; it is built into the type. We simply have to limit what types can be used as subtypes of **Ordered**. A version of our bubble sort that takes this approach is shown here.

```scala
def bubbleSort[A <: Ordered[A]](a: Array[A]) {
  for (i <- 0 until a.length; j <- 0 until a.length - 1 - i) {
    if (a(j + 1) < a(j)) {
      val tmp = a(j)
      a(j) = a(j + 1)
      a(j + 1) = tmp
    }
  }
}
```

You should take special note of the type bound on this example: [A <: Ordered[A]]. This is a recursive type bound where the bound refers back to the type that it is constraining. In this context we require the ability to build a recursive type bound because inside the sort, the code is doing a comparison between two objects of type A.

If you enter the sort that uses **Ordered** above into the computer and played with it, you will discover that it had some limitations you might not have been expecting. In particular, you could not use it to sort an array of **Ints**. The same is true of many of the other types that you might have tried to use it to sort. The reason for this is that **Ints** do not inherit from **Ordered**, thus they do not have an inherit ordering. This was done to keep the **Int** type simple.

This problem can be fixed by changing a single character in the code, though explaining what that character means and why it fixes the problem is a bit more involved. The fix is to replace the : with a % in the type bound.

```scala
def bubbleSort[A <% Ordered[A]](a: Array[A]) {
  for (i <- 0 until a.length; j <- 0 until a.length - 1 - i) {
    if (a(j + 1) < a(j)) {
      val tmp = a(j)
      a(j) = a(j + 1)
      a(j + 1) = tmp
    }
  }
}
```

The <% symbol says that you can use any type that is a subtype of, or has an implicit conversion to, a subtype of the specified type. This sort will work for **Array[Int]** or **Array[Double]** because the standard libraries in Scala define implicit conversions from **Int** and **Double** to other types that are subtypes of **Ordered[Int]** and **Ordered[Double]**, respectively.

Implicit conversions are an advanced feature that you have been using all the time, but which are generally invisible to you. They make life easier by converting one type to another without forcing you to do any work. They can make it hard to tell what code is really doing in some places though, and for that reason, there are strict rules on what implicit conversions are allowed to happen. These include only using implicit conversions that are currently in scope, and not applying conversion transitively.

4.4 End of Chapter Material

4.4.1 Summary of Concepts

- Polymorphic code is code that can work with many types. Universal polymorphism implies that it can work with an infinite number of types.

- Inclusion polymorphism is a form of universal polymorphism that comes from subtyping. Scala, like most `class`-based object-oriented languages, gets this through inheritance.

  - Subtypes cannot get direct access to `private` members of the supertype. The `protected` visibility is meant to address this, but it is not as secure as `private` because you have little control over what code might inherit from any `class` you write.

  - The `super` keyword refers to the part of the current object that holds the supertype and can be used to make calls on methods in the supertype that have been overridden in the current `class`.

  - In Scala, when a subtype overrides a method, the keyword `override` must come before the `def` keyword for the method declaration.

  - Anonymous classes are created when a call to `new` on a type is followed by a block of code.

  - Members and methods of a `class` can be left undefined. Such members are called `abstract`. If you have `abstract` members in a `class`, the `class` itself must be declared abstract. You cannot instantiate `abstract classes`.

  - A `class` can only inherit from one other `class`. There is a similar construct called a `trait` that allows for multiple inheritance. A `trait` is much like an `abstract class`.

  - Members and methods that should not be overridden can be declared `final`. You can also make whole `classes` `final` to prevent any other `classes` from extending them.

- Scala allows methods/functions, `classes`, and `traits` to have type parameters. These are passed in square brackets that come before any normal parameters in parentheses. This ability leads to parametric polymorphism, another form of universal polymorphism.

  - When a `class` takes a type parameter, the type must be specified at the point of instantiation with `new`. This is in contrast to parametric functions/methods which can typically infer the proper types of parameters based on provides value arguments.

  - You can place bounds on type parameters. Without bounds, the type parameters must be broadly assumed to be of the type `Any`.

4.4.2 Exercises

1. The `hourFromTime` method in the `ToDValues` class is not particularly robust. What are some situations where it could fail? Write a better replacement for it.

2. Draw out the inheritance hierarchies that you might design to represent the following types of objects.

   - Animal life
   - School/university organization
   - Accounts for a large bank (savings, checking, etc.)
   - Types of Facebook friends
   - Types of students at your school
   - Whatever else you can think of that makes a good example of an "is-a" relationship.

3. Implement one of your hierarchies from above with a few simple methods.

4.4.3   Projects

There are several different options here that you could pick from based on which of the projects you started in the last chapter. The general idea though is that you will build a hierarchy of types for things that are related to your project. You can also put in parametric polymorphism if it fits your project.

When you do this, there is something very significant to keep in mind. Just because two things fit the "is-a" requirement does not mean they should be modeled with inheritance. One alternative is to use a single type and have the differences be data driven. Whether you should use inheritance or just store values that provide the difference is best answered by whether the difference is truly a functional difference or a difference of magnitude. For example, in a game with different types of enemies or units, most of the different types should not be different subtypes because the only difference is one of things like strength or speed. However, when the difference is something like the behavior, that would require one to use a very different implementation of a particular method, and is therefore a reason to invoke inheritance.

1. In the first project on the MUD/text RPG, it was recommended that you create some character other than your player that wanders around the map. These other characters might play various roles in the final game. In a standard MUD, these would be the things that players do combat with. In many ways, these computer-controlled characters are just like players except that their movements and other actions are controlled by the computer instead of a human. This is a great potential use of inheritance. You might be able to come up with other uses as well.

2. One potential location for using inheritance in the web spider is with the types of data that you collect. You can make a supertype that represents data in general with abstract display capabilities. The details of this will depend on exactly what you are planning to do for your data collection.

3. For the graphical game project, the possibilities depend a lot on what you are doing. As with the MUD, it is possible that you will have computer-controlled entities that mirror functionality of the player. Even if that does not fit, it is likely that there are other types of entities, active or not, which go into the game that share functionality in many ways, but not completely. (Note that for networked games, the behavior of the local player and the remote player will be distinct, but that is something you do not need to deal with until chapter 11.)

Chapter 5

GUIs and Graphics

Odds are good that most of the programs you have used in your life have not had the text style interface that we have been creating with `println` and the various read methods. Most users work with programs that have GRAPHICAL USER INTERFACES, GUIs. The goal of this chapter is to introduce you to a GUI library in Scala, and show you how you can build GUIs and add our own graphics to them.

5.1   Project Analysis

To help illustrate principles of object-oriented design, we are going to follow the development of a single large program over many of the remaining chapters in this book. The program that we are going to develop is a drawing program. We haven't introduced it sooner because we couldn't implement much of it earlier without GUIs and graphics. This program is not going to be like Paint® or quite like any other "normal" drawing program you might have used. We will do a few things differently to make it so that this program can better illustrate key concepts through the rest of the book. A few things will be thrown in just for educational purposes, but in general, features will be included to produce a final application that is capable of things that you do not normally see in drawing programs, including animation.

This is going to be a fairly big piece of software, so as was discussed in chapter 2, we should start with the analysis and then work on the design before we try to implement anything. We are not going to try to figure out every possible option that could go into the software right now. We want to figure out enough that we have some direction to go in. We

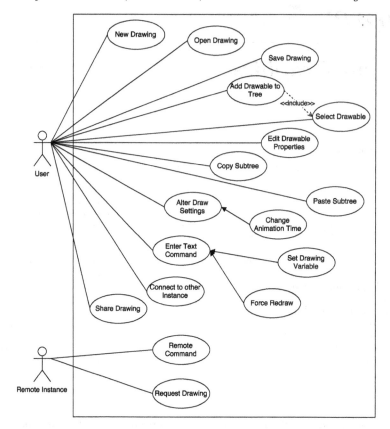

FIGURE 5.1: This is a use case diagram for our drawing program.

also want to identify where we might want to add functionality later so that as we move into the design phase, we can keep those in mind.

The program itself is a drawing program that has some extra features like networking, animation, and text input to issue commands. Figure 5.1 shows a use case diagram for our system. There are two actors in the diagram. The obvious one is the user who is running the program. This user has the ability to add elements to the drawing and edit elements of the drawing. They also have the ability to give text commands to the program. The less obvious parts of the diagram involves having two separate instances of the program talk to one another. This leads to use cases where the local user makes connections to other instances of the program and does things such as sending them parts of drawings or other forms of communication.

This diagram shows some arrows connecting different use cases. The solid line with the filled arrow indicates a generalization. This implies that one use case is a specific type of another use case. There is also a dotted arrow with an open head labeled <<include>>. That style of arrow indicates that whenever one use case occurs, another will be included as part of it. This diagram is sufficient for our current needs. It lays out the basic operations of the program, and its creation forces us to think about what it is we want the program to be able to do. You should go through the different options shown in that diagram, as they will help to inform the decisions that we make as we start to lay out a design.

Our next step is to just draw out what we want the GUI to look like. During this process, we will introduce a number of the elements that we can put into GUIs that will be elaborated on later in the chapter. Running through the use cases, we can determine a number of elements that our GUI needs to have. These include the following.

- A set of menus for basic operations.

- Tabs for switching between drawings.

- An area to display the drawing.

- An area that lists the elements in the drawing so that we can select specific elements from a graph.

- An area to see and edit the properties of things in the drawing.

- A sliding element for adjusting the time in an animation.

- A field to type in text commands.

- An area to display the output of text commands.

Figure 5.2 shows a simple line drawing of how we might lay out these various components. The bars that are drawn thicker will allow the user to adjust their position. The others will just take up whatever space is needed for things. Many of the different components have names in parentheses next to them like `TreeView` or `Canvas`. These are the names of classes that we will use to produce those different elements.

5.2 ScalaFX

Before we can go much further, we need to actually learn a bit about what goes into writing a GUI. At the time of writing, there are three different GUI libraries that are part of the Java libraries. We are going to use the newest of these, JavaFX, through a wrapper called ScalaFX that provides better support for Scala syntax. ScalaFX makes it easy to create modern-looking graphical user interfaces (GUIs) with sophisticated visual effects. Most of the GUI code currently written for the Java Virtual Machine is written using the Swing library, which was introduced early in the life of the JVM as a pure Java way to make GUIs that was more flexible than the original Abstract Windowing Toolkit (AWT). Unfortunately, Swing has some design elements that caused problems when trying to modernize and optimize it. For that reason, Oracle® is pushing new GUI application development on the JVM to move to JavaFX, and this book is following that recommendation.

Currently, ScalaFX is not part of the standard libraries in Scala. You can find information for the project at `http://www.scalafx.org/`. This includes links to the API documentation[1] and the GitHub repository. If you are using sbt (Scala Build Tool)[2] for your projects, you can easily add an extra line to the dependencies to get ScalaFX support. If you are using an IDE without using an sbt project, you can pull down the code from the repository to build a JAR file for the library. JAR files of ScalaFX for current versions of the library,

---

[1] At the time of this writing, the API documentation for ScalaFX is fairly minimal. That isn't a problem though, because ScalaFX is just a wrapper for JavaFX, and at the top of each page is a link to the corresponding JavaFX API page, which is generally very well documented. You can read the JavaFX documentation and then make the calls using the Scala style that is supported by ScalaFX.

[2] This is discussed in the online appendices at `http://www.programmingusingscala.net`.

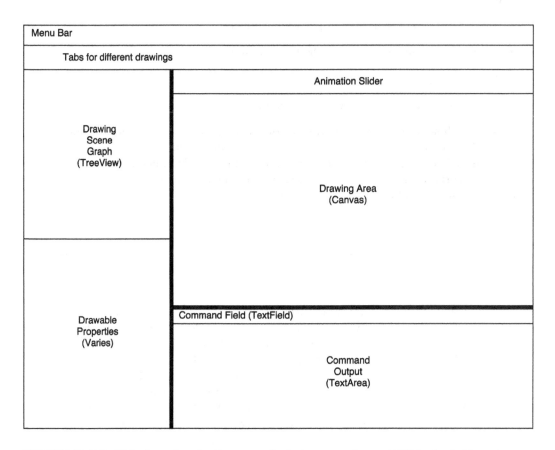

FIGURE 5.2: This is a simple diagram of what we want our GUI to look like when we are done with the project.

Scala, and Java are posted on the book website. You can download them, and add them to your path if you don't want to build the library yourself.

To make an application that runs a GUI using ScalaFX, simply make an `object` and have it `extend JFXApp`. So the simplest GUI you could write would be the following.

```
import scalafx.application.JFXApp

object FirstGUI extends JFXApp
```

If you run this, a small window with nothing in it will pop up.

Now we need to add elements into our GUI. JavaFX, and by extension ScalaFX, uses something called a scene graph to organize the different elements in a GUI. A scene graph is a tree data structure, most commonly found in graphical applications and libraries. The elements that can go into the scene graph are part of a large inheritance hierarchy. As a general rule, object-oriented GUI libraries make significant use of inheritance. Figure 5.3 shows a class diagram for many of the `classes`/`traits` that are used as part of the scene graphs in ScalaFX. A few `classes` have been left out of this diagram, but it is generally complete. The first thing that you probably notice is that it is large. ScalaFX has a lot of different pieces that allow you to build powerful GUIs. We clearly aren't going to cover every `class` shown here. Instead we will cover one or two elements of various branches of the hierarchy that we will be using for our project or that we believe you will use in your own projects. A more complete coverage of the various GUI elements is given in *Introduction to Programming and Problem Solving Using Scala* and the online videos associated with that book.

There are three classes at the top of this diagram: `Window`, `Scene`, and `Node`. As the name implies, the `Window` type and its subtypes represent windows, or the primary platforms that the GUI runs in.[3] We will concern ourselves primarily with the branch that goes down to `PrimaryStage`, as that is the main window for a desktop GUI.

The `Scene` class is a container that holds all the elements of a scene graph. The contents of the scene graph are all subtypes of `Node`, which make up the vast majority of figure 5.3. The significant branches that we will be dealing with are under `Shape`, `Pane`, and `Control`.

The following code shows a very simple GUI with the primary elements that you will wind up adding to any real GUI along with a button and an ellipse. Note that the code begins with a number of `import` statements. The ScalaFX library is organized in a number of different packages. To help you learn where things are, we fully specify all the paths in the `import` statements in our earlier examples. When you are using multiple elements from a package, you should feel free to use underscores as appropriate. The first `import` statement brings in all the members of an `object` called `Includes`. This provides a number of implicit conversions from JavaFX types to the appropriate ScalaFX types. We recommend that you put this at the top of all of your ScalaFX programs.

```
1  package guigraphics
2
3  import scalafx.Includes._
4  import scalafx.application.JFXApp
5  import scalafx.scene.Scene
6  import scalafx.scene.control.Button
7  import scalafx.scene.shape.Ellipse
8  import scalafx.scene.paint.Color
```

---

[3]JavaFX can also be used for mobile applications, which don't use windows in the same manner as desktop GUIs. This is part of the reason for the use of the term `Stage`, because calling it a window only applies to the desktop whereas `Stage` applies to both.

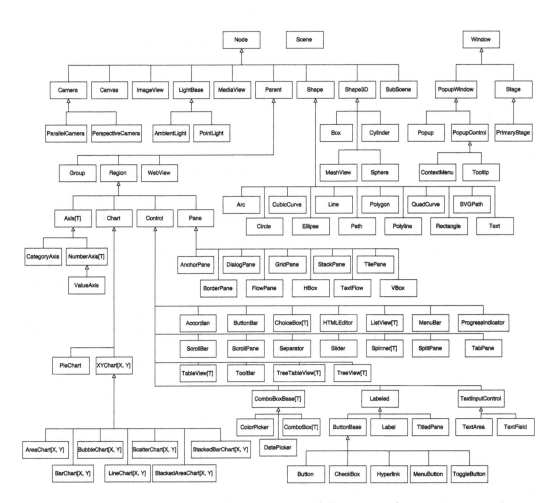

FIGURE 5.3: This class diagram shows the most of the classes/traits that are relevant to the ScalaFX scene graph.

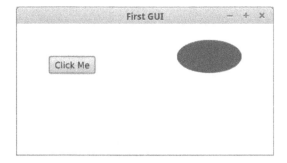

FIGURE 5.4: This is the window that pops up when you run the full version of FirstGUI.

```scala
 9
10  object FirstGUI extends JFXApp {
11    stage = new JFXApp.PrimaryStage {
12      title = "First GUI"
13      scene = new Scene(400,200) {
14        val button = new Button("Click Me")
15        button.layoutX = 50
16        button.layoutY = 50
17        val ellipse = Ellipse(300, 50, 50, 25)
18        ellipse.fill = Color.Red
19        content = List(button, ellipse)
20      }
21    }
22  }
```

If you run this program, you will see the window shown in figure 5.4. You can click the button, but it doesn't do anything. Also, if you resize the window, you will see that the elements just stay where they are. Using content[4] to add elements that have fixed positions is a simple approach to building GUIs, but it isn't the most user-friendly option when the sizes and resolutions of displays can vary dramatically. So we need to learn how to lay out our GUIs and make them interactive so that we can build our drawing program.

If you haven't worked with GUIs or graphics before, there is one very important aspect of the coordinates that appear on lines 15–17 that needs to be clarified. The origin of the coordinate system, $(0,0)$, occurs in the top left corner. The x-coordinate grows as you move to the right, as you would expect. So the left edge of the Button is 50 pixels from the left edge of the scene, and the center of the Ellipse is 300 pixels from the left edge of the scene. The y-coordinate grows as you move down, the opposite of what happens with the coordinate systems you are used to from math. So the top of the Button is down 50 pixels from the top of the scene, and the center of the Ellipse is also down 50 pixels from the top of the scene.

This inversion of the y-coordinate is not unique to ScalaFX or JavaFX, it is the standard behavior in most graphics libraries. While it might seem odd, it makes more sense when you consider that the first "graphical" displays were printed sheets of paper. The top line on the sheet was 0, the next line was 1, etc. So the y-coordinate grew going down the page.

---

[4]A container that holds elements.

Anonymous Classes

The lines that build the **stage** and the **scene** are doing something more than it might appear. When a use of **new** is followed by a name and curly braces, it creates something called an anonymous class. So the line **scene = new Scene(400, 200){** is not just instantiating a new instance of **Scene**, it is creating a new **class** that is a subtype of **Scene** and instantiating an instance of that new **class**. It is called an anonymous class because it has no name that you can refer to. You can only refer to its type as the supertype, in this case **Scene**.

This type of thing is done a lot in Scala and in Java as well. Fortunately, the syntax for anonymous **class**es is so natural in Scala that you rarely think of what you are doing as inheritance. Instead, you think of it more like you are making a specialized instance of some type that has some additional functionality.

5.3   Drawing Program Layout

Now we want to write code to lay out the GUI for figure 5.2. This requires significantly more code than the last example. We will present the whole program first, then go through it, explaining what is happening in different parts.

We start off with a lot of **import** statements to bring in the various **class**es that are being used as well as the implicit conversions in **scalafx.Includes**. All the code is then placed inside of an **object** that **extends JFXApp** and includes code like what we have seen before to set up the **stage** and **scene**.

```
 1  package guigraphics.v1
 2
 3  import scalafx.Includes._
 4  import scalafx.application.JFXApp
 5  import scalafx.geometry.Orientation
 6  import scalafx.scene.Scene
 7  import scalafx.scene.canvas.Canvas
 8  import scalafx.scene.control.Menu
 9  import scalafx.scene.control.MenuBar
10  import scalafx.scene.control.MenuItem
11  import scalafx.scene.control.ScrollPane
12  import scalafx.scene.control.SeparatorMenuItem
13  import scalafx.scene.control.Slider
14  import scalafx.scene.control.SplitPane
15  import scalafx.scene.control.Tab
16  import scalafx.scene.control.TabPane
17  import scalafx.scene.control.TextArea
18  import scalafx.scene.control.TextField
19  import scalafx.scene.control.TreeView
20  import scalafx.scene.layout.BorderPane
21
22  object DrawingMain extends JFXApp {
23    stage = new JFXApp.PrimaryStage {
```

```scala
24      title = "Drawing Program"
25      scene = new Scene(1000, 700) {
26        // Menus
27        val menuBar = new MenuBar
28        val fileMenu = new Menu("File")
29        val newItem = new MenuItem("New")
30        val openItem = new MenuItem("Open")
31        val saveItem = new MenuItem("Save")
32        val closeItem = new MenuItem("Close")
33        val exitItem = new MenuItem("Exit")
34        fileMenu.items = List(newItem, openItem, saveItem, closeItem, new
            SeparatorMenuItem, exitItem)
35        val editMenu = new Menu("Edit")
36        val addItem = new MenuItem("Add Drawable")
37        val copyItem = new MenuItem("Copy")
38        val cutItem = new MenuItem("Cut")
39        val pasteItem = new MenuItem("Paste")
40        editMenu.items = List(addItem, copyItem, cutItem, pasteItem)
41        menuBar.menus = List(fileMenu, editMenu)
42
43        // Tabs
44        val tabPane = new TabPane
45        tabPane += makeDrawingTab()
46
47        // Top Level Setup
48        val rootPane = new BorderPane
49        rootPane.top = menuBar
50        rootPane.center = tabPane
51        root = rootPane
52      }
53    }
54
55    private def makeDrawingTab(): Tab = {
56      // left side
57      val drawingTree = new TreeView[String]
58      val treeScroll = new ScrollPane
59      drawingTree.prefWidth <== treeScroll.width
60      drawingTree.prefHeight <== treeScroll.height
61      treeScroll.content = drawingTree
62      val propertyPane = new ScrollPane
63      val leftSplit = new SplitPane
64      leftSplit.orientation = Orientation.Vertical
65      leftSplit.items ++= List(treeScroll, propertyPane)
66
67      // right side
68      val slider = new Slider(0, 1000, 0)
69      val rightTopBorder = new BorderPane
70      val canvas = new Canvas
71      rightTopBorder.top = slider
72      rightTopBorder.center = canvas
73      val commandField = new TextField
74      val commandArea = new TextArea
75      commandArea.editable = false
76      val commandScroll = new ScrollPane
77      commandArea.prefWidth <== commandScroll.width
```

```
78        commandArea.prefHeight <== commandScroll.height
79        commandScroll.content = commandArea
80        val rightBottomBorder = new BorderPane
81        rightBottomBorder.top = commandField
82        rightBottomBorder.center = commandScroll
83        val rightSplit = new SplitPane
84        rightSplit.orientation = Orientation.Vertical
85        rightSplit.items ++= List(rightTopBorder, rightBottomBorder)
86        rightSplit.dividerPositions = 0.7
87
88        val topSplit = new SplitPane
89        topSplit.items ++= List(leftSplit, rightSplit)
90        topSplit.dividerPositions = 0.3
91
92        val tab = new Tab
93        tab.text = "Untitled"
94        tab.content = topSplit
95        tab
96      }
97    }
```

5.3.1   Menus

The new parts of this code are what is put in the **scene** on lines 26–51 and a method for making a tab for a drawing that comes on lines 54–95. Lines 26–41 set up the menus for the program. Most of this is variable declarations for a **MenuBar**, two **Menus**, and nine **MenuItems**. Line 34 adds the first five **MenuItems** to **fileMenu** and line 40 adds the other four to **editMenu**. Line 34 also adds a separator between the first three items and the last item to reduce the odds of a user accidentally exiting the program when they meant to save. Line 41 sets those two **Menus** to appear on the **menuBar**. It isn't until line 49 that the **menuBar** is actually added into the GUI when it is set to be the **top** element in **rootPane**.

In ScalaFX, a **MenuBar** is just like any other **Node**. They do not have to be placed at the top of windows. It is just customary to do so. If you change **top** to **bottom**, you can move the menu to the bottom of the window.

While it isn't illustrated here, you can also add sub-menus to a menu by making one of the **items** in a **Menu** be a **Menu**. There are also **classes** for **CheckMenuItem** and **RadioMenuItem** that support putting check boxes and radio buttons in menus.

5.3.2   Panes

Lines 43–51 set up the top-level structure of the GUI by adding a **TabPane** and the **MenuBar** into a **BorderPane** that is set to be the **root** of the **scene**. The **classes** whose names end in **Pane**, and a few others with different names, are used to lay out elements of a GUI. In figure 5.3, most of these components inherit from the **Pane** type, through there are a few additional ones that inherit from **Control**.

In the small example that we began with, we did not use any **Panes** or **root**. Instead, we hard coded the locations of the **Nodes** and set them to be the **content**. In ScalaFX, **content** is just a helper that creates a **Group**, adds the specified elements to that **Group**, then sets that **Group** to be the **root**. Using **Panes** to lay out GUIs produces results that behave better when the user changes things, such as resizing the window.

Line 44 creates a **TabPane**. As the name implies, this is a pane that has tabs across the top. Each tab has different contents. The user can select a tab to see its contents. Only one tab is added on line 45, and it uses another method to build that tab. This code is put in a separate method, because it will need to happen for each drawing when the user selects the menu items new or open. It also helps to make the code more readable, even without those options in place.

Line 48 makes a `BorderPane`. This type of pane has five different regions where `Nodes` can be added. They are called `top`, `bottom`, `left`, `right`, and `center`. The `BorderPane` is a good example to show the advantages of using `Panes` over simply placing `Nodes` by coordinates. When the size of a `BorderPane` changes, most `Nodes` inside of it will automatically resize to adjust to the new size. The top and bottom elements, if present, take the full width of the panel and as much vertical size as they want. The left and right elements, if present, take the remaining vertical height, and as much horizontal space as they want. The center element gets everything that is left over. The `rootPane` is used to put the `MenuBar` at the top of the window, and all the rest of the space goes to the `tabPane`.

Two `BorderPanes` are also used in the `makeDrawingTab` method. They are created on lines 69 and 80 to separate the `Slider` from the `Canvas` and the `TextField` from the `TextArea` respectively.

Another type of `Pane` that appears in the `makeDrawingTab` method is the `ScrollPane`. The `ScrollPane` gets to have a single content `Node`, and its purpose is to show scroll bars and allow the user to scroll around if the content needs more space than is available. The `drawingTree`, the properties, and the `commandArea` are all put in a `ScrollPane`, as each of these could become larger than the region of the screen used to display them.

The last type of pane used to lay out this GUI is the `SplitPane`. This allows multiple `Nodes` to be stacked vertically or horizontally with user-adjustable bars between them. Three of them are created in this GUI on lines 63, 83, and 88. After each one is created, we add items to it using `++=`. In this case, each of our `SplitPanes` has only two items. By default, the split between them would be in the middle. Lines 86 and 90 change the position of two of them so that the GUI pops up with roughly the appearance that we desire.

Figure 5.5 shows the wind that is produced when you run `DrawingMain`. Note how the panes cause the different elements to be laid out in the manner that we want.

5.3.3 Controls

Most of the GUI elements that are intended to interact with the user are in the `scalafx.scene.control` package and inherit from `Control`. This includes the `Nodes` associated with making menus and the interactive panes that we looked at above, as well as many others that provide the elements you are used to working with in GUIs. Currently, this example only has four other elements: `Slider`, `TextArea`, `TextField`, and `TreeView`.

At this point, each of these is simply created and added into the appropriate pane so that it appears in the correct area of the GUI. The more interesting aspects of controls come in when we make them interactive. We will return to that later, after we have done some of the design for implementing drawings.

The only other thing worth specifically mentioning at this point is the usage of the `<==` symbolic method on lines 59, 60, 77, and 78. This is a binding operator. Values in ScalaFX GUIs are generally stored as observable properties, and they have the very powerful feature that you can bind these values to one another, so that when one updates, other changes happen automatically. This is covered in more detail in section 5.5.1, but it is worth describing what is being done here.

The `TreeView` and the `TextArea` both start off with default sizes. If we put them directly inside of a `BorderPane` or a `SplitPane`, they would expand to fill the area that they were given by the pane that contained them. However, we have embedded these inside of

FIGURE 5.5: This is the window that pops up when you run `DrawingMain`.

`ScrollPane`s in anticipation of the situation where their contents do not fit in the visible area. The `ScrollPane` does not tell its contents how big they should be, it lets the contents determine their natural size, then creates scroll bars as needed. This leads to an odd appearance when the "natural" size of the `TreeView` or `TextArea` is smaller than the display for the `ScrollPane`. These bindings make it so that the preferred width and height of the `TreeView` and `TextArea` are set to be equal to the width and height of their enclosing `ScrollPane`s. If their contents force them to, they will grow to be larger than the preferred size, but they don't display smaller than the preferred size. So, they fill whatever area is visible in the `ScrollPane`.

5.4 Drawing Program Design

We have coded up the basic layout of our GUI for the drawing program, but it doesn't do anything. Before we can add real functionality, we need to think about the design a bit. Most of the GUI code goes in the main `object`, but what `class`es do we need in this project? Going back to the capabilities listed in the analysis, it stands out that we need something to represent a drawing. We can do that with a `class` called `Drawing`.

We also need a method of representing what goes into the drawing. The design of our drawing program is to use a scene graph for this. In a scene graph, there are elements that represent individual elements that can be drawn as well as elements that can contain multiple other elements. Generally, something in the system also has the ability to do transformations such as rotations. ScalaFX provides its own scene graph, but we want to implement our own for a variety of reasons. Some of those reasons have the educational goal of helping you understand scene graphs better and some reasons are practical such as being able to implement functionality that does not exist in ScalaFX's scene graph.

The scene graph in ScalaFX uses the `Node` type as the base for an inheritance hierarchy,

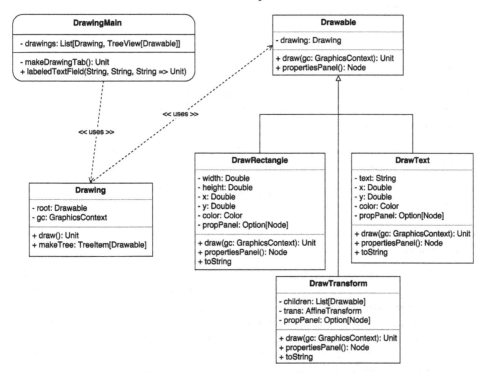

FIGURE 5.6: This is a first cut class diagram for our drawing project.

and everything that can go into the scene graph inherits from it. We will do something similar with an **abstract class** called Drawable. To start off, we will create just three types that inherit from Drawable. These are DrawRectangle, DrawText, and DrawTransform. These provide us with enough power capabilities that we can demonstrate the key ideas of the drawing program as well as most of the concepts that we want to cover for doing GUIs and graphics with ScalaFX.

Figure 5.6 shows a class diagram for this design. Note that this diagram was created after several iterations. When you sit down to start designing your solution, you shouldn't feel that you have to think of everything, and that all of it has to be perfect. Even seasoned professionals would find this is extremely difficult. The only time you can do that is if you have experience solving basically the same problem before. It is very common to iterate and improve over each iteration. Instead, you should think of as many features as you can, then try to implement your design. When you run into roadblocks or discover things that you hadn't considered, then just go back and adjust the design. Trying to get everything correct the first time through can lead to "paralysis by analysis" which leads to you never actually getting anything done. Especially as a novice programmer, you need to spend time thinking about design, but don't let it consume too much time before you start to actually convert your ideas into code.

5.5   Interactivity

Our previous implementation of `DrawingMain` lacked any form of user interactivity. For text applications, we get information from the user with calls to methods like `readInt` and `readLine`. That doesn't work for GUIs for a variety of reasons.

- There are multiple components from which you can get input. Each control can be interacted with independently.

- There are multiple devices that can potentially provide the input. In addition to the keyboard we have a mouse and potentially a touch screen.

- The `read` calls are blocking and stop everything else from happening while we wait for the user.

GUIs require a method of supporting interaction that gets around these issues. We have to be able to deal with the user interacting with different components in different ways and make sure that everything is responsive while it happens.

The general approach used to support interaction is to use event handlers. Any element that the user could interact with provides some ability for us to register code that will be executed when the user does some sort of action. We illustrate several examples in the following code, which shows the completed version of `DrawingMain` for this chapter.[5] After listing the full code, we will go through the details of various parts of it. The other `class`es in our implementation are shown later.

```scala
package guigraphics.v2

import scalafx.Includes._
import scalafx.application.JFXApp
import scalafx.geometry.Orientation
import scalafx.scene.Scene
import scalafx.scene.canvas.Canvas
import scalafx.scene.control.Menu
import scalafx.scene.control.MenuBar
import scalafx.scene.control.MenuItem
import scalafx.scene.control.ScrollPane
import scalafx.scene.control.SeparatorMenuItem
import scalafx.scene.control.Slider
import scalafx.scene.control.SplitPane
import scalafx.scene.control.Tab
import scalafx.scene.control.TabPane
import scalafx.scene.control.TextArea
import scalafx.scene.control.TextField
import scalafx.scene.control.TreeView
import scalafx.scene.layout.BorderPane
import scalafx.scene.paint.Color
import scalafx.event.ActionEvent
import scalafx.scene.control.TreeItem
import scalafx.scene.control.SelectionMode
import scalafx.scene.control.Label
```

---

[5]We will be adding more to this program, including the main `object`, in later chapters. Not all features can be implemented at this point, as we do not yet have the knowledge to implement all of them. Some that we could implement are being left out for now as we will learn better ways of dealing with them later.

```scala
26  import scalafx.scene.layout.Priority
27  import scalafx.scene.control.ComboBox
28  import scalafx.scene.control.Dialog
29  import scalafx.scene.control.ChoiceDialog
30
31  object DrawingMain extends JFXApp {
32    private var drawings = List[(Drawing, TreeView[Drawable])]()
33
34    stage = new JFXApp.PrimaryStage {
35      title = "Drawing Program"
36      scene = new Scene(1000, 700) {
37        // Menus
38        val menuBar = new MenuBar
39        val fileMenu = new Menu("File")
40        val newItem = new MenuItem("New")
41        val openItem = new MenuItem("Open")
42        val saveItem = new MenuItem("Save")
43        val closeItem = new MenuItem("Close")
44        val exitItem = new MenuItem("Exit")
45        fileMenu.items = List(newItem, openItem, saveItem, closeItem, new
              SeparatorMenuItem, exitItem)
46        val editMenu = new Menu("Edit")
47        val addItem = new MenuItem("Add Drawable")
48        val copyItem = new MenuItem("Copy")
49        val cutItem = new MenuItem("Cut")
50        val pasteItem = new MenuItem("Paste")
51        editMenu.items = List(addItem, copyItem, cutItem, pasteItem)
52        menuBar.menus = List(fileMenu, editMenu)
53
54        // Tabs
55        val tabPane = new TabPane
56        val (drawing, tree, tab) = makeDrawingTab()
57        drawings = drawings :+ (drawing, tree)
58        tabPane += tab
59
60        // Menu Actions
61        newItem.onAction = (ae: ActionEvent) => {
62          val (drawing, tree, tab) = makeDrawingTab()
63          drawings = drawings :+ (drawing, tree)
64          tabPane += tab
65        }
66        closeItem.onAction = (ae: ActionEvent) => {
67          val current = tabPane.selectionModel.value.selectedIndex.value
68          if (current >= 0) {
69            drawings = drawings.patch(current, Nil, 1)
70            tabPane.tabs.remove(current)
71          }
72        }
73        addItem.onAction = (ae: ActionEvent) => {
74          val current = tabPane.selectionModel.value.selectedIndex.value
75          if (current >= 0) {
76            val (drawing, treeView) = drawings(current)
77            val dialog = new ChoiceDialog("Rectangle", List("Rectangle", "Transform",
                  "Text"))
78            val d = dialog.showAndWait() match {
```

```scala
79              case Some(s) =>
80                val d: Drawable = s match {
81                  case "Rectangle" => new DrawRectangle(drawing, 0, 0, 300, 300,
                         Color.Blue)
82                  case "Transform" => new DrawTransform(drawing)
83                  case "Text" => new DrawText(drawing, 100, 100, "Text", Color.Black)
84                }
85                val treeSelect = treeView.selectionModel.value.getSelectedItem
86                def treeAdd(item: TreeItem[Drawable]): Unit = item.getValue match {
87                  case dt: DrawTransform =>
88                    dt.addChild(d)
89                    item.children += new TreeItem(d)
90                    drawing.draw()
91                  case d =>
92                    treeAdd(item.getParent)
93                }
94                if (treeSelect != null) treeAdd(treeSelect)
95              case None =>
96            }
97          }
98        }
99        exitItem.onAction = (ae: ActionEvent) => {
100         // TODO - Add saving before closing down.
101         sys.exit()
102       }
103
104       // Top Level Setup
105       val rootPane = new BorderPane
106       rootPane.top = menuBar
107       rootPane.center = tabPane
108       root = rootPane
109     }
110   }
111
112   private def makeDrawingTab(): (Drawing, TreeView[Drawable], Tab) = {
113     val canvas = new Canvas(2000, 2000)
114     val gc = canvas.graphicsContext2D
115     val drawing = new Drawing(gc)
116
117     // left side
118     val drawingTree = new TreeView[Drawable]
119     drawingTree.selectionModel.value.selectionMode = SelectionMode.Single
120     drawingTree.root = drawing.makeTree
121     val treeScroll = new ScrollPane
122     drawingTree.prefWidth <== treeScroll.width
123     drawingTree.prefHeight <== treeScroll.height
124     treeScroll.content = drawingTree
125     val propertyPane = new ScrollPane
126     val leftSplit = new SplitPane
127     leftSplit.orientation = Orientation.Vertical
128     leftSplit.items ++= List(treeScroll, propertyPane)
129
130     // right side
131     val slider = new Slider(0, 1000, 0)
132     val rightTopBorder = new BorderPane
```

```scala
133    val canvasScroll = new ScrollPane
134    canvasScroll.content = canvas
135    rightTopBorder.top = slider
136    rightTopBorder.center = canvasScroll
137    val commandField = new TextField
138    val commandArea = new TextArea
139    commandArea.editable = false
140    val commandScroll = new ScrollPane
141    commandArea.prefWidth <== commandScroll.width
142    commandArea.prefHeight <== commandScroll.height
143    commandScroll.content = commandArea
144    val rightBottomBorder = new BorderPane
145    rightBottomBorder.top = commandField
146    rightBottomBorder.center = commandScroll
147    val rightSplit = new SplitPane
148    rightSplit.orientation = Orientation.Vertical
149    rightSplit.items ++= List(rightTopBorder, rightBottomBorder)
150    rightSplit.dividerPositions = 0.7
151
152    // Top Level
153    val topSplit = new SplitPane
154    topSplit.items ++= List(leftSplit, rightSplit)
155    topSplit.dividerPositions = 0.3
156
157    // Tree Selection
158    drawingTree.selectionModel.value.selectedItem.onChange {
159      val selected = drawingTree.selectionModel.value.selectedItem.value
160      if (selected != null) {
161        propertyPane.content = selected.getValue.propertiesPanel()
162      } else {
163        propertyPane.content = new Label("Nothing selected.")
164      }
165    }
166
167    val tab = new Tab
168    tab.text = "Untitled"
169    tab.content = topSplit
170    tab.closable = false
171    (drawing, drawingTree, tab)
172  }
173
174  def labeledTextField(labelText: String, initialText: String, action: String =>
         Unit): BorderPane = {
175    val bp = new BorderPane
176    val label = Label(labelText)
177    val field = new TextField
178    field.text = initialText
179    bp.left = label
180    bp.center = field
181    field.onAction = (ae: ActionEvent) => action(field.text.value)
182    field.focused.onChange(if (!field.focused.value) action(field.text.value))
183    bp
184  }
185 }
```

Most of this code is the same as we had before. The first interactivity in this listing is on lines 60–102, where the code for menu items appears. `MenuItems` are like `Buttons`, and demonstrate the simplest form of interaction. They really only have one "action" that can happen with them, and that happens when the user selects a item from the menu. To make something happen when the user does this, we specify a function and set it to be the `onAction` property of the `MenuItem`. This code sets action handlers for `newItem`, `closeItem`, `addItem`, and `exitItem`.

The code shown here never uses the `ActionEvents` that are passed into the functions, but you could. Those events contain information related to what happened to cause the event. You can look in the API to see that there isn't all that much information in the `ActionEvent`, but later we will see that other event types contain information that is essential to most of the processing you will want to do with those events.

`onAction` is not the only property used to specify behaviors for interactivity. If you look in the API at the `Node class`, you will see a long list of methods that begin with `on` for handling many types of events. Most of these are for general keyboard, mouse, and touch interactions, which we will explore in section 5.7.

5.5.1 Properties, Observables, and Bindings

One of the significant features added in JavaFX is properties. These are objects that store values and include functionality for code to be notified when the value changes. You can think of a property as a value that emits events, much like the GUI itself. Most of the values stored in ScalaFX components are stored as properties. The impact of this can be seen in the handlers for the menu items with various calls to `value`. Consider the following line which is the first line in the handlers for both `closeItem` and `addItem`.

```
val current = tabPane.selectionModel.value.selectedIndex.value
```

A call to `selectionModel` returns an instance of `ObjectProperty[SingleSelectionModel [Tab]]`. Note that it isn't actually a selection model, it is an `ObjectProperty` wrapped around a selection model. To get to the actual value you have to call either the `value` or the `apply` method. Recall that the `apply` method can be called by putting parentheses after an object, so this line could be rewritten in the following way. You can decide which format you prefer.

```
val current = tabPane.selectionModel().selectedIndex()
```

Having to type `value` or put parentheses in many locations might seem like a pain, but it comes with a significant benefit. Lines 158–165 add interactivity with the `TreeView` so that when the user selects a `Drawable` from the tree, its properties are displayed in the `propertyPane`. This is done by setting code to execute when the selection changes. Properties have `onChange` methods that take a by-name argument that will be executed when the value changes.

The last method in `DrawingMain` is called `labeledTextField`. It is used by some of the subtypes of `Drawable` to reduce code duplication in those `classes`. As the name implies, it gives back a `Node` that has a `TextField` with a `Label`. The last argument is a function that is the action that will happen when the user stops editing a field. On lines 181 and 182, this action is registered to happen in two different situations. When the user hits the Enter key while editing a field, the `onAction` function is called, so line 181 sets the `onAction` property to a function that invokes our action.

If the user stops editing the field, we also need to process the modified value, even if they don't hit Enter. The term that is used to describe where keyboard input goes in a GUI is FOCUS. If you think of a form with multiple text fields, when you type, the key strokes only go to one of the fields. The one they go to is the one with focus. The `TextField` has a property called `focus`. Line 182 registers code to be executed when the focus changes. This code will be called both when focus is gained as well as when it is lost. For that reason, there is an `if` that makes it so that the action code only executes when the focus has been lost.

Properties represent single values. There is also a package called `scalafx.collections` that includes a number of collection types that begin with the word `Observable`. These are collections that behave like properties and will execute appropriate code when their contents change. We take advantage of this on line 89. The variable `item` refers to a `TreeItem` that is part of the `TreeView`. `TreeItem` has a member called `children` that we are adding a new element into. This might not look special, but because `children` is an observable collection, this change is automatically reflected in the GUI.

The `<==` operator that occurs on lines 122, 123, 141, and 142 is also related to properties. You can only do binding between operators. The `<==` operator does a one-way binding. As the symbol indicates, when the value on the right changes, the property on the left is automatically updated to reflect that change. Our examples are fairly simple as they just bind one property to another, but for numerical properties, you can do mathematical operations on the bindings as well. You could use the following line to keep a label centered horizontally in some parent panel.

```
label.layoutX = (parent.width - label.width)/2
```

There is also a two-way binding method that uses the symbol `<==>`. When this is used, changes to either the property on the left or the right of the symbol is reflected in a change to the other side. Note that some properties are read-only and can't be used with a two-way binding. Read-only properties can only appear on the right side of a one-way binding.

5.6 Graphics

There are two approaches to adding graphics to a ScalaFX GUI. The first is to add graphical elements into the scene graph and the second is to draw to a subtype of `Node` called `Canvas`. The drawing program utilizes the second option, but we want to discuss both briefly.

5.6.1 Shapes and Scene Graph

Back in figure 5.3 you might have noticed a type called `Shape` with twelve different subtypes, including `Line`, `Polygon`, and `Rectangle`. We made use of the `Ellipse` subtype in the code for `FirstGUI`. These different subtypes provide significant power in terms of what you can create. Unlike controls, you typically don't put your shapes in panes, instead, you specify their locations manually. You can use the `Group` type to group together multiple shapes into a single `Node`.

Each of the shapes has a number of different properties that you can set to specify how it is drawn. The `FirstGUI` example showed how the `fill` property could be set to a `Color`. The type of `fill` is actually `scalafx.scene.paint.Paint` and `Color` is one of three subtypes of `Paint`. The other two are `LinearGradient` and `RadialGradient`.

You can also adjust the style of line that a shape is drawn with. This is called the STROKE.

In addition to the `stroke` property, which allows you to set the color, you can set the `strokeWidth`, `strokeLineJoin`, `strokeLineCap`, and `strokeDashArray`. A full discussion of these options is beyond the scope of this chapter.

All Nodes, including Shapes, can also be given a transformation. Standard transformations include rotations, translations, scales, and shears. When you set a transform to a Group, it is applied to all the contents.

5.6.2   Canvas

The other approach to doing graphics is to add a Canvas to the scene graph, get a GraphicsContext from that Canvas, and call methods on the GraphicsContext that draw things. Lines 113 and 114 for DrawingMain make the Canvas and get the GraphicsContext. Then line 115 passes the GraphicsContext to the constructor for our Drawing. The Drawing class has a method called draw that uses the GraphicsContext to render our Drawable elements. Here is the code for Drawing.

```scala
1   package guigraphics.v2
2
3   import scalafx.scene.canvas.GraphicsContext
4   import scalafx.scene.control.TreeItem
5   import scalafx.scene.paint.Color
6
7   class Drawing(private var gc: GraphicsContext) {
8     private var root = new DrawTransform(this)
9
10    def draw(): Unit = {
11      gc.fill = Color.White
12      gc.fillRect(0, 0, 2000, 2000)
13      root.draw(gc)
14    }
15
16    def makeTree: TreeItem[Drawable] = {
17      def helper(d: Drawable): TreeItem[Drawable] = d match {
18        case dt: DrawTransform =>
19          val item = new TreeItem(d)
20          item.children = dt.children.map(c => helper(c))
21          item
22        case _ => new TreeItem(d)
23      }
24      helper(root)
25    }
26  }
```

This is a fairly simple class. In addition to the GraphicsContext, called gc, it stores a single DrawTransform that is called root. The draw method clears the drawing area by setting the fill on gc, then filling in a large rectangle. It finishes by calling the draw method on root.

There is one other method in Drawing that has nothing to do with graphics. The makeTree method is used to set up the TreeItems for the TreeView back in DrawingMain. It uses a recursive function called helper to go down through our scene graph, building one TreeItem[Drawable] for each Drawable that it finds. Note that the recursive call is inside of a call to children.map, so it happens on all of the children of the current node.

Just as there are options for how things are drawn on the shapes that you add to the scene graph, there are also options for how things are drawn on the `Canvas` using the `GraphicsContext`. We have seen the use of `fill`, which mirrors that for a shape. You can use `stroke` to set the color for lines, `lineWidth`, `lineCap`, and `lineJoin` to set properties for drawing lines.

If you look in the API, there are four types of methods that are used for drawing. One set begins with `fill`. As the name implies, these fill in whole areas. A second set begins with the word `stroke`. These draw lines for whatever shape is involved. The third set is three different versions of `drawImage`, which will render an `Image` to the `Canvas`. You can create instances of `scalafx.scene.image.Image` by passing in a URL for where the image is located. The last set of methods, which can be paired with the `fill` or `stroke` methods, all end in `to`. This set includes `moveTo`, `lineTo`, and `arcTo`. These allow you to build your own paths after a call to `beginPath`, and you can use `fillPath` or `strokePath` to render it.

With this brief discussion of how to draw things, we should look at the `Drawable` types, beginning with the **abstract class** for `Drawable`.

```scala
package guigraphics.v2

import scalafx.scene.canvas.GraphicsContext
import scalafx.scene.Node

abstract class Drawable(val drawing: Drawing) {
  def draw(gc: GraphicsContext): Unit
  def propertiesPanel(): Node
}
```

Hopefully this is the code you would expect to see based on the UML diagram. Each instance of any subtype of `Drawable` stores the `Drawing` that it is part of. It also has two abstract methods for `draw` and `propertiesPanel`. To see how these actually come out in code, consider the following implementation of `DrawRectangle`.

```scala
package guigraphics.v2

import scalafx.Includes._
import scalafx.scene.canvas.GraphicsContext
import scalafx.scene.paint.Color
import scalafx.scene.Node
import scalafx.scene.layout.VBox
import scalafx.scene.control.TextField
import scalafx.scene.control.ColorPicker
import scalafx.event.ActionEvent
import scalafx.scene.layout.Priority

class DrawRectangle(
    d: Drawing,
    private var _x: Double,
    private var _y: Double,
    private var _width: Double,
    private var _height: Double,
    private var _color: Color) extends Drawable(d) {

  private var propPanel: Option[Node] = None

```

```
23    override def toString = "Rectangle"
24
25    def draw(gc: GraphicsContext): Unit = {
26      gc.fill = _color
27      gc.fillRect(_x, _y, _width, _height)
28    }
29
30    def propertiesPanel(): Node = {
31      if (propPanel.isEmpty) {
32        val panel = new VBox
33        val xField = DrawingMain.labeledTextField("x", _x.toString, s => { _x =
                s.toDouble; drawing.draw() })
34        val yField = DrawingMain.labeledTextField("y", _y.toString, s => { _y =
                s.toDouble; drawing.draw() })
35        val widthField = DrawingMain.labeledTextField("width", _width.toString, s =>
                { _width = s.toDouble; drawing.draw() })
36        val heightField = DrawingMain.labeledTextField("height", _height.toString, s
                => { _height = s.toDouble; drawing.draw() })
37        val colorPicker = new ColorPicker(_color)
38        colorPicker.onAction = (ae: ActionEvent) => {
39          _color = colorPicker.value.value
40          drawing.draw()
41        }
42        panel.children = List(xField, yField, widthField, heightField, colorPicker)
43        propPanel = Some(panel)
44      }
45      propPanel.get
46    }
47  }
```

This **class** takes in multiple arguments. The first one is the **Drawing**, which is simply passed through to the constructor for **Drawable**. As was discussed in chapter 4, this argument is neither a **val** nor a **var**, and it is not used in any way inside of the **class**. The other five arguments are all **private vars** that hold information about the rectangle.

Line 21 declares another **private var** called **propPanel**, which is an **Option[Node]** that stores the properties panel, if one has been built. This is an **Option** because we don't want to actually construct the panel unless it is used. In the **propertiesPanel** method, we check if it is **None** and if so, we build a new panel, otherwise, we just return the value we stored previously.[6]

Line 23 overrides the default implementation of **toString**. This is important because the **TreeView** uses a call to **toString** when displaying a **Drawable** in the tree. If we don't **override** this method, the output that the user sees is quite odd, so we make sure to override it in each of the different implementations of **Drawable**.

The **draw** method on our **DrawRectangle** is extremely simple. It sets the fill color, then fills in a rectangle with the appropriate location and size. The **propertiesPanel** method is a bit more complex. The role of all of the **propertiesPanel** methods is to build a **Node** that has GUI controls that allow the user to change the properties of that **Drawable**. For the rectangle, this means that we need to be able to change the five values for the location, size, and color. The first four values are all of type **Double**, and the code uses the

---

[6]This style of making a value, where it isn't constructed until it is used, is called "lazy" evaluation. It is a useful technique when there is an expensive operation that might not need to happen. Scala has a language feature of **lazy vals** that is discussed in the on-line appendices which can do this type of thing automatically.

`DrawingMain.labeledTextField` method that we saw earlier. Hopefully this makes it clear why we wanted to make that into a method. Each call to `labeledTextField` would require roughly eight lines of code instead of the one line that it needs the way it is written.

The last argument in each call to `labeledTextField` is the most interesting one. Recall that this is a function that is invoked each time the user stops editing the `TextField`. In this code, we see that each one sets the appropriate member data, then calls `drawing.draw()` so that the drawing is updated with the new value.

The last element in the properties panel is one that is used to set the color of the rectangle. This is something that could be challenging to do, but it is greatly simplified by the fact that ScalaFX includes a `ColorPicker`. We simply create an instance of `ColorPicker` giving it the current color as the initial value, then we register an `onAction` handler. That handler function is called whenever the user selects a new color.

The code for `DrawText` is very similar to that for `DrawRectangle` except that there is a `String` for the text instead of a width and a height. This code includes the text as part of the `toString` method as well.

```scala
package guigraphics.v2

import scalafx.Includes._
import scalafx.scene.canvas.GraphicsContext
import scalafx.scene.paint.Color
import scalafx.scene.Node
import scalafx.scene.layout.VBox
import scalafx.scene.control.ColorPicker
import scalafx.event.ActionEvent

class DrawText(
    d: Drawing,
    private var _x: Double,
    private var _y: Double,
    private var _text: String,
    private var _color: Color) extends Drawable(d) {

  private var propPanel: Option[Node] = None

  override def toString = "Text: "+_text

  def draw(gc: GraphicsContext): Unit = {
    gc.fill = _color
    gc.fillText(_text, _x, _y)
  }

  def propertiesPanel(): Node = {
    if (propPanel.isEmpty) {
      val panel = new VBox
      val textField = DrawingMain.labeledTextField("Text", _text, s => { _text = s;
          drawing.draw() })
      val xField = DrawingMain.labeledTextField("x", _x.toString, s => { _x =
          s.toDouble; drawing.draw() })
      val yField = DrawingMain.labeledTextField("y", _y.toString, s => { _y =
          s.toDouble; drawing.draw() })
      val colorPicker = new ColorPicker(_color)
      colorPicker.onAction = (ae: ActionEvent) => {
        _color = colorPicker.value.value
```

```
36        drawing.draw()
37      }
38      panel.children = List(textField, xField, yField, colorPicker)
39      propPanel = Some(panel)
40    }
41    propPanel.get
42  }
43 }
```

The last `Drawable` subtype is the `DrawTransform`. Before we talk about the `DrawTransform`, we should probably discuss transforms. This is another setting that you can have on the things you draw in ScalaFX. In general, most 2D graphics libraries support AFFINE TRANSFORMS. These are transforms that preserve parallel lines. There are four types of affine transforms: translation (moving thing in x or y), rotation, scaling (can be done independently in x and y), and shearing (think of pushing on the top of a square to make a rhombus). Any combination of affine transforms is also an affine transform.

The `DrawTransform` class is a bit different from the other two, though it still has the methods that are required by `Drawable`. In addition, it has methods called `addChild` and `removeChild` along with a `private var` called `_children`. This is because the `DrawTransform` is the one `Drawable` that we use to group together other `Drawables`. In the ScalaFX scene graph, all subtypes of `Parent` can play this role.

```
1  package guigraphics.v2
2
3  import scalafx.Includes._
4  import scalafx.scene.Node
5  import scalafx.scene.canvas.GraphicsContext
6  import scalafx.scene.layout.VBox
7  import scalafx.scene.control.ComboBox
8  import scalafx.event.ActionEvent
9  import scalafx.scene.transform.Rotate
10 import scalafx.scene.transform.Translate
11 import scalafx.scene.transform.Scale
12 import scalafx.scene.transform.Shear
13
14 class DrawTransform(d: Drawing) extends Drawable(d) {
15   private var _children = List[Drawable]()
16   private var propPanel: Option[Node] = None
17   private var transformType = DrawTransform.Translate
18   private var value1 = 0.0
19   private var value2 = 0.0
20
21   def children = _children
22
23   override def toString = "Transform"
24
25   def addChild(d: Drawable): Unit = {
26     _children ::= d
27   }
28
29   def removeChild(d: Drawable): Unit = {
30     _children = _children.filter(_ != d)
31   }
32
33   def draw(gc: GraphicsContext): Unit = {
```

```
34    gc.save()
35    transformType match {
36      case DrawTransform.Rotate => gc.rotate(value1)
37      case DrawTransform.Translate => gc.translate(value1, value2)
38      case DrawTransform.Scale => gc.scale(value1, value2)
39      case DrawTransform.Shear => gc.transform(1.0, value1, value2, 1.0, 0.0, 0.0)
40    }
41    _children.foreach(_.draw(gc))
42    gc.restore()
43  }
44
45  def propertiesPanel(): Node = {
46    if (propPanel.isEmpty) {
47      val panel = new VBox
48      val combo = new ComboBox(DrawTransform.values.toSeq)
49      combo.onAction = (ae: ActionEvent) => {
50        transformType = combo.selectionModel.value.selectedItem.value
51        drawing.draw()
52      }
53      combo.selectionModel.value.select(transformType)
54      val v1Field = DrawingMain.labeledTextField("x/theta", value1.toString, s => {
            value1 = s.toDouble; drawing.draw() })
55      val v2Field = DrawingMain.labeledTextField("y", value2.toString, s => {
            value2 = s.toDouble; drawing.draw() })
56      panel.children = List(combo, v1Field, v2Field)
57      propPanel = Some(panel)
58    }
59    propPanel.get
60  }
61 }
62
63 object DrawTransform extends Enumeration {
64   val Rotate, Scale, Shear, Translate = Value
65 }
```

Both the **draw** and **propertiesPanel** methods make use of the enumeration defined in the companion object. Recall from chapter 4 that enumerations are used to represent small sets of possible values. This code is written so that each instance of **DrawTransform** can be either a rotate, a scale, a shear, or a translate. To combine them, you have to create multiple **DrawTransforms** and have one be a child of the other. Given this design choice, we need to have a way to represent values selected from one of those four options and nothing else. That is a perfect example of where we would want to use an enumeration. The **transformType** member has this type, and is given an initial value of **DrawTransform.Translate**.

The **draw** method, really has three main things going on. There is a **match** on the **transformType** that does the appropriate transformation on the **GraphicsContext**. For rotations, translations, and scaling, there are methods in **GraphicsContext** that apply those transformations. Since shearing is a less common operation, we have to use a more general method called **transform** to create a shear. This method can actually do any combination of scaling, shearing, and translating. The values that we pass as 1.0 are scaling arguments and those with the value 0.0 are translation arguments.

After the transform is applied, line 41 runs through all the children and draws them using **foreach**. The more interesting lines are the first and last lines of the **draw** function. These take advantage of a nice feature of the ScalaFX **GraphicsContext** that allows you to save off the draw settings, then bring them back later. In this case, after we have drawn the

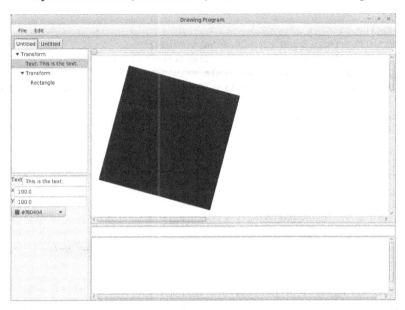

FIGURE 5.7: This is the window that pops up when you run the more complete version of `DrawingMain`.

children, we need to go back to using the transform that had been in place before we applied the current transform. So the call to `gc.save()` saves off the current transform, along with fill and stroke information, then the call to `gc.restore()` at the end brings those values back so that anything drawn after this `DrawTransform` won't have this transform in place.

In the `propertiesPanel` method, we use a `comboBox` to display the different types of transformations the user can choose from. Note the `DrawTransform.values.toSeq` that is passed in at construction. The `Enumeration` type provides a member called `values` that gives us a collection of the different possible values. Here we see another benefit of using the Enumeration over other approaches we might have taken. If we were to add or remove a value from the Enumeration, very little would have to be changed elsewhere. Only the `match` would require updating cases to adjust to the change.

We have now seen code that completes all of our current design. This code can be run, and the result could be something that looks like figure 5.7. Here we have a `DrawText` under the top `DrawTransform`, which is set to be a rotation. There is also a second `DrawTransform` that does a translate and has a `DrawRectangle` under it.

5.7   Keyboard and Mouse Input

While the drawing program allowed us to illustrate many of the features of GUIs and graphics that you will need for the rest of this book, there are a few key elements that it doesn't hit on sufficiently. One of those is getting input from the keyboard and the mouse in ways other than interacting with standard GUI elements. If the user clicks a button, you want your code to use `onAction`. However, if you have some shape or image that you want to respond to the mouse being clicked, then there is no `onAction`.

The `Node` type defines a fairly large set of methods that begin with `on` for handling basic input. Methods that start with **onKey** handle keyboard input, methods that start with **onMouse** handle mouse input, and methods that start with other things like **onDrag** and **onSwipe** support a touch interface.

The following code shows the use of **onMouseMoved** and **onKeyPressed** in an application that uses the scene graph for drawing. The goal is to have an image follow the mouse around and a box that can be moved with the arrow keys.

```scala
package guigraphics

import scalafx.Includes._
import scalafx.application.JFXApp
import scalafx.scene.Scene
import scalafx.scene.image.ImageView
import scalafx.scene.shape.Rectangle
import scalafx.scene.input.MouseEvent
import scalafx.scene.paint.Color
import scalafx.scene.input.KeyEvent
import scalafx.scene.input.KeyCode

object KeyboardMouseInput extends JFXApp {
  stage = new JFXApp.PrimaryStage {
    title = "Keyboard and Mouse Input"
    scene = new Scene(600, 600) {
      val img = new ImageView("file:Mimas_Cassini.jpg")
      val box = Rectangle(0, 0, 30, 30)
      box.fill = Color.Green
      content = List(img, box)
      onMouseMoved = (me: MouseEvent) => {
        img.x = me.x
        img.y = me.y
      }
      onKeyPressed = (ke: KeyEvent) => {
        ke.code match {
          case KeyCode.Up => box.y = box.y.value - 2
          case KeyCode.Down => box.y = box.y.value + 2
          case KeyCode.Left => box.x = box.x.value - 2
          case KeyCode.Right => box.x = box.x.value + 2
          case _ =>
        }
      }
    }
  }
}
```

Lines 17 and 18 create an **ImageView** and a **Rectangle** that are set as the contents of the scene. We then assign **onMouseMoved** for the scene to a function that changes x and y for the image to values of those fields in the **MouseEvent**. The last element in this code is an assignment to the **onKeyPressed** for the scene. The key-handling functions take a **KeyEvent** that has information such as which key the event is for. This code does a **match** on the **code** in the **KeyEvent**. It has cases for the arrow keys, which alter the appropriate values for the box.

5.8   Animation

Keyboard and mouse handlers are fine when all the actions you want to perform can be motivated by the user. If you want things to happen without the user doing anything, you need to do something different. ScalaFX has various animations that we won't cover in the `scalafx.animation` package. We will only consider the `scalafx.animation.AnimationTimer` class, because it is generally applicable to many different problems.

The following code demonstrates using an `AnimationTimer` combined with mouse and keyboard input. While the last program used `Nodes` in the scene graph, this one uses a `Canvas` to do the graphics. The goal is to have a box that is controlled by the arrow keys and a "swarm" of circles that follow the mouse pointer around, moving at different speeds to chase the mouse. To simplify things, this code uses the `Vect2D class` that appears on page 97 in chapter 3.

```scala
package guigraphics

import scalafx.Includes._
import scalafx.application.JFXApp
import scalafx.scene.Scene
import scalafx.scene.canvas.Canvas
import scalafx.scene.layout.BorderPane
import scalafx.animation.AnimationTimer
import oodetails.Vect2D
import scalafx.scene.paint.Color
import scalafx.scene.input.MouseEvent
import scalafx.scene.input.KeyCode
import scalafx.scene.input.KeyEvent

object AnimationExample extends JFXApp {
  case class Swarmer(p: Vect2D, speed: Double)
  private var swarm = List.tabulate(20)(i => Swarmer(Vect2D(300, 300), (i * 5 + 20)
      * 0.02))
  private var mouse = Vect2D(300, 300)
  private var box = Vect2D(0, 0)
  private var upPressed = false
  private var downPressed = false
  private var leftPressed = false
  private var rightPressed = false
  private val speed = 25.0
  stage = new JFXApp.PrimaryStage {
    title = "Animation Example"
    scene = new Scene(600, 600) {
      val border = new BorderPane
      val canvas = new Canvas(600, 600)
      val gc = canvas.graphicsContext2D
      border.center = canvas
      root = border

      canvas.width <== border.width
      canvas.height <== border.height

```

```
37    canvas.onMouseMoved = (me: MouseEvent) => {
38      mouse = Vect2D(me.x, me.y)
39    }
40
41    canvas.onKeyPressed = (ke: KeyEvent) => {
42      ke.code match {
43        case KeyCode.Up => upPressed = true
44        case KeyCode.Down => downPressed = true
45        case KeyCode.Left => leftPressed = true
46        case KeyCode.Right => rightPressed = true
47        case _ =>
48      }
49    }
50    canvas.onKeyReleased = (ke: KeyEvent) => {
51      ke.code match {
52        case KeyCode.Up => upPressed = false
53        case KeyCode.Down => downPressed = false
54        case KeyCode.Left => leftPressed = false
55        case KeyCode.Right => rightPressed = false
56        case _ =>
57      }
58    }
59
60    var lastTime = 0L
61    val timer = AnimationTimer(time => {
62      if (lastTime == 0) lastTime = time
63      else {
64        val interval = (time - lastTime) / 1e9
65        lastTime = time
66        swarm = swarm.map(s => s.copy(p = s.p + (mouse - s.p) * (interval *
                 s.speed)))
67        if (upPressed) box += Vect2D(0, -interval * speed)
68        if (downPressed) box += Vect2D(0, interval * speed)
69        if (leftPressed) box += Vect2D(-interval * speed, 0)
70        if (rightPressed) box += Vect2D(interval * speed, 0)
71      }
72      gc.fill = Color.White
73      gc.fillRect(0, 0, canvas.width.value, canvas.height.value)
74      gc.fill = Color.Red
75      for (Swarmer(p, _) <- swarm) {
76        gc.fillOval(p.x, p.y, 20, 20)
77      }
78      gc.fill = Color.Green
79      gc.fillRect(box.x, box.y, 20, 20)
80    })
81    timer.start
82    canvas.requestFocus()
83    }
84  }
85
86 }
```

Line 16 declares a **case class** called **Swarmer** that is used to represent the dots that chase the mouse. Line 17 declares a **var List** that contains 20 instances of **Swarmer**, each located at the same spot, but with different speed values. Line 18 declares a **var** to keep track of the last location of the mouse. Line 19 declares a `var` to store the location where the box will be drawn. Lines 20–23 define `Boolean` `vars` that tell us if a particular key is currently being held down. Line 24 declares `speed`, which is how fast the box should move in pixels per second.

Lines 28–32 create a `BorderPane`, a `Canvas`, and the `GraphicsContext` of that `Canvas`, then puts the `Canvas` in the `BorderPane` and sets the `BorderPane` as the root of the scene graph. The reason for including a `BorderPane` is on lines 34 and 35. The `Canvas` type does not naturally change size with the container that holds it the way many other subtypes of `Node`, like `BorderPane`, will do. We create the `BorderPane` so that we can bind the size of the `Canvas` to the `BorderPane` so that the `Canvas` will change size when the window is resized.

Lines 37–39 set up the mouse handling. Similar to the example above, we just set a value to remember where the mouse moved to. Lines 41–58 create the key handlers for both pressing and releasing keys. As with the previous example, these are looking for arrow keys, but when they execute, all they do is set the `Boolean` values appropriately.

Lines 61–79 define the `AnimationTimer`. The code in the `AnimationTimer` happens as frequently as the computer is able to call it. The one argument passed in when we build an `AnimationTimer` is a function of type `Long => Unit`. The argument is a measure of time in nanoseconds, so if there were a one-second delay between calls, the value in the second call would be 10^9 larger than the value in the first one. For many applications, including this one, we need to know how much time passes between calls. That is why line 60 declared `lastTime` to be a `Long` with value 0.

Inside the timer function, several things happen. It begins with an `if` that checks whether this is the first call or not. The value of `time` passed in is likely to be extremely large, so we don't want to use that for calculating a time between calls. Instead, the first time we just set `lastTime = time`. For all calls after the first one, lines 64–70 execute. These lines are responsible for moving the various things that we are drawing. It starts by calculating `interval`, which is the number of seconds since the last tick of the timer. That interval is used to update the positions of all of the elements in the swarm and, if any keys are pressed, to change the location of the box.

There are several reasons why the value of `box` is updated in the timer instead of in the keyboard handlers. If you run the last program you will note that if you press an arrow key and hold it, the box took one quick step, then paused and then took off quickly. If you pressed a second key, it didn't move diagonally; instead it showed the same behavior in the other direction. When you run this example, you will find that the box moves smoothly at a constant speed as soon as you press a key. In addition, holding multiple keys has the behavior that you would really like it to. The other reason for this approach is that the movements of the swarm and the box are now correlated with one another. That doesn't matter much for this program, but there are many situations, such as games, where it would.

The last part of the timer draws everything to the `Canvas`. The last example used scene graph `Node`s for the drawing. As a result, the graphics were updated automatically. Using a `Canvas`, we have to manually draw everything that we want to have appear. Whether this is a benefit or a hindrance depends on the application that you are creating.

The last two lines of code before we close the curly braces are simple and easy to forget, but they are very important. Line 81 starts the timer. If you leave out this line, nothing will be drawn and all you will see is an empty window. Line 82 is a bit more subtle, it requests focus for the `Canvas`. Earlier we introduced the term "focus" and used the example of a form with multiple text fields. Key strokes only type in one of them at a time. The same goes for the lower-level keyboard input discussed in this section. The previous example put the keyboard listener on the scene, which gets focus by default, so we didn't have to request it. In this example we opted to put the handlers on the `Canvas`. If you forget to include this line in this example, things will be drawn and the swarm will follow the mouse, but the keys won't move the box because without that line, the keystrokes go to the scene, and the `Canvas` never gets them.

5.9   Other Elements of ScalaFX

There are a number of other elements of ScalaFX that are not touched on at all in this chapter. There are other forms of animation and a variety of effects. You can also add media in the form of video and sounds. ScalaFX includes `Node` types that will display web pages and draw various forms of charts. It also has support for 3D graphics. Unfortunately, those are beyond the scope of what we can introduce in this chapter. The book website includes links to some additional resources and the interested reader is strongly encouraged to explore whichever of these capabilities appeals to him/her.

5.10   End of Chapter Material

5.10.1   Summary of Concepts

- The current recommended GUI library to use for Java is JavaFX, which is used through a wrapper called ScalaFX that provides better support for Scala syntax. To make an application that runs a GUI using ScalaFX, simply make an `object` and have it `extend JFXApp`.

- JavaFX, and by extension ScalaFX, use something called a scene graph to organize the different elements in a GUI. A scene graph is a tree data structure, most commonly found in graphical applications and libraries.

- The origin of the coordinate system, $(0,0)$, occurs in the top left corner. The x-coordinate grows as you move to the right. The y-coordinate grows as you move down.

- There are many different GUI elements that can be included in your GUIs. Some that were covered in this chapter including `MenuBars`, `MenuItems`, `TabPanes`, `BorderPanes`, `ScrollPanes`, `SplitPanes`, `Sliders`, `Canvases`, `TreeViews`, `TextFields`, and `TextAreas`.

- The general approach used to support interaction in GUIs is to use event handlers.

- The `<==` operator is a symbolic method that does a one-way binding. Values in ScalaFX GUIs are generally stored as observable properties, and they have the very powerful feature that you can bind these values to one another, so that when the value on the right changes, the one on the left is automatically updated. There is also a two-way binding method that uses the symbol `<==>`. When this is used, changes to either the property on the left or the right of the symbol is reflected in a change to the other side.

- There are two approaches to adding graphics to a ScalaFX GUI. The first is to add graphical elements into the scene graph and the second is to draw to a `Canvas`.

- Shapes have a number of different properties that you can set to specify how they are drawn. Not only can a `fill` property be set, you can also adjust the style of line that a shape is drawn with using the `stroke`, `strokeWidth`, `strokeLineJoin`, `strokeLineCap`, and `strokeDashArray` properties.

- All `Nodes`, including `Shapes`, can also be given a transformation. Standard transformations include rotations, translations, scales, and shears. When you set a transform to a `Group`, it is applied to all the contents.

- There are four types of methods that are used for drawing to a `Canvas`.
  - One set begins with `fill` and these fill in whole areas.
  - A second set begins with the word `stroke` and these draw lines for whatever shape is involved.
  - A third, which contains different versions of `drawImage`, renders an `Image` to the `Canvas`.
  - The last set of methods, which can be paired with the `fill` or `stroke` methods, all end in `to` and allow you to build your own paths.

- The `Node` type defines a fairly large set of methods that begin with `on` for handling basic input. Methods that start with `onKey` handle keyboard input, methods that start with `onMouse` handle mouse input, and methods that start with other things like `onDrag`, `onSwipe`, and some other options support a touch interface.

- If you want things to happen without the user doing anything, you can consider using the `scalafx.animation.AnimationTimer` class.

5.10.2   Exercises

1. Using three `ComboBoxes` and a `TextField`, set up a little GUI where the user makes simple math problems. The `ComboBoxes` should be in a row with the first and third having numbers and the middle one having math operators. When the user picks a different value or operation, the `TextField` should update with the proper answer.

2. Write a simple GUI that displays your name and major when a button is clicked. You should have two buttons, one called "Show Info", which displays your name and major when clicked, and one called "Quit".

3. Write a GUI application that calculates a car's gas mileage. The program's window should have a place that allows the user to enter the number of gallons of gas the car holds and the number of miles it can drive on a full tank of gas. When the user clicks a button, the application should display the number of miles that the car may be driven per gallon of gas.

4. Madame Maxine's Day Spa provides the following services:
   - Hair Cut — $50.00
   - Facial — $100.00
   - Eye Brow Wax — $25.00

- Manicure — $35.00
- Pedicure — $45.00
- Massage — $85.00

Write a GUI application with check buttons that allow the user to select any or all of these services. When the user clicks a button, the total charges should be displayed.

5. Write a GUI for a simple text editor. Use a `TextArea` for the main editing field and put in menu options to open and save files.

6. Write a tic-tac-toe game with a 3x3 grid of buttons. Have something print when each button is clicked.

7. Make a GUI that has one `Slider` and one `TextField`. When the user moves the `Slider`, the value it is moved to should appear in the `TextField`.

8. Write a GUI that has two `RadioButtons` and a `ListView`. The `RadioButtons` should be labeled "Uppercase" and "Lowercase" for uppercase letters and lowercase letters. You can put whatever list of strings you want in the `ListView` initially. When the state of a `RadioButton` is altered, the `ListView` should be changed to show the values exclusively in that case.

9. This exercise can extend the scheduling exercise started at 2. For this exercise you will write a GUI that lets users build schedules of courses. It should read information on the courses that are being offered from a file. It needs to have GUI elements that let you pick courses you want to take and put them into a different `ListView`. You should also be able to remove courses from that `ListView`. Give the user menu options to save and load the built schedules. You do not have to deal with conflicts in times or even duplicates of courses right now. Just let the user build a schedule from options. Consider how to prevent different types of conflicts and do that if you want an extra challenge.

   You should also add a graphical representation to the schedule. The goal is to have what looks like a fairly standard "week view" of the schedule. You will have a column for each day, and the courses should be displayed as labeled rectangles that span the section of the day that the course would occupy. This makes it easier to see when there are overly long blocks of class time or too many small breaks spread around.

   How you specify times is left up to you. You could give each time in the schedule a unique integer number. If you do that, you need to hard code the times and days for those time slots. The alternative is to add functionality so that the user can enter specific times and days. Taking that route will require more complex code in building to make sure that no two courses are scheduled for the same time.

10. This exercise can extend the recipe exercise found in chapter 3. For this exercise you should convert the functionality of the text menu to a GUI. You should decide the exact format, but you might consider using a `ListView` or a `Table` to display information related to the pantry and recipe items. You will need to have menu options for saving and loading. Like the text menu-based version, users should be able to add items to their pantries, add recipes to the recipe books, check if they can make a recipe, and tell the program that they have enough ingredients to make a recipe and have the proper amount reduced from their pantry contents. You should also add the ability to have images associated with ingredients and recipes. Lastly, you should also add in directions for how to make the recipe in the GUI. The `image` class in ScalaFX has the ability to read from files or URLs to get your images.

11. This exercise extends the music library exercise found in chapter 3. Make a GUI program that displays a music database. Use a GUI to display all the song information as well as allow the user to narrow down their selection. Also, provide menu options to add and remove song information.

    You should also show cover art for albums. Put that into your solution so that when albums (or possibly songs) are selected, the cover art for the album is displayed. Cover art can be stored in the same directory as the data file. The **image** class in ScalaFX has the ability to read from files or URLs to get your images.

5.10.3   Projects

This chapter described a new project that will be followed through the rest of the text and there will be new lines of projects at the end of chapters too. There are several different projects at the end of this chapter that are developed through the rest of this book. Information on these and how different projects in each chapter are related to different end projects as well as some sample implementations can be found at the book's website. once again, note that code you write now is not likely to survive all the way to the end of the book. You are going to learn better ways of doing things and when you do, you will alter the code that you have written here.

1. Editing rooms for your MUD project using a text editor in a basic text file can be challenging and is error prone. For this reason, it could be helpful to have a GUI that lets you edit the rooms. It can display a **ListView** of rooms that you can select from and then other options for setting values on the selected room. You should also include options to create new rooms or delete rooms. Note that when you insert or delete, the indices of all the rooms after that one change.

2. If you have been working on the web spidering project option, you can build a GUI to do minimal display of the information from those files. If you are feeling like you want to take on a bigger task, you can also include code to graphically represent the links between pages using either a **Canvas** or by positioning elements in the scene graph and connecting them with lines.

3. If you are working on the graphical game project, then at this time you should put together a GUI for your game. Utilizing what you now know about GUIs and graphics, you should get code up where you can have the player (who is at the computer the application is running on) move something around that is visible in the GUI. Your game could have an **AnimationTimer** to make actions occur as well and take input from the user in the form of key or mouse events.

    For an added challenge, you can make a more complex graphical game that includes computer-controlled players with more significant AI.

4. This is a new project line. By the end of the semester you will have the ability to parse, evaluate, and manipulate functions. These are critical aspects of graphical mathematical software like Mathematica®. As such, one project option you could do is to make a Mathematica-like worksheet program. In the end, this program will have the ability to plot functions and do basic symbolic manipulations of those functions. It should also have some programmability in a simple language. All of this will be inside of a GUI. For now you should build a GUI, make a **class** for a **Worksheet**, and see if you can get it to plot some preprogrammed functions that you can select between in the GUI using a **ComboBox** or a **ListView**.

Because you will have a worksheet that can have multiple different types of things in it, that makes a great possibility for an inheritance hierarchy. At the very least, you should have commands and plots. However, you might have a lot more possibilities when you think about it more.

Another element you could add to this project that uses inheritance is to implement the command pattern to provide undo and perhaps redo capabilities [6]. This was discussed in chapter 4. This works by creating a top-level command type that has two methods that might be called something like `execute` and `undo`. The user does something that should change the worksheet in some way; instead of directly changing the worksheet, you build a subtype of the command type with an `execute` method that does the action and an `undo` method that undoes it. You pass that object to a main control object and it executes the command. You can keep track of the commands to provide undo and redo capabilities.

5. This is a new project line. Another project option you could work on is an image processing program somewhat similar in nature to Photoshop®. The goal is to have the ability to build an image from multiple layers as well as a number of different filters for processing images and some tools for creating images. The result will not be as complex as Photoshop® itself, but it will be significantly more complex and functional than Paint®.

   For now you can write something on the order of Paint®. Give it the ability to load and save files as well as a number of different drawing tools for things like lines, rectangles, and similar primitives as well as color selections.

   An obvious use of an inheritance hierarchy is the different things that can be drawn. Other elements that are options for the user to choose from can probably go into this as well. What is more, you can also implement the command pattern, described in the project above, to help make this work and to provide the undo and redo capabilities.

6. The simulation workbench project will be dealing with different types of simulations. Each one will be a bit different, but they will all need to have the ability to run, draw, provide analysis data, and perhaps be paused and restarted. Create a supertype that can provide this type of functionality, then make your simple N-body gravity simulation be a subtype. You can then put the program into a GUI with graphics that will draw the particles moving around.

7. If you are working on the stock portfolio management project, now is the time to put together a GUI interface for your application. Provide displays for the information and menus to manage the portfolio, and maybe add buttons to buy and sell stocks. You will need to have menu options for saving and loading portfolios. You can also investigate how to plot trends and create graphs that depict the distribution of stock, stock price fluctuation, and comparisons of initial and final investment values.

8. For this exercise you should convert the functionality of the text menu in your movies application to a GUI. You should decide the exact format, but you might consider using a `ListView` or a `Table` to display information related to the movies. You will need to have menu options for saving and loading. You could also add the ability to have images associated with movies. There are a lot of movies in the list, and it might be impractical to have images for all, but you could consider having a featured selection section that displays movies you want to currently promote. The `image` class in ScalaFX has the ability to read from files or URLs to get your images.

9. Now that you know how to do graphics, you can actually create turtle-based graphics

for your L-system project. You can do this with a `Canvas` using the various path building methods or with the path capabilities in `scalafx.scene.shape`.

You should include controls for the user to specify the start string, the productions, the original line length, the turning angle, and buttons to advance the string by applying productions. Note, the `String` in an L-system grows exponentially in length. As a result, you probably want to have the length of the turtle move for an `F` or `f` get exponentially shorter. Start with the value provided by the user and divide it by a certain factor each time the system is updated. You could have the user specify that factor. For the sample grammars we gave previously, 3 is a good factor, but it varies based on the grammar.

Chapter 6

Other Collection Types

Back in chapter 1 we talked about the most basic sequence collections in Scala, the `Array` and the `List`. These collections are subtypes of the `Seq trait`, and they share a large number of different methods that we have used to great effect. The sequences are not the only type of collection. The Scala collections library is far more extensive than we have seen so far. In this chapter we will examine the inheritance hierarchy of the entire collections library and look at three other types of collections in detail.

6.1 The scala.collection Packages

At this point you have likely seen in the API that there are a number of packages that start with `scala.collection` including one called simply `scala.collection`. A class diagram showing the inheritance relationships in this top level package is shown in figure 6.1. At the top of the hierarchy is the `Traversable[A]` type. This is a **trait** that underlies the rest of the hierarchy and can be used to represent anything that can run through its contents.

Below `Traversable[A]` is the `Iterable[A]` trait. As the name implies, this type, and its subtypes, have the ability to provide `Iterators`. Anything that is `Iterable[A]` will have a method called `iterator` that will give you an `Iterator[A]`. Though not shown in the figure, the `Iterator[A]` is also in `scala.collection` and it inherits from a type called `TraversableOnce[A]`. Both of these types represent collections that are consumed as they are read.

There are three main subtypes of `Iterable[A]` in the Scala libraries. These are `Seq[A]`, which we have worked with previously, `Set[A]`, and `Map[A, B]`. The latter two allow you

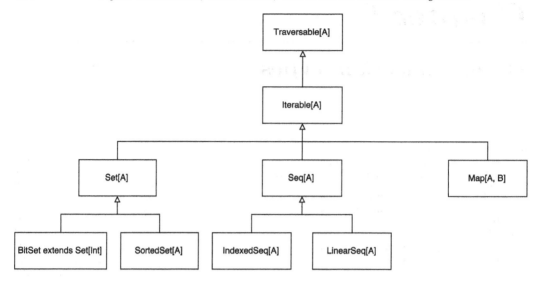

FIGURE 6.1: This is a UML diagram of the main types in the `scala.collection` package. The collection types we have dealt with, including `List[A]` and `Array[A]`, have been under the `Seq[A]` part of the hierarchy.

to run through their contents; hence they are subtypes of `Iterable[A]`, but they handle their contents in a very different way than does a sequence. The sections below will go into the details of those types.

In this top package there are two subtypes of `Set` and `Seq`. The `BitSet` type is optimized to work with sets of `Int`. It stores one bit for each integer. If the value is in the set, the bit will be on, otherwise it will be off. The `SortedSet` type is a specialized version of `Set` that works when the contents have a natural ordering.

The two subtypes of `Seq` are `IndexedSeq` and `LinearSeq`. These two types differ in how the elements can be efficiently accessed. A `List` is an example of a `LinearSeq`. To get to an element using a `LinearSeq` type, you have to run through all the elements before it. On the other hand, an `IndexedSeq` allows you to quickly get to elements at arbitrary indexes, as is the case with `Array`. Each of these types implements their methods in a different way, depending on what is the most efficient way to get to things.

These types are implemented in two different ways in the Scala libraries based on whether the implementations are immutable or mutable. To keep these separate, there are packages called `scala.collection.immutable` and `scala.collection.mutable`.

6.1.1   scala.collection.immutable

The `scala.collection.immutable` package stores various types of collections that are all immutable and which have been optimized so that common operations can be done efficiently with them. This means that "updates" can be made without making complete copies. Of course, because these collections are immutable, no object is really updated, instead, a new, modified version is created. To be efficient, this new object will share as much memory as possible with the original. This type of sharing is safe only because the objects are immutable. A simple class diagram showing the inheritance relationships between the main classes and traits in this package are shown in figure 6.2.

The top part of the hierarchy looks the same as that shown in figure 6.1 for the `scala.collection` package. It begins with `Traversable` and `Iterable` and then has three

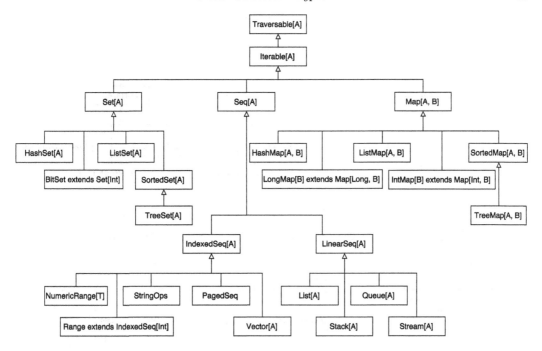

FIGURE 6.2: This is a UML diagram of the main types in the `scala.collection.immutable` package. This image is significantly simplified. If you look in the API you will see that most of the types inherit from multiple other types.

main subtypes in the form of **Set**, **Seq**, and **Map**. The difference is that those three types have many more subtypes. We will not go into detail on most of these, but it is worth noting a few. First there is **List**. This is the immutable sequence that we have been using up to this point. You can see here that it is a **LinearSeq**. This means that you cannot efficiently jump to random elements. We will see exactly why that is the case in chapter 12.

If you want an immutable sequence that has efficient indexing, your default choice would generally be **Vector**. The other immutable subtypes of **IndexedSeq** include **Range**, which we saw in chapter 1, and a type called **StringOps**. The **StringOps** type provides Scala collection operations on **String**s. Many of the methods that you have been using with **String**s are not defined on the **java.lang.String** type. Instead, there is an implicit conversion from **java.lang.String** to **scala.collection.immutable.StringOps**. That is how you can do things like this.

```scala
val (upper,lower) = "This is a String.".filter(_.isLetter).partition(_.isUpper)
```

The **java.lang.String** type does not actually have methods called **filter** or **partition**. You can verify this by looking in the Java API. When Scala sees this line of code, it sees that those methods are missing and looks for an implicit conversion that will give a type that has the needed methods. In this case, the conversion will be to **StringOps**.

There are a number of different types listed under **Map** as well. Two of these are optimized specifically to work with **Int** and **Long**. For most applications, the **HashMap** will probably be ideal. The details of **HashMap** are covered in chapter 22. We will cover possible implementations of the **TreeMap** in chapter 16.

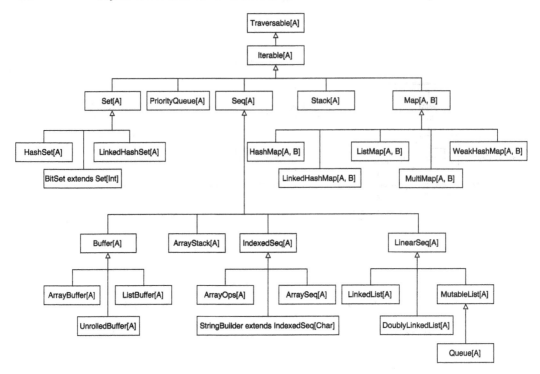

FIGURE 6.3: This is a UML diagram of the main types in the `scala.collection.mutable` package. This image is significantly simplified. If you look in the API you will see that most of the types inherit from multiple other types.

6.1.2   scala.collection.mutable

The last package under `scala.collection` that we will concern ourselves with right now is the `scala.collection.mutable` package. As the name implies, this package includes collections that are mutable. This is the most extensive of the three packages. It is worth scanning through the API to look at the various types in this package and to see what they are used for. A UML class diagram of this package is shown in figure 6.3.

The fact that these types are mutable implies that they include methods that will change their values. When using them, you need to try to remember that the "standard" methods, those we have been using regularly to this point, are still non-mutating methods. For example, `filter` does not change the contents of a collection, even if the collection is mutable. Instead, the mutable types introduce a few additional methods that do perform mutations. We'll talk about these a bit more below.

It is also worth noting that one of the mutable types is `ArrayOps`. This class is much like `StringOps`. In truth, the `Array` type that we have been using is `scala.Array`. The `ArrayOps class` has some additional methods that require an implicit conversion to use.

Now we will look at the `Set`, `Map`, and `Buffer` types. These types can provide a significant boost to your programming and make it much easier to do certain tasks.

6.2 Sets

The concept of a set comes from math. There are two main differences between the **Set** type and the Seq type. The most important one is that a **Set** does not allow duplicate values. If you add a value to a **Set** and that value is already there, it does not add a new value and the size of the resulting **Set** is the same as you had before. The other difference is that the order of elements in a **Set** is not guaranteed. [1] We can see both of these effects in the following sequence of REPL commands.

```
scala> Set(1,2,3)
res0: scala.collection.immutable.Set[Int] = Set(1, 2, 3)

scala> res0 + 4
res1: scala.collection.immutable.Set[Int] = Set(1, 2, 3, 4)

scala> res1 + 2
res2: scala.collection.immutable.Set[Int] = Set(1, 2, 3, 4)

scala> res1 + -10
res3: scala.collection.immutable.Set[Int] = Set(-10, 4, 1, 2, 3)
```

This begins with a set that has the values 1, 2, and 3. As with the **List** and **Array** types, Scala figures out that this is a **Set[Int]**. You can also see from the inferred type that by default we get an immutable **Set**.

You can add to a **Set** with +, and you get back a new **Set** that has the element after the plus sign added in. The second statement gives us back a **Set** that includes 4 in addition to the 1, 2, and 3. In the third command we add a 2 to this new **Set**. The result looks the same as the **Set** before the 2 was added. This is because the **Set** cannot contain duplicate elements.

The last statement adds the value -10 into the **Set** from **res1**. This value goes into the resulting **Set** because there had not been a -10 previously. However, the result shows that order is not preserved in **Sets**. Adding the -10 into the **Set** not only put it in an odd place at the beginning, it also altered the location of the 4 in the **Set**. From this you can see that you want to use **Sets** when you care about the uniqueness of elements, not the order in which you get them.

The fact that the order can change means that "indexing" into a **Set** in the manner we have done with an **Array** or a **List** does not have much meaning. You can perform that type of operation on a **Set**, but the meaning is a bit different. Consider the following examples.

```
scala> res3(2)
res4: Boolean = true

scala> res3(5)
res5: Boolean = false

scala> res3(-10)
res6: Boolean = true
```

---

[1]If you happen to be using a **SortedSet**, then the order of the objects will match the natural ordering of the type, but this is more the exception than the rule.

This type of operation on a `Set` does not give you back a value at a particular location. Instead, it returns a `Boolean` that tells you whether or not the specified value is in the `Set`. This works equally well if the `Set` contains values other than `Int`s. Consider the following.

```scala
scala> val strSet=Set("Jen","Jason","Julie")
strSet: scala.collection.immutable.Set[String] = Set(Jen, Jason, Julie)

scala> strSet("Jason")
res7: Boolean = true

scala> strSet("Bill")
res8: Boolean = false
```

In this example, the `Set` contains `String`s and the "indexing" is done by passing in `String`s. The meaning is the same either way. The return value is true if the argument passed in is in the `Set` and false if it is not. To see why this works this way, go look in the Scala API at one of the versions of `Set`. They all have a type parameter called `A` and they define the following method.

```scala
def apply (elem: A): Boolean
```

Remember that `apply` is a special method that is invoked when an object is used with function call notation. So `f(5)` in Scala gets expanded to `f.apply(5)`. Here the `apply` takes one argument, `elem`, which must have the type `A`. That is the type parameter of the `Set` as a whole, so you have to pass in an object of the type that the `Set` contains. This method then returns to you a `Boolean` telling you whether or not that value is in the `Set`.

This operation of looking to see if elements are members is the most significant operation on a `Set`, and for that reason this is typically a fast operation. Whereas determining if an element is in a `Seq` using methods like `contains` or `indexOf` is a fairly slow operation that must run through all of the elements.

6.2.1   Running through Sets

In addition to the standard higher-order methods like `map`, `filter`, and `foreach`, that are available for both `Set` and the sequences that we have worked with previously, the `Set` type works perfectly well with `for` loops.

```scala
scala> for(v <- res3) println(v)
1
2
-10
3
4
```

On the other hand, using a `while` loop to run through a `Set` is more challenging, because you cannot index by position. If you want to use a `while` loop with a `Set`, you will have to get an `Iterator` for the set with the `iterator` method and then go through that with the `while` loop calling `hasNext` in the condition and `next` in the body. Keep in mind that the order in which the loop runs through the `Set` values is not something you can predict. It does not even have to match the order in which the values are displayed when the `Set` is converted to a `String` for printing.

6.2.2 Mutable vs. Immutable

The examples above showed the default immutable `Set` type. Like the `List` type that we saw previously, an immutable `Set` does not change after it is created. When we add elements to the `Set` with the + operator, we get back a new `Set` and the original `Set` is unaltered. We can demonstrate this by calling back up the original `Set`.

```scala
scala> res0
res9: scala.collection.immutable.Set[Int] = Set(1, 2, 3)
```

All of the operations that were done have not altered it in any way.

As we saw earlier, there is another `Set` type in the `scala.collection.mutable` package. You could get to this type using the fully specified name of `scala.collection.mutable.Set`. However, doing so not only requires a lot of typing, it also makes code difficult to read. You could `import scala.collection.mutable.Set` and then refer to it as just `Set`. The problem with this approach is that you can no longer easily refer to the immutable `Set` type. A recommended style for dealing with this is to `import scala.collection.mutable`. The packages in Scala are truly nested. As a result, this import will allow us to refer to a mutable `Set` by the name `mutable.Set` and the immutable set as just `Set`.

All the operations that you can do with an immutable `Set` are available on a mutable `Set` as well. Above we used the + operator on the immutable `Set` and it works equally well on the mutable `Set`.

```scala
scala> import scala.collection.mutable
import scala.collection.mutable

scala> val mset = mutable.Set(1,2,3)
mset: scala.collection.mutable.Set[Int] = Set(1, 2, 3)

scala> mset + 4
res10: scala.collection.mutable.Set[Int] = Set(1, 4, 2, 3)

scala> mset
res11: scala.collection.mutable.Set[Int] = Set(1, 2, 3)
```

This code shows you the `import` statement, the declaration of a mutable `Set` after the `import`, and the creation of a new `Set` by adding an element to the old one. Just using + does not change the mutable `Set`. The way in which the mutable `Set` differs from the immutable version is that it includes additional operators and methods.

To see this, we will call one of those extra operators. On a mutable `Set` we can use `+=` the same way we might with an `Int` that was declared as a `var`.

```scala
scala> mset += 4
res12: mset.type = Set(1, 4, 2, 3)

scala> mset
res13: scala.collection.mutable.Set[Int] = Set(1, 4, 2, 3)
```

You can see from this that the use of `+=` actually changes the original `Set`. The `+=` does not exist on the immutable `Set` type. If you try to call it, you will get a syntax error.[2] The list of methods/operators on the mutable `Set` that mutates them includes the following for `Set[A]`.

---

[2]You can call `+=` on a `var` immutable set, but then you are assigning to the `var`, not changing the `Set`. The original `Set` will still be around if you have another name you can use to reach it.

- `+=(elem : A)` - Adds the specified element to the `Set`.

- `-=(elem : A)` - Removes the specified element from the `Set`. Does nothing if it was not there initially.

- `++=(xs : TraversableOnce[A])` - Adds all the elements from a collection into the `Set`. The `TraversableOnce` type is the most basic type for Scala collections so this can be used with any of the collections in Scala. `Traversable` inherits from `TraversableOnce`.

- `--=(xs : TraversableOnce[A])` - Removes all the elements in `xs` from the `Set`.

- `add(elem : A) : Boolean` - Adds the specified element to the `Set`. Returns true if it was not there before the call and false otherwise.

- `clear():Unit` - Removes all elements from the `Set`.

- `remove(elem : A) : Boolean` - Removes the specified element from the `Set` and returns true if it was there to remove or false otherwise.

- `retain(p : (A) => Boolean) : Unit` - Removes all elements that do not satisfy the predicate from the `Set`. This is like a mutating version of `filter`.

Mutable `Set`s can also be updated using assignment and indexing, which calls the `update` method. The index is the value you are adjusting in the `Set` and the assignment is either to true or false. To see how this works, consider the following examples.

```scala
scala> mset(7) = true

scala> mset(2) = false

scala> mset
res14: scala.collection.mutable.Set[Int] = Set(1, 4, 7, 3)
```

The first line adds 7 to the `Set` while the second removes 2. This works because the `scala.collection.mutable.Set` type defines an `update` method that looks like this.

```scala
def update (elem: A, included: Boolean): Unit
```

So the line `mset(7)=true` gets expanded to `mset.update(7,true)`.

Earlier in this section it was mentioned that the standard methods, like `map` and `filter`, will do the same thing on a mutable set as they do on an immutable `Set`. This was very intentional. The same operations on subtypes should always have the same outward appearance and significant effects. This reduces the chances of programmers making silly mistakes. If you want to mutate the `Set` you have to call one of the specific methods for doing so. For example, you would call `retain` instead of `filter`. If you accidentally try to use one of these operations on an immutable type, the code will not compile. If the operations with the same name had different behaviors, it would be very easy for the programmer to introduce logic errors because he/she believed an object was one type when it was actually another, or for bugs to be introduced later on by a programmer changing the specific type that is assigned to a variable or passed into a method.

6.2.3   Using a Set

To see how the `Set` type can be useful we will write a function that tells us all the unique words in a text file. The function will take the name of the file as an argument and return a `Set[String]` that has all the words in lowercase. As with any function of any significance, this can be done in many ways. We are going to do it making heavy use of the Scala collection methods. The fact that we are using the collection methods makes it more sensible to use a `Source` than a `Scanner`. Here is code that will do this.

```
1  def uniqueWords(fileName: String): Set[String] = {
2    val source = io.Source.fromFile(fileName)
3    val words = source.getLines.toSeq.flatMap(_.split(" +")).
4      map(_.filter(_.isLetter).toLowerCase).toSet
5    source.close()
6    words
7  }
```

A function is defined using the signature that was described above. This function includes only four statements on lines 2–5. The first and third statements in the function open the file and close it. The last statement is just an expression for the `Set` that was created from the source.

The second statement is where all the work is done and is worth looking at in detail. This line declares the variable `words` and makes it equal to a `Set` of the words. To do that it gets the lines from the source and turns them into a sequence. The `toSeq` call makes types work out better with `flatMap`. As you might recall, `getLines` returns an `Iterator`. The `Iterator` type is a collection that you can only pass through once. It is consumed during that pass. By bumping it up to a sequence we get something that works better with the `Array[String]` that we get from the `split` method.

After the lines are converted to a sequence there is a call to `flatMap`. This is one of the standard higher-order methods on collections. We have not used it much, so it is worth taking a minute to discuss. As you can tell from the name, `flatMap` is closely related to `map`. The `map` method takes each element of a collection, applies some function, and makes a new collection that has the results of that function application. The `flatMap` does basically the same thing only it expects that the function passed in is going to produce collections. Instead of returning a collection of collections, as `map` would do, `flatMap` will flatten out all of the inner collections into one long collection. If you use `map` on the sequence of lines and use the same function argument, you will get a `Seq[Array[String]]`. Each element of the sequence will contain an `Array` of the words from that line. However, that is not quite what we want. We want a simple sequence of words, not a sequence of `Array`s of words. This is what `flatMap` does for us. It puts each of the `Array`s end to end in the new sequence so that we get a `Seq[String]`.

If we knew that the file only included simple words in lowercase we could stop there. However, most text files will include capitalization and punctuation as well. This would mess up the `Set` because "Hi" is not equal to "hi" and both would appear in the result when we want only one. Similarly, the word "dog" would be different than "dog." or "dog," if the word happened to appear directly before a punctuation mark. The second line of the second statement includes logic to deal with this. Each of the words that the file was broken into is passed through a call to `map`. That call to `map` filters out any non-letter characters and converts the resulting `String`s to lowercase. The filtering of non-letter characters is perhaps too aggressive because it removes things like apostrophes in contractions, but the result it provides is sufficiently correct for our purposes.

The last thing that happens in the second statement is that the sequence is converted to a `Set`. This gives us the type we desired and removes any non-unique elements by virtue of being a `Set`.

6.3 Maps

Another collection type that is extremely useful, probably more so than the **Set** in most programs, is the **Map**. Like the **Set**, the term **Map** has origins in math. A mapping in mathematics is something that takes values from one set and gives back values from another set. A function in math is technically a mapping. The **Map** collection type is not so different from this. It maps from value of one type, called the key type, to another type, called the value type. The key and value types can be the same, but they are very often different.

To help you understand **Maps**, let's look at some examples. We will start with a statement to create a new **Map**.

```scala
scala> val imap = Map("one" -> 1, "two" -> 2, "three" -> 3)
imap: scala.collection.immutable.Map[String,Int] = Map(one -> 1, two -> 2, three
    -> 3)
```

The **Map** type takes two type parameters. In this example, they are **String** and **Int**. The values passed in to build the **Map** are actually tuples where the key is the first element and the value is the second element. In this example we used the -> syntax to build the tuples. You can only create a tuple with two elements using ->. It was added to the Scala library specifically so that **Maps** could be created in this way. You can read this as making a **Map** where the **String** "one" is associated with the **Int** 1 and so on.

When you index into a **Map** you do so with value of the key type. What you get in return is the corresponding value object.

```scala
scala> imap("two")
res0: Int = 2
```

If you like, you can think of a **Map** as a collection that can be indexed by any type that you want instead of just the integer indexes that were used for sequences. This works because the **Map** type defined an **apply** method like this.

```scala
def apply (key: A): B
```

The types **A** and **B** are the type parameters on **Map** with **A** being the key type and **B** being the value type. If you try to pull out a key that is not in the map, you will get an exception.

In addition to allowing any type for the key, the **Map** also has the benefit that it does not require any particular "starting" value for the indexes. For this reason, you might sometimes want to build a map that uses an **Int** as the index. Consider this example.

```scala
scala> val imap2 = Map(10 -> "ten", 100 -> "hundred", 1000 -> "thousand")
imap2: scala.collection.immutable.Map[Int,String] = Map(10 -> ten, 100 -> hundred,
    1000 -> thousand)
```

Technically you could build a sequence that has the values "ten", "hundred", and "thousand" at the proper indexes. However, this is very wasteful because there are many indexes between 0 and 1000 that you will not be using and they would have to store some type of default value. For numeric types that would be 0, for reference types it would be **null**.

As with the **Set** type, you can add new key, value pairs to the **Map** with the + operator.

```scala
scala> imap + ("four" -> 4)
res1: scala.collection.immutable.Map[String,Int] = Map(one -> 1, two -> 2, three
    -> 3, four -> 4)
```

The key and value need to be in a tuple, just like when the original `Map` was built. The parentheses in this sample are required for proper order of operation. Otherwise the `+` happens before the `->` and the result is a (`String`,`Int`) that will look rather unusual.

You can also use the `-` operator to get a `Map` with fewer elements. For this operation, the second argument is just the key you want to remove. You do not have to have or even know the value associated with that key.

```
scala> imap - "two"
res1: scala.collection.immutable.Map[String,Int] = Map(one -> 1, three -> 3)
```

Here we remove the value 2 from the `Map` by taking out the key "two".

As with the `Set` type, there are mutable and immutable versions. As you can see from the examples above, the default for `Map`, like for `Set`, is immutable. To use the mutable version you should follow the same recommendation as was given above and import `scala.collection.mutable` and then refer to it as `mutable.Map`.

6.3.1 Looping through a Map

You can use the higher-order methods or a `for` loop to run through the contents of a `Map`, but there is a significant difference between this and the other collections we have looked at. This is because the contents of the `Map` are effectively tuples. To illustrate this, we'll look at a simple `for` loop.

```
scala> for(tup <- imap) println(tup)
(one,1)
(two,2)
(three,3)
```

Each element that is pulled off the `Map` is a tuple. This has significant implications when using higher-order methods like `map`. The argument to `map` is a function that takes a single argument, but that argument is a tuple with a key-value pair in it. Consider the following code.

```
scala> imap.map(tup => tup._2*2)
res3: scala.collection.immutable.Iterable[Int] = List(2, 4, 6)
```

Because the argument, `tup`, is a tuple, we have to use the _2 method to get a value out of it. The _1 method would give us the key. The alternative is to make the body of the function longer and use a pattern in an assignment to pull out the values. For many situations this is not significantly better.

Running through a map works much better with a `for` loop because the `for` loop does pattern matching automatically. This syntax is shown here.

```
scala> for((k,v) <- imap) yield v*2
res4: scala.collection.immutable.Iterable[Int] = List(2, 4, 6)
```

This syntax is not significantly shorter for this example, though it can be in others, and it is easier for most people to read. The real advantage is that you can easily use meaningful variable names. This example used `k` and `v` for key and value, but in a full program you could use names that carry significantly more meaning.

You might have noticed that both the example with `map` and with the `for` loop gave back `Lists`. This is also a bit different as we are used to seeing this type of operation give us back the same type that was passed in. The problem is that to build a `Map` we would need key, value pairs. The way the libraries for Scala are written, you will automatically get back a `Map` any time that you use a function that produces a tuple.

```scala
scala> imap.map(tup => tup._1 -> tup._2*2)
res5: scala.collection.immutable.Map[String,Int] = Map(one -> 2, two -> 4, three
    -> 6)

scala> for((k,v) <- imap) yield k -> v*2
res6: scala.collection.immutable.Map[String,Int] = Map(one -> 2, two -> 4, three
    -> 6)
```

These two examples show how that can work using either `map` or the `for` loop with a `yield`.

6.3.2   Using Maps

The example we will use for illustrating use of a `Map` is closely related to the one we used for the `Set`. Instead of simply identifying all the words, we will count how many times each one occurs. This function will start off very similar to what we did before. We'll start with this template.

```scala
def wordCount(fileName: String): Map[String, Int] = {
  val source = io.Source.fromFile(fileName)
  val words = source.getLines.toSeq.flatMap(_.split(" +")).
    map(_.filter(_.isLetter).toLowerCase)
  val counts = ???
  source.close()
  counts
}
```

Here `words` will just be a sequence of the words and we will use it to build the counts and return a `Map` with those counts. We will do this last part in two different ways.

The first approach will use a mutable `Map`. The logic here is that we want to run through the sequence of words and if the word has not been seen before, we add a count of 1 to the `Map`. If it has been seen before, we add the same key back in, but with a value that is one larger than what we had before. The code for this looks like the following.

```scala
def wordCount(fileName: String): mutable.Map[String, Int] = {
  val source = io.Source.fromFile(fileName)
  val words = source.getLines.toSeq.flatMap(_.split(" +")).
    map(_.filter(_.isLetter).toLowerCase)
  val counts = mutable.Map[String, Int]()
  for (w <- words) {
    if (counts.contains(w)) counts += w -> (counts(w) + 1)
    else counts += w -> 1
  }
  source.close()
  counts
}
```

Note that the `for` loop to run through the words here occurs after the declaration of `counts`. Because `counts` is mutable, this approach works. You should also note that the result type for the function has been altered to be a `mutable.Map[String,Int]`. If you leave this off, you have to add a call to `.toMap` at the end, after `counts`, so that the counts will be converted to an immutable type. This distinction between mutable and immutable types is very significant for safety in programs. In general, there could be other references to the

`Map` you are given that are retained in other parts of the code. If it were possible for you to get a mutable `Map` when you were expecting an immutable one, those other parts of the code could alter the `Map` when you were not expecting it. This would lead to logic errors that are very hard to track down.

Using the mutable version is fairly logical and can have some efficiency advantages, but as was just alluded to, mutable values come with risks. For this reason, we will write a second version that uses an immutable `Map` and the `foldLeft` method. The `foldLeft` method is a more complex method we have not really dealt with in detail. However, it is ideal for what we are doing here as it effectively does what we wrote previously, but without the need for a loop, `vars`, or mutable types. The `foldLeft` method runs through the collection from left to right and applies a function to each element, compiling a result as it goes. The function is curried and the first argument list is a single value that tells what the compiled value should start off as. In this case, we are compiling a `Map` so we will start off with an empty one. The code we want to apply is very much like what was in the `for` loop previously. This is what it all looks like when put together.

```
def wordCount2(fileName: String): Map[String, Int] = {
  val source = io.Source.fromFile(fileName)
  val words = source.getLines.toSeq.flatMap(_.split(" +")).
    map(_.filter(_.isLetter).toLowerCase)
  val counts = words.foldLeft(Map[String, Int]())((m, w) => {
    if (m.contains(w)) m + (w -> (m(w) + 1))
    else m + (w -> 1)
  })
  source.close()
  counts
}
```

The function that is passed into `foldLeft` takes two arguments. The first is the compiled value so far and the second is the value that is being operated on. In this case the compiled value is a `Map` and we use the variable m to represent it. The value being passed in is the next word that we want to count and is represented by the name w as it was in the `for` loop previously.

The body of the function passed into `foldLeft` as the second argument list differs from what was inside the `for` loop previously in two ways. First, instead of referring to the mutable `Map` `counts`, we use the argument name m. Second, we do not use `+=`. Instead we just use `+`. The `+=` operator is not defined for the immutable `Map` and what we are doing here is building a new immutable `Map` that is passed forward to be used with the next word.

If you really want to push things a little harder you can get rid of the declaration of `words` all together. After all, the only thing it is used for is as the argument to `foldLeft` to build `counts`. Doing this and getting rid of some curly braces that are not technically needed produces the following code.

```
def wordCount3(fileName: String): Map[String, Int] = {
  val source = io.Source.fromFile(fileName)
  val counts = source.getLines.toSeq.flatMap(_.split(" +")).
    map(_.filter(_.isLetter).toLowerCase).
    foldLeft(Map[String, Int]())((m, w) =>
      if (m.contains(w)) m + (w -> (m(w) + 1))
      else m + (w -> 1))
  source.close()
  counts
}
```

This code is compact, but whether you really want to do things this way or not is debatable. The introduction of the variable `words` did add extra length to the code, but for most readers it probably also made things more understandable.[3]

Having any version of `wordCount` implicitly gives us a `Set` of the words as well. This is because the `Map` type has a method called `keySet` that returns a `Set` of all of the keys used in the `Map`. Because each key can only occur once, nothing is lost when you get the keys as a `Set`. For efficiency reasons you would probably want to keep the `uniqueWords` function around if you were going to be using it often instead of replacing calls to it with `wordCount("file.txt").keySet` because the counting of the words does extra work that is not needed to produce the `Set`.

A more general use case for `Map`s is using them as a fast way to look up groups of data like `case class`es by a particular key value. An example of this would be looking up an instance of a `case class` that represents a student by the student's ID or perhaps their unique login number. The real power of the `Map` type comes from the fact that the key and value types can be anything that you want.

6.4   Buffers

The last of the new collection types that we will look at in detail in this chapter is the `Buffer`. The `Buffer` type is like an `Array` that can change size. As a result, the `Buffer` type is implicitly mutable. There is no immutable `Buffer`. If you are using other mutable types, the rule of importing `scala.collection.mutable` and referring to a `Buffer` as `mutable.Buffer` will still work well and it makes it perfectly clear to anyone reading the code that you are using a mutable type. However, because the `Buffer` type is only mutable, you could also consider doing an `import` of `scala.collection.mutable.Buffer` and simply calling it a `Buffer` in your code.

There are several subtypes of `Buffer` in the Scala libraries. The two most significant ones are `ArrayBuffer` and `ListBuffer`. The former uses an `Array` to store values. The latter uses a structure called a linked list. The `List` type in Scala is also a linked list, though it is not directly used by the `ListBuffer` because the `ListBuffer` needs to be mutable. We will use the `ArrayBuffer` in this chapter. The nature of the `ListBuffer` and how it is different will become more apparent in chapter 12.

You can create a `Buffer` like this.

```
scala> val buf = mutable.Buffer(1,2,3,4,5)
buf: scala.collection.mutable.Buffer[Int] = ArrayBuffer(1, 2, 3, 4, 5)
```

You can also use some of the methods we learned about for `Array` and `List` like `fill` and `tabulate`.

```
scala> val rbuf = mutable.Buffer.fill(10)(math.random)
rbuf: scala.collection.mutable.Buffer[Double] = ArrayBuffer(0.061947605764430924,
    0.029870283928219443, 0.5457301708447658, 0.7098206843826819,
    0.8619215922836797, 0.5401420250956313, 0.6249953821782052,
    0.1376217145656472, 0.26995766937532295, 0.8716257556831167)
```

---

[3]There is another higher-order method on sequences called `groupBy` that could make this code even shorter. The `groupBy` method takes a function with a single argument of the key type. Let's say that function has the type `A => B`. Calling `groupBy` produces a `Map[B, List[A]]`. Using this, one could say `val counts = words.groupBy(w => w).map(t => t._1 -> t._2.length)`.

You can index into a `Buffer` and change values the same way you did for an `Array`.

```scala
scala> rbuf(3)
res0: Double = 0.7098206843826819

scala> buf(3)=99

scala> buf
res1: scala.collection.mutable.Buffer[Int] = ArrayBuffer(1, 2, 3, 99, 5)
```

The way in which the `Buffer` type differs from the `Array` type is that it can easily grow and shrink. There are a whole set of methods that take advantage of this. Here are some of them.

- `+=(elem: A): Buffer[A]` - Append the element to the `Buffer` and return the same `Buffer`.

- `+=:(elem: A): Buffer[A]` - Prepend the element to the `Buffer` and return the same `Buffer`.

- `++=(xs: TraversableOnce[A]): Buffer[A]` - Append the elements in `xs` to the `Buffer` and return the same `Buffer`.

- `++=:(xs: TraversableOnce[A]): Buffer[A]` - Prepend the elements in `xs` to the `Buffer` and return the same `Buffer`.

- `-=(elem: A): Buffer[A]` - Remove the element from the `Buffer` and return the same `Buffer`.

- `--=(xs: TraversableOnce[A]): Buffer[A]` - Remove all the elements in `xs` from the `Buffer` and return the `Buffer`.

- `append(elem: A): Unit` - Append the element to the `Buffer`.

- `appendAll(xs: TraversableOnce[A]): Unit` - Append the elements in `xs` to the `Buffer`.

- `clear(): Unit` - Remove all the elements from the `Buffer`.

- `insert(n: Int, elems: A*): Unit` - Insert the specified elements at the specified index.

- `insertAll(n: Int, elems: Traversable[A]): Unit` - Insert all the elements in `elems` at index `n`.

- `prepend(elems: A*): Unit` - Prepend the elements to this `Buffer`.

- `prependAll(xs: TraversableOnce[A]): Unit` - Prepend all the elements in `xs` to this `Buffer`.

- `remove(n: Int, count: Int): Unit` - Remove `count` elements starting with the one at index `n`.

- `remove(n: Int): A` - Remove the one element at index `n` and return it.

- `trimEnd(n: Int): Unit` - Remove the last `n` elements from this `Buffer`.

- `trimStart(n: Int): Unit` - Remove the first `n` elements from this `Buffer`.

Most of these should be fairly self-explanatory. The only methods that you might question are the ones that involve symbol characters. There are two things of interest here. The first is that they all return the `Buffer` that they are called on. The reason for this is that it allows you to string them together to append or prepend multiple elements in a row. Consider this example with appending.

```scala
scala> buf += 6 += 7
res2: buf.type = ArrayBuffer(1, 2, 3, 99, 5, 6, 7)
```

Here both 6 and 7 are appended in a single line.

The other operations that might seem odd are +=: and ++=:. These operations prepend to the `Buffer`. What is interesting about them is their usage. Here is an example.

```scala
scala> 0 +=: buf
res3: buf.type = ArrayBuffer(0, 1, 2, 3, 99, 5, 6, 7)
```

The value to be prepended is to the left of the operator and the `Buffer` is on the right. In some ways that makes sense given where it is in the `Buffer` after the prepending. We have seen something like this before. The cons operator on a `List` is : : and it prepends the element to the `List`. That operator also ends with a colon. This is one of the more esoteric rules of Scala. Any symbolic operator method that ends in a colon is right associative instead of being left associative the way most operators are. This means that the object the operator works on is to the right of the operator and the operations are grouped going from the right to the left.

6.5 Collections as Functions

The different collection types all inherit from `PartialFunction` which inherits from the standard function types. This means that you can use collections in places where you would normally be expected to use functions. For example, the `Seq[A]` type will work as a `Int => A`. Similarly, a `Map[A, B]` can be used as a `A => B`.

A simple example usage of this is to take every third element of an `Array` by mapping over a `Range` with the desired indexes and using the `Array` as the function argument.

```scala
scala> val nums = Array.tabulate(30)(i => 2*i)
nums: Array[Int] = Array(0, 2, 4, 6, 8, 10, 12, 14, 16, 18, 20, 22, 24, 26, 28,
    30, 32, 34, 36, 38, 40, 42, 44, 46, 48, 50, 52, 54, 56, 58)

scala>  val everyThird = (nums.indices by 3).map(nums)
everyThird: scala.collection.immutable.IndexedSeq[Int] = Vector(0, 6, 12, 18, 24,
    30, 36, 42, 48, 54)
```

In this way, applying `map` with a collection as the argument gives us the ability to do arbitrary indexing into some other collection.

6.6 Project Integration

There are quite a few places where we can use these different collections. In the previous chapter the children in the **DrawTransform** were held in a **var List**, but this would probably be better done with a **mutable.Buffer**. This would make it easier to include code that adds and removes children using **+=** and **-=**. In this section we are going to implement the command processing, which we did not do earlier. To do this, we are going to use the **Map** type. We will also refactor the way in which new drawables are added to our drawings.

6.6.1 Commands

The command processor is a part of the code that could be implemented with inheritance. It would be possible to make a type called **Command** that we have other types inherit from. That type of construct would make logical sense, but might not be what we really need. One aspect you need to learn about inheritance is not to use it if it is not required. To know if we need it, we have to do some analysis.

You might wonder why we would want to put a text-based command processor into a drawing program. The reason is similar to why you would want to include a command line interface in an OS that has a GUI. There are certain operations that simply work better when they are typed in than they would when you have to point and click them. For now we will start with some simple examples that we can use as use cases for the command tool. There are three commands that we will start with:

- echo - prints whatever follows the command,

- add - should be followed by a bunch of space separated numbers and prints the sum,

- refresh - takes no arguments and causes the drawing to refresh.

So the user might type in "echo Hello, World!", "add 1 2 3 4 5", or "refresh". The first two take arguments and the only result of calling them is that something is displayed in the output. The last one is very different in that it does not take arguments or display anything. Instead, it needs to cause the **Drawing** object that is displayed to repaint itself.

To make this happen in the code, we will need to set the **onAction** of the **TextField** for the user input. That action needs to take the line that is typed in and identify the first word, then based on what that word is, appropriate functionality needs to be invoked.

One way to implement this functionality is to have each command include a function of the form **(String,Drawing) => Any**. The first parameter of this function will be the rest of the line after the command. The second parameter is the **Drawing** object the command was invoked on. This is needed so that a command like refresh can cause something to happen associated with the drawing. The return value is anything we want to give back. For most things this object will be converted to a **String** and appended to the output.

Having these functions is not quite enough. We also need to have the ability to quickly identify which of these functions should be invoked based on the first word of the input. This is a perfect usage of a **Map**. More specifically, we want to use a **Map[String, (String,Drawing)=>Any]**. This is a **Map** from each command **String** to the function that command should invoke.

Using this approach, we can declare an object that contains the command functionality. An implementation that includes the three commands described above might look like the following.

```scala
package mapset.drawing

object Commands {
  private val commands = Map[String, (String, Drawing) => Any](
    "add" -> ((rest, d) => rest.trim.split(" +").map(_.toInt).sum),
    "echo" -> ((rest, d) => rest.trim),
    "refresh" -> ((rest, d) => d.draw()))

  def apply(input: String, drawing: Drawing): Any = {
    val spaceIndex = input.indexOf(' ')
    val (command, rest) = if (spaceIndex < 0) (input.toLowerCase(), "")
        else (input.take(spaceIndex).toLowerCase(), input.drop(spaceIndex))
    if (commands.contains(command)) commands(command)(rest, drawing) else "Not a
      valid command."
  }
}
```

This object contains an `apply` method that is used to invoke various commands. It takes two arguments. The first is the full input from the user and the second is the `Drawing` object for calls to be made back on. This method finds the location of the first space, assuming there is one, then breaks the input into a command part and the rest of the input around that space. If there is no space, the entire input is the command. Once it has split the input, it checks to see if the command is one that it knows.

The `Map` called `commands` is used to identify known commands and to invoke the command. A call to `contains` tells us if the command is a known command. If it is, then the `apply` method is called using function notation. If not, an appropriate `String` is returned.

To make this code function, we need to update `DrawingMain` so that when the `TextField` for commands is created, an `onAction` handler is set that will execute the command and alter the text in the associated `TextArea`. The following lines need to be added after `commandField` and `commandArea` have both been declared.[4]

```scala
commandField.onAction = (ae:ActionEvent) => {
  val text = commandField.text.value
  if(text.nonEmpty) {
    commandArea.text = (commandArea.text + "> "+text+"\n"+Commands(text,
        drawing)+"\n").value
    commandField.text = ""
  }
}
```

This code sets a function to be executed when the user hits Enter in the `TextField`. The function executes the command and appends the result to the command area, then clears the text in the `TextField` to accept the next command.

The primary advantage of the design using a `Map` is that it makes it very easy to add new commands. Simply add a new member to the `command Map` and you are done. For these commands, the code for the function could be done in line. Longer functions could easily be written as methods in the `Commands` object. If needed, the functionality for commands could even be placed in other parts of the code, as long as it is reachable from `Commands` or through the `Drawing` instance that is passed in.

---

[4]A full version of `DrawingMain` with this and other alterations is shown below.

6.6.2 Adding Drawables

There is another place where we might use a collection in an interesting way to clean up the code so that it is more extensible and less error prone. This is in the code to add a new `Drawable`. The code we had in chapter 5 looked like this.

```
addItem.onAction = (ae: ActionEvent) => {
  val current = tabPane.selectionModel.value.selectedIndex.value
  if (current >= 0) {
    val (drawing, treeView) = drawings(current)
    val dialog = new ChoiceDialog("Rectangle", List("Rectangle", "Transform",
        "Text"))
    val d = dialog.showAndWait() match {
      case Some(s) =>
        val d: Drawable = s match {
          case "Rectangle" => new DrawRectangle(drawing, 0, 0, 300, 300,
              Color.Blue)
          case "Transform" => new DrawTransform(drawing)
          case "Text" => new DrawText(drawing, 100, 100, "Text", Color.Black)
        }
        val treeSelect = treeView.selectionModel.value.getSelectedItem
        def treeAdd(item: TreeItem[Drawable]): Unit = item.getValue match {
          case dt: DrawTransform =>
            dt.addChild(d)
            item.children += new TreeItem(d)
            drawing.draw()
          case d =>
            treeAdd(item.getParent)
        }
        if (treeSelect != null) treeAdd(treeSelect)
      case None =>
    }
  }
}
```

This code is not all that long, but it contains one significant part that is particularly error prone and that could lead to irritating errors later in development as more `Drawable` types are added to the project. The weak point in this code is that the string literals in the `match` cases have to be exactly equal to the strings in the `ChoiceDialog`. Imagine what happens over time as more options are added. You have to add a string to `options`, then add a new case with the same string. If you ever have a typo in a string, the code will continue to compile just fine, but selecting the option with the typo will not add anything new to the drawing.

An alternate approach is to do something very similar to what we did with the commands. We can group the names with the functionality that makes the `Drawable` objects. We can make a `Map` with keys that are the names of the types and values that are functions to create new `Drawables`. Such a `Map` could be put at the top of `DrawingMain` and might look like the following.

```
private val creators = Map[String, Drawing => Drawable](
  "Rectangle" -> (drawing => new DrawRectangle(drawing, 0, 0, 300, 300,
      Color.Blue)),
  "Transform" -> (drawing => new DrawTransform(drawing)),
  "Text" -> (drawing => new DrawText(drawing, 100, 100, "Text", Color.Black)))
```

Once this has been defined, we can use it in action for the menu item as shown below.

```scala
addItem.onAction = (ae: ActionEvent) => {
  val current = tabPane.selectionModel.value.selectedIndex.value
  if (current >= 0) {
    val (drawing, treeView) = drawings(current)
    val dtypes = creators.keys.toSeq
    val dialog = new ChoiceDialog(dtypes(0), dtypes)
    val d = dialog.showAndWait() match {
      case Some(s) =>
        val d = creators(s)(drawing)
        val treeSelect = treeView.selectionModel.value.getSelectedItem
        def treeAdd(item: TreeItem[Drawable]): Unit = item.getValue match {
          case dt: DrawTransform =>
            dt.addChild(d)
            item.children += new TreeItem(d)
            drawing.draw()
          case d =>
            treeAdd(item.getParent)
        }
        if (treeSelect != null) treeAdd(treeSelect)
      case None =>
    }
  }
}
```

Line 5 uses the **creators Map** to get all the strings that we want to display in the **ChoiceDialog**. Line 9 uses the string the user selected to look up the proper function in the Map, then passes it the **drawing** to instantiate the appropriate type.

The real benefit of this approach is that the names and the code to create them are grouped together in such a way that new options can be added with very little chance of the programmer messing things up. All that has to be done is to add another line to the **creators Map**. Having the string and the type appear on the same line means that a simple visual inspection can tell you whether or not they have been set up correctly. In addition, the handler code does not need to be changed in any way when we make new subtypes of **Drawable**.

To help you see how the pieces fit together, here is a full version of **DrawingMain** with the changes from this chapter included.

```scala
package mapset.drawing

import scalafx.Includes._
import scalafx.application.JFXApp
import scalafx.geometry.Orientation
import scalafx.scene.Scene
import scalafx.scene.canvas.Canvas
import scalafx.scene.control.Menu
import scalafx.scene.control.MenuBar
import scalafx.scene.control.MenuItem
import scalafx.scene.control.ScrollPane
import scalafx.scene.control.SeparatorMenuItem
import scalafx.scene.control.Slider
import scalafx.scene.control.SplitPane
import scalafx.scene.control.Tab
import scalafx.scene.control.TabPane
```

```scala
import scalafx.scene.control.TextArea
import scalafx.scene.control.TextField
import scalafx.scene.control.TreeView
import scalafx.scene.layout.BorderPane
import scalafx.scene.paint.Color
import scalafx.event.ActionEvent
import scalafx.scene.control.TreeItem
import scalafx.scene.control.SelectionMode
import scalafx.scene.control.Label
import scalafx.scene.layout.Priority
import scalafx.scene.control.ComboBox
import scalafx.scene.control.Dialog
import scalafx.scene.control.ChoiceDialog

object DrawingMain extends JFXApp {
  private var drawings = List[(Drawing, TreeView[Drawable])]()

  private val creators = Map[String, Drawing => Drawable](
    "Rectangle" -> (drawing => new DrawRectangle(drawing, 0, 0, 300, 300,
        Color.Blue)),
    "Transform" -> (drawing => new DrawTransform(drawing)),
    "Text" -> (drawing => new DrawText(drawing, 100, 100, "Text", Color.Black)))

  stage = new JFXApp.PrimaryStage {
    title = "Drawing Program"
    scene = new Scene(1000, 700) {
      // Menus
      val menuBar = new MenuBar
      val fileMenu = new Menu("File")
      val newItem = new MenuItem("New")
      val openItem = new MenuItem("Open")
      val saveItem = new MenuItem("Save")
      val closeItem = new MenuItem("Close")
      val exitItem = new MenuItem("Exit")
      fileMenu.items = List(newItem, openItem, saveItem, closeItem, new
          SeparatorMenuItem, exitItem)
      val editMenu = new Menu("Edit")
      val addItem = new MenuItem("Add Drawable")
      val copyItem = new MenuItem("Copy")
      val cutItem = new MenuItem("Cut")
      val pasteItem = new MenuItem("Paste")
      editMenu.items = List(addItem, copyItem, cutItem, pasteItem)
      menuBar.menus = List(fileMenu, editMenu)

      // Tabs
      val tabPane = new TabPane
      val (drawing, tree, tab) = makeDrawingTab()
      drawings = drawings :+ (drawing, tree)
      tabPane += tab

      // Menu Actions
      newItem.onAction = (ae: ActionEvent) => {
        val (drawing, tree, tab) = makeDrawingTab()
        drawings = drawings :+ (drawing, tree)
        tabPane += tab
```

```scala
70       }
71       closeItem.onAction = (ae: ActionEvent) => {
72         val current = tabPane.selectionModel.value.selectedIndex.value
73         if (current >= 0) {
74           drawings = drawings.patch(current, Nil, 1)
75           tabPane.tabs.remove(current)
76         }
77       }
78       addItem.onAction = (ae: ActionEvent) => {
79         val current = tabPane.selectionModel.value.selectedIndex.value
80         if (current >= 0) {
81           val (drawing, treeView) = drawings(current)
82           val dtypes = creators.keys.toSeq
83           val dialog = new ChoiceDialog(dtypes(0), dtypes)
84           val d = dialog.showAndWait() match {
85             case Some(s) =>
86               val d = creators(s)(drawing)
87               val treeSelect = treeView.selectionModel.value.getSelectedItem
88               def treeAdd(item: TreeItem[Drawable]): Unit = item.getValue match {
89                 case dt: DrawTransform =>
90                   dt.addChild(d)
91                   item.children += new TreeItem(d)
92                   drawing.draw()
93                 case d =>
94                   treeAdd(item.getParent)
95               }
96               if (treeSelect != null) treeAdd(treeSelect)
97             case None =>
98           }
99         }
100      }
101      exitItem.onAction = (ae: ActionEvent) => {
102        // TODO - Add saving before closing down.
103        sys.exit()
104      }
105
106      // Top Level Setup
107      val rootPane = new BorderPane
108      rootPane.top = menuBar
109      rootPane.center = tabPane
110      root = rootPane
111    }
112  }
113
114  private def makeDrawingTab(): (Drawing, TreeView[Drawable], Tab) = {
115    val canvas = new Canvas(2000, 2000)
116    val gc = canvas.graphicsContext2D
117    val drawing = new Drawing(gc)
118
119    // left side
120    val drawingTree = new TreeView[Drawable]
121    drawingTree.selectionModel.value.selectionMode = SelectionMode.Single
122    drawingTree.root = drawing.makeTree
123    val treeScroll = new ScrollPane
124    drawingTree.prefWidth <== treeScroll.width
```

```scala
125   drawingTree.prefHeight <== treeScroll.height
126   treeScroll.content = drawingTree
127   val propertyPane = new ScrollPane
128   val leftSplit = new SplitPane
129   leftSplit.orientation = Orientation.Vertical
130   leftSplit.items ++= List(treeScroll, propertyPane)
131
132   // right side
133   val slider = new Slider(0, 1000, 0)
134   val rightTopBorder = new BorderPane
135   val canvasScroll = new ScrollPane
136   canvasScroll.content = canvas
137   rightTopBorder.top = slider
138   rightTopBorder.center = canvasScroll
139   val commandField = new TextField
140   val commandArea = new TextArea
141   commandArea.editable = false
142   commandField.onAction = (ae:ActionEvent) => {
143     val text = commandField.text.value
144     if(text.nonEmpty) {
145       commandArea.text = (commandArea.text + ">
              "+text+"\n"+Commands(text,drawing)+"\n").value
146       commandField.text = ""
147     }
148   }
149   val commandScroll = new ScrollPane
150   commandArea.prefWidth <== commandScroll.width
151   commandArea.prefHeight <== commandScroll.height
152   commandScroll.content = commandArea
153   val rightBottomBorder = new BorderPane
154   rightBottomBorder.top = commandField
155   rightBottomBorder.center = commandScroll
156   val rightSplit = new SplitPane
157   rightSplit.orientation = Orientation.Vertical
158   rightSplit.items ++= List(rightTopBorder, rightBottomBorder)
159   rightSplit.dividerPositions = 0.7
160
161   // Top Level
162   val topSplit = new SplitPane
163   topSplit.items ++= List(leftSplit, rightSplit)
164   topSplit.dividerPositions = 0.3
165
166   // Tree Selection
167   drawingTree.selectionModel.value.selectedItem.onChange {
168     val selected = drawingTree.selectionModel.value.selectedItem.value
169     if (selected != null) {
170       propertyPane.content = selected.getValue.propertiesPanel()
171     } else {
172       propertyPane.content = new Label("Nothing selected.")
173     }
174   }
175
176   val tab = new Tab
177   tab.text = "Untitled"
178   tab.content = topSplit
```

```
179     tab.closable = false
180     (drawing, drawingTree, tab)
181   }
182
183   def labeledTextField(labelText: String, initialText: String, action: String =>
          Unit): BorderPane = {
184     val bp = new BorderPane
185     val label = Label(labelText)
186     val field = new TextField
187     field.text = initialText
188     bp.left = label
189     bp.center = field
190     field.onAction = (ae: ActionEvent) => action(field.text.value)
191     field.focused.onChange(if (!field.focused.value) action(field.text.value))
192     bp
193   }
194 }
```

6.7 End of Chapter Material

6.7.1 Summary of Concepts

- The Scala collections libraries go far beyond the sequences like `Array` and `List` that we have looked at previously. There is a significant type hierarchy in related different parts of the hierarchy.

- The `Set` type is based on the mathematical concept of a set. All values in the `Set` are unique and order does not matter. The advantage over a sequence is that the check for membership is fast. There are mutable and immutable `Set` types.

- A `Map` is a collection that allows you to store and look up values using an arbitrary key type. This allows you to store and index values by `Strings`, tuples, or anything else. The type used for keys should really be immutable. There are mutable and immutable `Map` types.

- The `Buffer` is the ultimate mutable sequence type. Values can be mutated like an `Array`, but you are also able to insert or remove elements in a way that alters the size.

6.7.2 Exercises

1. Implement the following function that will build a `Map` from any sequence of a type with a function that can make keys from values.

   ```
   def buildMap[A, B](data:Seq[A], f: A => B): Map[B, A]
   ```

2. Write code to represent a deck of cards. Use the following as a starting point.

   ```
   case class Card(suite: String, value: Char)
   class Deck {
   ```

```
private val cards = mutable.Buffer[Card]()

def dealHand(numCards: Int): List[Card] = ...
def shuffle() { ... }
def split(where: Int) { ... }
}
```

Note that `cards` is a `mutable.Buffer`, so dealing a hand removes cards and the other two methods should change the ordering of elements.

3. In math discussions of probability, it is common to use a "bag of marbles" with different colored marbles as a mental image for illustrating certain concepts. The "bag" is not a `Seq` because there is no ordering. In Scala, it is better modeled as a `Set`.

Write a little script to make a `Set[Marble]` and try using these two definitions of `Marble`.

```
case class Marble(color: String)
class Marble(val color: String)
```

For each one, try making a set of three red and two blue marbles. How do those implementations differ? Why do you think that is?

4. Write code that creates several `Maps` that will display course information. First, create a `Map` that contains course numbers and the room numbers where the courses meet. Next, create a `Map` that contains the instructors and the courses that they teach. Lastly, create a `Map` that contains the meeting times that each course meets. Let the user enter a course number and then display the course's instructor, room number, and meeting time.

5. A standard example of a `Map` is a telephone book. The key is the name. The value would be a `case class` with telephone number, address, etc. You should make a GUI to display the telephone book. For the first cut you can make it so that there can only be one person with a given name and use a `Map[String,Person]`.

If you want a bit of a challenge, make it so that there can be multiple people with the same name. For that you could use a `Map[String,mutable.Buffer[Person]]`.

6. Create a text file that contains all of the states in the United States and their associated capitals. Then write code that loads this information into a `Map`. Use this to create a simple "State Capital" quiz. You should randomly pick a state and ask the user to enter the name of the capital.

6.7.3 Projects

The project ideas for this chapter all have you adding functionality to what you have done before using one of the new types that was introduced.

1. If you are working on the MUD project, one obvious use of the `Map` type is to look up commands that a user types in to get functions that execute the required actions. You could also consider using a `Map` for any other place where something will need to be indexed by a value other than a simple number. For example, items or characters in a room could be indexed by their name, with some care for the fact that you

should allow multiples. The ability to have multiple values could be provided by a `Map[String,mutable.Buffer[Item]]`.

Your program is also keeping track of things like all the rooms. Originally you implemented this using an `Array[Room]` and the links between rooms were stored as `Ints`. You should change it so that each room has a unique keyword associated with it and all the rooms are stored as a `Map[String, Room]`. Change the exits so that they store `Strings` for their destinations instead of `Ints`.

You could also change your code to use a `mutable.Buffer` to store things like the items in a room or the players inventory as those vary over time.

2. If you are working on the web spider, you should probably keep a `Buffer` of the data that you have collected from the various pages. In addition, you definitely need a way to not repeat visiting pages and perhaps to associate data with the page that it came from. The former can be nicely done using a `Set` as it can efficiently tell you if a value is already a member or not. If you want to have the ability to see what data was associated with a particular URL, you could use a `Map` with the URL as the key. The fact that the keys of a `Map` are themselves a `Set`, means that you do not need a separate `Set` if you use a `Map` in this way.

3. The broad variety of possibilities for graphical games, networked or not, makes it hard to predict the ways in which these new collections might be useful. However, the vast majority of possibilities will have entities in them whose numbers vary and could be implemented well with a `Buffer`. You might also find uses for `Maps` for efficient look-up or `Sets` to efficiently demonstrate uniqueness.

4. The math worksheet project will have commands, much like the MUD. At this point, that makes a clear case for using a `Map`. In addition, the worksheet itself is going to be a sequence of different elements that will grow as the user adds things, or possibly shrink when the user takes things away. After reading this chapter, that should sound like a `Buffer`.

5. The Photoshop® project is filled with different aspects that can be nicely implemented with `Buffers`. For example, one of the standard features of software like Photoshop® is to include multiple layers that can be worked on independently. The elements that are drawn on the layers could also be stored in `Buffers`.

6. If you are working on the stock portfolio management project you may want to provide the users with a way of accessing information quickly. A `Map[String, Stock]` would be useful for looking up stock information quickly when the user types in a stock symbol.

7. If you are working on the movies project you can definitely provide some quick lookups by doing `Map[String, Movie]` or `Map[String, Actor]`, etc.

8. If you are working on the simulation workbench project, it makes sense to use a `Buffer` for different simulations that are running at one time. You could link that to a `TabbedPane` that shows the different simulations.

Having a `Buffer` of particles could be interesting for some simulations, but probably not the gravity ones that you have right now.

Another way to add more collections to this project is to add another simulation type altogether. One that you can consider including is a particle-mesh for doing cloth simulations. The way this works is that you have a number of particles that each have a small amount of mass. They are connected with "springs" so that if they move too close to one another or two far away, there is a restoring force. In many ways, this works just like the gravity simulation, the only difference is that the forces only occur between particles that are connected in the mesh, and the force is not a $1/r^2$ type of force.

The most basic implementation of this would use a standard Hook's Law with $F_s = -kx$, where k is the spring constant and x is the offset from the rest length of the spring. The force is directed along the segment joining the particles. A grid of particles with springs between each particle and its neighbors in the same row and column provides a simple setup that you can test. To make it dynamic, consider fixing one or two particles so that forces cannot move them. You might also add a somewhat random forcing to mimic wind.

A more accurate simulation can be obtained by the addition of extra, weaker springs. These would go between each particle and the ones immediately diagonal from it on the grid, as well as those two down along the same row or column. These additional springs give the fabric some "strength" and will make it bend in a more natural way. The simple grid configuration can completely fold over with an unnatural crease.

Chapter 7

Stacks and Queues

The previous chapter extended the depth of your knowledge in libraries so that you have the capability to do more inside of your programs. Now we are going to switch gears and start looking at what goes into constructing data structures and writing some basic libraries. Most of the time you are writing programs, you will use pre-existing libraries. After all, they have been written by experienced people who put a lot of time and effort into tuning them so that they are flexible and efficient.

The fact that you will typically use libraries written by others might tempt you to believe that you really do not need to know what is going on inside of these data structures. On the contrary, understanding the inner working of data structures is an essential skill for any competent developer. One reason for this is that you often have many different options to choose from when you select a data structure and picking the right one requires knowing certain details. For example, so far we have treated the `List` and `Array` types almost interchangeably when we need a `Seq`, with the exception that the `List` type is immutable. In chapter 12 we will see that the two are much more different and that even when we just focus on linked lists, there are many options to choose from.

Knowing how different data structures work and how to implement them is also required if you ever have to implement your own. This might happen for one of several reasons. One is that you are developing for a system where you are required to use a lowerlevel language that does not come with advanced collections libraries. Another is that you have a problem with very specific requirements for which none of the provided structures are really appropriate. If you do not understand and have experience writing data structures, you will almost certainly lack the ability to write an efficient and correct one for a specific task. These are the types of skills that separate passable developers from the really good ones.

In this chapter, we will start our exploration with some of the simplest possible data structures, the stack and the queue. We will also write basic implementations of a stack and a queue that are based on an `Array` for storage and which use parametric polymorphism so that they can be used with any type. Particular attention will be paid to how well these implementations perform, and what we have to do to make sure they do not slow down when larger data sets are being used.

7.1 Abstract Data Types (ADTs)

Implementations of the stack and the queue are found in Scala's collections library because they hold or collect data. They could also be referred to using the term data structure. However, the most accurate term for them is ABSTRACT DATA TYPE (ADT). The "abstract" part has the same meaning as the usage we first saw in chapter 4. An ADT specifies certain operations that can be performed on it, but it does not tell us how they are done. The *how* part can vary across implementations, just as the implementation of abstract methods varies in different subtypes.

Different implementations can have different strengths and weaknesses. One of the learning goals of this book is for you to know the capabilities of different ADTs as well as the strengths and weaknesses of different implementations so that you can make appropriate choices of which is appropriate to use in different situations.

7.2 Operations on Stacks and Queues

What defines a specific ADT is the set of methods that have to be included in it along with descriptions of what they do. Most implementations will go beyond the bare minimum and add extra methods to improve usability, but it is those required methods that matter most. In the case of both the stack and the queue ADT, there are only two methods that are absolutely required. One adds elements and the other removes them.

In the case of the stack, these methods are commonly called **push** and **pop**. For the queue they are called **enqueue** and **dequeue**. These ADTs are so remarkably simple that the removal methods, **pop** and **dequeue**, do not take arguments. You cannot specify what you are taking out. What makes the stack and the queue different from one another is how each selects the item to give you when something is removed. In a stack, the object that is removed by **pop** is the one that was most recently added. We say that it follows a "Last In, First Out" (LIFO) ordering. Calls to **dequeue**, conversely, remove the item that was least recently added with **enqueue**. The queue has a "First In, First Out" (FIFO) ordering.

The names of these ADTs and the methods on them are meant to invoke a mental image. In the case of a stack, you should picture a spring-loaded plate holder like you might see in a cafeteria. When a new plate is added, its weight pushes down the rest of the stack. When a plate is removed, the stack pops up. Following this image, you only have access to the top plate on the stack. Generally these devices prevent you from randomly pulling out any plate that you want.

In many English-speaking parts of the world, the term "queue" is used where Americans use the term "line". When you go to buy tickets for an event, you do not get in a "line", instead you get in a "queue". The queue ADT is intended to function in the same way that a service line should. When people get into such a line, they expect to be serviced in the order they arrive. It is a first-come, first-serve type of structure. The queue should work in this same way with the `dequeue` method pulling off the item that has been waiting in the queue the longest.

7.3 Real Meaning of O

Hopefully your previous exposure to computer science introduced you to the concept of O order. It is used to describe how many operations various algorithms must perform when working on inputs of different sizes. You might have seen it used to describe the relative costs of searches or sorts. We want to explore this concept a bit more deeply here because we are going to use it a lot in discussing the performance of different ADTs and we will also use it to put constraints on what we will accept for implementations of certain ADTs.

While we often care about how long something will take to run in units of time, this really is not a very good way to characterize the performance of programs or parts of programs. The run time can vary from one machine to another and do so in odd ways that are very particular to that hardware. A more stable definition is to consider one or more operations of interest, and look at how the number of times those operations are executed varies with the size of the input set we are working with. Imagine that we have some part of a program that is of interest to us, it might be a single function/method or the running of an entire program. We would run the program on various inputs of different sizes and count how many times the operations of interest occur. We could call the input size n and the number of operations at a particular input size $g(n)$. We say that $g(n)$ is $O(f(n))$, for some function $f(n)$, if $\exists m, c$ such that $c \times f(n) > g(n)$ $\forall n > m$. The best way to see what this means is graphically. Figure 7.1 shows how $g(n)$ and $c \times f(n)$ might look in a plot.

There are some significant things to notice about this definition. First, O is only an upper bound. If $g(n)$ is $O(n)$, then it is also $O(n^2)$ and $O(n^3)$. In other words, the upper bound of an algorithm's complexity describes how the algorithm's execution time changes with a change in the amount of data being processed in the operation. This means that as the amount of data is increased, a more complex algorithm will often take an increasingly long time to execute. In addition, the O behavior of a piece of code only really matters when n gets large. Technically, it is an asymptotic analysis. It deals with the situation where $n \to \infty$. In general this is very significant because so much of what we do on computers deals with large sets of data, and we are always trying to push them to work with more information. However, if you have a problem where you know the size will be small, it is quite possible that using a method with an inferior O behavior could be faster.[1]

You will generally ignore all constant factors c in the definition of O because they have little impact on overall performance. So $2n$ and $100n$ are both $O(n)$, even though the former is clearly better than the latter. Similarly, while $100n$ has a better order than $3n^2$ ($O(n)$ vs. $O(n^2)$), the latter is likely to be more efficient for small inputs.

The O notation also ignores all lower-order terms. The functions $n^3 + 2n^2 + 100$ and $n^3 + \log n$ are both $O(n^3)$. It is only the fastest, growing term that really matters when n gets large. Despite all of the things that technically do not apply in O, it is not uncommon for people publishing new algorithms to include coefficients and lower order terms as they can be significant in the range of input size that individuals might be concerned with.

---

[1] An example of this is the fact that insertion sort is really hard to beat on small data sets, even though sorts like quicksort and merge sort have much better asymptotic performance and are much more efficient on large data sets.

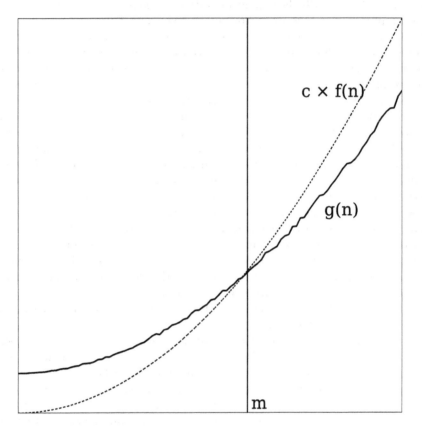

FIGURE 7.1: This figure shows a graphical representation of the meaning of O. The solid line represents $g(n)$, which would be a measure of the amount of computation done for an input size n. The dotted line is the function, $c \times f(n)$, where we want to say that $g(n)$ is $O(f(n))$. This plot labels a position m, above which the dotted curve is always above the solid curve.

7.4 $O(1)$ Requirement

The fact that the stack and queue are so very simple allows us to place a strong limit on the methods that we put into them. We are going to require that all methods we put into our implementations be $O(1)$ on average. That is to say that the amount of work they do should remain constant and not depend on the number of values that have been stored. To understand why we want such a stringent limit, consider what would happen to the speed of a full program if a stack or queue grew to having n elements. In order to get there, and then make that useful, you have to have at least $O(n)$ calls to both the adding and removing methods. Combine the fact that orders are multiplied with a slower implementation that is $O(n)$, and we would get whole program performance that is $O(n^2)$. This is prohibitively expensive for many applications.

The requirement of $O(1)$ performance will only be for the AMORTIZED COST. The amortized cost is the cost averaged over many calls. This means that we are willing to have an occasional call take longer and scale with the size of the data structure as long as this happens infrequently enough that it averages to a constant cost per call. For example, we can accept a call with an $O(n)$ cost as long as it happens only once every $O(n)$ times.

For most ADTs, it is not possible to impose such a strict performance requirement. It is only the simplicity of the stack and the queue that make this possible. For others we will have to accept $O(n)$ behavior for some of the operations or we will be happy when we can make the most expensive operation have a cost of $O(\log n)$.

7.5 Array-Based Stack

Having laid out the groundwork, it is now time to write some code. We will start with the stack ADT, and our implementation in this chapter will use **Arrays** to store the data. For usability, we will include **peek** and **isEmpty** methods on both our stack and queue in addition to the two required methods. The **peek** method simply returns the value that would be taken off next without taking it off. The **isEmpty** method returns **true** if the stack or queue has nothing in it and **false** otherwise.

We start by defining a **trait** that includes our methods. This **trait** has a type parameter so that it can work with whatever type the user wants and provide type safety so that objects of the wrong type cannot be added.

```
package stackqueue.adt

/**
 * This trait defines a mutable Stack ADT.
 * @tparam A the type of data stored
 */
trait Stack[A] {
  /**
   * Add an item to the stack.
   * @param obj the item to add
   */
  def push(obj: A)
```

```
/**
 * Remove the most recently added item.
 * @return the item that was removed
 */
def pop(): A

/**
 * Return the most recently added item without removing it.
 * @return the most recently added item
 */
def peek: A

/**
 * Tells whether this stack is empty.
 * @return true if there are no items on the stack, otherwise false.
 */
def isEmpty: Boolean
}
```

Scaladoc comments are added here for the **trait**. They will be left off of implementations to conserve space, but your own code should generally include appropriate Scaladoc comments throughout. This code should be fairly straightforward. If there is anything worthy of questioning, it is the fact that **pop** is followed by parentheses while **peek** and **isEmpty** are not. This was done fitting with style recommendations that any method that causes a mutation should be followed by parentheses, while methods that simply return a value, and which could potentially be implemented with a **val** or **var**, should not. This makes it clear to others what methods have side effects and allows flexibility in subtypes.

This **trait** effectively defines the ADT in Scala. Once we have it, we can work on a **class** that **extends** the **trait** and provides concrete implementations of the methods. Before we show the code for such an implementation, we should pause to think about how it might work and what it would look like in memory. A possible view of this is shown in figure 7.2. For illustration purposes, this stack stores **Ints**. We have an array that stores the various values as well as an integer value called top that stores the index where the next element should be placed in the stack. Each time a value is pushed, it is placed in the proper location and top is incremented. When a value is popped, top is decremented and the value at that location is returned.

Though this figure does not show it, we also need to handle the situation where enough elements have been pushed that the **Array** fills. In this situation, a larger **Array** needs to be allocated and the values should be copied over so that we have room for more elements.

Code that implements this view of a stack can be seen here.

```scala
 1  package stackqueue.adt
 2
 3  import scala.reflect.ClassTag
 4
 5  class ArrayStack[A: ClassTag] extends Stack[A] {
 6    private var top = 0
 7    private var data = new Array[A](10)
 8
 9    def push(obj: A) {
10      if (top >= data.length) {
11        val tmp = new Array[A](data.length * 2)
12        Array.copy(data, 0, tmp, 0, data.length)
13        data = tmp
```

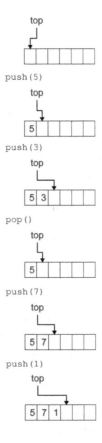

FIGURE 7.2: This figure shows a graphical representation of an `Array`-based stack and how it changes over several operations. The stack stores data in the `Array` and uses an integer index to keep track of the position of the top.

```scala
14      }
15      data(top) = obj
16      top += 1
17    }
18
19    def pop(): A = {
20      assert(!isEmpty, "Pop called on an empty stack.")
21      top -= 1
22      data(top)
23    }
24
25    def peek: A = data(top - 1)
26
27    def isEmpty: Boolean = top == 0
28  }
```

We will run through this beginning at the top. The `class` is called `ArrayStack` and it extends `Stack`. The `ArrayStack` takes a type parameter called `A` that is passed through to `Stack`. This type parameter is modified in a way we have not seen before; it is followed with `: ClassTag`. This is required for us to make an `Array` of `A` using `new Array[A](...)`. The details of why this is needed are explained below.

Inside the `class` on lines 6 and 7 we have two `private var` declarations for `top` and the `Array` called `data`. These are both `private` because they are mutable, and having outside code make changes to them would cause the stack to have inappropriate behavior. The implementation of `pop(): A` decrements `top` and returns the data that is there. Both `peek` and `isEmpty` are even simpler. All of these methods are clearly $O(1)$ because they do math on an integer and direct access on an `Array`, two operations that occur in constant time.

The only complex method is `push(obj: A)`. This method ends with two lines that actually place the object in the `Array` and increment `top`. The first part of the method is a check to make sure that there is space to place the object in the `Array`. If there is not, a new `Array` is made and the contents of the current `Array` are copied into the new one, then the reference `data` is changed to point to the new `Array`.[2] Everything here is $O(1)$ except the copy. That line takes $O(n)$ time because it has to move $O(n)$ values from one `Array` to another.

The obvious question is, will this average out to give us an amortized cost of $O(1)$? For this code, the answer is yes, but it is very important to see why, as it would have been very easy to write a version that has an amortized cost of $O(n)$. The key is in how we grow the `Array`. In this code, every time we need more space, we make the `Array` twice as big. This might seem like overkill, and many would be tempted to instead add a certain amount each time. Perhaps make the `Array` 10 slots bigger each time. The following table runs through the first six copy events using these two approaches. At the bottom it shows generalized formulas for the mth copy event.

---

[2]Instead of using the `for` loop, the contents of the `Array` could be copied with the line `Array.copy(data, 0, tmp, 0, data.length)`.

Resize Event	data.length*2		data.lenth+10	
	Size	Total Work	Size	Total Work
1	10	10	10	10
2	20	30	20	30
3	40	70	30	60
4	80	150	40	100
5	160	310	50	150
6	320	630	60	210
...	...	...	...	...
m	10×2^m	$10 \times (2^{m+1} - 1)$	$10m$	$5m(m+1)$

To get the amortized cost, we divide the total work by the size because we had to call push that number of times to get the stack to that size. For the first approach we get

$$\frac{10 \times (2^{m+1} - 1)}{10 \times 2^m} = \frac{2^{m+1} - 1}{2^m} = 2 - \frac{1}{2^m} \approx 2 \in O(1).$$

For the second approach we get

$$\frac{5m(m + 1)}{10m} = \frac{m + 1}{2} \in O(m).$$

Neither the multiple of 2, nor the addition of 10 is in any way magical. Growing the array by a fixed multiple greater than one will lead to amortized constant cost. Growing the array by any fixed additive value will lead to an amortized linear cost. The advantage of the fixed multiple is that while the copy events get exponentially larger, they happen inversely exponentially often.

ClassTags (Advanced)

To understand **ClassTags**, it helps to have some understanding of the history of Java as the nature of the Java Virtual Machine (JVM) plays a role in their existence. The parallel to Scala-type parameters in Java is generics. Generics did not exist in the original Java language. In fact, they were not added until Java 5. At that point, there were a huge number of JVMs installed around the globe and it was deemed infeasible to make changes that would invalidate all of them. For that reason, as well as a motivation to keep the virtual machine fairly simple, generics were implemented using something called "type erasure". What this means is that the information for generics was used at compile time, but was erased in the compiled version so that at run time it was impossible to tell what the generic type had been. In this way, the code that was produced looked like what would have been produced before generics were introduced.

This has quite a few effects on how Java programs are written. It comes through in Scala as a requirement for having a **ClassTag** on certain type parameters. The **ClassTag** basically holds some of the type information so that it is available at run time. To understand why we need this, we should go back to the expression that was responsible for us needing a **ClassTag**: new **Array[A](10)**. This makes a new **Array[A]** with 10 elements in it and sets the 10 elements to appropriate default values. It is the last part that leads to the requirement of having a **ClassTag**. For any type that is a subtype of **AnyRef**, the default value is **null**. For the subtypes of **AnyVal**, the default varies. For **Int**, it is 0. For **Double** it is 0.0. For **Boolean** it is **false**. In order to execute that expression, you have to know at least enough about the type to put in proper default values.[3] Type erasure takes away that information from the run time. The `ClassTag` allows Scala to get back access to it.

It is also worth noting that our implementation of the stack technically does not actually remove the value from the array when `pop()` is called. This remove is not really required for the stack to execute properly. However, it can improve memory usage and performance, especially if the type `A` is large. Unfortunately, that would require having a default value to put into the array in that place, which we have just discussed it hard for a completely general type parameter.

7.6 Array-Based Queue

Like the stack, the code for the `Array`-based queue begins by setting up the ADT as a completely abstract `trait`. This `trait`, which takes a type parameter so that it can work with any type, is shown here.

```
package stackqueue.adt

/**
 * This trait defines a mutable Queue ADT.
 * @tparam A the type of data stored
 */
trait Queue[A] {
  /**
   * Add an item to the queue.
   * @param obj the item to add
   */
  def enqueue(obj: A)

  /**
   * Remove the item that has been on the longest.
   * @return the item that was removed
   */
  def dequeue(): A

  /**
   * Return the item that has been on the longest without removing it.
   * @return the most recently added item
   */
  def peek: A

  /**
   * Tells whether this queue is empty.
   * @return true if there are no items on the queue, otherwise false.
   */
  def isEmpty: Boolean
}
```

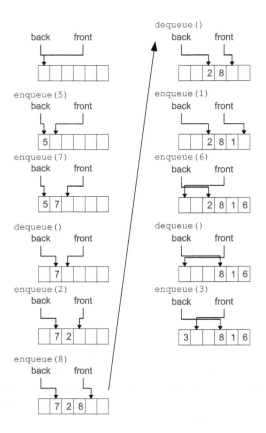

FIGURE 7.3: This figure shows a graphical representation of an `Array`-based queue and how it changes over several operations. The queue stores data in the `Array` and uses integer indexes to keep track of the front and back.

As with the stack, this has only the signatures of the methods along with comments describing what they do.

The queue ADT can also be implemented using an `Array`. Figure 7.3 shows how this might look for several operations. Unlike the stack, the queue needs to keep track of two locations in the `Array`: a front and a back. The back is where new elements go when something is added to the queue and the front is where values are taken from when they are removed. Both indexes are incremented during the proper operations. When an index moves beyond the end of the `Array`, it is wrapped back around to the beginning. This allows us to continually add and remove items without having to grow the `Array` or copy elements around. Both of those would either require $O(n)$ operations and prevent our total performance from being $O(1)$ or would lead to unbounded memory consumption, even when the queue never has more than a few elements in it.

This figure does not show what happens when the queue gets full. At that point a bigger `Array` is needed and values need to be copied from the old `Array` into the new one. The details for the queue are not quite as straightforward as they were for the stack. Figure 7.4 demonstrates the problem and a possible solution. Doing a direct copy of elements is not appropriate because the wrapping could result in a situation where the front and back are inverted. Instead, the copy needs to shift elements around so that they are contiguous. The

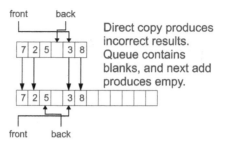

Direct copy produces incorrect results. Queue contains blanks, and next add produces empy.

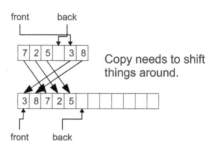

Copy needs to shift things around.

FIGURE 7.4: This figure shows a graphical representation of an `Array`-based queue when it needs to grow on an enqueue. The queue is seen as empty if `front==back`. For that reason, the grow operation should happen when the last slot will be filled. In addition to allocating a bigger `Array`, the copy operation needs to move elements of the `Array` around and alter the location of front and back.

values of front and back need to be adjusted appropriately as well. There are many ways that this could be done. Figure 7.4 demonstrates the approach that we will take in the code here where the front is moved back to the beginning of the `Array`.

This graphical description of implementing an `Array`-based queue can be implemented in code like the following.

```
package stackqueue.adt

import scala.reflect.ClassTag

class ArrayQueue[A: ClassTag] extends Queue[A] {
  private var front, back = 0
  private var data = new Array[A](10)

  def enqueue(obj: A) {
    if ((back + 1) % data.length == front) {
      val tmp = new Array[A](data.length * 2)
      for (i <- 0 until data.length - 1) tmp(i) = data((i + front) % data.length)
      front = 0
      back = data.length - 1
      data = tmp
    }
    data(back) = obj
    back = (back + 1) % data.length
  }
```

```
  def dequeue(): A = {
    assert(!isEmpty, "Dequeue called on an empty queue.")
    val ret = data(front)
    front = (front + 1) % data.length
    ret
  }

  def peek: A = data(front)

  def isEmpty: Boolean = front == back
}
```

This code is very similar to what we had for the stack. The only element that you might find odd is that the wrapping from the end of the `Array` back to the beginning is done with a modulo. The wrap in dequeue and the one at the end of enqueue could easily be done with an `if`. For the other two, the wrapping needs to be part of an expression; this makes the modulo far more effective. If you are not happy using modulo, you are more than welcome to rewrite this code using `if` statements. We expect that once you have done so, you will see why we prefer modulo.

All the operations have an amortized cost of $O(1)$. As with `push` on the stack, the `enqueue` method will be $O(n)$ on the calls where it has to grow. However, those are rare enough that they average out to constant time.

7.7 Unit Tests

As you are well aware by this time, writing code that compiles does not mean that you have code that is correct. There can still be runtime and logic errors present in the code. In order to find these, you need to test the code. It is quite possible that up to this point you have been using manual tests. This is where each test has to be run explicitly by the developer to test the code. In addition, some of the tests would print values that had to be verified for correctness by the programmer. In large-scale development, these approaches fall short of helping us to produce reliable code. In this section we will address this by introducing the concept of automated UNIT TESTS.

A unit test is a test that checks the functionality of some particular unit of a program. Typically it is something like a single class or even some methods in a class. The unit test is intended to be run automatically and, as such, does not rely on programmer interaction to determine if the test passed or failed. That logic should be built into the test. This allows unit tests to be run frequently. The idea is that every time you make changes to the code, you rerun all the unit tests to make sure that the changes did not break anything.

There are many types of unit testing frameworks. Some, such as ScalaTest, ScalaCheck, and Specs, have been written specifically for Scala. In this book we will use JUnit, the primary unit testing framework for Java. The reasons for this choice include the fact that JUnit is broadly used, it is easy to use, and it has support built into Eclipse and other IDEs. In many ways, ScalaTest is a more powerful framework, but it is not used as broadly, and at this point it benefits you more to see a basic implementation of how unit testing is done.

7.7.1 Setup

If you are using Eclipse, you can add JUnit support to your project by right-clicking on the project and selecting Build Path > Add Libraries. You will see an option for JUnit listed there. You can find documentation for JUnit at `http://www.junit.org/`. This includes a JavaDoc API that you will likely find helpful.

7.7.2 Writing Tests

Unlike tests that we would put in a `main` method in a companion object, unit tests are put in separate files and even in separate packages or projects. We will take the approach of putting all test code in a parallel package hierarchy that begins with `test`. As the `ArrayStack` and `ArrayQueue` were in the `stackqueue.adt` package, our tests will be in `test.stackqueue.adt`. To make the tests we simply write classes with methods and annotate our test methods with the `@Test` annotation from JUnit. For that to compile, you should do an import for `org.junit._`. Every method that you put the `@Test` annotation on will be run as part of the unit testing.

Inside of the test methods, you need to have calls to JUnit `assert` methods. There are methods such as `assertTrue`, `assertFalse`, and `assertEquals`. Typically, tests are written building up from those that test very simple requirements to more significant ones. To understand how simple, consider the following examples.

```
@Test def emptyOnCreate {
  val stack = new ArrayStack[Int]
  assertTrue(stack.isEmpty)
}

@Test def nonEmptyOnPush {
  val stack = new ArrayStack[Int]
  stack.push(5)
  assertFalse(stack.isEmpty)
}

@Test def pushPop {
  val stack = new ArrayStack[Int]
  stack.push(5)
  assertEquals(5,stack.pop)
}
```

The first two only check that a new stack is empty and that after something has been pushed it is not empty. It is not until the third test that we even pop the stack and check that we get back the correct value.

You should notice that all of these methods begin with the same line. Code duplication like this is frowned on in testing as much as it is in normal code development. However, because JUnit runs every test separately, you have to use an annotation to make sure that the code you want to run happens before each and every test. The annotation for this purpose is `@Before`. If you have code that needs to follow every test to clean up after something that happens in the "before" method, there is also an `@After` annotation. You can see these used, along with a more complete set of tests for the `ArrayStack`, in this code.

```
package test.stackqueue.adt

import org.junit._
import org.junit.Assert._
```

```scala
import stackqueue.adt._

class TestArrayStack {
  var stack: Stack[Int] = null

  @Before def initStack {
    stack = new ArrayStack[Int]
  }

  @Test def emptyOnCreate {
    assertTrue(stack.isEmpty)
  }

  @Test def nonEmptyOnPush {
    stack.push(5)
    assertFalse(stack.isEmpty)
  }

  @Test def pushPop {
    stack.push(5)
    assertEquals(5, stack.pop)
  }

  @Test def pushPopPushPop {
    stack.push(5)
    assertEquals(5, stack.pop)
    stack.push(3)
    assertEquals(3, stack.pop)
  }

  @Test def pushPushPopPop {
    stack.push(5)
    stack.push(3)
    assertEquals(3, stack.pop)
    assertEquals(5, stack.pop)
  }

  @Test def push100Pop100 {
    val nums = Array.fill(100)(util.Random.nextInt)
    nums.foreach(stack.push(_))
    nums.reverse.foreach(assertEquals(_, stack.pop))
  }
}
```

These tests are incomplete. It is left as an exercise for the student to come up with and implement additional tests.

With these tests written, now we just need to run them and make certain that our implementation passes all the tests. If you select a run option for that file in Eclipse, you should have an option for "JUnit Test". When you select that, you should see something similar to what is shown in Figure 7.5. This shows the six tests shown above all passing. When everything passes, we get a green bar. If any tests fail, there would be a red bar and the tests that fail would be shown in red. You can click on failed tests to find out what caused them to fail. Tests can fail because of an assert or because they throw an exception when they should not.

FIGURE 7.5: This is what Eclipse shows when a JUnit test is run and all tests are passed. A green bar means that everything is good. If you get a red bar, then one or more tests have failed and they will be listed for you.

Similar tests can be written for our queue. They might look like the following.

```scala
package test.stackqueue.adt

import org.junit._
import org.junit.Assert._
import stackqueue.adt._

class TestArrayQueue {
  var queue: Queue[Int] = null

  @Before def initQueue {
    queue = new ArrayQueue[Int]
  }

  @Test def emptyOnCreate {
    assertTrue(queue.isEmpty)
  }

  @Test def nonEmptyOnEnqueue {
    queue.enqueue(5)
    assertFalse(queue.isEmpty)
  }

  @Test def enqueueDequeue {
    queue.enqueue(5)
    assertEquals(5, queue.dequeue)
  }

  @Test def enqueueDequeueEnqueueDequeue {
    queue.enqueue(5)
    assertEquals(5, queue.dequeue)
    queue.enqueue(3)
    assertEquals(3, queue.dequeue)
  }

  @Test def enqueueEnqueueDequeueDequeue {
    queue.enqueue(5)
    queue.enqueue(3)
    assertEquals(5, queue.dequeue)
    assertEquals(3, queue.dequeue)
```

```
  }

  @Test def enqueue100Dequeue100 {
    val nums = Array.fill(100)(util.Random.nextInt)
    nums.foreach(queue.enqueue(_))
    nums.foreach(assertEquals(_, queue.dequeue))
  }
}
```

You can check these as well and see that our **Array**-based implementation of a queue passes all of these tests.

7.7.3 Test Suites

The purpose of automated testing is to make it easy to run all the tests regularly during development. Having the ability to run all of your tests on a regular basis makes you more confident when you go to make changes to your code. If you run the tests and they pass, then make a change and one or more fail, you know exactly what caused the failure and it will be easier to locate and diagnose. If you make that change and everything passes, you feel a certain level of confidence that the change you made did not break anything.

Unfortunately, going through every test class and running it would be just as much of a pain as going through a large number of objects that have a **main** defined for testing. To consolidate your test runs, you can make a TEST SUITE. This lets you specify a group of different **classes** that contain test code that should be run together. To build a test suite you make a **class** and annotate it with the **@RunWith** and **@Suite.SuiteClasses** annotations. Here is an example of a suite that contains the two tests we just wrote as well as a third that appears later in this chapter.

```
package test.stackqueue

import org.junit.runner.RunWith
import org.junit.runners.Suite

@RunWith(classOf[Suite])
@Suite.SuiteClasses(Array(
  classOf[test.stackqueue.adt.TestArrayStack],
  classOf[test.stackqueue.adt.TestArrayQueue],
  classOf[test.stackqueue.util.TestRPNCalc]))
class AllTests {}
```

Note that the **class** does not actually contain anything at all.[4] It is the annotations that tell JUnit all the work that should be done here. As we add more test **classes**, we will also add them to our suite.

7.7.4 Test-Driven Development

Unit tests are a central part of an approach to development called "Test-Driven Development" (TDD). This style of development is highlighted by the idea that you write tests for code before you write the actual code. In a TDD environment, a programmer who wants to write a new **class** would create the **class** and put in blank methods which have just

---

[4]It is tempting to remove the curly braces here, but doing so causes problems for the JUnit test suite loader.

enough code to get the `class` to compile. Methods that return a value are set to return **???** in Scala or default values such as `null`, 0, or `false` in Java. No logic or data is put in.

After that, the developer writes a basic test for the `class` and runs the test to see that it fails. Then just enough code is added to make that test pass. Once all existing tests pass, another test is written which fails and this process is repeated until a sufficient set of tests have been written to demonstrate that the `class` must meet the desired requirements in order to pass them all.

While this textbook does not follow a TDD approach, there are some significant benefits to it. The main one is that it forces a developer to take small steps and verify that everything works on a regular basis. Students are prone to write large sections of code without testing that they even compile, much less run. This approach to software development inevitably leads to many headaches. You need to focus on a small task, write the code for it, and then make sure that it works. TDD also gives you a feeling of security when you are making changes because you generally feel confident that if you make a change which breaks something, it will immediately show up as a red bar in testing. This can be important for novice programmers who often leave large sections of code in comments because they are afraid to delete them in case something goes wrong. For this reason, the literature on TDD refers to tests as giving you the "courage" to make changes.

There is a lot more to unit testing that we have not covered here. Interested readers are encouraged to look at the JUnit website or look into Scala-based testing frameworks such as ScalaTest. Full development needs more than just unit testing as well. Unit tests, by their nature, insure that small pieces of code work alone. To feel confident in a whole application, you need to test the parts working together. This is called integration testing and it too is done on a fairly regular basis in TDD, but it involves a different set of tools that we do not cover in this book.

7.8 RPN Calculator

A classic example usage of a stack is the Reverse Polish Notation (RPN) calculator. In RPN mode, operations are entered after the values that they operate on. This is also called post-fix notation. The RPN representation of "2 + 3" is "2 3 +". While this might seem like an odd way to do things, it has one very significant advantage, there is no need for parentheses to specify order of operation. Using standard notation; also called in-fix notation, the expression "2 + 3 * 5" and "5 * 3 + 2" are both 17 because multiplication occurs before addition, regardless of the order things are written. If you want the addition to happen first, you must specify that with parentheses, "(2 + 3) * 5". Using post-fix notation, we can say "3 5 * 2 +" to get 17 or "2 3 + 5 *" if we want the addition to happen first so we get 25. This lack of parentheses can significantly reduce the length of complex expressions which can be appealing when working on such things.

The evaluation of a post-fix expression can be done quite easily with a stack that holds the numbers being worked with. Simply run through the expression, and for each element you do one of the following. If it is a number, you push that number onto the stack. If it is an operator, you pop two elements from the stack, perform the proper operation on them, and then push the result back on the stack. At the end there should be one value on the stack which represents the final result. We could implement this basic functionality using the following `object`.

```scala
object RPNCalc {
```

```scala
def apply(args: Seq[String]): Double = {
  val stack = new ArrayStack[Double]
  for(arg <- args; if (arg.nonEmpty)) arg match {
    case "+" => stack.push(stack.pop + stack.pop)
    case "*" => stack.push(stack.pop * stack.pop)
    case "-" =>
      val tmp = stack.pop
      stack.push(stack.pop - tmp)
    case "/" =>
      val tmp = stack.pop
      stack.push(stack.pop / tmp)
    case x => stack.push(x.toDouble)
  }
  stack.pop
}
}
```

The arguments are passed in as a sequence that is traversed using a **for** loop. The subtract and divide operators are slightly different from those for addition and multiplication simply because they are not commutative; the order of their operands matters. This code also assumes that anything that is not one of the four operators is a number.

There are a few options that would be nice to add to this basic foundation. For example, it would be nice to have a few other functions beyond the standard four arithmetic operations, such as trigonometric functions and a square root. It would also be nice to have support for variables so that not every value has to be a literal numeric value. An **object** that supports these additions is shown here.

```scala
package stackqueue.util

import stackqueue.adt.ArrayStack

object RPNCalc {
  def apply(args: Seq[String], vars: collection.Map[String, Double]): Double = {
    val stack = new ArrayStack[Double]
    for (arg <- args; if (arg.nonEmpty)) arg match {
      case "+" => stack.push(stack.pop + stack.pop)
      case "*" => stack.push(stack.pop * stack.pop)
      case "-" =>
        val tmp = stack.pop
        stack.push(stack.pop - tmp)
      case "/" =>
        val tmp = stack.pop
        stack.push(stack.pop / tmp)
      case "sin" => stack.push(math.sin(stack.pop))
      case "cos" => stack.push(math.cos(stack.pop))
      case "tan" => stack.push(math.tan(stack.pop))
      case "sqrt" => stack.push(math.sqrt(stack.pop))
      case v if (v(0).isLetter) => stack.push(try { vars(v) } catch { case ex:
          NoSuchElementException => 0.0 })
      case x => stack.push(try { x.toDouble } catch { case ex:
          NoSuchElementException => 0.0 })
    }
    stack.pop
  }
}
```

This has added the three main trigonometric functions along with a square root function. These operations pop only a single value, apply the function, and push back the result. This code also includes an extra argument of type `Map[String, Double]` to store variables by name. In this case, the `scala.collection.Map` type is used. This is a supertype to both the mutable and immutable maps. This code does not care about mutability so this provides additional flexibility. Lastly, some error handling was added so that if a value comes in that is neither a number, nor a valid variable, we push 0.0 to the stack. This is done using `try/catch` as an expression. If the `try` succeeds, it has the value of the block. If it throws an exception that is caught by a `case` in the `catch`, then the value from that expression is used.

Before we can proclaim that this code works, we need to test it.[5] Here is a possible test configuration.

```
package test.stackqueue.util

import org.junit._
import org.junit.Assert._
import stackqueue.util.RPNCalc

class TestRPNCalc {
  @Test def basicOps {
    assertEquals(5, RPNCalc("2 3 +".split(" "), null), 0.0)
    assertEquals(6, RPNCalc("2 3 *".split(" "), null), 0.0)
    assertEquals(3, RPNCalc("6 2 /".split(" "), null), 0.0)
    assertEquals(1, RPNCalc("3 2 -".split(" "), null), 0.0)
  }

  @Test def twoOps {
    assertEquals(20, RPNCalc("2 3 + 4 *".split(" "), null), 0.0)
    assertEquals(3, RPNCalc("2 3 * 3 -".split(" "), null), 0.0)
    assertEquals(5, RPNCalc("6 2 / 2 +".split(" "), null), 0.0)
    assertEquals(0.25, RPNCalc("3 2 - 4 /".split(" "), null), 0.0)
  }

  @Test def vars {
    val v = Map("x" -> 3.0, "y" -> 2.0)
    assertEquals(20, RPNCalc("2 x + 4 *".split(" "), v), 0.0)
    assertEquals(3, RPNCalc("y 3 * x -".split(" "), v), 0.0)
    assertEquals(5, RPNCalc("6 2 / y +".split(" "), v), 0.0)
    assertEquals(0.25, RPNCalc("x y - 4 /".split(" "), v), 0.0)
  }

  @Test def specials {
    val v = Map("pi" -> 3.14159, "x" -> 3.0, "y" -> 2.0)
    assertEquals(0, RPNCalc("pi cos 1 +".split(" "), v), 1e-8)
    assertEquals(math.sqrt(2) + 3, RPNCalc("y sqrt x +".split(" "), v), 1e-8)
    assertEquals(1, RPNCalc("pi 2 / sin".split(" "), v), 1e-8)
    assertEquals(0.0, RPNCalc("x y y 2 / + - tan".split(" "), v), 1e-8)
  }
}
```

As with the earlier tests, these are fairly minimal. There are literally an infinite number of

---

[5]It is worth noting that there was a bug in this code as originally written by the authors that was caught by these tests.

tests that could be run to verify the RPN calculator. This set simply picks a few different tests related to the different types of functionality the calculator is supposed to support.

True unit testing adherents would prefer that each **assertEquals** be split into a separate method because none depends on the results before it. This is not just a style issue either; there are valid functionality benefits to that approach. JUnit will run all the test methods independently, but each test method will stop when an assert fails. So if the processing of "2 3 +" were to fail, we would not get results from the other three lines of that test as it is written now. That makes it harder to know why it failed. If "2 3 +" fails and all the other single operator calls pass, we can feel fairly confident the problem is with the handling of "+". As it is written now, we will not get that extra information. The only reason for using the format shown here is that it takes less than half as many lines so the code is more readable in printed format.

This **RPNCalc** object can also be integrated into the drawing program as a command. To make use of the variables, we add an extra **val** to the **Drawing** class, called **vars**, that is a **mutable.Map[String, Double]**. Then we add two commands called "rpn" and "set". The first invokes the **RPNCalc object** with the specified arguments. The second allows you to set variables in the **Map** for that drawing. Here is code for the modified version of **Commands**

```scala
package stackqueue.drawing

import stackqueue.util.RPNCalc

object Commands {
  private val commands = Map[String, (String, Drawing) => Any](
    "add" -> ((rest, d) => rest.trim.split(" +").map(_.toInt).sum),
    "echo" -> ((rest, d) => rest.trim),
    "refresh" -> ((rest, d) => d.draw()),
    "rpn" -> ((rest,d) => (RPNCalc(rest.trim.split(" +"), d.vars))),
    "set" -> ((rest,d) => {
      val parts = rest.trim.split(" +")
      d.vars(parts(0)) = parts(1).toDouble
      parts(0)+" = "+parts(1)
    }))

  def apply(input: String, drawing: Drawing): Any = {
    val spaceIndex = input.indexOf(' ')
    val (command, rest) = if (spaceIndex < 0) (input.toLowerCase(), "")
      else (input.take(spaceIndex).toLowerCase(), input.drop(spaceIndex))
    if (commands.contains(command)) commands(command)(rest, drawing) else "Not a
      valid command."
  }
}
```

Thanks to the flexibility of the original design, there is very little effort required for adding two additional commands. Adding the **vars** member to **Drawing** is also simple at this point. We just add the following line to the file.

```scala
private[drawing] val vars = mutable.Map[String, Double]()
```

Note the visibility here is specified as **private[drawing]**. This uses the option to specify a level of visibility that was introduced on page 93. Putting **drawing** in brackets after **private** allows the other code in the **drawing** package to access the variable, but keeps it hidden from all other parts of the code.

7.9   Breadth-First Search

One of the most common activities done on computers is searching. The term search can have many meanings, depending on the context, but in all of them, the idea is that there are many possibilities, and we need to identify those that meet a certain criteria. It is possible that you have previously explored the idea of searching for data in a sequence like an `Array`. In this section, we want to consider a different type of searching. We want to search through a number of different possible solutions to a problem. This is a broad area so we will focus on something that everyone should be able to identify with—finding a way through a maze.

There are two broad approaches to solving a problem like this. They are called breadth-first search and depth-first search. We will consider breadth-first search here, because it is a good example use of a queue. We will return to depth-first search in several later chapters. The idea of breadth-first search is that given some starting condition, you try everything that is one step away from that position, followed by everything that is two steps away, etc. In a depth-first search, you would take one path all the way down many steps, then backtrack and try other paths, branching off as needed.

We will create a type called `DrawMaze` to fit this into our drawing program. As the name implies, we are going to make this a `Drawable` so that we can see the maze and the shortest path to it displayed graphically. Here is the code for the full class.

```scala
package stackqueue.drawing

import collection.mutable
import scalafx.Includes._
import scalafx.scene.layout.VBox
import scalafx.scene.canvas.GraphicsContext
import scalafx.scene.Node
import scalafx.scene.paint.Color
import scalafx.scene.control.Button
import scalafx.event.ActionEvent
import stackqueue.adt.ArrayQueue

class DrawMaze(d: Drawing) extends Drawable(d) {

  private val maze = Array(
    Array(0, 1, 0, 0, 0, 0, 0, 0, 0, 0),
    Array(0, 1, 0, 0, 1, 0, 1, 0, 1, 0),
    Array(0, 1, 0, 1, 1, 0, 1, 0, 1, 0),
    Array(0, 1, 0, 0, 1, 0, 1, 0, 1, 0),
    Array(0, 1, 1, 0, 1, 0, 1, 0, 1, 0),
    Array(0, 0, 0, 0, 0, 0, 1, 0, 1, 0),
    Array(0, 1, 1, 1, 1, 1, 1, 0, 1, 1),
    Array(0, 1, 0, 0, 0, 0, 0, 0, 0, 0),
    Array(0, 1, 0, 1, 1, 1, 1, 0, 1, 0),
    Array(0, 0, 0, 1, 0, 0, 0, 0, 1, 0))

  private val gridWidth = 20
  private val offsets = Seq((0, -1), (1, 0), (0, 1), (-1, 0))

  private var drawVisited = Set[(Int, Int)]()
  private var startX = 0
```

```
32    private var startY = 0
33    private var endX = 9
34    private var endY = 9
35
36    private var propPanel: Option[Node] = None
37
38    override def toString = "Maze"
39
40    def draw(gc: GraphicsContext): Unit = {
41      gc.fill = Color.Black
42      gc.fillRect(-1, -1, 2 + maze(0).length * gridWidth, 2 + maze.length * gridWidth)
43      for (j <- maze.indices; i <- maze(j).indices) {
44        if (maze(j)(i) == 0) {
45          gc.fill = if (drawVisited(i -> j)) Color.Green else Color.White
46        } else {
47          gc.fill = Color.Black
48        }
49        gc.fillRect(i * gridWidth, j * gridWidth, gridWidth, gridWidth)
50      }
51    }
52
53    def propertiesPanel(): Node = {
54      if (propPanel.isEmpty) {
55        val panel = new VBox
56        val bfs = new Button("Breadth First Search")
57        bfs.onAction = (ae: ActionEvent) => {
58          breadthFirstShortestPath() match {
59            case None => drawVisited = Set()
60            case Some(lst) => drawVisited = lst.toSet
61          }
62          drawing.draw()
63        }
64        panel.children = List(bfs)
65        propPanel = Some(panel)
66      }
67      propPanel.get
68    }
69
70    def breadthFirstShortestPath(): Option[List[(Int, Int)]] = {
71      val queue = new ArrayQueue[List[(Int, Int)]]
72      queue.enqueue(List(startX -> startY))
73      var solution: Option[List[(Int, Int)]] = None
74      val visited = mutable.Set[(Int, Int)]()
75      while (!queue.isEmpty && solution.isEmpty) {
76        val steps @ ((x, y) :: _) = queue.dequeue
77        for ((dx, dy) <- offsets) {
78          val nx = x + dx
79          val ny = y + dy
80          if (nx >= 0 && nx < maze(0).length && ny >= 0 && ny < maze.length &&
81              maze(ny)(nx) == 0 && !visited(nx -> ny)) {
82            if (nx == endX && ny == endY) {
83              solution = Some((nx -> ny) :: steps)
84            }
85            visited += nx -> ny
86            queue.enqueue((nx -> ny) :: steps)
```

```
86          }
87        }
88      }
89      solution
90    }
91
92  }
```

The maze itself is declared as a 2D array on lines 15–25. The locations with a 0 are open and the location, with a 1 are walls that you cannot go through. Lines 31–34 define the locations to treat as the start and the end of the maze. They are initially set so the start is at the top left and the end is at the bottom right.

The `draw` method is mostly a `for` loop that runs through the maze drawing squares. The colors of the squares are determined by whether a position is open or a wall. For open positions, a `Set` called `drawVisited` is checked to see if it was part of the solution so that we can see where the solution went. The `propertiesPanel` method adds a `Button` that has an `onAction` handler which calls `breadFirstShortestPath`, sets the value of `drawVisited`, and tells the drawing to redraw.

The part of this code that we care most about in this section is the last method, `breadthFirstShortestPath`. It does not take any arguments because it uses values stored in the `class`. It produces an `Option[List[(Int, Int)]]` that will hold the path for the solution, if one exists. Line 71, the first line in the method, creates an instance of our `ArrayQueue` and calls it `queue`. The next line adds the starting location to this queue. The contents of the queue are lists of x, y coordinate tuples. Each new location that is checked is consed to the front of the list from the step before it, so the "current location" is always the head of the list.

Line 73 declares a `var` called `solution` that will hold the shortest path if one is found. If there is no solution, the initial value of `None` will still be there when the end of the method is reached. Line 74 declares a `mutable.Set` called `visited` whose contents are x, y positions. As the name implies, this is a set of all the places that have been reached. This is a vital piece of the algorithm. We have to include some way of preventing the code from going backward over places it has already been. A `Set` is efficient for this purpose, because all we are doing is adding things to it and then checking if certain values are contained in it. Those are the exact operations that `Sets` perform well.

Line 75 starts the `while` loop that does the real work of the method. It is supposed to keep going as long as the queue is not empty and we have not yet found a solution. The breadth-first approach to searching works well for shortest path because it deals with every location that is n steps from the start point in order. So the first time that you reach a location, you know that you have gotten there in the fewest number of possible steps. Problems that do not have this type of property are poorly suited for breadth-first search.

Inside the `while` loop on line 76 there is a call to `dequeue` that is assigned to a reasonably complex pattern. The pattern is `steps @ ((x, y) :: _)`. Recall that the `@` symbol is used in patterns when we want to give something a name, but also break it down further with a more detailed pattern. So the `List` that we dequeue is given the name `steps`. The pattern to the right of the `@` is a `List` matching pattern by virtual of the `::`. This gives the names `x` and `y` to the values in the head of the list, and does not give any name to the tail of the list as the underscore matches anything without giving it a name.

After the dequeue, there is a `for` loop that goes through the members of `offsets`, which was declared up on line 28. It is a sequence of tuples for x, y offsets for going up, right, down, and left, in that order. Having this loop prevents us from having to duplicate code four times for testing the different directions we can move from the current position.

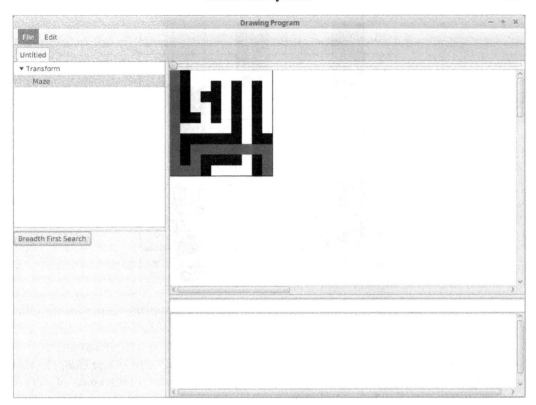

FIGURE 7.6: This is the drawing program after adding a maze and executing a breadth-first solution.

After introducing the `vals` `nx` and `ny` to store the location that is one step away from `x`, `y`, line 80 checks that the `nx`, `ny` location is in the bounds of the maze, not on a wall, and not a place that we have already reached. If that is the case, we can go there. Line 81 checks if we have gotten to the end of the maze. If so, it stores the path to get there in `solution`. Line 84 adds this location to the `visited Set` and line 85 enqueues the path into the queue.

By adding a line to create a `DrawMaze` in the declaration of `creators` in `DrawingMain` we can see this code in action in our program. Figure 7.6 shows what it looks like. Note that there are a lot of features that could be added to the GUI here. We could make the maze editable along with the starting and ending locations. Those enhancements are left as exercises for the reader.

Now that we have stepped through the code, we should make sure that you understand what it is doing. This is probably easiest to do with a maze that has no walls at all. We start at some position, which is put on the queue, then taken off in the `while` loop. The `for` loop will add the four adjacent locations (up, right, down, left) to the queue. Then each of those will be taken off in order and their neighbors which have not yet been visited will be added to the queue. Figure 7.7 shows a graphical representation of how this might work in a maze with few walls where we start near the center and have an end at the top left corner. Each square is labeled with the number of elements that would be in `steps`. So the starting location has a 1. Its 4 neighbors have a 2. The eight squares that are adjacent to those with the 2s are labeled 3. At that point, some walls begin to get in the way, but the general procedure is the same; anything that touches a 3 would get a 4. Those that touch a 4 get a 5 and so on.

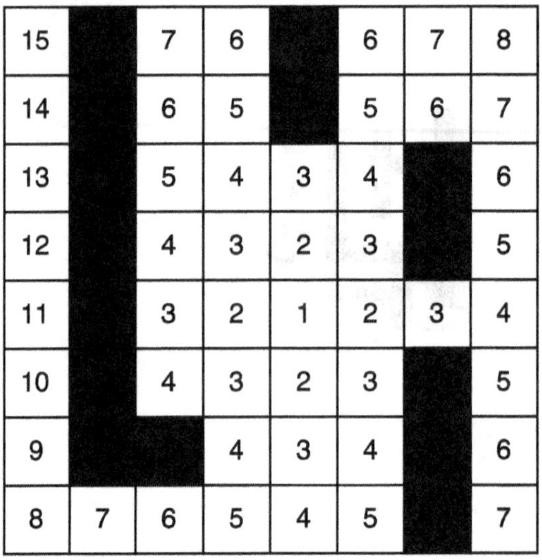

15		7	6		6	7	8
14		6	5		5	6	7
13		5	4	3	4		6
12		4	3	2	3		5
11		3	2	1	2	3	4
10		4	3	2	3		5
9			4	3	4		6
8	7	6	5	4	5		7

FIGURE 7.7: This figures shows in a graphical way what happens when a breadth-first search is run.

We close by noting that this is an efficient procedure because in the worst case, like that shown in the figure, each square can only be visited once. So the total amount of work is $O(n)$ in the number of squares in the maze.

Breadth-first is not always efficient or even useful though. Consider the example of trying to find the longest path through the maze that does not cross itself. That type of problem needs a different method. We will see how we might approach that problem in chapter 15.

7.10   End of Chapter Material

7.10.1   Summary of Concepts

- The term Abstract Data Type, ADT, is used to describe a programmatic structure that has well-defined operations, but that does not specify how those operations are carried out. This concept links back to the idea of separation of interface and implementation.

- A stack is a simple ADT with two basic operations, **push** and **pop**. The **push** operation adds an element to the top of the stack. The **pop** operation removes the most recently added element. This order of content removal is called Last In First Out, LIFO.

- A queue is another simple ADT with **enqueue** and **dequeue** operations. The **enqueue** adds an element to a queue while the **dequeue** takes out the element that has been in the queue the longest. This ordering scheme for removal of elements is called First In First Out, FIFO.

- Order analysis helps us understand the way in which algorithms behave as input sizes get larger. The O notation ignores constant multiples and lower-order terms. It is only really relevant when the input sizes get large.

- Due to their simplicity, any reasonable implementation of a stack or a queue should have operations that have an $O(1)$ amortized cost.

- One way to implement the stack ADT is using an **Array** to store the data. Simply keep an index to the **top** of the stack in the **Array** and build up from 0. When the **Array** is full, it should be grown by a multiplicative factor to maintain $O(1)$ amortized cost over copying operations.

- It is also possible to implement a queue with an **Array**. The implementation is a bit more complex because data is added at one end of the queue and removed at the other. The **front** and **back** are wrapped around when they hit the end so the **Array** stays small and no copies are needed for most operations. Like the stack, when the **Array** does have to grow, it should do so by a multiplicative factor.

- Testing is a very important part of software development. Unit tests are tests of the functionality of small units of code, such as single classes. The **JUnit** framework provides a simple way of building unit tests that can be run regularly during software development.

7.10.2 Exercises

1. Add extra functions to the RPN calculator.

2. Write more tests for the JUnit tests that are shown in the book.

3. For this exercise you should implement a few possible versions of both the queue and the stack and then perform timing tests on them to see how well they scale. You will include the implementation that we developed in this chapter plus some slightly modified implementations to see how minor choices impact the final speed of the implementation.

7.10.3 Projects

As you can probably guess, the projects for this chapter have you adding stacks and/or queues to your code. Many of them have you doing path searching using breadth-first searches.

1. If you have been working on the MUD project, for this iteration you should add a new command called "shortpath" that takes a single argument for the keyword of a room. When the user executes this command, you will print out the directions they should move in from their current location to get to that room on the shortest path. Use a breadth-first search with your own queue implementation for this.

2. In a good website, pages should be reachable in a small number of clicks. If you have been working on the web spider project, you have probably built up something called a graph where each of your pages knows what other pages it links to. Each link requires the user to do a click. Write a breadth-first shortest path that will tell the user the smallest number of click it takes to get from one page to another on the site you are working with.

3. If your graphical game includes things that move around and have barriers to motion, then it often makes sense for them to be able to find the shortest path from one place to another. Depending on the game, this might be something that the player does or something that certain types of enemies do. Use your own queue to implement a breadth-first search and use that to create the proper type of motion for an entity in your game.

4. Back in chapter 5, project 4 on the math worksheet, it was suggested that you could integrate polymorphism into your project by using the command pattern. One of the main uses of the command pattern is to implement undo and redo capabilities. A stack is an ideal data structure for doing this. You have the main application keep track of two stacks. One is the undo stack, which you push commands onto as they are executed. When the user selects to undo something, you pop the last command on the undo stack, call its undo method, then push it to a second stack, which is a redo stack. When a user selects redo, things move in the opposite direction. Note that any normal activity should probably clear the redo stack.

    The stack that you use for this will probably be a bit different from the default implementation. For one thing, you need the redo stack to have a clear method. In addition, you might not want unlimited undo so you might want the stack to "lose" elements after a certain number have been pushed. It is left as an exercise for the student to consider how this might be implemented in a way that preserves $O(1)$ behavior.

5. In chapter 5, project 5 suggested that you add the command pattern to the Photoshop® project in order to implement undo and redo capabilities. As was described in the project above, this is best done with two stacks. For this iteration you should do for Photoshop® what is described above for the math worksheet.

6. The stock portfolio management project could also benefit from an implementation of the command pattern as described above in the math worksheet project. You will want to provide the users with the ability to undo or redo a stock purchase. We can assume that a decision to buy or sell a stock is an order for you, the management company, to perform that action for them thus the actual buy/sell can be undone or redone.

7. The movie project could benefit from the ability to undo and redo as well. Refer to the description of the command pattern and how it could be implemented above in the math worksheet project. You should allow the user to undo and redo changes to their list of movies that they want to watch and their favorites list.

8. So far the projects for the simulation workbench have had you working with N-body simulations like gravity or a cloth particle-mesh. Now we want to add a fundamentally different type of simulation to the mix, discrete event simulations. These are discussed in more depth in chapter 13, but you can create your first simple one here, a single-server queue.

    The idea of a single-server queue is that you have people arriving and getting in line to receive service from a single server. The people arrive at random intervals and the server takes a random amount of time to process requests. When people arrive and the server is busy, they enter a queue. The simulation can keep track of things like how long people have to wait and the average length of the queue.

    The key to a discrete event simulation is that all you care about are events, not the time in between. In this system, there are only two types of events, a customer arriving and a service being completed. Only one of each can be scheduled at any given time. To model arrivals you will use an exponential distribution and to model service times you will use a Weibull distribution. The following formulas can be used to generate numbers from these distributions, where r is the result of a call to `math.random`.

- $a(\lambda)_{exponential} = -\frac{\ln(1-r)}{\lambda}$
- $a(k,\lambda)_{Weibull} = \lambda(-\ln(1-r))^{\frac{1}{k}}$

For the Weibull, use $k = 2$. The λ values determine the average time between arrivals or services. You should let the user alter those.

The simulation should start at $t = 0$ with no one in the queue. You generate a value from the exponential distribution, $a_{exponential,1}$ and instantly advance to that time, when the first person enters. As no one is being served at that time, they go straight to the server. You also generate another number from the exponential distribution, $a_{exponential,2}$ and keep that as the time the next person will enter. The first person going to the server causes you to generate a service time, $a_{Weibull,1}$ for when that person will finish the service, and you store that.

You then instantly advance to the lower of the two event times. So either a second person arrives and enters the queue, or the first person is serviced and the server becomes free with no one present. Each time one person arrives, you schedule the next arrival. Each time a person goes to the server, you schedule when they will be done. Note that the generated times are intervals so you add them to the current time to get when the event happens.

Let the user determine how long to run the simulation. Calculate and display the average wait time, the average line length, and the fraction of time the server is busy.

9. For the L-system project, if you have not yet implemented the '[' and ']' characters in your strings, and you should do so now. These store the state of the turtle and pop it back off so that you can make branching structures. As the term "pop" implies, this can be nicely done with a stack. Add this functionality to your drawing capabilities. Every time you get to a '[' you should push information to the stack about where your turtle is and where it is facing. When you get to a ']' the stack should be popped and the turtle should go back to that state.

Chapter 8

Multithreading and Concurrency

Our programs so far have been written based on the basic assumption that only one thing is happening at any given time. The instructions in the program execute one after another and control might jump from one method to another, but at any given time you can point to a single line of code and say that is what is happening at that time. This type of sequential processing of instructions is called a thread. Our programs so far have been written to utilize a single thread of execution. Such programs do not take full advantage of modern hardware. In this chapter we will learn how to make our programs use multiple threads. Multithreading can be used to simplify logic in some places, but most of the time we will use it to make our programs run faster or be more responsive.

8.1 The Multicore and Manycore Future

For many decades after the development of the integrated circuit, computers got consistently faster because of a combination of two factors. The most significant is related to *Moore's Law*. This was a prediction made by Gordon Moore at Intel that the number of transistors that could be etched on a chip would double roughly every two years. This growth was enabled by technological improvements that allowed the transistors to be made at smaller and smaller scales. Associated with this was the ability to push those transistors to higher frequencies.

Figure 8.1 shows clock frequencies, number of transistors, and core counts for x86 processors released by Intel and AMD between 1978 and 2011. The log scales on the vertical axes make it so that exponential growth displays as straight lines. A dotted line shows an exponential fit to the transistor count data. From this you can see that Moore's Law has held fairly well. Transistor counts have grown exponentially with a doubling time fairly close to the predicted two years.[1]

From the beginning of the figure until the 2000–2005 time frame, the clock frequencies of chips grew exponentially as well. The chips available in 1978 and 1979 ran at 5 MHz. By 2000 they had crossed the 1 GHz speed. Another dotted line shows an exponential fit for the clock speeds through 2000. As you can see though, clock speeds stalled at a few GHz and have not gone up much since just after 2000. This was a result of problems with heat dissipation. Above these speeds, silicon-based chips need special cooling to keep them running stably.

Much of the speed boost that computers got until 2005 came from increases in clock frequency. When that ended, chip manufactures had to look for an alternate way to improve the speed of chips. After all, no one buys new computers so they can run programs slower than the old one. So beginning in 2005, commodity computer chips started going in a different direction.[2] The direction they went is shown by the line with triangle markers in figure 8.1. Instead of using additional transistors primarily to add complexity to the data paths for single instructions or just adding more cache, chip makers started making chips that had multiple cores.

When a chip has multiple cores, it can be executing more than one instruction at any given time. We normally say that code can be executed in parallel. Unfortunately, a program written in the way we have done so far will not automatically take advantage of multiple cores. Instead, it has to be rewritten so that it will run faster on these newer computers. This is a significant change from what had been happening when processors were getting higher clock speeds. The significance of this change was described in an article titled "The Free Lunch Is Over" [12]. The "free lunch" that is referred to is getting more speed from new processors without programmers having to do any work. To take full advantage of newer processors, programs have to be multithreaded. In this chapter we will begin to explore exactly what that means.

---

[1] The low outlying points in 2008 and 2011 are chips specifically aimed at low-energy, mobile computing. This is why they have low transistor counts compared to other chips being made at the same time.

[2] High-end server chips started doing this a few years earlier.

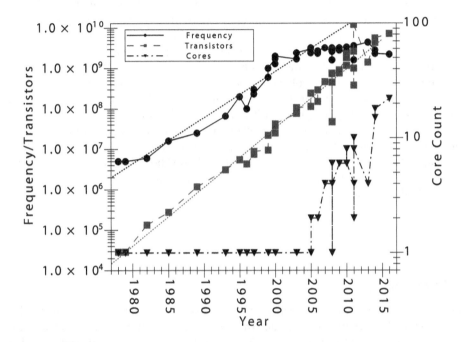

FIGURE 8.1: This figure shows frequencies, transistor counts, and core counts for x86 processors produced between 1978 and 2016. The log scale on the plot makes it so that exponential curves appear as straight lines. The dotted lines show fits to the full transistor count data and the frequency data through 2000. Clock speeds shown are the highest release for each chip.

8.2 Threads (Conceptual)

When you imagine one of your programs running, you can picture the computer moving an arrow from one line to the next. The arrow points to the line that is currently being executed. When you call a function/method, the arrow jumps to that other part of the code, then jumps back when the function/method returns. The term we use for this behavior is a "thread". The programs we have written so far use a single thread of execution. We start this way because it is easier, but the current nature of computer hardware makes it important for you to understand the concept of multiple threads early on.

The idea of a multithreaded program is that you have two or more of those arrows pointing at different lines and moving through the program at the same time. These different threads can be run on different cores, giving us the ability to more completely utilize modern processors. Just like when multiple people cooperate on a task, there has to be some coordination between the individuals to get the maximum benefit. There are also typically limits to how many people can productively work together on a particular task. If you want to dig a small hole, you probably cannot efficiently use more than two or maybe three people. However, if you want to dig a large trench, you can involve hundreds, if not thousands, of people productively. These same limitations can apply to programs.

There are many ways of using threads to produce parallelism in Scala. We will start off in this chapter with the ones that you should probably be using in your Scala programs. In the next chapter we will explore lower-level constructs in the Java libraries that underlie what we talk about here.

8.3 Futures

When you want to cause one calculation or task to be done in a separate thread in Scala, you should probably use the `scala.concurrent.Future` trait and a companion object. The companion object has an `apply` method that takes an argument that is passed by name, which we will often refer to as a by-name argument, that contains the code you wish to execute. It starts with code running in a separate thread and returns an instance of `Future` to represent an ongoing calculation. As the name implies, a `Future` is an object that represents a value that is in the process of being computed and might not be ready to be used yet. It is a parametric type, and the type parameter is the type of the value that is produced by the calculation. The following code is a simple demonstration of using a `Future` in a manner that makes it clear when you run it that the code is running in a separate thread.

```scala
import scala.concurrent.Future
import scala.concurrent.ExecutionContext.Implicits.global

object SimpleExample extends App {
  Future {
    for(i <- 1 to 26) {
      println(i)
      Thread.sleep(10)
    }
  }
```

```
  for(c <- 'a' to 'z') {
    println(c)
    Thread.sleep(10)
  }
}
```

Notice the `imports` at the top. The first one should make sense because we want to use the name `Future` instead of `scala.concurrent.Future` in our code. The second import does something very different though. The `apply` method on the `Future` companion object is actually curried with a second argument list that takes an instance of `ExecutionContext`. That argument is labeled as `implicit`. The second `import` brings an `implicit ExecutionContext` into scope so that you do not have to worry about it. You might feel like this is a really long name to remember. Fortunately, if you leave that line out, the error message you will get specifically says that you might want to add that `import`, so if you forget, it tells you what to do. You just have to actually take the time to read the error message.[3]

If you run this code, it will print alternating numbers and letters. The numbers are being printed in the body of the `Future` while the characters are being printed by code after the `Future`. If this was plain sequential code, all the numbers would come before all the letters. The calls to `Thread.sleep` are needed to slow things down a bit. `Thread` is a `class` in Java that we will talk more about in the next chapter, but which has a few methods, like `sleep`, that are generally useful. The argument to `sleep` is a length of time in milliseconds. So the pauses inserted here are only 1/100th of a second, and it is unlikely that you would even notice them. Without the pause, one loop inevitably finishes before the other and we do not see the interleaving.

If you run this program several times, you might notice that the output can vary.[4] For example, one run might give you perfect interleaving with 1 followed by 'a'. However, in another run, you might see 1 and 2 print before 'a'. This is a hint to the source of many challenges with writing multithreaded code; you do not have control over thread scheduling. With single-threaded programs, if you run the program twice and give it the same input, it will have the same behavior. We like this feature, because determinism helps us make certain that our programs are correct. Once you introduce a second thread, you do not know exactly when that second thread will start execution or how fast it will get through things relative to the first thread. As a result, your programs can become non-deterministic. That is something we will try to control with the use of proper threading libraries, like `Future`s.

8.3.1 Using `foreach`, `map`, and `flatMap`

Our first example showed that the code in a `Future` executed in a separate thread, but it did not give the `Future` a name or do anything with it. The power of the `Future` type and the way in which it can help us to avoid the pitfalls of multithreading are best achieved when we utilize the instances of `Future` and the methods that they have. A `Future` should be used in a situation where some operation might take a while, and your current thread can do other productive things while that code executes and before it needs the answer. Demonstrating this at an introductory level can be a challenge given that computers are so very fast. It can be challenging to come up with calculations that take any amount of

---

[3]Remember, any method that has a parameter list that takes a single argument can have that argument put in curly braces instead of parentheses. This is particularly useful when the argument is passed-by-name, and will be a block of code with multiple statements.

[4]This depends on details of your computer, but adjusting the sleep time can allow you to see variations on any computer.

time that is noticeable by the user. There are some operations that are still slow on modern computers that we can utilize for these examples. These include reading from files or pulling data over a network. We will make our examples using the latter.

When you go to websites or run an app on your phone that uses data, there are good odds that the program you are using is communicating with other machines to get data using what is called a RESTful API. These are standards for communicating data that respond to HTTP requests. You might recognize the acronym HTTP from the first part of the URLs that are used by web browsers. RESTful APIs specify how you can make requests that will respond with particular pieces of information. When you make a call to one of these APIs, there is a lot of work that goes on. Your computer sends out an HTTP request, which is received by the machine specified in the URL. That machine does whatever processing is required and responds. Depending on how far away the other machine is across the network, the response time can easily be long enough for people to notice.

We will make our examples using the MetaWeather API. You can read the specifications for this API at `https://www.metaweather.com/api/`.[5] You can test a query by simply putting the URL in your browser's address bar. For example, the URL `https://www.metaweather.com/api/location/search/?query=san` does a location query asking for all of the known locations that begin with "san". If you try this link, you will get back the following text.

```
[{"title":"Santa Cruz","location_type":"City","woeid":2488853,"latt_long":
"36.974018,-122.030952"}, {"title":"San Francisco","location_type":"City",
"woeid":2487956,"latt_long":"37.777119, -122.41964"}, {"title":"San Diego",
"location_type":"City","woeid":2487889,"latt_long":"32.715691,-117.161720"},
{"title":"San Jose","location_type":"City","woeid":2488042,"latt_long":
"37.338581,-121.885567"}, {"title":"San Antonio","location_type":"City",
"woeid":2487796,"latt_long":"29.424580,-98.494614"}]
```

This lists all the different locations along with information about the type of location, the ID number for that location, and the latitude and longitude of that location. There is a fair bit of punctuation in here as well. This is using a standard format called JSON that is broadly used for communicating information on the web. We'll come back to JSON in a bit, but first let's consider a brief Scala program that can read this same data and display it using a Future.

```scala
1   import scala.concurrent.Future
2   import scala.concurrent.ExecutionContext.Implicits.global
3
4   object WeeksWeather extends App {
5     println("Starting app.")
6     val fut = Future {
7       io.Source.fromURL("https://www.metaweather.com/api/location/search/?query=san").
8         mkString
9     }
10    fut.foreach(s =>{
11      println(s)
12    })
13    println("About to sleep")
14    Thread.sleep(5000)
```

---

[5]Many RESTful APIs require authentication, which is a complication that we do not want to deal with in this book. MetaWeather was chosen in part because it has a fairly straightforward API, and it does not require authentication. A number of exercises are given using other APIs that also do not require authentication.

```
15    println("Done")
16  }
```

We begin with the same **imports** as before. There are **println** calls on lines 5, 13, and 15 just to help you see what is going on. Lines 6–9 create a **Future** called **fut** that does the API call using **io.Source.fromURL**. This call is similar to **io.Source.fromFile**, which we have used a number of times previously, but as the name implies, it reads the data from a URL instead of a file. The call to **mkString** simply takes the result and builds one long string from it. The type of **fut** is **Future[String]** because the code inside produces a **String**. Lines 10-12 call **foreach** on **fut**. This usage of **foreach** is completely analogous to similar calls on collections, but there is only one value, and that **foreach** call does not happen until the result of the **Future** is ready. Unlike calls to **foreach** on regular collections, the **foreach** on a **Future** does not block the current thread; instead, it schedules the function passed to it to be called when the **Future** that is is called on completes. Line 14 pauses the main thread for five seconds because when the main thread finishes, the application closes, even if there are still **Futures** executing. Once the application closes, all of the **Futures** will then stop executing. Since the loading from the URL takes a while, it never has a chance to complete if we leave this out.

The previous paragraph mentioned that **foreach** does not "block". It is worth taking a moment to discuss what exactly we mean by this. A BLOCKING CALL is one that stops the progress of the current thread until it has been completed. Most of the methods we have worked with are of this variety, but the most obvious are methods like **readLine**. When **readLine** is called, the execution of that thread simply halts until the user types something in. If the user takes 20 minutes to type something, the program will sit there for 20 minutes doing nothing until the user hits the Enter key. There are times when this is the desired behavior, but in a large number of situations, especially when a program has many users and needs to respond quickly to them, this type of behavior would be very detrimental to performance and the user experience. One of the main reasons for using **Futures** is to push this type of blocking behavior into other threads so that the program can continue to respond as quickly as it has the required information.

The long string of information in the JSON format has the information that we want, but it is not in a format that we can use very easily. The fact that JSON is broadly used means that there are lots of libraries for parsing strings and turning it into other formats. There are many of these just for the Scala language, but most of them would require you to download and install additional libraries. There happens to be one in the standard Scala library, so we will use that one.[6]

If you look back at the JSON data, you see that the whole thing is in square brackets, with a comma-separated list of things in curly braces. Inside of each set of curly braces are pairs of keys and values. The pairs are separated by commas and the keys and values are separated by colons. The brackets represent a JSON array, which will parse to a **List[Any]** using the standard library. Each set of curly braces represents a JSON object which parses to a **Map[String, Any]**. The presence of **Any** in the parse results makes it a bit challenging to deal with. Instead of a **println** in the **foreach**, we could put in code to parse to JSON and find the value of "woeid" for a particular city. That functionality is shown in the following code.

```
1  import scala.concurrent.Future
2  import scala.concurrent.ExecutionContext.Implicits.global
```

---

[6]While the "parser combinator" library is part of the standard Scala library, it is in a separate module that you might have to add to your project in some IDEs. You can find it in the "lib" directory of your Scala installation.

```scala
import scala.util.parsing.json.JSON

object WeeksWeather extends App {
  println("Starting app.")
  val f1 = Future {
    io.Source.fromURL("https://www.metaweather.com/api/location/search/?query=san").
      mkString
  }
  f1.foreach(jsonStr => {
    val json = JSON.parseFull(jsonStr)
    json.get match {
      case lst: List[Map[String, Any]] =>
        lst.find(m => m("title") == "San Antonio") match {
          case Some(map) =>
            map("woeid") match {
              case d:Double => println("woeid = "+d.toInt)
              case _ => println("Wrong type")
            }
          case None =>
            println("City not found.")
        }
      case _ =>
        println("Not expected")
    }
  })
  println("About to sleep")
  Thread.sleep(5000)
  println("Done")
}
```

The code that has changed in this example is from lines 11 to 25 along with an extra import on line 3. As you can probably tell from reading it, the import is for the JSON parser. This parser is used on line 11 with the call to JSON.parseFull. For this JSON, that gives us back an Option[List[Map[String, Any]]], but if you look at the type of the json variable you will find that it is just a Option[Any]. The reason for this is that we do not know exactly what string value we will get until we run the program. The top level could have curly braces instead of square brackets or might just be a string or a number. Even with square brackets, the contents could be a mix of values. It is perfectly legal to have a JSON string with the value [1.25, "testing", "first": "word", "second": 42]. This would parse to Some(List(1.25, testing, Map(first -> word, second -> 42.0))). This is a List with a Double, and String, and a Map in it. That has to be represented as a List[Any].[7]

The fact that we cannot know the types in advance means that we need to do matches on them. Line 12 assumes that we have a Some. If we somehow got None for the Option, the call to get would throw an exception. That is matched against two cases. The first one is a type of List[Map[String, Any]] and the second is anything else. When you put this code in your IDE and compile it, you will get a warning on line 13 that says the pattern is unchecked since it is eliminated by erasure. The term erasure here refers to something

---

[7]The reason for this is found in the name JSON, which stands for JavaScript Object Notation. JavaScript is a dynamically typed language. That means that it does not check types until you run the program. In a sense, the language does not have a concept of types. It is the runtime that deals with types. In JavaScript, and other dynamically typed languages, it is like everything is of an Any type that allows you to call whatever methods you want. If you happen to call something it does deal with, it crashes during runtime.

that happens on the JVM. Type parameters are not preserved in compiled code. Recall from chapter 7 that we had to put in a `ClassTag` when creating `Arrays` that held our type parameter for this reason. To get rid of this warning, we would have to match on a `List[Any]`, then inside of that, match on the other types we believe are in the list as we run through the list. Since we know the format of the response for this call, we can take this shortcut and feel safe about it.

Inside of the match for the `List[Map[String, Any]]` we have a find that looks for the `Map` with a key of "title" that has a value of "San Antonio". If one is found, we pull off the value for the key "woeid". This is a key that is used to specify locations in other searches to this API.

As with collections, the call to `foreach` does stuff, it does not produce a value. As we just said, the "woeid" value is useful for subsequent calls. In collections, if we wanted to apply a function and get back values, we would have used `map` or `flatMap`. Those work for `Futures` as well. And just as calling `map` on a `List` gives back a new `List`, calling `map` on a `Future` gives back a new `Future`.

With a few minor changes, we can make it so that the previous code gives back an `Option[Int]` instead of printing values. If we do that and change the `foreach` to a `map`, the result is a new `Future[Option[Int]]` that we can call `map`, `flatMap`, or `foreach` on to do more work.

8.3.2 Working in Parallel

This example really did not need a `Future`. We were only doing one thing. We did not have anything else happening in the main thread. In fact, we had to have it sleep so that it did not terminate. In order to demonstrate the real power of `Futures`, we need to have a number of these potentially time-consuming calls that we could do at the same time. For example, we could take the "woeid" that we just got and use it to look up the weather information on several different days. We will do that for five different days as that should be sufficient to show what we want without taking up too much bandwidth on the server. The code to do this is shown here.

```scala
package multithreading1

import scala.concurrent.Future
import scala.concurrent.ExecutionContext.Implicits.global
import scala.util.parsing.json.JSON

object WeeksWeather extends App {
  println("Starting app.")
  val f1 = Future {
    io.Source.fromURL(
        "https://www.metaweather.com/api/location/search/?query=san").mkString
  }
  val f2 = f1.map(jsonStr => {
    val json = JSON.parseFull(jsonStr)
    json.get match {
      case lst: List[Map[String, Any]] =>
        lst.find(m => m("title") == "San Antonio") match {
          case Some(map) =>
            map("woeid") match {
              case d: Double => Some(d.toInt)
              case _ => None
            }
```

```scala
23              case None =>
24                  None
25          }
26        case _ =>
27            None
28      }
29    })
30    val days = 1 to 5
31    val f3 = f2.flatMap(optInt => optInt match {
32      case Some(woeid) =>
33        val dayFutures = days.map(day => Future {
34          io.Source.fromURL(
35            s"https://www.metaweather.com/api/location/$woeid/2016/5/$day/").mkString
36        })
37        Future.sequence(dayFutures)
38      case None => Future { Nil }
39    })
40    f3.foreach(lst => lst.foreach(println))
41    println("About to sleep")
42    Thread.sleep(10000)
43    println("Done")
44  }
```

The primary changes are between lines 30 and 40. We start by defining the days of the month that we are going to request information for. Next, we do a `flatMap` on the `Future` that holds the "woeid" value. The reason for using a `flatMap` is because the function inside is going to produce a `Future`, and we do not want a `Future[Future[_]]`.[8] Inside the `flatMap` we start by doing a `match` on the `Option[Int]`. If we have a value, it gets the name `woeid` and do the processing to get the weather data. Otherwise we just give back a `Future` with an empty `List`.

Lines 33–36 map over `days` and each one produces a `Future[String]`. This means that `dayFutures` is actually a `Seq[Future[String]]`. This is not really what we want. We want the data from all of those loads to complete before we do some other task. We do not want to do different things for each one. So we need to switch the `Seq[Future[String]]` to a `Future[Seq[String]]`. This is a fairly common thing to want to do. For that reason, there is a method called `sequence` in the `Future` companion object that converts a collection of `Futures` to a `Future` of that collection type. This new `Future` does not complete until all the `Futures` in the original sequence have completed. We call this method on line 37.

The code ends with a `foreach` that prints the JSON return values for those five queries. The JSON returned for a query on a day is significantly longer than that produced for a location search, but not really all that much more complex. It has a JSON array with a bunch of JSON objects in each, and each JSON object has keys and values that provide information for a point in time and the weather conditions at that time. We are not going to spend time parsing values out of it here because our focus is on the `Futures`, not JSON parsing. Interested readers can figure out how to pull more information from the MetaWeather API responses.

---

[8]While you do sometimes want a `List[List[_]]` or an `Array[Array[_]]` for storing 2D data, you never want a `Future[Future[_]]`. It really is not a meaningful thing. It is similar to how you never really want an `Option[Option[_]]`. For both `Future` and `Option`, one level is sufficient to express what you want.

Methods in the Future Object

This section has used the `apply` method of the `Future` companion object to create new `Future` instances. We also just saw how you can use the `sequence` method to convert a collection of `Future`s into a single `Future` around a collection of the values. There are other methods in the `Future` companion object that can be very helpful for various situations when you have collections of `Future`s. We briefly introduce them here. Note that all of these operate on collections of `Future`s and they all produce `Future`s, so none of them are blocking.

The `sequence` method is actually a special case of the more general `traverse` method. The `traverse` method is curried and the first argument list takes a single collection. Unlike `sequence`, the collection does not have to be a collection of `Future`s. The second argument list takes a function that takes the type in the collection and produces a `Future` of some type. Like `sequence`, it produces a `Future` wrapped around a collection. In this case, the values in the collection will be the results of `Future`s produced by the function in the second argument.

You can think of the `sequence` method as taking a collection of `Future`s and producing a `Future` that effectively waits until all the input `Future`s are done. There is also a method called `firstCompletedOf` that does what the name implies. It takes a collection of `Future`s and gives back a `Future` that will complete when the first one is done and provide that value. This can be handy if you have several ways of getting the information you want, but do not know which will be fastest, so you try all of them in separate `Future`s.

There is a `find` method that is like the `find` method on collections. It takes a collection of `Future`s and a predicate. It produces a `Future` with an `Option` of the type in the original collection of `Future`s. The result is the first of the `Future`s in the input collection that finishes and satisfies the predicate. So in a way it combines the functionality of `firstCompletedOf` with `filter`-like capabilities.

Finally, there are `reduceLeft` and `foldLeft` methods. As with the other methods, these have a first argument list that is a collection of `Future`s. They run through that collection using a provided function to aggregate the results and produce a `Future` of the final result. As with the `foldLeft` on collections, this `foldLeft` takes an argument for the "zero" of the aggregation. Note that in this case that means it is a curried method with three argument lists: the collection, the zero value, and the aggregating function.

8.3.3 Futures with for Loops

One detail of the `for` construct in Scala that you might not recall from earlier is that when you put a `for` in your code, it actually gets converted to the appropriate combination of calls to `foreach`, `map`, `filter`, and `flatMap`. While we typically use the term "for loop", it is more appropriate to call them "for comprehensions". The advantage of this approach is that the `for` in Scala can work with more than just collections. They can work with any type that has the appropriate methods. That happens to include `Future`s as well as `Option`s.

Many people find the `for` syntax easier to read than using all the methods directly. Here is a segment of code that creates three `Future`s that each call different methods that might take a while to run. They are followed by a `for` comprehension that gets the values from each of those three futures, then combines them and gives back a fourth `Future` with the combined result.

```scala
val f1 = Future { slowMethodProducingString() }
val f2 = Future { slowMethodProducingInt() }
val f3 = Future { slowMethodProducingListInt() }
val f4 = for {
  str <- f1
  i <- f2
  lst <- f3
} yield {
  combineResults(str, i, lst)
}
```

The declaration of `f4` could have been done with the following line of code.

```scala
val f4 = f1.flatMap(str => f2.flatMap(i => f3.map(lst => combineResults(str, i,
    lst))))
```

Readers can decide which they prefer, but it should be noted that in the `for` version, you do not have to think about whether you are using `map` or `flatMap`.

Some readers might wonder about the details of the `for` expression here, in particular, the use of curly braces instead of parentheses after `for`. Scala allows you to use either. The difference is that when you use parentheses, you have to put semicolons between the different elements, and when you use curly braces you can just use line feeds. So you could rewrite the `for` as `for (str <- f1; i <- f2; lst <- f3)`.

You might be tempted not to declare `f1`, `f2`, and `f3` in this example. Instead, you could put that code to the left of the `<-` symbols in the `for` as shown here.

```scala
val f = for {
  str <- Future { slowMethodProducingString() }
  i <- Future { slowMethodProducingInt() }
  lst <- Future { slowMethodProducingListInt() }
} yield {
  combineResults(str, i, lst)
}
```

This is shorter, but there was actually a reason for putting the declarations outside. As soon as we create the `Future`, it can start processing, and since these `Future`s do not depend on one another, we start them all running at essentially the same time. Then the body of the `for` will happen as soon as they are all done. In this modified version, the second `Future` is not created until the first is done and the third is not created until the second finishes. The result is that we do not get any parallelism at all. A general rule is that you should create your `Future`s as soon as you have the information you need to do so. That way they can get to work earlier and finish sooner.

One more example of using a `for` comprehension with `Future`s is to convert the `WeeksWeather` application to using one. An approach to doing that is shown below.

```scala
1  package multithreading1
2
3  import scala.concurrent.Future
4  import scala.concurrent.ExecutionContext.Implicits.global
5  import scala.util.parsing.json.JSON
6
7  object WeeksWeatherFor extends App {
```

```scala
 8    println("Starting app.")
 9    val days = 1 to 5
10    val future = for {
11      jsonStr <- loadURL("https://www.metaweather.com/api/location/search/?query=san")
12      Some(woeid) <- getWoeid(jsonStr)
13      daysJson <- getDaysJSON(woeid, days)
14    } {
15      daysJson.foreach(println)
16    }
17    println("About to sleep")
18    Thread.sleep(10000)
19    println("Done")
20
21    def loadURL(url: String): Future[String] = Future {
         io.Source.fromURL(url).mkString }
22
23    def getWoeid(jsonStr: String): Future[Option[Int]] = Future {
24      val json = JSON.parseFull(jsonStr)
25      json.get match {
26        case lst: List[Map[String, Any]] =>
27          lst.find(m => m("title") == "San Antonio") match {
28            case Some(map) =>
29              map("woeid") match {
30                case d: Double => Some(d.toInt)
31                case _ => None
32              }
33            case None =>
34              None
35          }
36        case _ =>
37          None
38      }
39    }
40
41    def getDaysJSON(woeid: Int, days: Seq[Int]): Future[Seq[String]] = {
42      val dayFutures = days.map(day =>
           loadURL(s"https://www.metaweather.com/api/location/$woeid/2016/5/$day/"))
43      Future.sequence(dayFutures)
44    }
45  }
```

The first thing that you should notice about this version is that the logic has been broken up into methods. You could make a strong argument that this is a very good thing, and we probably should have done it to start with. Here we do not build the Futures at the beginning because we cannot do so. Each one takes the result of the one before it as an argument, so these have to be done sequentially.

One of the standard advantages of **for** comprehensions, pattern matching, helps simplify this code as well. The middle generator does a pattern match on the Option and only gets executed if a "woeid" is found. This eliminates a **match** from the earlier code. If the middle generator does not find a "woeid" (returns None), all the lines of code within the **for** loop that follow this middle generator are skipped. This is like the Python **continue** statement. The last thing to note is that every one of the methods at the bottom of the singleton **object** returns a Future. This is essential. Think of it this way, if you were using collections with a **for**, you could not have one of the generators have a simple Int on the right-hand side of a generator; they all have to be collections. You also cannot mix collections and `Futures`. If the first generator uses a `Future`, then all the rest must do the same.

The last thing to point out is how the final step in the first code was done with a `foreach` and that code has now moved into the body of the `for` with no `yield` on line 14. If the last step had been a call to `map`, we would use a `yield`. However, `foreach` does not produce a value, which is the behavior we get when we do not use `yield`.

8.3.4 Getting a Result

So far, all the calls that we have looked at for `Futures` have been non-blocking. This is intentional. Blocking calls slow things down when you have a heavy load. For that reason, the creators of the Scala `Futures` library specifically tried to prevent people from causing threads to block with `Futures`. This contrasts the `Future` type in many other languages which will include a simple method that will block the current thread until the result is ready.

Sometimes you simply will not be able to figure out a way to do what you want without somehow blocking the current thread to wait until some `Future` has completed and you can get the value. In those extreme situations, you can use `Await.result`.[9] Here is what a sample call would look like.

```scala
import scala.concurrent.Await
import scala.concurrent.duration._
...
  val answer = Await.result(future, 10.seconds)
...
```

The first argument is the `Future` that you want the result of. The second argument is how long you are willing to wait for it. By including the contents of `scala.concurrrent.duraction`, we get conversion from numeric types to durations like the `seconds` method used here. If the `Future` takes longer than that duration, the call throws a `java.util.concurrent.TimeoutException`. If you are willing to wait forever, which you probably should not be as your user likely is not, you can use a duration of `Duration.Inf`.

---

blocking

Both the operating system and the Scala libraries include code that handles the task of thread scheduling. This is the process of figuring out which threads should be running at any given time. This can be a very challenging problem and it is fortunate that we do not have to do it ourselves. However, we can do some things to help the thread scheduler.

In particular, if you know that some piece of code in a `Future` is going to potentially block and take a long time to complete, you can communicate that to the scheduler by putting it in a block of code after a call to `scala.concurrent.blocking`. A simple example of such code would be user input methods. If for some reason you were reading a word from the user inside of a `Future`, you should consider doing the following.

```scala
import scala.concurrent.blocking
```

---

[9] Early on you might be tempted to use this quite a bit. Try to avoid it. Spend time thinking of other ways. As you gain experience, you will probably find that you never have to use this approach.

```
...
  val word = blocking { io.StdIn.readLine() }
...
```

This helps the thread scheduler in the Scala libraries know that the user might take a while to type something in, and that it might be worth doing other things in the meantime. Keep in mind that your computer can do billions of operations every second, so even the reaction time of a human is a long time for a processor.

8.4   Parallel Collections

Futures are not your only option for creating parallelism in Scala. Most of the time, Futures are used to make applications more responsive, but it can be challenging to get significant speed improvements using them. A lot of applications that need a speed boost have time-consuming sections where they are doing the same operations on large collections of data. This type of DATA PARALLELISM is well addressed with parallel collections, which are generally the ideal tool for speeding up applications that are not performing fast enough. These are collections that have their basic operations performed in parallel. Not only do they do their operations in parallel, they have been written in such a way that they will automatically load-balance. If one thread finishes its own work load, it will steal work from another thread that has not yet finished. For this reason, if your problem can be solved using parallel collections, you should probably use them. Not only will they be easier to write, they will probably be more efficient than anything you are going to take the time to write yourself.

The structure of the parallel collections largely mirrors the structure of the normal collections with a `scala.collection.parallel` package that has subpackages for **immutable** and **mutable** types. The parallel collections are not subtypes of the normal collections we have used to this point. Instead, they have mutual supertypes that begin with **Gen** for general. The reason for this is that the original collections had an implicit contract to go through their elements in order and in a single thread. A lot of the code we have written did not depend on this, but some of it did. The general types explicitly state that they do not have to preserve that behavior.

The parallel collections have names that begin with **Par**. Not all of the types we used previously have parallel equivalents. This is because some types, for example the **List**, are inherently sequential. If you think back to chapter 6, we saw that there was a whole set of types under the **LinearSeq** type that had linear access behavior. None of those types can be efficiently implemented in parallel because of that. Figure 8.2 shows a UML class diagram of the parallel collections.

To convert from a normal collection to a parallel collection, call the **par** method on the standard collection. To convert back, call the **seq** method on the parallel collection. These calls are efficient for types that have a parallel equivalent. You can convert a **Range**, a **Vector**, an **Array**, a **HashSet**, or a **HashMap** from sequential to parallel and back in O(1) time. For other types, the conversion will be less efficient as it has to build a completely new collection. Consider the following examples from the REPL.

```
scala> (1 to 10).par
```

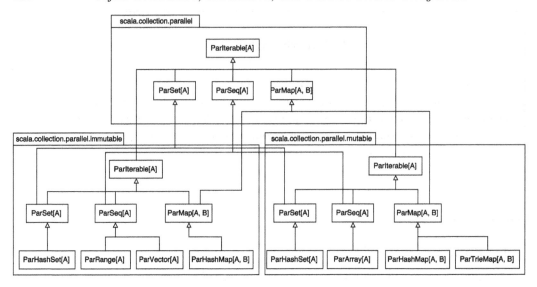

FIGURE 8.2: This UML diagram shows the main types in the three different parallel collection packages.

```
res0: scala.collection.parallel.immutable.ParRange = ParRange(1, 2, 3, 4, 5, 6, 7,
    8, 9, 10)

scala> Array(1,2,3,4).par
res1: scala.collection.parallel.mutable.ParArray[Int] = ParArray(1, 2, 3, 4)

scala> List(1,2,3,4).par
res2: scala.collection.parallel.immutable.ParSeq[Int] = ParVector(1, 2, 3, 4)
```

The first two calls are able to simply wrap the sequential version in a parallel collection. The last one has to convert the `List` over the a `ParVector`. The choice of `ParVector` in this situation is made because preserving immutability is the primary consideration.

The parallel collections have all the normal methods that you are used to using. Given this, you can write a parallel version of factorial using `BigInt` with the following code.

```
def parFact(n:BigInt) = (BigInt(1) to n).par.product
```

If you compare runtimes of this implementation to other versions that you might write that are not parallel, you will find that this version is faster and scales better because it does not have as much inherently sequential code.

8.4.1   Parallel `for` Loops

The fact that the parallel collections implement parallel versions of `map`, `filter`, `flatMap`, and `foreach` means that they can be used with `for` loops and those loops will automatically happen in parallel. To help illustrate this, we will define a reasonably slow method for calculating Fibonacci numbers. These are sequences of numbers where each element is the sum of the previous two.

```
def fib(n: Int): Int = if (n < 2) 1 else fib(n - 1) + fib(n - 2)
```

This method produces a sequence like 1, 1, 2, 3, 5, 8, 13, ... Now we can calculate quite a few of these in a `for` loop. To make that parallelism clear, we will calculate larger values and have the `for` loop count down for n values instead of up. That way the first iteration of the loop is doing the most work.

```
for(n <- (30 to 15 by -1).par) println(fib(n))
```

If we remove the call to `par` and run this as a sequential loop, the values print out from largest to smallest. However, if you run this code, the values print out in very different orders, though the biggest values are nearly always close to the end because they take the longest to calculate. Note that if we had a `yield` instead of a `println`, you would not see this effect because the results are always put back together in their proper order in the resulting collection when you use `yield`. The `println`, on the other hand, prints out the value as soon as it has been calculated.

This approach of using parallel collections in `for` loops is likely to be the way you use them most frequently. When you do this, there is one main thing that you have to remember: the actions of one iteration through the loop cannot depend on what happened in an earlier iteration. That behavior is only safe with sequential loops. Also, as we just saw, side effecting parallel `for` loops might exhibit behavior that seems odd if you are not considering the fact that it is running in parallel. Having your loops produce values with `yield` often produces safer code that is easier to work with.

8.4.2 Parallel `reduce`, `fold`, and `aggregate`

While most of the methods on parallel collections are ones we have used on the sequential collections, there is one method that is very significant for parallel collections called **aggregate** that we have not talked about with sequential collections, because it is not as useful in that context. In many ways, you can think of this method as being like a parallel version of the `fold` or `reduce` methods. The fact that you specifically call either `foldLeft`, `foldRight`, `reduceLeft`, or `reduceRight` tells you that those operations are sequential in their nature. A parallel version has to be able to run through the elements in whatever order it wants. Different threads will be running through different parts of the collection at the same time. For parallel collections, there are `fold` and `reduce` methods that will do those operations in an order-independent manner.

The `reduce` method is a full replacement for `reduceLeft` or `reduceRight` in situations where the order in which elements are processed does not matter. The same is not true of `fold`. To see why, compare the signatures of `fold` and `foldLeft` from `ParIterable[T]` in the API.

```
def fold [U >: T] (z: U)(op: (U, U)  U): U
def foldLeft [S] (z: S)(op: (S, T)  S): S
```

The `fold` method is more restricted in the types that it can use. While `foldLeft` could result in any type that you want, `fold` must result in a type that is a supertype of the type the collection contains. The reason for this restriction is that when a `fold` is done in parallel, you can have one thread apply `op` to one section of the collection while another thread does the same for a different section. To get your final answer, you have to be able to combine the results of the two threads. The combining operation only works with the same function if both parameters for the function have the same type as the output of that function.

The **aggregate** method addresses this shortcoming, but doing so requires two separate operators. Here is the signature of the method.

```scala
def aggregate [S] (z: S)(seqop: (S, T)  S, combop: (S, S)  S): S
```

The `seqop` is equivalent to the `op` that is passed into `foldLeft`. The `combop` is used to combine the results of processing the different pieces. Any time when you would use `foldLeft` or `foldRight` with a result type that is distinctly different from the type stored in the collection and the processing order is not critical to getting the correct answer, you can use `aggregate` instead.

A simple example that uses the ability to have different types is summing up the length of all the `Strings` in a long `Array[String]`. The following code does this with `aggregate`.

```scala
val words = makeReallyBigArrayOfStrings()
val sumLengths = words.par.aggregate(0)((lenSum, str) => lenSum + str.length,
    (len1, len2) => len1 + len2)
```

In this example, that type of the collection is `String` and the type of the result is `Int`, two completely unrelated types. The first argument list to `aggregate` is the base value to use when putting things together. Since we are summing, we use 0. The second argument list takes two functions. The first function is used to sum up the length of separate groups of `Strings`. The second function takes two variables and then has some method for combining them. In this example, the second function takes two `Ints` that were the sums of different groups of `Strings` and adds them together. This version uses the "rocket" notation with variable names to make it explicit what is going on. In general usage, it would probably be more common to use the underscore notation.

```scala
val sumLengths = words.par.aggregate(0)(_ + _.length, _ + _)
```

A more complex example of this is finding both the minimum and maximum values in a collection. Consider the following code that generates a parallel array of 1000 random doubles then finds the min and max values using `aggregate`.

```scala
val data = ParArray.fill(1000)(math.random)
val (min, max) = data.aggregate((data(0), data(0)))(
    (mm, n) => (n min mm._1, n max mm._2),
    (mm1, mm2) => (mm1._1 min mm2._1, mm1._2 max mm2._2))
```

It would have been much shorter to simply write `data.min` and `data.max`, but each of those calls has to run through the full set of data. This approach only runs through the data once to find both the minimum and maximum values. Given the overhead in creating tuples it is not clear that this is a benefit when we are simply comparing `Doubles`, but if the values we were comparing were found via a complex calculation on the data, which would be what you could get from a method like `maxBy`, this approach might be beneficial.

8.4.3 When to Use Parallel Collections

Given how easy they are to write, it might be tempting to make every `for` loop in your program use a parallel collection. There are two reasons why you should not do this. First, it does not work for everything. As mentioned earlier, a parallel collection is unsafe if one iteration of the loop depends on the result of the previous iteration. We will see other specific examples of when doing things in parallel is not safe in the next section.

There are also times when doing things in parallel actually slows the application down. At the beginning of the chapter we mentioned the idea of multiple people working to build a hole. For a small hole, the effort of organizing 100 people would probably take longer than just digging the hole, but for a large trench, more people will definitely be worth the effort.

The same applies to everything done in parallel. Splitting up the work takes some effort. It is possible that the effort of splitting things up and giving tasks to different threads can be larger than the original task's effort. In that situation, parallel processing will definitely be slower.

The general rule of thumb for all optimization is that to really know if it helps, you have to run tests. Going parallel is likely to speed things up if you have a lot of work to do and that work breaks nicely into tasks that are each big enough that they take significantly longer than the effort of scheduling them. However, how much benefit you get depends on many factors, not the least of which is the machine being used. If you want to experiment some with that, the following method might come in handy.

```scala
def timeCode[A](body: => A): A = {
  for(i <- 1 to 100) body
  val start = System.nanoTime()
  for(i <- 1 to 99) body
  val ret = body
  println((System.nanoTime() - start) / 1e9 / 100)
  ret
}
```

Note that this method runs whatever code is passed to it 200 times and measures the time for the last hundred, then prints the average run time of that code. It also gives back the value of the last execution. The short explanation for this is that the JVM only optimizes code after it has been called a reasonable number of times. The hope here is that the first 100 calls will get it optimized. Then we do another 100 because the actual amount of time varies. Clearly you would not use this exact approach for code that takes a long time to execute. It is left as an exercise for the reader to add to this method so it is easy to adjust the number of "warm-up" runs and the number of timing runs.

8.5 Race Conditions

Futures and parallel collections make it very easy to run code in multiple threads. Unfortunately, that apparent simplicity can mask the many problems that can occur when you are using threads. The challenges of multithreading are the reason it is not done unless it is really needed. The two main problems that one runs into are called RACE CONDITIONS and DEADLOCK. In this section, we will consider race conditions. After we learn how we might deal with race conditions, we will look at deadlock.

The term race condition is used to describe a situation where the outcome of a program can depend on the order in which threads are scheduled. Typically, if you run a program with the same inputs, you expect to have the same outputs. This deterministic nature is part of what makes computers so useful in the everyday world. Programmers do not have full control over what threads are running at any given time. You have the ability to do things like make a thread sleep for a period of time, but in a program with many threads, you do not get to explicitly say which ones will be running at any given time. Using the human analogy, you typically do not have the fine-grained control of telling each and every person exactly where to be and what to be doing. If they are digging a trench, you cannot say exactly when each shovel will cut into the ground. If you tried to exert such control, it would probably slow things down in the end and defeat the purpose of having many people/threads working on the task.

Simply lacking detailed control does not lead to race conditions. Race conditions happen when you combine that lack of control with shared mutable state. You are probably aware that most of the time your computer is running many different programs. These different programs are often called processes and they can utilize multiple cores as well. The difference between different processes and different threads is that each process gets its own chunk of memory that it can work with, but threads share access to the same memory. As a first example of this, consider the following rather meaningless example.

```scala
var cnt = 0
for(i <- (1 to 1000000000).par) cnt += 1
println(cnt)
```

We make a `var` that starts at zero and increment it by one a billion times. Clearly, the print should output one billion. If you take out the call to `par` that is exactly what will be printed. However, with the call to `par` in there, this will print out a number much less than one billion. How much less depends on the computer you run it on. The more core you have, the farther off you are likely to be. On a workstation that supports 48 threads, we have seen values under 5 million.

To help explain how this happens and to give a meaningful context to the problem, we will consider a slightly stripped-down version of our bank account from chapter 2. We have taken out the `Customer`, which is not significant here.

```scala
package multithreading1

class Account(
    private var _balance: Int,
    val id: String) {

  def balance = _balance

  def deposit(amount: Int): Boolean = {
    if (amount > 0) {
      val oldBalance = _balance
      val newBalance = oldBalance + amount
      _balance = newBalance
      true
    } else false
  }

  def withdraw(amount: Int): Boolean = {
    if (amount > 0 && amount <= _balance) {
      val oldBalance = _balance
      val newBalance = oldBalance - amount
      _balance = newBalance
      true
    } else false
  }
}
```

The lines that had been `_balance += amount` and `_balance -= amount` have been rewritten in a longer way using local variables because this is what the code will really be expanded to when it is compiled and it makes it easier to see where the problem arises. The problem here would arise with parallel calls to `deposit` or `withdraw`. Here is what the loop above might look like in this context.

```
val account = new Account(0, "01234567")
for(i <- (1 to 1000000000).par) account.deposit(1)
println(account.balance)
```

This code should deposit one billion pennies into the account, but as you can guess by now, it does not do so.[10]

To see how these types of problems occur, imagine that you make a withdrawal at an ATM at the same time as your paycheck is being deposited. Both of these activities read from and write to the _balance member of the object that represents your account in lines 11–13 for deposit and lines 20–22 for withdraw. The result at the end depends on the exact relative time at which these lines execute. If one of them starts and finishes completely before the other, everything will be fine. The race condition occurs when both are executing those critical lines at overlapping times.[11] In that situation, the one that executes _balance = newBalance last will have its results stored. The other one will have the work that it did wiped out. You might not mind this if your withdrawal is not recorded and you get your money for free. However, you might have a problem with losing a whole month's salary because you happened to visit the ATM at the wrong time.

Hopefully you can see now how race conditions can cause problems. Specific timing is required for the error to manifest. You might run a program for a long time before two critical segments of code ever happen simultaneously. This does not make them unimportant. Instead, it just makes them hard to demonstrate and debug. Indeed, that type of bug you get from a race condition is one of the hardest to diagnose and fix because it can be so rare and hard to reproduce.

Note again that race conditions only happen when at least one thread is writing to memory. In many places in this book, we have argued that there are benefits to using a functional approach in your programming and using immutable values as much as possible. This should probably be the strongest reason yet. If all of your data is immutable, it is impossible to have race conditions. Indeed, the advent of multicore chips has been a major factor in the rise in popularity of functional languages and the addition of functional features to languages that were created as functional languages.

8.6 Synchronization

The first approach to dealing with race conditions is SYNCHRONIZATION. When a block of code is synchronized, it means that when that block executes it first checks a MONITOR on the specified object. If the monitor is unlocked, that thread acquires the lock on the monitor and executes the body. If the monitor is locked by another thread, the thread blocks until the lock is released by that thread. This is done by calling the synchronized method in AnyRef. That method has the following signature.

```
def synchronized[T](body: => T): T
```

Note that the body is passed by name and the result type is the type parameter. This makes it so that you can write whatever code you might have normally written as the argument

---

[10]If only it were that easy to get $10,000,000 into your bank account.

[11]Note that this is a problem even if you do not have multiple cores. In that situation, if you have multiple threads, one thread stops executing for a while and is swapped out for another one. So the problem would involve one thread giving up control in the middle of this type of calculation and then its execution is not recorded.

after the call to **synchronized**. The fact that it is defined in **AnyRef** means that you can call this inside of any **class**, **object**, or **trait** declaration, and you often do not have to specify the object it is being called on as it will implicitly be called on **this**. You do have to be careful to keep in mind what instance object it is being called on though, as that will be the instance the monitor is locked on, and using the wrong one can lead to incorrect results.

As a first example, let us look at how we would alter the bank account class to be synchronized so that we do not have to worry about deposits and withdrawals happening at the same time.

```
package multithreading1

class AccountSync(
    private var _balance: Int,
    val id: String) {

  def balance = _balance

  def deposit(amount: Int): Boolean = synchronized {
    if (amount > 0) {
      val oldBalance = _balance
      val newBalance = oldBalance + amount
      _balance = newBalance
      true
    } else false
  }

  def withdraw(amount: Int): Boolean = synchronized {
    if (amount > 0 && amount <= _balance) {
      val oldBalance = _balance
      val newBalance = oldBalance - amount
      _balance = newBalance
      true
    } else false
  }
}
```

As you see, all we did here was put a call to **synchronized** around the full method for **deposit** and **withdraw**. That is all that is required for that example.

At this point you might be wondering why you do not synchronize everything to avoid race conditions. The reason is that synchronization has a cost in execution speed. Checking and setting locks on monitors takes real time. In addition, it prevents multiple threads from operating. If you really did synchronize everything, you would likely make the program run very slowly with only a single thread active most of the time and have very good odds of deadlock. Instead, you have to be careful with how you synchronize things.

To illustrate this, consider our first example with a **for** loop that simply increments the variable **cnt**. It is tempting to change that to look like the following.

```
for(i <- (1 to 1000000000).par) synchronized { cnt += 1 }
```

Technically, this works and it gets the right answer. However, it also runs many times slower than the original sequential code. That happens because this basically is sequential code. Each execution of **cnt += 1** has to happen independently of the others. Since that is all the work there is, nothing can happen in parallel. Unlike the straight sequential code though, this code also breaks up the work, schedules it for threads, and most importantly, checks and sets the monitor lock before each increment, then releases it after the increment.

There are other ways of doing this type of thing that can give a speed boost and always provide the correct answer. We will consider a number of them in the next chapter, but before we close this section, we will look at an approach that just uses methods in the parallel collection. The use of counting up by one made a good example earlier because that way we knew the answer, but it is not really useful. If we instead sum up the values from calling some unspecified code, the usefulness might be more apparent.[12] With a race condition, that code would look like this.

```scala
var cnt = 0L
for (i <- (1 to 1000000000).par) cnt += someFunction(i)
println(cnt)
```

Since all we are doing is adding things up, we could `yield` the values from the function and then sum them all up as is done here.

```scala
val cnt = (for (i <- (1 to 1000000000).par) yield someFunction(i)).sum
```

Of course, a `for` loop just gets converted to various higher-order methods, so if you prefer, you can also write the following.

```scala
val cnt = (1 to 1000000000).par.map(someFunction).sum
```

The `sum` method on parallel collections is an efficient way to do a parallel sum, so you are likely to find that either of these last two code approaches runs faster than the sequential code.

If you have something more complex than a sum, product, minimum, or maximum you cannot simply use the built-in methods that calculate those things. It might take a bit more thought, but that is when the **reduce**, **fold**, or **aggregate** methods can come into play.

8.7   Deadlock

There is another reason to not simply put everything in **synchronized** blocks. It turns out that over-synchronization has a tendency to cause the other primary error associated with multithreading, deadlock. In addition to using **synchronized**, you can do custom locks, and you can manually have some threads wait for other threads to finish before they move on to further tasks. In our hole digging analogy, you probably should not start filling the trench with water until after all the digging is done and everyone has gotten out. We will explore how you can do those things in the next chapter, but we can describe how you get deadlock with just **synchronized**.

What happens in deadlock is that you have different threads that are waiting on each other and none of them can progress until one or more of the others finishes. As a result, they all just sit there and whatever work they were supposed to be doing never happens. Consider the following example methods.

```scala
def method1(a: A, b: B): Unit = {
  a.synchronized {
```

---

[12]Note that we are making `cnt` of type `Long` because the number could get too big to be stored in an `Int`.

```
    // Do stuff - 1
    b.synchronized {
      // Do Stuff - 2
    }
    // Do Stuff - 3
  }
}

def method2(a: A, b: B): Unit = {
  b.synchronized {
    // Do stuff - A
    a.synchronized {
      // Do Stuff - B
    }
    // Do Stuff - C
  }
}
```

Now imagine that one thread calls `method1` while a second thread calls `method2` and they are passed the same objects for `a` and `b`. As soon as `method1` starts, it acquires the monitor on `a` and does some work. Similarly, `method2` will acquire the monitor on `b`, then do some work. Before releasing the locks they have, they try to acquire the locks they do not have yet. This can cause a deadlock, because `method1` cannot get the monitor for `b` until `method2` is done and `method2` cannot get the monitor for `a` until `method1` is done. The result is that they sit around waiting for one another forever and those two threads simply stop.

You might argue that this is a contrived example and that you would never put such alternating locks into real code. In reality though, the inner lock would probably be located in some function/method that is called by the outer lock, so it will not be at all obvious that such an alternating lock structure is present in the code. As with race conditions, deadlock is very intermittent and often extremely challenging to debug, perhaps even harder than race conditions. If a race condition occurs, you generally know it because something is output that you know is wrong. That gives you some idea of where to look. If your program has a deadlock, the whole program or parts of it simply stop running, and you generally get very little information about where or why it happened.

So how do you save yourself from spending hours tracking down bugs created by race conditions or deadlock? One simple approach is to use the features of the libraries discussed in this chapter and avoid mutable data. If you do not use mutable data, you do not need to have explicit synchronization or locking, so you cannot have race conditions. What about when you need to get different threads to agree about when certain things should happen like the example earlier of not filling the trench with water until all the workers were done digging? It turns out that we have already seen how to do that. Consider the following pseudocode.

```
val workers = Vector.fill(100)(new Worker)
val futures = workers.map(w => Future { w.digTrench() })
Futures.sequence(futures).foreach(ws => fillWithWater())
```

Here we make 100 workers and have each of them do their work digging in separate `Futures`. The call to `sequence` gives us back a single `Future` that will not complete until all the workers are done digging. The call to `foreach` schedules the filling to happen only after the digging is done.

Alternately, you could just use a parallel collection.

```
val workers = Vector.fill(100)(new Worker)
for (w <- workers.par) w.digTrench()
fillWithWater()
```

The parallel collection methods automatically block until all of the tasks are done, so the call to `fillWithWater()` cannot happen until all the workers are done.

Of course, this example ignores the details of the digging, which certainly seems like a mutating process, though you could probably come up with functional methods of doing it. In the next section, we will see another approach that allows us to protect mutability by encapsulating it in a way that only allows one thread to touch the mutable data. That might be a more appropriate model for digging a trench.

8.8 Multithreading in GUIs

Prior to this chapter, your code has been written using a single thread. However, in reality, any code that uses a GUI is inherently multithreaded, though ScalaFX does a good job of hiding this from you. One way that you can identify the multithreading is to print the name of the current thread, which you can get with `Thread.currentThread().getName`. If you print this in a normal application, it will print out `main`. However, if you do this in a ScalaFX application, it will print `JavaFX Application Thread`. This primary thread created by JavaFX handles all the events in the GUI and is responsible for doing all the drawing. By putting all of these activities in a single thread, they prevent race conditions.

The fact that a single thread is responsible for handling all events and repaints means that you have to be careful what you do in that thread or you can stop all responses from the GUI. The following code demonstrates that fact.

```
package multithreading1

import scalafx.Includes._
import scalafx.application.JFXApp
import scalafx.scene.Scene
import scalafx.scene.control.Button
import scalafx.event.ActionEvent

object ThreadBlockGUI extends JFXApp {
  stage = new JFXApp.PrimaryStage {
    title = "GUI Thread Demo"
    scene = new Scene(100, 50) {
      val button = new Button("Click Me")
      root = button
      button.onAction = (ae: ActionEvent) => Thread.sleep(10000)
    }
  }
}
```

If you click the button, it blocks the event thread from doing anything else for ten seconds. If you run this program, you can click the button and see the effects. One effect is that the button will "stay down". More interesting is seeing what happens if you try to close or resize the window during the 10 seconds after clicking the button. If you try to close, it will not actually happen until the end of the ten seconds. If you resize the window, the window will not repaint itself during that time either.

To see this same effect on a larger scale, add the following line into the **commands** map in the **Commands** object in the drawing project code.

```
"freeze" -> ((rest,d) => Thread.sleep(rest.trim.toInt*1000))
```

Now you can run the drawing program and enter the command "freeze 10". After doing this, nothing you do in the GUI will take effect for ten seconds. This command will be useful in the next section as well to see how blocking the event thread impacts other things that happen in the GUI.

Much of this chapter has focused on the challenges associated with using threads. Those challenges remain for the threads associated with a GUI. As you just saw, keeping the event thread busy too long can lead to an unresponsive GUI. It is also risky to have other threads altering mutable data that is used by the event thread as that can lead to race conditions. This combination makes things a bit tricky because you should have any significant work load happen in a separate thread, but you should not have that other thread mutate values that are used for painting or event handling. Instead, you should have the other thread schedule certain things to happen back in the main event thread.

The way you do this scheduling in ScalaFX is with the **Platform.runLater** method. It takes a single, by-name argument that is the code that should be executed in the event thread. So if you run some code in a different thread using a **Future**, parallel collections, or some other approach, and you find that you need to do something that modifies the GUI, you should put the code that modifies the GUI in a block after **Pltaform.runLater**.

While the "freeze" command is a rather contrived way of stalling the GUI, it is possible that we will create other commands in the future that require a lot of processing and we will not want those to stall out our GUI while they are processing. For that reason, it would probably be better to run commands in a separate thread. However, when the commands are done, we have to update the text in the **commandArea**. That is an action that should happen in the primary event thread to prevent race conditions.

Here is the code that we have been using to run the commands.

```
commandField.onAction = (ae:ActionEvent) => {
  val text = commandField.text.value
  if(text.nonEmpty) {
    commandArea.text = (commandArea.text + ">
      "+text+"\n"+Commands(text,drawing)+"\n").value
    commandField.text = ""
  }
}
```

The **Commands(text,drawing)** part is what could be time consuming, so it really should be run in a **Future**. Consider the following alternative code.

```
commandField.onAction = (ae: ActionEvent) => {
  val text = commandField.text.value
  if (text.nonEmpty) {
    commandField.text = ""
    val future = Future { Commands(text, drawing) }
    future.foreach(result => Platform.runLater {
      commandArea.text = (commandArea.text+"> "+text+"\n"+result+"\n").value
    })
  }
}
```

This code puts the potentially blocking code into a **Future**. We then use **foreach** to schedule the update to the **commandArea** after that calculation has completed. The key thing to note is the use of **Platform.runLater** in the **foreach**. This is needed because the code in **foreach** is also done in a separate thread.

To really see the impact of this code, you need to run **DrawingMain** then try using the "freeze" command again. What you will find is that the GUI no longer freezes up. If you do "freeze 10" you can keep using the GUI and nothing will happen for ten seconds. At the end of the ten seconds, the output area will show the proper feedback to indicate that the command completed. You can even type in other commands while a freeze command is working. If the other commands finish before the end of the freeze time, their feedback in the output panel will come before that of freeze.

Making this change means that we need to be more careful about what is happening in the **Commands object**. In particular, because the commands will now be run in a separate thread, if a command does anything directly to alter the GUI, we need that to be done in a call to **Platform.runLater**. Currently, we only have one command that falls into this category, and that is **refresh**. To make the code safe, that command should be updated to the following because the call to **d.draw()** potentially does a lot of things to the **Canvas**, which is part of our GUI.

```
"refresh" -> ((rest, d) => Platform.runLater { d.draw() } ),
```

This might seem like a minor thing, but making significant changes to a GUI, such as drawing to a **Canvas** from another thread, can cause some very significant bugs that are extremely challenging to diagnose. We have seen situations where parts of the **Canvas** just go blank and subsequent updates do not draw anything, even though the rest of the code continues to run. What makes it worse is that when this happened there was no stack trace printed or other feedback indicating what might have caused the error.

8.9 Multithreaded Mandelbrot (Project Integration)

Now that you have seen many of the different possibilities for multithreading as well as the challenges that it poses, let us write a simple example that uses multithreading for our project. The code that we will write here will draw a Mandelbrot set with the user able to specify the range to draw on the real and imaginary axes as well as the size of the region to draw. We will put this in a new class called **DrawMandelbrot** that extends **Drawable**.

If you do a web search for the Mandelbrot set, you will see a lot of pretty pictures. Figure 8.3 shows what it will look like in our implementation. In many ways, the pretty picture is what we are interested in producing with the drawing program, but to code it, we need to have a slightly deeper understanding. The Mandelbrot set is a fractal set of points in the complex number plane characterized by the mapping $z_{n+1} = z_n^2 + c$, where $z_0 = 0$ and c is a point in the plane. So given any complex value, c, there is a different sequence of z_i values that are generated. For some points, the sequence will diverge to infinity. For others, it will stay bound close to zero. For our purposes here, it is not the math we are concerned with, but making the math happen in parallel while being translated to an image that we can draw.

Calculating the Mandelbrot set is an EMBARRASSINGLY PARALLEL problem. This means that it can be broken into separate pieces that can be solved completely independently. So you can put each piece in a different thread and they do not have to talk to one another.

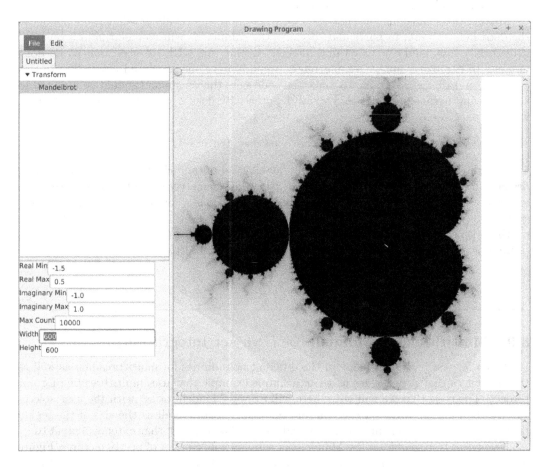

FIGURE 8.3: This is a screen capture of the drawing program with a Mandelbrot image.

This makes it an ideal candidate for solving with the parallel collections library. This code shows just such a solution.

```scala
package multithreading1.drawing

import scalafx.Includes._
import scalafx.scene.Node
import scalafx.scene.canvas.GraphicsContext
import scalafx.scene.layout.VBox
import scalafx.scene.paint.Color
import scalafx.scene.image.WritableImage
import scala.concurrent.Future
import scala.concurrent.ExecutionContext.Implicits.global
import scalafx.application.Platform

class DrawMandelbrot(d: Drawing) extends Drawable(d) {
  private var (rmin, rmax, imin, imax) = (-1.5, 0.5, -1.0, 1.0)
  private var (width, height) = (600, 600)
  private var maxCount = 100
  private var img = new WritableImage(width, height)
  private var propPanel: Option[Node] = None

  startDrawing()

  override def toString() = "Mandelbrot"

  def draw(gc: GraphicsContext): Unit = {
    gc.drawImage(img, 0, 0)
  }

  def propertiesPanel(): Node = {
    def checkChangeMade[A](originalValue: A, newValue: => A): A = {
      try {
        val nv = newValue
        if (originalValue == nv) originalValue else {
          startDrawing()
          nv
        }
      } catch {
        case nfe:NumberFormatException => originalValue
      }
    }
    if (propPanel.isEmpty) {
      val panel = new VBox
      panel.children = List(
        DrawingMain.labeledTextField("Real Min", rmin.toString(), s => { rmin =
            checkChangeMade(rmin, s.toDouble) }),
        DrawingMain.labeledTextField("Real Max", rmax.toString(), s => { rmax =
            checkChangeMade(rmax, s.toDouble) }),
        DrawingMain.labeledTextField("Imaginary Min", imin.toString(), s => { imin
            = checkChangeMade(imin, s.toDouble) }),
        DrawingMain.labeledTextField("Imaginary Max", imax.toString(), s => { imax
            = checkChangeMade(imax, s.toDouble) }),
        DrawingMain.labeledTextField("Max Count", maxCount.toString(), s => {
            maxCount = checkChangeMade(maxCount, s.toInt) }),
```

```
48      DrawingMain.labeledTextField("Width", width.toString(), s => { width =
           checkChangeMade(width, s.toInt) }),
49      DrawingMain.labeledTextField("Height", height.toString(), s => { height =
           checkChangeMade(height, s.toInt) }))
50    propPanel = Some(panel)
51  }
52  propPanel.get
53  }
54
55  private def mandelIter(zr: Double, zi: Double, cr: Double, ci: Double) = (zr * zr
       - zi * zi + cr, 2 * zr * zi + ci)
56
57  private def mandelCount(cr: Double, ci: Double): Int = {
58    var ret = 0
59    var (zr, zi) = (0.0, 0.0)
60    while (ret < maxCount && zr * zr + zi * zi < 4) {
61      val (tr, ti) = mandelIter(zr, zi, cr, ci)
62      zr = tr
63      zi = ti
64      ret += 1
65    }
66    ret
67  }
68
69  private def startDrawing(): Unit = Future {
70    val image = new WritableImage(width, height)
71    val writer = image.pixelWriter
72    for (i <- (0 until width).par) {
73      val cr = rmin + i * (rmax - rmin) / width
74      for (j <- 0 until height) {
75        val ci = imax - j * (imax - imin) / height
76        val cnt = mandelCount(cr, ci)
77        writer.setColor(i, j, if (cnt == maxCount) Color.Black else
78          Color(1.0, 0.0, 0.0, math.log(cnt.toDouble) / math.log(maxCount)))
79      }
80    }
81    Platform.runLater {
82      img = image
83      drawing.draw()
84    }
85  }
86 }
```

Many of the aspects of this code are the same as other `Drawables` that we have written previously. We have member data at the top as well as methods overriding `toString`, `draw`, and `propertiesPanel`. The calculations for and drawing of the Mandelbrot set happen in three helper methods at the bottom of the `class`. You can explore the details of the math in the `mandelIter` and `mandelCount` methods, but they are not particularly relevant to our learning goals here. `startDrawing` is more interesting because it deals with threading.

The part of this code that makes the calculation happen in parallel is the call to `par` on line 72. There is not a call to `par` in the inner loop on line 74 based on the assumption that whole columns make a fairly good level at which to break things up for threads. The code in lines 73 through 79 basically calculate the values for each pixel in each row of a column. By placing a call to `par` in line 72, we are allowing the application to run all the calculations needed for each column independently using multithreading. A `WritableImage` is used so that we can set individual pixels based on the calculation of how many iterations it takes for a point to diverge.

The first version of this code that we wrote put the contents of `startDrawing` inside of the `draw` method. That is fine when the value of `maxCount` is 100. However, if you go up to 10000, which gives more refinement to the edges of the set and is really needed if you zoom in, the drawing takes long enough that it can stall the GUI. For that reason, we decided to use a `Future`, and pull the code that actually creates the drawing out of the `draw` method. This way, the rendering calculations can begin as soon as any change is made, before the image is even needed, and the GUI does not stall at all. When the calculation of the image is done, a call to `runLater` is made, which will set the member variable `img` and tell the drawing that it needs to be redrawn.

It is tempting to add extra synchronization to lines 25 and 82 where the `img` member is used because line 82 alters the value in code that is in a `Future`. However, we make the assumption that the `draw` method will only be run by the event thread for ScalaFX, if it were run in a different thread that would be a bug because it alters the GUI. Since the code in `runLater` gets scheduled for that same thread, that is sufficient synchronization to prevent a race condition.

Note that `startDrawing` is called in two places in this code. The first time is on line 20. This call simply starts making an initial image that can be displayed using the original values. The second time is on line 33, which is inside of a little helper method[13] called `checkChangeMade` that is nested in `propertiesPanel`. This method holds code that needs to be executed every time the user changes a property of the Mandelbrot set drawing. It takes a type parameter, because some of the properties are `Int`s and others are `Double`s. Note that the new value is passed by-name. This allows us to do error handling in the function. The entire body of the function is a `try` expression. It starts by evaluating `newValue` and giving it the name `nv`. If you look at lines 43–49, you will see that all the new values come from code that invokes `toInt` or `toDouble`. If the user types in something that is not of the right format, such as "abc", those calls will throw `NumberFormatExceptions`. Thanks to the way that pass-by-name works, that would happen on line 31 instead of down in each of the handlers. If the code that produces the new value does not throw an exception, the new value is checked against the old one. If it has changed, we start rendering again and return the new value. If it did not change, we give back the original value. If an exception is thrown calculating the new value, we also give back the original value.

It is worth taking a minute to reflect on what the `propertiesPanel` method would look like if we did not have `DrawingMain.labeledTextField` or `checkChangeMade` and instead, wrote out the full code for creating each `TextField` with the `Label` and event handling code. Each one of the seven properties would require 10–15 lines of code to set up instead of the one line used to build each one here. That one method would be as long as the entire `class` is currently. This is a nice example of how we can use the different features of Scala to reduce code duplication. This code specifically utilizes passing functions, pass-by-name, and type parameters to provide powerful abstractions to minimize duplication.

Once you have this code, you can add a line to the `DrawingMain` for a creator that builds a `DrawMandelbrot`, then run the program and add a Mandelbrot set. The result should look like figure 8.3.

---

[13]A helper method is a method that helps another method. It is used to organize code to perform smaller tasks within the method. It is not meant to becalled by outside methods.

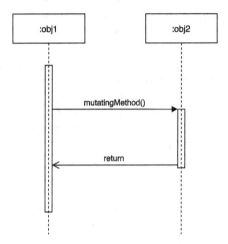

FIGURE 8.4: This figure shows a single method call between two objects in a basic sequence diagram. The key thing to note is that the thread goes with the call. So when you call methods, the thread weaves around and between objects.

8.10   Actors and Akka

Another approach to doing multithreading in a safe manner is the ACTOR MODEL. This approach was first explored in 1973, but as was the case with many approaches to parallelism, there was not a strong motivation to study it until the advent of multi-core chips made everything capable of parallel computing.

The object model that we have been learning about allows you to protect your data by encapsulating it inside of objects. These objects should only allow appropriate operations to be performed on the data. The protection comes from the fact that other parts of code cannot get access to `private` data. However, when you call methods, threads go with them. Figure 8.4 shows a UML sequence diagram that illustrates this. UML sequence diagrams show multiple objects spread out horizontally with time advancing as you move down vertically. This figure shows that some method in `obj1` is executing and it makes a call to `mutatingMethod()` on `obj2`. When that method is done, the method returns a value and the thread of control comes with it.

Unfortunately, as we saw with the bank account example, the standard model of object-orientation does not do anything to protect you from race conditions. Figure 8.5 shows how race conditions can occur as a sequence diagram. Here there are two threads that are originally executing code in `obj1` and `obj3` respectively. Each of them makes a call to a method of `obj2`. As a result, we have a period of time when there are two threads executing in `obj2`. So if those methods happen to be using the same mutable memory, and either one of them is writing to it, a race condition is likely to occur.

A simple way to view the actor model is that each object gets a thread and a message inbox, and we call it an actor. In reality, actor systems do not make a full thread for each actor, but they do make sure that each actor is only using one thread at a time. Instead of calling methods on your actors, you send them messages. The messages are queued up in their inbox and they handle them in the order that they are received. This approach insures that each actor is only processing a single message at a time. You can see this represented as a sequence diagram in figure 8.6.

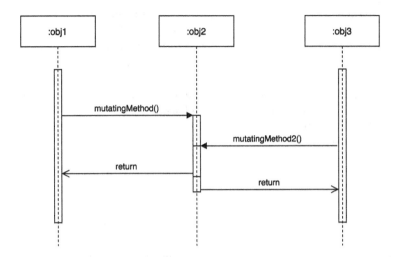

FIGURE 8.5: This figure shows three objects and two method calls in a basic sequence diagram that illustrates how a race condition occurs. Even if the objects properly encapsulate and protect their values, you still have to use synchronization to prevent race conditions.

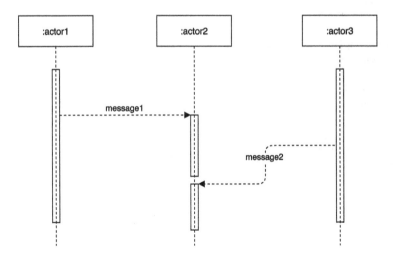

FIGURE 8.6: This figure shows three actors and two messages in a basic sequence diagram that illustrates how the actor model prevents race conditions. Each actor has only a single thread, and messages are processed sequentially, so as long as mutable state is confined to a single actor, you cannot have race conditions.

The parallelism of the actor model comes from the fact that a system can have multiple actors, and all the actors can be doing things at the same time. In many ways, the actor model is more similar to a group of people working on a project together. Each person only does one thing at a time, but many of them can be working simultaneously. There will be times when some actors do not have messages to process, but as long as there are as many actors with messages to process as you have cores in your machine, you can feel your hardware fully busy.

Indeed, this analogy to humans can be useful in building actor systems. When you think about how to set up an actor-based system and what types of messages you want, a good approach is to try to imagine that the actors are people sending emails or other electronic messages between one another.

8.10.1   Akka

The primary actor system for Scala is called Akka.[14] The Akka library is robust and broadly used. The Scala libraries originally included their own actor library, but it was flawed in a number of ways and has been deprecated. Instead, developers are encouraged to use Akka. You can find information on Akka, including downloads and APIs, at `http://akka.io/`. You should go there and potentially download the latest version so that you can add it to your path.[15]

Let us look at a very simple example program that uses Akka actors. The following code is an application that has a nested **class** called **SimpleActor** that extends **akka.actor.Actor** and shows how to use it.

```scala
package multithreading1.actorexamples

import akka.actor.Actor
import akka.actor.ActorSystem
import akka.actor.Props

object SimpleExample extends App {
  class SimpleActor extends Actor {
    def receive = {
      case s: String => println("String "+s)
      case i: Int => println("Int "+i)
    }
  }
  val system = ActorSystem("SimpleExample")
  val actor = system.actorOf(Props[SimpleActor], "FirstActor")
  actor ! "Hi"
  actor ! 42
  actor ! 'a'
  system.terminate()
}
```

The **Actor trait** has one abstract method called **receive** which we have to implement. The **receive** method defines which messages your actor can handle. It is a PARTIAL FUNCTION, which we write in Scala as curly braces with various **cases** inside, much like what you do with a **match** or a **catch**. The term partial function means that it does not have to work

---

[14]Akka can also be used in Java and other JVM languages. They are working on a .NET port of the library as well.

[15]The examples in this book require the akka-actor and config JAR files that you will find in the lib/akka directory of the install. If you are using an IDE and not sbt, you can add these to your build path.

with every possible input. In this example, we have `cases` for `String` and `Int`. If the actor receives one of those, a simple message is printed. Note that the actor `class` could be in its own file, and it should be for larger projects, but it is nested in the application in our small examples for brevity.

After the declaration of the `Actor class`, line 14 creates and `ActorSystem` and calls it `system`. Every application written with Akka needs to have a system, and most of the time there should only be one. When we create the `ActorSystem` we pass it a `String` for a human readable name.

Once we have a system, we can use it to make an actor. Line 15 calls `system.actorOf` and gives the result the name `actor`. There are several interesting things to note about this line. First, the value returned by `actorOf` is not actually an `Actor`. Instead, it is an `ActorRef`. This is a safety mechanism in Akka. We'll come back to this in a bit after going through the rest of this code.

There are two arguments to `actorOf`. The first one is of the type `akka.actor.Props`. This argument tells `actorOf` what type of `Actor` it is building and how to build it. The usage here is the simple form where we do not have to pass arguments to the constructor for `Actor`. We will see the alternate form where we can pass values in a later example. You have to use `Props` to create an instance of any `Actor` in Akka. If you try to simply call `new SimpleActor` without using `Props`, it will throw an exception. The second argument to `actorOf` is a `String` with the name you want for your actor instance. The names of actors have to be unique. If you leave this argument off, Akka will generate a unique name for you, but we will see later that there are reasons why you might want to know the names of your actors.

Lines 16–18 send messages to `actor`. This is being done with an exclamation point. The use of an exclamation point was chosen to mirror the syntax of Erlang, a functional language that is completely based on the actor model. In Scala, this is not some odd language syntax, there is just a method called `!` in the `ActorRef class` that sends messages. There is also a method called `tell` that can be used for this purpose, but we will generally use `!`. Note that the exclamation point is often read as bang in code because it is a lot fewer syllables.

It is important to note that sending messages does not block. So after the message of `"Hi"` is sent, the message of `42` can be sent before the `println` for the first message happens.[16] The print statements in this example will always happen in the same order with `String Hi` printing before `Int 42` because messages that go to the same actor are queued and handled in order. As we will see later though, messages that go to different actors can be processed in a different order than the order that they are sent.

The first and second messages send types that the `receive` method in `SimpleActor` has a `case` for, so each of those print out. The third message sends a `Char`, which the `SimpleActor` does not have code to handle. As it happens, this is not a problem. When an `Actor` gets a message that it does not know how to handle, it just moves on to the next message. The messages themselves are of type `Any`. This means that Akka is not doing any static type checking. The determination of whether a message does something happens at run time when it is received and checked against the partial function. While some people view this as a strength of Akka, there are enough people who appreciate type safety that there is a project called Akka Typed, which is aimed at creating a type safe alternative. That extension is still in the early stages at the time of this writing and is likely to evolve, so we do not cover it.

While the messages can be of any type, using standard types like `String` and `Int` as we did in this example is not recommended. Those types are too common. We would like

---

[16]In reality, you do not know whether the print will happen before or after the next message is sent because that is up to the thread scheduler.

two things from the types of messages that we pass around. They should be meaningful and immutable. Providing meaning in the code makes it easier for programmers to figure out what is going on and to fix things when something goes wrong. The reason for being immutable is less obvious, but far more important: sending mutable messages can cause race conditions. The reason for this is that when one actor sends a message, it could keep a reference to it. If it does, then the actor that receives the message will also have a reference to it, and that means we have two actors that use two separate threads that can potentially access the same mutable memory. The simple way to achieve both of these goals is to use `case class`es that store immutable data for messages. We will do this in future examples.

The last line in the application calls `system.terminate`. As you can probably tell from the name, this method tells the `ActorSystem` that you are done with it and that it should shut things down. **Earlier versions of Akka used shutdown instead of terminate, but that method has now been deprecated.** Back in section 8.3, we saw that the thread for a `Future` does not keep the application alive when the main thread terminates. To make sure that our `Future`s had completed their work, we inserted calls to `Thread.sleep`. The `ActorSystem` is the opposite. It will keep an application running forever unless we terminate it. We will see later that we do have to be careful calling `terminate`, as doing so does stop all the actors, even if they still have messages that we want them to process.

We mentioned earlier that the `actorOf` method gives back an `ActorRef`, not an `Actor` for safety reasons. When you communicate with `Actor`s, you are supposed to do it by passing messages, not calling methods. If you got an actual reference to an `Actor` you could call methods on it, which could potentially induce race conditions and eliminate the whole reason for using the actor model. Instead, you get an `ActorRef`, which is an instance object that wraps around an `Actor` and includes code for handling the inbox and messages among other things. To make this clearer, consider this slightly different form of `SimpleActor` that includes a mutable value and a method that mutates it.

```scala
class SimpleActor extends Actor {
  private var counter = 0

  def receive = {
    case s: String => println("String "+s); count()
    case i: Int => println("Int "+i); count()
  }

  def count(): Unit = counter += 1
}
```

The idea here is that we want to count up how many messages our actor has handled. This works great and is free of race conditions as long as we are only processing messages one at a time and no other code can call `count` from the outside. The first requirement is a general property of the actor model. The second requirement is something that Akka helps enforce by giving us an `ActorRef`. If Akka gave us back an instance of `SimpleActor`, we could accidentally call `count` from a different thread and create a race condition. We could further protect ourselves by making `count` and any other methods, other than `receive`, that we put into our `Actor`s, be `private[this]`. Doing so allows the Scala language to enforce that they are never called by anything other than the current instance of the object.

8.10.2   Actors Communicating

The next step up in complexity is to show how actors can communicate with one another. To demonstrate this, we will use a simple example with two actors that count down. So one actor will print a number, then the other actor will print the next one, and they will go back and forth until they get to zero. For this example, we will break out the `Actor class` into its own file and set up the code in the way that we would with larger applications. The following code shows the file with the implementation of the `Actor` type.

```scala
package multithreading1.actorexamples

import akka.actor.Actor
import akka.actor.ActorRef

class CountDownActor extends Actor {
  import CountDownActor._
  def receive = {
    case StartCounting(n, o) =>
      println(n)
      o ! Count(n - 1)
    case Count(n) =>
      if (n > 0) {
        println(n)
        Thread.sleep(1000)
        sender ! Count(n - 1)
      } else {
        context.system.terminate()
      }
  }
}

object CountDownActor {
  case class StartCounting(n: Int, partner: ActorRef)
  case class Count(n: Int)
}
```

The file includes both a `class` and the companion `object`. The companion `object` holds the two `case classes` that we are using for our message types. The use of the companion object here is a style choice. The `case classes` that are used as message types have to go somewhere. By always putting the message types that an actor knows how to handle in the companion object for that `class`, makes it easier for people reading the code to quickly known what they should be able to send in to this type of actor without looking through a lot of code. One of the message types is used to initiate the counting, and the second one is used for the standard counting down. The `import` statement on line 7 allows us to refer to the `case classes` using just the `class` name. Without that, we would have to say things like `case CountDownActor.Count(n) =>`.

The `StartCounting` message includes both the number that they are counting from, as well as the other actor involved in the counting. Note that the type that is used to refer to the actor is `ActorRef`. This will always be the case. You should virtually never have a reference to an `Actor` or a subtype of `Actor` because that would allow you to call methods that might induce race conditions.[17] When the `receive` method gets an instance of `StartCounting`, it prints that value and then sends a `Count` message to the other actor with a value one less than what it just printed.

---

[17] We could get by with a single message type if we used `tell` instead of `!` in our application. We did not do that, as it is important for readers to see how actors are passed by `ActorRef`. It is left as an exercise for the reader to make that simplification to this code.

The `Count` message type only holds an `Int` with the current value. The `case` for it in `receive` first checks if the value has reached zero. If it has not, it prints the value, pauses a bit, and then sends a message back to where this one came from. Notice the use of `sender` for the response message. `sender` is a method defined in `Actor` that provides the `ActorRef` that sent the most recently received message. Having access to that is why the normal `Count` message does not need to include an `ActorRef`. If the value in the `Count` message is zero or less, then the actor makes a call to `context.system.terminate`. `context` is a `val` in `Actor` of the type `ActorContext` that can be used to get access to a lot of general Akka functionality, including the `system` that the actor is running in.

We make use of `CountDownActor` in an application called `CountDown` that is shown here.

```
1   package multithreading1.actorexamples
2
3   import akka.actor.ActorSystem
4   import akka.actor.Props
5
6   object CountDown extends App {
7     val system = ActorSystem("CountDownExample")
8     val actor1 = system.actorOf(Props[CountDownActor], "Actor1")
9     val actor2 = system.actorOf(Props[CountDownActor], "Actor2")
10
11    actor1 ! CountDownActor.StartCounting(10, actor2)
12  }
```

As you can see, this code is fairly simple. It builds an `ActorSystem`, then makes two actors, giving them different names. The last thing it does is send a message to `actor1` telling it to start doing a count down from 10 with `actor2`. If you run this program, you will see it print 10 and 9 in quick succession. There will then be a one-second pause for each of the following numbers. One second after printing the value 1, the program terminates.

For this application, the call to `terminate` is in an actor, not at the end of the application. If you put a call at the end of the application, you will get a very different behavior. The values 10 and 9 will print, then there will be an error printed telling you that a message went to the "dead letter office" and the application will stop. The dead letter office is a construct in Akka where messages go when the actor they were sent to is not available to get them. In this case, the terminate message tells all the actors in the system to stop accepting new messages while one of the actors is executing `Thread.sleep`. It wakes up and sends a message to the other actor who can no longer receive it. That leads to the error printout and this is why we call `terminate` in an actor at the point when the countdown is done.

8.10.3   The Ask Pattern

There were two key differences between figures 8.5 and 8.6. Earlier we focused on the first one, which is that the queueing of messages and only having one thread executing in each actor prevents race conditions. However, you might have also noticed that there was no line providing a return value in figure 8.6. That has to be the case because when you pass a message it is non-blocking. The current thread does not wait around for a response, it just continues. This is actually the most challenging thing to get used to when you start using the actor model.

This does not mean that no information can come back. As we just saw in the countdown example, when you send a message to an actor, it can respond by sending a message back to sender. However, that message will be processed in a completely different invocation of `receive`. Most of the time that is fine. The response type can often provide enough information that you do not really need to know what you were doing when you sent the original message. There are some situations though when it really is nicer to get a "return" value back.

The fact that passing messages does not block means that getting back a value has to be handled in a different way. If you send a message and you really need to know at that point in the code what the response is, you can use the ask pattern. The ask pattern allows you to "ask" an actor for something. This is still a non-blocking call, but while sending a message with ! returns `Unit`, when you use ask you get back a `Future[AnyRef]`. You can use that `Future` instance to execute code as soon as you get back a response. The actor that receives an ask message does not get any indication that the message was sent with ask instead of a normal send. It is expected that when it handles that message type, it will send a message back to `sender` and the contents of that message will complete the `Future`.

There are two methods of doing an ask. One is with `akka.pattern.Patterns.ask`. The other is using ? in place of ! when sending a message. To demonstrate this, here is an `Actor` `class` that stores a name. It has two types of functionality. We can ask it for its name or have it ask another actor for its name.

```
 1  package multithreading1.actorexamples
 2
 3  import scala.concurrent.duration._
 4
 5  import akka.actor.Actor
 6  import akka.actor.ActorRef
 7  import akka.pattern._
 8  import akka.util.Timeout.durationToTimeout
 9
10  class AskActor(val name: String) extends Actor {
11    implicit val timeout = durationToTimeout(1.seconds)
12    implicit val ec = context.system.dispatcher
13    import AskActor._
14    def receive = {
15      case AskName =>
16        sender ! name
17      case AskOf(o) =>
18        (o ? AskName).foreach(n => println(s"They said their name is $n."))
19    }
20  }
21
22  object AskActor {
23    case object AskName
24    case class AskOf(otherActor: ActorRef)
25  }
```

The companion `object` in this file has two message types. Note that the first one is declared as a `case object` instead of a `case class`. This is because it does not take any arguments.

When an actor gets the `AskName` message, it simply sends its `name` back to the `sender`. Note that `name` is just a plain `String`. We said before that we should not send messages with these types to actors. We are willing to relax that rule for the response to an `ask`.

When an actor gets the `AskOf` message, it uses the ? operator to ask the provided `ActorRef` its name. That produces a `Future[AnyRef]` that we call `foreach` on. So as soon as the other actor responds, the current actor will print out a message that includes their name. The use of ? requires the definition of an `implicit` value for how long it should wait and it must be of the `akka.util.Timeout` type. The declaration on line 11 satisfies this requirement. If the other actor does not respond in the one second that we set as the timeout, an exception will be thrown.

As we saw previously, `Futures` need an `ExecutionContext` that can give them a thread to work with. In the examples with just `Futures`, we always did an `import` of `scala.concurrent.ExecutionContext.Implicits.global`. There are other execution contexts though, and one happens to be built into each `ActorSystem` that is used to schedule the actors. Line 12 is a declaration that tells the `Futures` in this code to use the `system.dispatcher` as their `ExecutionContext`. We will tend to do that in our actor examples under the assumption that using a single `ExecutionContext` for all the threading could have benefits compared to two separate ones where each does not know what the other is doing.

We can see how this type of actor is used in the following code that sets up an application.

```scala
package multithreading1.actorexamples

import scala.concurrent.duration._

import akka.actor.ActorSystem
import akka.actor.Props
import akka.pattern.Patterns.ask
import akka.util.Timeout.durationToTimeout

object AskPattern extends App {
  val system = ActorSystem("SimpleExample")
  val bob = system.actorOf(Props(new AskActor("Bob")), "BobActor")
  val jill = system.actorOf(Props(new AskActor("Jill")), "JillActor")
  implicit val ec = system.dispatcher

  jill ! AskActor.AskOf(bob)

  val nameFuture = ask(jill, AskActor.AskName, 1.seconds)

  nameFuture.foreach { s =>
      println("name = "+s)
  }

  Thread.sleep(100)
  system.terminate()
}
```

As we have seen before, we create an `ActorSystem` and two `Actors` which we have named `bob` and `jill`. The first operation we do is to send `jill` a message to ask the name of `bob`. That message is sent, then the code immediately uses the `ask` method to send another message asking the name of `jill`. The `Future` that is the result of that `ask` is given the name `nameFuture`. We once against use `foreach` to print a response once `nameFuture` has been completed.

The `ask` method was used here instead of `?` for two reasons. The first is that we want you to see how it is used. The second is that the `?` needs an `implicit` sender, and we did not want to complicate this code by adding another `implicit` declaration. Also note that the `ask` takes an explicit timeout, so we do not have to declare an `implicit` one. As a general rule, `ask` is almost certainly easier to use outside of `Actors` and is probably what you should prefer in general.

The code ends with a short `sleep` and a `terminate`. For this code, the print statements can occur in either order, so there is no clear last operation that we can put the call to `terminate` after. They both happen very quickly though, so only a short pause is needed before we can be certain that it is safe to shut things down.

Keep in mind that the `Future` you get back from `?` or `ask` is the same one we talked about earlier in this chapter, with all the same capabilities. This example does the most basic usage of `foreach` to keep things simple, but you certainly are not limited to that. All the other capabilities of `Future`s are at your disposal.

One thing to note is that there is a general recommendation that you should not use `ask` all that much. That is part of why they made it harder to use `ask` or `?` than to use `!`. It is generally considered better if your design can work by just sending messages that go to `receive` in the standard way. This is not just a style recommendation either. The next section will describe how there are certain bugs that you can create when you start using `Future`s in your `Actor`s that cannot exist otherwise.

8.10.4 Be Careful with `sender` and `this`

Special care must be taken when you are using `Future`s inside of `Actor`s. The functions that you create with the rocket or underscore syntax in Scala are also sometimes called CLOSURES. This is because in addition to the code in the functions, they also remember the values that are used inside of them. This is generally described by saying they encompass or "close over" those declarations. In doing this, they can keep those declarations around longer than their normal scope. There are two main ways in which this becomes a problem with `Actor`s and closures in `Future`s. One type of problem arises when you close over `sender` and another arises when you close over the current object, `this`.

The problem with closing over `sender` comes from the fact that `sender` is a `def`, and it gives back the `ActorRef` for where the most recent message came from. A `Future` can live longer than the current message. So if the closure in a `Future` that is in an `Actor` uses `Sender` directly, and that code is executed after another message has been received, the value of `sender` could be the wrong one.

Fortunately, there is a simple way of dealing with this potential problem. If you are going to use `sender` inside of a `Future`, declare a `val` that you set equal to `sender` before the `Future`, then use that `val` in the `Future` instead of sender. So you might have a line like `val s = sender` before any `Future` is created. Then inside of the `Future` code you use `s` instead of `sender`.

The other problem stems from the fact that the code in a `Future` is running in a different thread, but it is inside the `Actor`, so you could accidentally call methods and cause mutations in the `Future` thread while a message is being processed that is working with the same memory. If you do this, you will create a race condition, which is one of the main things we were trying to avoid by moving to the actor model.

Here is another recommendation that is fairly simple, do not mutate values in the `Future`s that are inside of `Actor`s. If something needs to be mutated, send a message to that `Actor` to have it done instead of doing it directly in the `Future`. If this is something that code outside the `Actor` should not be able to do, you can make the message type `private` in the companion `object`. An `Actor` can send a message to itself using `self`, which is a `val` in `Actor` that refers to the `ActorRef` of the current `Actor`.

So to summarize, you have to be careful with the functions you build in `Future`s that are inside of `Actor`s. If you are going to use `sender`, make a `val` outside of the `Future` and use it instead. Also, never mutate values in the `Actor` inside of a `Future`. If you need to mutate something, send a message to `self` to make it happen. Otherwise you are likely to create a race condition.

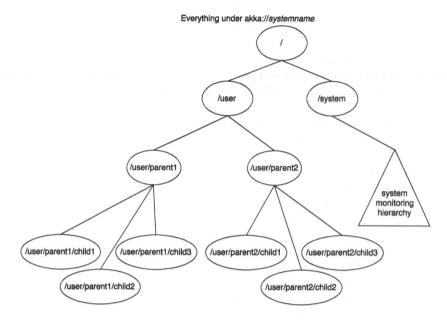

FIGURE 8.7: This figure shows the organization of the actor hierarchy that we are building in the example in this section.

8.10.5   The Actor Hierarchy

To this point we have just been creating a few actors in our application using the **system** and we have not really worried about the relationship between the actors. In large applications it is helpful to be able to organize actors into logical groups. As we will see in the next section, these logical groups help with error handling as well.

The way in which actors are organized is a hierarchy. Each actor can have other actors below it. Figure 8.7 shows a graphical representation of the hierarchy that we are going to build in our example in this chapter. All the actors that we create wind up being under the **/user** side of the hierarchy. There is also a system side with actors that are created specifically to run the Akka system.

Where an actor goes in the hierarchy depends on how it is created. So far we have created all of our actors using **system.actorOf**. When an actor is created using a call on system, it goes directly below /user. You can also create actors inside of other actors using **context.actorOf**. You can see an example of this in the following code segment, which gives the code for a **class** called **ParentActor**.

```scala
package multithreading1.actorexamples

import akka.actor.Props
import akka.actor.Actor

class ParentActor extends Actor {
  override def preStart() = {
    context.actorOf(Props[ChildActor], "child1")
    context.actorOf(Props[ChildActor], "child2")
    context.actorOf(Props[ChildActor], "child3")
  }

```

```
13     import ParentActor._
14     def receive = {
15       case Cascade =>
16         println("Parent")
17         context.children.foreach(_ ! ChildActor.SimpleMessage)
18     }
19   }
20
21   object ParentActor {
22     case object Cascade
23   }
```

You can see on lines 8–10 there are three calls to `context.actorOf` that create instances of `ChildActor`. The code for that is below. These three calls are inside of a method called `preStart`. You can tell from the **override** keyword that this is a method that is part of `Actor`. It is one of several methods that are invoked as part of the actor lifecycle, which we will discuss in detail in the next section. As the name implies, this method is called before the actor is started. That means that this code will execute before any messages can be received.

The `ParentActor` only responds to one message called `Cascade`. When this message is received, the actor prints a message and then sends a message to all of its children. We did not do anything to store the children in `preStart`. The Akka system does that for us. You can see on line 17 the usage of `context.children`, which is an `Iterable[ActorRef]`.

The `ChildActor` type, which is shown below, should be fairly straightforward at this point with the exception of two new things on lines 10 and 12 that are described below the code.

```
1    package multithreading1.actorexamples
2
3    import akka.actor.Actor
4
5    class ChildActor extends Actor {
6      import ChildActor._
7      def receive = {
8        case ToParent =>
9          println("Child sending Cascade to parent")
10         context.parent ! ParentActor.Cascade
11       case SimpleMessage =>
12         println("Child Simple = "+self.path)
13     }
14   }
15
16   object ChildActor {
17     case object ToParent
18     case object SimpleMessage
19   }
```

The `ChildActor` responds to two different types of messages. The `ToParent` message causes it to print something so we can tell where it is, then it sends a `Cascade` message up to its parent. It does this using `context.parent`, which is of type `ActorRef`.

The second message is `SimpleMessage`. When the `ChildActor` receives this message, it prints out a notification that includes showing its "path". The path of an actor is a full specification of where it is in the hierarchy. It starts with "akka://" followed by the system name. The other elements are what appear in figure 8.7. When you put them all together you get paths like "akka://HierarchyExample/user/parent2/child3". We will use these paths in the application code below.

We said before that actors have to have unique names. This is not quite true. They actually have to have unique paths. In this example, there are two actors with the name "child1". However, one of them is under "parent1" and one of them is under "parent2", so they have unique paths.

The full application sets up a system and then makes two instances of `ParentActor`. As we saw, each one of those makes three instances of `ChildActor`. So by line 10, we have created eight different actors.

```scala
package multithreading1.actorexamples

import akka.actor.ActorSystem
import akka.actor.Props

object HierarchyExample extends App {
  val system = ActorSystem("HierarchyExample")
  val actor1 = system.actorOf(Props[ParentActor],"parent1")
  val actor2 = system.actorOf(Props[ParentActor],"parent2")

  val c1 = system.actorSelection("akka://HierarchyExample/user/parent1/child1")
  c1 ! ChildActor.ToParent
  val c2 = system.actorSelection("/user/parent2/child1")
  c2 ! ChildActor.ToParent

  Thread.sleep(100)
  system.terminate()
}
```

The interesting parts of this example are on lines 11 and 13 where we look up actors using their paths with `system.actorSelection`. On line 11 we use the full path, which happens to be a URL, for the first child of the first parent. The value `c1` is not an `ActorRef`. It is an `ActorSelection`. While an `ActorRef` wraps around a single instance of `Actor`, an `ActorSelection` might resolve to one `ActorRef`, to many, or to none. In this case, it should represent a single `Actor`, but if we had given it a path that did not match any of the `Actors` in our system, then the `ActorSelection` would not be selecting any `Actors`. You can also put wildcards of * and ? in a path the way you do in operating system command lines. This allows an `ActorSelection` to represent many `Actors`. If you do this, then sending a message to the `ActorSelection`, like we do on lines 12 and 14, will do a broadcast to all of the `Actors` in that selection. If you need to get an `ActorRef` by path, the `ActorSelection` has a method called `resolveOne` that takes a timeout and gives back a `Future[ActorRef]` which you can use to get one `Actor` from the selection, assuming it has one.

Note that line 14 uses a shorter version of a path. The path specifies work like URLs or files in a directory system. If you use `context.actorSelection`, you can specify paths that are relative to the current `Actor`. So inside of "child1" you could use the path `"../child2"` to find its sibling named "child2".

A sample run of this application produced the following output.

```
Child sending Cascade to parent
Child sending Cascade to parent
Parent
Parent
Child Simple = akka://HierarchyExample/user/parent2/child3
Child Simple = akka://HierarchyExample/user/parent1/child1
```

```
Child Simple = akka://HierarchyExample/user/parent1/child3
Child Simple = akka://HierarchyExample/user/parent2/child2
Child Simple = akka://HierarchyExample/user/parent1/child2
Child Simple = akka://HierarchyExample/user/parent2/child1
```

Note the ordering of the print statements. You can tell from the "Child Simple ..." lines that there is some arbitrariness to the order. That is because the Actors are working on separate threads, and they process messages as the thread scheduler allows. The order of the prints is not completely arbitrary though. One "Child sending ..." message has to come before each "Parent" message, and a "Parent" message must come before the three "Child Simple" messages of the children of that parent. Other than that, the ordering is completely arbitrary, though you are likely to find that the thread scheduler spreads work fairly uniformly and you are not likely to see a "Child Simple ..." message ever come before the "Child sending ..." message associated with the other parent.

8.10.6   Error Handling and the Actor Lifecycle

One of the key reasons why many people like actor systems is that they are designed to be highly resilient. The primary user of the Erlang language, which is completely based on the actor model, was telecommunication companies. The telecoms are very sensitive to availability. Having a large section of a city lose phone service because of a bug in software is generally not acceptable. The key to actor systems being highly available is that they are **built to fail**. This means that they are constructed with the knowledge that things will go wrong. Part of the system will crash. That is simply unavoidable. What really matters is how quickly you can come back up and get things running again when that happens.

The feature of actor systems that makes them highly reliable is that the hierarchy we just saw provides a built-in structure for supervision so that when something does fail, there is another piece in the system that can be notified and it can do something about the failure to keep the system as a whole running smoothly.

To illustrate this, we are going to do another example that is very much like the last one with parents and children, but we will focus on the issues of supervision and actor lifecycle. We will start with supervision. A parent in the actor hierarchy is, by default, responsible for supervising its children.[18] When you write an Actor class that is going to have children, you can specify how it will handle their errors by overriding the supervisorStrategy val. There are two main strategies that you can pick from, OneForOneStrategy and AllForOneStrategy. If you are using the OneForOneStrategy, when one child actor has an error, your decision of how to handle it applies to only one child. So, if a child misbehaves, only that one child is dealt with. If you are using the AllForOneStrategy, then the decision made for that one child is applied to all the children at that time.[19]

When something goes wrong with a child, the supervisorStrategy gets the exception that caused the problem. Based on the details of that exception, it can decide to respond in one of four ways.

Resume tells the system to just let the child continue processing. It is like a parent telling their small child who has fallen that they are okay and can just get up and keep going.

Restart tells the system that the child is no longer usable, but that it should be replaced with a new instance. Since instantiating humans is far more time consuming than instantiating actor objects, there is not a strong human analogy here.

---

[18]It is possible to change this so that some other actor has the supervision responsibility.

[19]Life is not always fair for child actors.

Stop tells the system that the child is no longer usable and that it should not be replaced. In real life this would be a rather severe punishment that is illegal in most countries.

Escalate tells the system that the current actor does not know what to do with this problem, so it should be sent up to the next supervisor up the hierarchy. The parent is not certain what to do, so they go ask the grandparent. Unlike the human analogy, in Akka the decision of the next supervisor up will be applied not just to the child, but to the supervisor who decided to escalate the error.

To see what all of this looks like in code, here is an implementation of a parent actor that sets a supervision strategy to restart all children.

```scala
package multithreading1.actorexamples

import akka.actor.Props
import akka.actor.Actor
import akka.actor.OneForOneStrategy
import akka.actor.SupervisorStrategy._

class LifecycleParent extends Actor {
  override def preStart() = {
    context.actorOf(Props[LifecycleChild], "child1")
  }
  def receive = {
    case m =>
      println("Parent")
      context.children.foreach(_ ! LifecycleChild.SimpleMessage)
  }
  override val supervisorStrategy = OneForOneStrategy(loggingEnabled = false) {
    case ex: Exception =>
      println("Child had an exception")
      Restart
  }
}
```

Like the previous parent actor, this one creates children in `preStart`. We only create one here to keep the output simpler and easier to understand. The `receive` method has one case that will do the same thing for any message that it gets. Note that this really is not a style you want to emulate, but it works for this simple example.

The interesting code in this example is on lines 17–21 where we create the `supervisorStrategy`. The `OneForOneStrategy` is a `case class` with default values for all the arguments of the primary argument list. Here we pass in one argument by name to turn the logging off. It is also curried, and the second argument is a partial function that determines what to do for different possible exceptions. You can see in this case that all exceptions are treated the same, and when one happens we print a message and then tell the system to restart that child.

Akka provides a fair bit of control over what happens at different points in the lifecycle of an actor. We have seen the `preStart` method that gets called before the actor begins accepting messages. There are also methods called `postStop`, `preRestart`, and `postRestart` that we can `override`. The `postStop` method is called after the actor has been stopped and is no longer being sent messages. Both `preStart` and `postStop` are always called on all actors. The `preRestart` and `postRestart` are only called on methods that had an exception and were told to restart by their supervisor.

We can see these methods defined in the following implementation of a child actor that illustrates the actor lifecycle.

```
package multithreading1.actorexamples

import akka.actor.Actor

class LifecycleChild extends Actor {
  println("Making a child "+self.path)

  import LifecycleChild._
  def receive = {
    case Throw =>
      println("Child Dying")
      throw new Exception("Something bad happened.")
    case SimpleMessage =>
      println("Child Simple = "+self.path)
  }
  override def preStart() = {
    super.preStart
    println("Prestart")
  }
  override def postStop() = {
    super.postStop
    println("Poststop")
  }
  override def preRestart(reason: Throwable, message: Option[Any]) = {
    super.preRestart(reason, message)
    println("Prerestart "+message)
  }
  override def postRestart(reason: Throwable) = {
    super.postRestart(reason)
    println("Postrestart")
  }
}

object LifecycleChild {
  case object Throw
  case object SimpleMessage
}
```

This actor knows how to deal with two types of messages. The first one tells it to throw an exception. This allows us to activate the supervision behavior and see the methods associated with a restart being called. A second message type simply prints a message that includes the path of the actor.

Lines 16–31 are the declarations of the lifecycle methods. We do not have particular code for these, so we simply call the versions in the supertype, which is something you should generally do, and then print a message so we know that they happened. The restart methods take additional arguments that provide information about why the restart occurred.

When a restart occurs, the **preRestart** method is called after **postStop** on the actor that is being restarted. The **postRestart** method is called on the new actor, after the restart has occurred and after **preStart** has been called on that new actor. When you view all these calls in order, this ordering seems rather natural. Where it can be confusing is when you look at a single instance of an actor because **preRestart** is one of the last methods that is called and **postRestart** is one of the first methods that is called, but the **postRestart** is called on a different instance that is created later.

To see how all of this works together, we have a little application that makes one parent, looks the child up by its path, then sends that child some messages.

```scala
package multithreading1.actorexamples

import akka.actor.ActorSystem
import akka.actor.Props

object SupervisorExample extends App {
  val system = ActorSystem("SupervisorExample")
  val actor1 = system.actorOf(Props[LifecycleParent], "parent1")

  val c1 = system.actorSelection("akka://SupervisorExample/user/parent1/child1")
  c1 ! LifecycleChild.Throw
  Thread.sleep(10)
  c1 ! LifecycleChild.SimpleMessage

  Thread.sleep(100)
  system.terminate()
}
```

Line 11 tells the child actor to throw an exception. This invokes the supervision strategy of the parent, which tells the system to restart the child. Line 12 does a very brief **sleep**, then line 13 sends another message. Note that we continue to use **c1**, even though the actor was restarted. This is another advantage of using **ActorRef**s over having direct references to **Actor**s.[20] When an actor is restarted, we make a new instance of the **Actor** type. If we had a reference to the original **Actor**, it would now refer to a dead actor that we are not allowed to use. However, when a restart happens, the **ActorRef** stays the same, it just changes what instance of **Actor** it is wrapping around.

The short call to **sleep** on line 12 is actually quite significant. During the restart, there is a brief period of time when there is not a valid instance of **Actor** in the **ActorRef**. There is a period of time when we have the old instance after it has been stopped, and another period when we have the new instance before it has been started. If we send a message to the **ActorRef** during this period of time, the message will get bounced to the dead letter office instead of going to the actor. To prevent that from happening in this code, the short **sleep** call was inserted.

If you run this program, you get the following output. The fact that we only have one parent and one child, and all the messages are sequential means that this output is actually deterministic.

```
Making a child akka://SupervisorExample/user/parent1/child1
Prestart
Child Dying
Child had an exception
Poststop
Prerestart Some(Throw)
Making a child akka://SupervisorExample/user/parent1/child1
Prestart
Postrestart
Child Simple = akka://SupervisorExample/user/parent1/child1
Poststop
```

---

[20]Technically **c1** is an **ActorSelection**, but this capability works equally well with an **ActorRef**.

8.10.7   Routers (Project Integration)

Akka provides you the capability to break up work among different actors however you wish. When you have a lot of different tasks, there are certain approaches to breaking up the work that are used frequently. For this reason, Akka includes types called routers that implement these standard strategies for breaking up work among a group of actors so that developers do not have to rewrite code for doing them.

Here are a number of different strategies for breaking up work that are implemented in the `akka.routing` package. For all of these, the general behavior is that a message is sent to the router, and the router then forwards that message to one or more actors based on the desired distribution strategy.

Round-robin forwards messages to each actor under the router in order. Once every one has gotten one message, it starts back at the beginning and repeats this process.

Random simply picks a random member and forwards the message to it.

Smallest mailbox forwards the message to the actor with the smallest mailbox. Remember that mailboxes queue up, so if messages are coming in faster than they can get processed, the actors could have multiple messages in their inboxes.

Balancing forwards the message to the actor who has been least busy. This might seem similar to smallest mailbox, but the results can be quite different if some messages are processed faster than others.

Broadcast routing will forward the message to all the actors that are members of that router.

Scatter/gather first completed forwards the message to all members like a broadcast, but this distribution style replies with the first answer that it gets.

Consistent hashing forwards the message to one member based on a function that involves some aspect of the message. The function makes sure that the same member of the router gets all messages that are associated in some way. For example, if you were processing orders for some product, you might want all orders for the same product to go to the same actor. This approach would do that.

The classes in `akka.routing` that have names that start with a distribution strategy and end with `Group` or `Pool`. The `Groups` allow you to manage their members while `Pools` manages their own members. Not all distribution styles include both. For our example, we will use a `BalancingPool`, which does not have an equivalent `Group`.

For this example, we are not going to write a little standalone application like we did for the previous section. Instead, we are going to create a new **Drawable** and integrate this with our drawing project. In section 8.9 we added a **Drawable** that would draw a type of fractal called the Mandelbrot set. In this section, we will create a very similar **Drawable** that draws a very similar family of fractals called Julia sets. The math for the Julia set uses the same equation as for the Mandelbrot set, but instead of making c the value of the current point in the plane, we use a single value of c for every point, and z_0 is the value of the location in the plane. So instead of having one Julia set, there are an infinite number of them, one for every point in the complex number plane. Figure 8.8 shows the drawing program with the result of what we are going to write.[21]

---

[21]There is a remarkable relationship between the Mandelbrot sets and the Julia sets. If you pick a point in the complex plane, if that point belongs to the Mandelbrot set, then the corresponding Julia set is connected. If the point is not in the Mandelbrot set, then the corresponding Julia set is not connected. Not all Julia sets look interesting, only those that are fairly near the boundary of the Mandelbrot set.

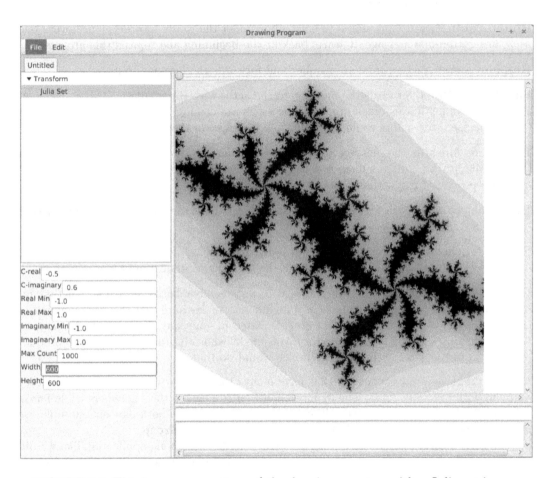

FIGURE 8.8: This is a screen capture of the drawing program with a Julia set image.

The difference between this example and what we did in section 8.9 is that instead of breaking up the work with parallel collections and making it behave nicely with Futures, we are going to do everything with actors, including a router.

As with all of our previous examples, we cannot start using actors until we have created an ActorSystem. We really only need one ActorSystem in the application, so we do not want to create it in our Drawable. Instead, we will add the following line to the top of DrawingMain next to the declaration of drawings.

```
val system = ActorSystem("DrawingSystem")
```

This is a public val because we want other parts of the project to be able to access the system so that they can use actors. Adding an ActorSystem to the program has the side effect that the application will no longer terminate when we close the window, because the system is keeping the progam alive. We can fix that by adding the following line in the block of code that sets up the stage.

```
onCloseRequest = (we:WindowEvent) => system.terminate()
```

With this, we will terminate the system when the user clicks to close the window. Note that we do not need to add this to the "Exit" option in the menus because the call to sys.exit shuts down everything, including the actor system.

With that in place, we can write the DrawJulia class. The code for this class is shown here.

```scala
1  package multithreading1.drawing
2
3  import scalafx.Includes._
4  import scalafx.scene.Node
5  import scalafx.scene.canvas.GraphicsContext
6  import scalafx.scene.layout.VBox
7  import scalafx.scene.paint.Color
8  import scalafx.scene.image.WritableImage
9  import scalafx.scene.image.PixelWriter
10 import scala.concurrent.Future
11 import scala.concurrent.ExecutionContext.Implicits.global
12 import scalafx.application.Platform
13 import akka.actor.Props
14 import akka.routing.BalancingPool
15 import akka.actor.Actor
16 import akka.actor.ActorRef
17
18 class DrawJulia(d: Drawing) extends Drawable(d) {
19   private var (rmin, rmax, imin, imax) = (-1.0, 1.0, -1.0, 1.0)
20   private var (creal, cimag) = (-0.5, 0.6)
21   private var (width, height) = (600, 600)
22   private var maxCount = 100
23   private var img = new WritableImage(width, height)
24   private val juliaActor = DrawingMain.system.actorOf(Props(new JuliaActor),
        "JuliaActor")
25   private val router = DrawingMain.system.actorOf(BalancingPool(5).props(Props(new
        LineActor)), "poolRouter")
26   private var propPanel: Option[Node] = None
27
28   juliaActor ! MakeImage
29
```

```scala
30    override def toString() = "Julia Set"
31
32    def draw(gc: GraphicsContext): Unit = {
33      gc.drawImage(img, 0, 0)
34    }
35
36    def propertiesPanel(): Node = {
37      def checkChangeMade[A](originalValue: A, newValue: => A): A = {
38        try {
39          val nv = newValue
40          if (originalValue == nv) originalValue else {
41            juliaActor ! MakeImage
42            nv
43          }
44        } catch {
45          case nfe: NumberFormatException => originalValue
46        }
47      }
48      if (propPanel.isEmpty) {
49        val panel = new VBox
50        panel.children = List(
51          DrawingMain.labeledTextField("C-real", creal.toString(), s => { creal =
                checkChangeMade(creal, s.toDouble) }),
52          DrawingMain.labeledTextField("C-imaginary", cimag.toString(), s => { cimag
                = checkChangeMade(cimag, s.toDouble) }),
53          DrawingMain.labeledTextField("Real Min", rmin.toString(), s => { rmin =
                checkChangeMade(rmin, s.toDouble) }),
54          DrawingMain.labeledTextField("Real Max", rmax.toString(), s => { rmax =
                checkChangeMade(rmax, s.toDouble) }),
55          DrawingMain.labeledTextField("Imaginary Min", imin.toString(), s => { imin
                = checkChangeMade(imin, s.toDouble) }),
56          DrawingMain.labeledTextField("Imaginary Max", imax.toString(), s => { imax
                = checkChangeMade(imax, s.toDouble) }),
57          DrawingMain.labeledTextField("Max Count", maxCount.toString(), s => {
                maxCount = checkChangeMade(maxCount, s.toInt) }),
58          DrawingMain.labeledTextField("Width", width.toString(), s => { width =
                checkChangeMade(width, s.toInt) }),
59          DrawingMain.labeledTextField("Height", height.toString(), s => { height =
                checkChangeMade(height, s.toInt) }))
60        propPanel = Some(panel)
61      }
62      propPanel.get
63    }
64
65    private def juliaIter(zr: Double, zi: Double, cr: Double, ci: Double) = (zr * zr
          - zi * zi + cr, 2 * zr * zi + ci)
66
67    private def juliaCount(z1r: Double, z1i: Double): Int = {
68      var ret = 0
69      var (zr, zi) = (z1r, z1i)
70      while (ret < maxCount && zr * zr + zi * zi < 4) {
71        val (tr, ti) = juliaIter(zr, zi, creal, cimag)
72        zr = tr
73        zi = ti
74        ret += 1
```

```
75        }
76      ret
77    }
78
79    case object MakeImage
80    case class Line(row: Int, imaginary: Double)
81    case class LineResult(row: Int, colors: Array[Color])
82
83    class JuliaActor extends Actor {
84      private var image: WritableImage = null
85      private var writer: PixelWriter = null
86      private var lineCount = 0
87      def receive = {
88        case MakeImage =>
89          image = new WritableImage(width, height)
90          writer = image.pixelWriter
91          lineCount = 0
92          for (i <- 0 until height) {
93            val imag = imax - i * (imax - imin) / width
94            router ! Line(i, imag)
95          }
96        case LineResult(r, colors) =>
97          lineCount += 1
98          for (i <- colors.indices) {
99            writer.setColor(i, r, colors(i))
100          }
101          if (lineCount >= height) {
102            Platform.runLater {
103              img = image
104              drawing.draw()
105            }
106          }
107      }
108    }
109
110    class LineActor extends Actor {
111      def receive = {
112        case Line(r, imag) =>
113          sender ! LineResult(r, Array.tabulate(width)(i => {
114            val real = rmin + i * (rmax - rmin) / width
115            val cnt = juliaCount(real, imag)
116            if (cnt == maxCount) Color.Black else
117              Color(1.0, 0.0, 0.0, math.log(cnt.toDouble) / math.log(maxCount))
118          }))
119      }
120    }
121  }
```

For the most part, this code looks very much like the code for `DrawMandelbrot`. There are a number of name changes to reflect the fact that it is making Julia sets, but only a few significant changes.

Lines 20, 51, and 52 are new additions that handle the fixed value of c that specifies what Julia set we are drawing. The more interesting changes are the ones that deal with actors. Lines 24 and 25 create the actors that we are going to use for rendering. There is a `juliaActor` that is responsible for drawing whole images. Then there is a router of type

`BalancingPool` that holds five actors of the type `LineActor`. The `Actor` classes and their messages are on lines 79–120.

The `JuliaActor` takes two types of messages. The first one tells it to make a new image. This includes a loop that runs through all of the rows of the image and for each row sends a message to the router of type `Line`. The other message type handled by `JuliaActor` is `LineResult`. This has information about one row of data, including the row number and the colors of all the pixels on that row. Then this message is received, the code does a loop to set the colors of all of those pixels.

There is also a `var` called `lineCount` in `JuliaActor` that is used so that we know when the whole image is done. Every time we make a new image, this is set back to zero. Each time a `LineResult` is received, it is incremented by one. When it gets up to the height of the image, we can do the `Platform.runLater` call that had appeared in `DrawMandelbrot` after the parallel for loop was done.

The `LineActor` handles one type of message that gives it a row number and the imaginary value for that row. It uses `Array.tabulate` to run through the width of that row and calculate all the individual counts and corresponding colors. The resulting `Array[Color]` is sent back as part of a `LineResult` to the `sender`.

The last change in moving to the actor model is that lines 28 and 41 no longer call `startDrawing`, as was done in `DrawMandelbrot`. Instead, those lines send `MakeImage` to `juliaActor` to initiate the calculation of a new image.

With the `DrawJulia class` completed, we can add the following to `DrawingMain` so that the user can add Julia sets to the drawings and you can test this out.

```
"Julia Set" -> (drawing => new DrawJulia(drawing))
```

8.10.8   More to Akka

There is a lot more to Akka that we have not discussed in this chapter including testing as well as remote actors and clusters that can communicate across computers. Our goal here has been to hit the main topics that you need in order to write applications using Akka for actor parallelism as well as to give you the knowledge you will need to do projects through the rest of this book.

GPUs

Multicore chips are not the only example of computer hardware moving toward parallelism. Another significant trend is the use of the Graphics Processing Unit (GPU) to help with general purpose computing. This is commonly called GPGPU. Higher-end GPUs are typically found on separate graphics cards, though there are integrated graphics on the motherboard. Beginning around 2010, it became more common to find GPU elements etched onto the same silicon as CPUs.

The idea of doing general purpose calculations on a graphics card might seem odd at first. To see why it works, you need to know something about how graphics cards work. Graphics is basically a lot of math. Early GPUs were built specifically to do just the math that was needed with graphics. They did this in the fastest way possible and did not adhere to floating point number standards because you do not need many digits of accuracy for building an image that is on the order of 1000 pixels across.

Over time, the desire for more realistic graphics capabilities led to the creation of pixel shaders. These allowed programmers to write limited logic that could be run for each pixel in an image or on the screen. This was a significant break from earlier designs as it allowed true programmable logic on the GPU. At first this logic was still very heavily focused on graphics-related calculations, but the fact that a GPU could do floating-point math extremely fast tempted people into trying to use pixel shaders for other purposes.

Since then, GPUs have become more flexible and more programmable. The support for GPGPU became a selling point, so hardware manufacturers put in proper IEEE floating-point arithmetic. First this was only in single precision, but later double precision was added as well. This has culminated in cards built specifically for GPGPU. These cards have lots of memory and one or more GPU chips in them, but they often do not have any graphics output. They are designed and sold just to support GPGPU style calculations.

While multicore chips have a few cores and there are plans to take them to tens or even a few hundreds, GPUs will support thousands or tens of thousands of threads. You have to break up a problem into a lot of pieces to keep a GPU busy. In addition, the threads on a GPU are not as fully capable as those on the cores of a CPU. Typically GPU threads need to be run in groups that are all doing very nearly the same thing to different data.

At this time there are three main interfaces to support GPGPU style programming: CUDA®, OpenCL®, and OpenACC®. CUDA is an API created by NVIDIA® which only runs on NVIDIA GPUs. OpenCL and OpenACC, as the names imply, are open standards that supports GPU and other high thread count computation models. There is currently preliminary support for OpenCL in Scala through the ScalaCL package. ScalaCL works in large part through an augmented set of collections that split the work up on the GPU instead of over multiple cores like the normal parallel collections do.

If you major in computer science, GPGPU is probably a topic that should be on your radar and all indications are that CPUs with integrated GPUs will become common. As was shown at the beginning of this chapter, it was not long ago that most programs did not need multithreading because most computers did not require it to run at full speed. A similar change could be in the works for GPUs and programmers will have to have some understanding of what is happening in the hardware to write programs that take full advantage of the resources they have.

8.11 End of Chapter Material

8.11.1 Summary of Concepts

- Multicore processors have become the norm and core counts are growing exponentially without increases in chip clock speed. This means that programs have to include parallelism to take full advantage of the hardware. Multithreading is one of the primary ways to doing this.

- When implementing multithreading in Scala, one method you could use is the `scala.concurrent.Future` **trait** and companion object. A `Future` is an object that represents a value that is in the process of being computed and might not be ready to be used.

- Through the Java libraries, Scala programs have access to basic thread operations.

  - One of the main problems with threads is that they all access the same memory. That means that if different threads are mutating parts of the memory, you can have race conditions where the outcome of the computation depends on the details of thread handling.

  - Another problem you run into with threads is that if you start pausing them to improve their cooperation, you run the risk of making a number of threads freeze up in a state called deadlock where they are all waiting around for the others to finish.

  - The primary way of dealing with race conditions is through synchronization. Critical blocks of code where values are mutated can be synchronized. Only one thread can be executing code that is synchronized on a particular object at any given time. Unfortunately, synchronization is a bit slow and too much of it can lead to deadlock.

- A blocking call is one that stops the progress of the current thread until it has been completed.

- The `Future.sequence` method takes a collection of `Future`s and produces a `Future` that effectively waits until all the input `Future`s are done.

- There is a `Future.find` method that takes a collection of `Future`s and a predicate. It produces the first of the `Future`s in the input collection that finishes and satisfies the predicate.

- The `Future` object has `reduceLeft` and `foldLeft` methods. These have a first argument list that is a collection of `Future`s. They run through that collection using a provided function to aggregate the results and produce a `Future` of the final result.

- You can use `Await.result` to block the current thread and wait until some `Future` has completed.

- Scala makes certain types of parallelism simple with parallel collections. These collections have methods that split the work across multiple threads and automatically load balance. Calling `par` on a regular collection gives you a parallel version. Calling `seq` on the parallel version gives you back a sequential one. These calls are efficient for types that have a parallel equivalent.

- When using parallel collections, there is one main thing that you have to remember, the actions of one iteration through the loop cannot depend on what happened in an earlier iteration. That behavior is only safe with sequential loops.

- One of the biggest benefits to using a functional approach in your programming and using immutable values as much as possible is the fact that if all of your data is immutable, it is impossible to have race conditions.

- Any code that uses a GUI is inherently multithreaded. GUIs also have to deal with the challenges of deadlock and race conditions. Keeping the event thread busy too long can lead to an unresponsive GUI. Allowing threads to alter mutable data that is used by the event thread as that can lead to race conditions. A way that you can control scheduling to help prevent these problems in ScalaFX is with the `Platform.runLater` method.

- Akka is the primary library for actor parallelism on the JVM.

  - Simply put, an actor is an object that gets a thread and a message inbox. The Akka systems make sure that each actor is only using one thread at a time, assuming you do not make your own `Futures`. Instead of calling methods on your actors, you send them messages. The messages are queued up in their inbox and they handle them in the order that they are received. This approach insures that each actor is only processing a single message at a time.

  - Actors are useful for designing a whole parallel system. When you create an actor, the names of actors should be unique.

  - You have to be careful with the functions you build in `Futures` that are inside of `Actors`. If you are going to use `sender`, make a `val` outside of the `Future` and use it instead. Also, never mutate values in the `Actor` inside of a `Future`. If you need to mutate something, send a message to `self` to make it happen. Otherwise you are likely to create a race condition.

  - In large applications it is helpful to be able to organize actors into logical groups. The way in which actors are organized is a hierarchy. Each actor can have other actors below it. Where an actor goes in the hierarchy depends on how it is created.

  - One of the key reasons why many people like actor systems is that they designed to be highly resilient. The feature of actor systems that makes them highly reliable is that the actor hierarchy provides a built-in structure for supervision so that when something fails, there is another piece in the system that can be notified and it can do something about the failure to keep the system as a whole running smoothly.

  - Akka provides you the capability to break up work among different actors however you wish. Because of this, Akka includes types called routers that implement various standard strategies for breaking up work among a group of actors.

  - There are a number of different strategies for breaking up work that are implemented in the `akka.routing` package. For all of these, the general behavior is that a message is sent to the router, and the router then forwards that message to one or more actors based on the desired distribution strategy.

- Scala provides many options for adding multithreading to your code. It can be challenging to determine which one to use in different contexts. Here are some simple rules of thumb.

  - If the problem automatically brings to mind a for loop or higher-order methods like `map` and `filter` and the things happening in one iteration do not depend on another, a parallel collection is probably best.

  - If you are designing a large system with many parts that are all able to happen in parallel, then you should strongly consider using an actor-based system for the full application.

  - If it does not match one of those two, use a `Future`.

8.11.2 Exercises

1. Compare and contrast futures, parallel collections, and the actor model in regards to multithreading.

2. Earlier in this chapter a method called `timecode` was presented that runs whatever code is passed to it 200 times and measures the time for the last hundred then prints the average run time of that code. You should add to this method so it is easy to adjust the number of "warm-up" runs and the number of timing runs.

3. Earlier in this chapter, the code for `CountDownActor` was produced, but it requires two message types. We could get by with a single message type if we used `tell` instead of `!` in our application. We did not do that, as it is important for readers to see how actors are passed by `ActorRef`, however for this exercise, you should make that simplification to this code.

4. Make a different version of `CountDownActor` that allows three actors to participate in the count down.

5. There are some other open APIs that you can use to find information. Create parallel requests for these APIs using `Futures` like we did in the chapter for MetaWeather. Consider accessing Google Books or Google Locations. You also may find additional sources here at `https://shkspr.mobi/blog/2014/04/wanted-simple-apis-without-authentication/`.

8.11.3   Projects

1. It you are working on the MUD/text RPG, you can change up your code to use Akka and be completely actor based. This is a good early step for making it multiplayer in a few chapters. Any element that contains mutable data should be considered as a possible actor. It is obvious to make a player an actor. You probably think of making the any non-player characters into actors as well. In addition, the type that represents your room is inevitably mutable as well, so it should probably be an actor as well.

   You should also consider making actors that manage other groups of actors. For example, you might have an actor that manages all the rooms as well as one that will manage all the players.

   If you do not plan to make your MUD actor based, you should instead work on making a thread safe queue. When you add networking, there will have to be some threading and you will need to add players from one thread to another. A thread safe queue will be a good way for you to do this.

2. There is no specific project for web spidering in this chapter, but you will be putting in multithreading capabilities in this project in the next chapter.

3. If you are writing a networked graphical game, you will be forced to put in multithreading to deal with networking. However, you can start preparing for that now. You can put in code to handle two players from the keyboard where the different players get messages from a single keyboard handler and a main thread runs the primary game logic. You might also be able to think of ways to split up the game logic across multiple threads.

4. For a non-networked game, the threading will primarily be done to help with the computer-controlled entities in the game. Care must be taken to avoid race conditions, but you should look for intelligent ways to split up the work across different threads. This could include grouping active entities if they have complex decision-making needs. It could also be just breaking things up by type. Note that the second option generally does not scale to high core counts as well.

5. If you are doing the Math worksheet problem, you have quite a few options for parallelism. Worksheets do not have to be linear. If one section of a worksheet does not depend on the results of another, they can be processed in parallel. In addition, complex calculations can often be done in parallel by working on different data values using different threads.

6. The Photoshop® project runs into some challenges with multithreading because most of the work that it is doing involves graphics and, as was described in the introduction to these projects, normal graphics does not generally benefit from multithreading. It is probably worth you spending a little bit of time to test this hypothesis though. You could render layers to separate `Image`s using separate threads. If you make the scenes complex enough, you can really see if multithreading provides any benefit.

   There are some other things that could be done which would definitely benefit from multithreading. The methods to get and set colors using `PixelReader` and `PixelWriter` on a `WritableImage` to not use normal graphics operations. They work with direct access to the memory behind the image. For this reason, any operations that you can do using only those methods will benefit from multithreading. An example of something that can be done this way is filters. Many effects that you might want to apply to an image can be done with only the RGB values of a pixel and those around it.

   You might want to add some more complex and interesting elements to the things you can add to drawings. That could include fractals like the Mandebrot shown in this chapter, Julia sets, L-systems, or even ray-tracing elements. All of those could benefit from some or all parts being done in parallel.

7. The simulation workbench is also a plentiful area for parallelization because it is processor intensive and can be arranged in ways that do not conflict. This does not mean it is easy to do efficiently though. Using the simple gravity simulation, devise at least three different ways of calculating the forces in parallel, then do a speed comparison between those approaches and your original code.

8. In order to populate your stock data at the beginning of this project you needed to manually download CSV files from Yahoo®Finance. However, now you should build URLs and then request these in separate threads. This will allow you to more easily maintain current stock information. You also add a feature to make it possible for the user to have the option to reload stock information.

9. In order to populate your movie data at the beginning of this project you needed to manually download files from various sites. However, now you should build URLs, and make requests in order to automatically update your movie information. If you are getting your information from different URLs then request these in separate threads. This will allow you to more easily maintain current movie information. You also add a feature to make it possible for the user to have the option to reload movie information.

10. There are various approaches that you could take to make the L-systems utilize parallelism. In theory, you could use any of the three approaches from this chapter, though they are probably not as well suited for actors. Try some and pick the one that works best.

Chapter 9

Low-Level Multithreading and Java Libraries

In the last chapter, we introduced the key concepts of multithreading and showed the primary libraries that you should use when you are multithreading in Scala. The use of Futures, parallel collections, and Akka actors provides a safe way to do the vast majority of parallel tasks. So why do we have another chapter on parallel programming? There are two main reasons for having this chapter, which focuses on the threading features of the Java languages. First, some of these libraries are the lower-level underpinnings of all the Scala libraries, and they represent the type of parallelism that will be available in a broader variety of languages.[1] The second reason is that the parallel collections in Java are recommended for use in Scala as well, and there are use cases where they are an easier way of doing things when using Futures and parallel collections.

9.1   Basic Threads

The creators of Java saw ,back in the early 1990s, that multithreading was going to be significant. As such, they added a number of threading support features to the core of the language including the `java.lang.Thread` class and elements associated with it. The `Thread class` spawns a new thread to perform a calculation. It lacks the additional functionality that you get from Futures. The fact that it always launches a new thread is significant. We did not deal with this detail in the last chapter, but there is a fair bit of overhead to creating threads. Each thread has to have memory set aside for its call stack

---

[1] At the time of this writing, most languages do not have the breadth and variety of support for multi-threading provided by Scala.

along with other setup aspects. The reason for the `ExecutionContext` that was required when you create a `Future` was to oversee the creation of threads. If you create 1000 `Futures`, you will almost certainly have a lot fewer threads because the `ExecutionContext` has the ability to manage a small pool of threads and give them tasks when new `Futures` are created. By contrast, when you do **new Thread**, you always create a new thread. There are times when you might feel that you need this, but most of the time, you are probably better off letting library code handle that decision.

We will start with a reasonably simple example that shows you how you create and use the `Thread class`. This is an application that will ask your age and count down from 10 to 1. If you do not answer before the countdown ends, it will terminate without you answering. If you answer, it prints a message based on whether or not you are 18 years of age and stops counting.

```scala
package multithreading2

import io.StdIn._

object ReadTimeout extends App {
  val countThread = new Thread {
    override def run: Unit = {
      for (i <- 10 to 1 by -1) {
        println(i)
        Thread.sleep(1000)
      }
      println("Time's up.")
      sys.exit(0)
    }
  }
  println("Enter your age.")
  countThread.start()
  val age = readInt()
  if (age < 18) println("Sorry, you can't come here.")
  else println("Welcome.")
  sys.exit(0)
}
```

A new thread is created on lines 6–15 using **new** and the type `Thread`, and overriding the `run` method. You can also make a new `Thread` and pass it an instance of the `Runnable` type which must have a `run:Unit` method defined. Beginning in Java 2.12 you can use a function literal in place of Runnable.[2] In the `run` method we place the code that we want the other thread to execute. In this case it has a for loop that counts down from 10 to 1. After each number is printed, there is a call to `Thread.sleep(1000)` that pauses the thread for a second. After the loop, it tells the user that time is up and calls `sys.exit(0)`. This call forcibly causes a Scala program to terminate. The 0 is the exit code. By convention, a value of 0 indicates normal termination. Some other value would indicate an error.

Creating a new `Thread` object does not cause it to start working. For that to happen we have to call the **start** method. In this case, that is done on line 17 after the user has been prompted. We originally called **start** before the `println`, but because of the randomness of thread scheduling, this led to occasional executions where the "10" would print before we even asked the user their name. After the thread is started, there is a `readInt` call and

---

[2]Java 8 introduced lambda expressions to the Java language, but instead of being their own type, they can be used in places where the code needed one abstract method to be implemented, like `Runnable`. Scala 2.12 added the ability to use Scala lambdas with Java libraries in places where Java would allow a lambda.

a conditional followed by another call to `sys.exit`. The last call to `exit` might seem odd to you. We do not normally tell our program to stop when it gets to the end, but in this case we need to. An executing `Thread` will keep the process alive in the same way that an `ActorSystem` does. If we leave out the call to `exit` at the end, the user can enter their value and get the response, but the thread continues to count down and tells them that time is up well after they did their input before the program actually stops.

The use of `sys.exit` is pretty heavy-handed. It works in this simple example, but it is easy to imagine situations where you do not want to terminate the entire execution of a program. The call to `sys.exit` in the thread is needed here because `readInt` is a blocking method that we cannot get out of once we call it. The call to `sys.exit` at the end is needed because we did not include a mechanism that allows us to tell the thread to stop early. Let us try to see how we could get rid of both of these.

We will start with the call at the end of the script because it is easier to deal with. There are two problems involved here. One is that we need some way for the code outside the thread to "talk" to the code inside the thread so that it can know when the user has answered. The second is that the `for` loop in Scala does not allow you to break out in the middle of it. If you want to have the possibility of terminating early, you need to use a `while` loop. The condition in the `while` loop should say that we keep counting while a counter variable is greater than zero and the user has not answered. This second part tells us how to get the two parts to "talk". We simply introduce a `Boolean` variable that keeps track of whether the user has answered. The `while` loop stops if they have. The line after reading the input should change this variable to say they have answered. A possible implementation of this looks like the following.

```scala
package multithreading2

import io.StdIn._

object ReadTimeout2 extends App {
  var answered = false
  val countThread = new Thread {
    override def run():Unit = {
      var i = 10
      while (!answered && i > 0) {
        println(i)
        Thread.sleep(1000)
        i -= 1
      }
      if (!answered) {
        println("Time's up.")
        sys.exit(0)
      }
    }
  }
  println("Enter your age.")
  countThread.start()
  val age = readInt()
  answered = true
  if (age < 18) println("Sorry, you can't come here.")
  else println("Welcome.")
}
```

Enter this code and see that it behaves the same as the earlier version.

Note that we also altered the way that we are building the **Thread**. Instead of passing in a **Runnable**, we are making a anonymous class that is a subtype of **Thread** and overriding the **run** method in that. The astute reader might also notice that there is technically a race condition here because **answered** is being utilized in two threads, and one of them is modifying the value. While there is a race going on here, because the **while** loop checks the value of **answered** each time through, this race condition does not cause incorrect behavior.

This version still contains one call to **sys.exit**. This one is a little more difficult to get rid of. The problem is that **readInt** is a blocking method. In this case, it blocks until the user has typed in an appropriate input value. That is not acceptable in this situation because we want to make it so that the countdown thread can prevent the read from happening, and more importantly blocking, if it gets to the end before the user enters their value. An approach to this is shown here.

```scala
package multithreading2

import io.StdIn._

object ReadTimeout3 extends App {
  var answered = false
  var timeUp = false
  val countThread = new Thread(() => {
    var i = 10
    while (!answered && i > 0) {
      println(i)
      Thread.sleep(1000)
      i -= 1
    }
    if (!answered) {
      println("Time's up.")
      timeUp = true
    }
  })
  println("Enter your age.")
  countThread.start()
  while (!timeUp && !Console.in.ready) {
    Thread.sleep(10)
  }
  if (!timeUp) {
    val age = readInt()
    answered = true
    if (age < 18) println("Sorry, you can't come here.")
    else println("Welcome.")
  }
}
```

This code introduces another mutable **Boolean** variable that keeps track of whether time has expired. At the end of the countdown this variable is set to be true. This code uses a third approach to making a **Thread** passing in a function literal that serves as the **run** method.

The more interesting part of this code occurs right after the prompt for the user's age. Here you find a **while** loop like this.

```scala
while (!timeUp && !Console.in.ready) {
  Thread.sleep(10)
```

```
}
```

This loop will stop under one of two conditions, either time must run out, or `Console.in` must be ready. To understand this second condition requires some digging through the API. The object `scala.Console` is what you read from when you do calls like `readInt` or what you write to when you do `println`. The input comes through an object called `Console.in`, which is a `java.io.BufferedReader`. We will talk more about the various classes in `java.io` in the next chapter, but for now you should know that this type has a method called `ready` that will tell you when you can read from it without having it block.

Inside the loop is another call to `Thread.sleep`. This is here simply to prevent the `while` loop from doing a "busy wait" and making the processor do a lot of extra work for no real reason. It will still check if the user has input something or if time has run out 100 times each second. This is more than sufficient for the type of user interaction that we have here. After the loop, the same code from before appears inside an `if` so that it will not happen if the loop stopped as a result of time running out.

9.1.1 Wait/Notify

The `synchronized` method can help you to prevent race conditions, but it is not generally sufficient to coordinate the behaviors of different threads. `Futures` and Akka actors have mechanisms that help you do this in safe ways. In the basic thread libraries this is done with `wait`, `notify`,[3] and `notifyAll`. You should probably only use these if you had made the decision to go down to the level of using `Thread`, but even if you are not doing that, there is some benefit to knowing how things work at the lowest level as the other mechanisms are often built on top of these, so we will discuss these briefly.

As with the `synchronized` method, `wait`, `notify`, and `notifyAll` are methods in `AnyRef` and can be called on any object of that type. The `wait` method does exactly what the name implies, it causes the current thread to stop executing and wait for something. That something is for it to be wakened by a call to `notify` or `notifyAll` in another thread. All of these calls require that the current thread hold the lock to the monitor for the object the method is called on. In addition, `notify` and `notifyAll` only wake up threads that were set to wait on the same object. You will normally see these calls in synchronized blocks. The call to `wait` releases the lock on that monitor. It will be reacquired when the thread wakes back up.

To see how this can work, consider this example that has a number of different threads that use `wait` to count off in turns with no more than one thread active at any given time.

```scala
 1  package multithreading2
 2
 3  object WaitCounting extends App {
 4    val numThreads = 3
 5    val threads = Array.tabulate(numThreads)(i => new Thread {
 6      override def run():Unit = {
 7        println("Start "+i)
 8        for (j <- 1 to 5) {
 9          WaitCounting.synchronized {
10            WaitCounting.wait()
11            println(i+" : "+j)
12            WaitCounting.notify()
13          }
```

---

[3]Use of `notify` is typically discouraged because you have no control over which thread it will wake up. This has a tendency to lead to deadlock.

```
14        }
15      }
16    })
17    threads.foreach(_.start)
18    Thread.sleep(1000)
19    println("First notify.")
20    synchronized { notify() }
21 }
```

Each thread has a `for` loop that counts to 5. In the loop it synchronizes on the `WaitCounting` object and calls `wait`. Once the thread is wakened by another thread, it will print out the thread number and the loop counter value. Then it notifies some other thread. After that it immediately goes back to sleep as the next time through the loop leads to another call to `wait`. After the threads are built, they are started and the main thread sleeps for one second before calling `notify`.

The call to `sleep` makes sure that all the threads have had time to start up and call `wait` before the first call to `notify` is made. If you leave this out, it is possible that the first call to notify will happen before the threads have started. A more rigorous method of accomplishing this would be to have a counter that increments as each thread starts.

Each call to `notify` wakes up a single thread that is waiting on the `WaitCounting` object so this works like dominoes with the prints running through the threads in whatever order `notify` happens to wake them up. The inability to predict what thread will be woken up is a problem. The real problem arises because it is possible that in later development some other thread might start waiting on this same object and that thread might not follow the same rule of printing then calling `notify`. That can lead to deadlock. In this example, if you simply add the following line to the bottom after the call to `notify`, you can create deadlock.

```
synchronized { wait() }
```

This causes a problem because now when one of the other threads calls `notify`, it wakes up the main thread instead of one of the printing threads. The main thread does not do anything after it wakes up though, so we are left with printing threads that remain asleep forever.

The lack of predictability in `notify` has led to the standard rule that you should not use it. The recommended style is to have all calls to `wait` appear in `while` loops that check some condition and to always use `notifyAll`. The following code shows this rule in practice.

```
1  package multithreading2
2
3  object WaitCountingSafe extends App {
4    val numThreads = 3
5    var handOff = Array.fill(numThreads)(false)
6    val threads = Array.tabulate(numThreads)(i => new Thread {
7      override def run {
8        println("Start "+i)
9        for (j <- 1 to 5) {
10          WaitCountingSafe.synchronized {
11            while (!handOff(i)) {
12              WaitCountingSafe.wait()
13            }
14            handOff(i) = false
15            println(i+" : "+j)
16            handOff((i + 1) % numThreads) = true
```

```
17        WaitCountingSafe.notifyAll()
18      }
19    }
20  }
21  })
22  threads.foreach(_.start)
23  Thread.sleep(1000)
24  println("First notify.")
25  handOff(0) = true
26  synchronized { notifyAll() }
27 }
```

Now each thread has its own **Boolean** flag telling it whether another thread has handed it control. If that flag is false, the thread will go back to waiting. In this way we regain control over things and reduce the odds of deadlock. Indeed, adding the same call to **wait** at the end of this code will not cause any problems at all.

9.1.2   Other Thread Methods

There are a number of other methods in **Thread** that are worth knowing about. We have already seen the **sleep** method, which causes a thread to stop processing for a specified number of milliseconds. Similar to **sleep** is the **yield** method. Calling **Thread.'yield'** will make the current thread give up control so that if another thread is waiting it can take over. The backticks are required in this call because "yield" is a keyword in Scala. You use it with the **for** loop to make it into an expression. Using backticks allows you to have Scala interpret tokens as normal even if they are keywords that are part of the language.

The **dumpStack** method will print out the current stack trace. This can be helpful if you are running into problems in a certain method and you are not certain what sequence of calls got you there. You can also use **Thread.currentThread()** to get a reference to the **Thread** object that represents the current thread.

If you have a **Thread** object that you either created or got through a call to **Thread.currentThread()**, you can call other methods on that thread. The **join** method is one of the more useful ones. This is a blocking call that will stop the current thread until another thread has completed. The following code shows how it might be used.

```
1  package multithreading2
2
3  object ThreadJoining extends App {
4    val threads = for(i <- 1 to 10) yield {
5      new Thread {
6        override def run():Unit = {
7          // Do something here
8        }
9      }
10   }
11   threads.foreach(_.start())
12
13   // This will cause the main thread to pause until the other threads are done.
14   threads.foreach(_.join())
15   // Do other stuff that needs the threads to have finished.
16 }
```

Lines 4–10 make the threads and line 7 should be replaced with whatever work you want them to do. Line 11 starts the threads. You have to remember to do this as threads do not start themselves. Then line 14 will block until all of those threads have completed their work.

This example highlights the two main pitfalls of standard Java **Thread**s. The first is that they do not produce values. So if the code in the **run** method had been doing some calculation that produced a result, we would have to write our own logic to get those values. The second is that we really prefer not to block, and standard **Thread**s make that more challenging. If you are tempted to write this code, you should probably consider the following version using **Future**s instead.

```scala
package multithreading2

import scala.concurrent.Future
import scala.concurrent.ExecutionContext.Implicits.global

object FutureJoining extends App {
  val futures = for (i <- 1 to 10) yield Future {
    // Do something here
  }
  Future.sequence(futures).foreach(values => {
    // Do other stuff that needs the futures to have finished.
  })
}
```

One other pair of **Thread** methods that is worth mentioning are the **setPriority** and **getPriority** methods. Each thread has an integer priority associated with it. Higher-priority threads will be preferentially scheduled by the thread scheduler. If you have work that needs to happen in the background and is not critical or should not impact the function of the rest of the system, you might consider making it lower priority. You can do this with a thread in a **Future** by calling **Thread.currentThread** to get the thread you are running in. You probably should not do this much, and if you do, remember to set the priority back to what it started at when the task is done because **Future**s potentially share threads made by the **ExecutionContext**, and you do not want some later task operating at lower priority when it should not be.

There are also a number of methods for **java.lang.Thread** that you can find in the API which are deprecated. This implies that they should no longer be used. This includes **stop**, **suspend**, and **resume**. These methods are fundamentally unsafe and should not be used. It might seem like a good idea to be able to tell a thread to stop. Indeed, the writers of the original Java library felt that was the case. However, experience showed them that this was actually a very bad thing to do. Forcibly stopping a thread from the outside will often leave things in unacceptable conditions. A simple example of this would be if you kill a thread while it is in the middle of writing to a file. This not only leaves the file open, it could very well leave the file in an inconsistent state that will make it unusable later.

Scala will give you warnings if you use deprecated library calls. You should pay attention to these, find them in your code, and look in the API to see how they should be handled. In the case of stopping threads, the proper technique is to have long running threads occasionally check **Boolean** flags that tell them if they should continue running. If the flag has been changed, they should terminate in an appropriate manner. This is what we did in the **ReadTimeout3** example application above.

9.2 Concurrency Library

While the Java libraries are still catching up to the Scala libraries in terms of making threading easier and safer, they have been upgraded significantly since the original release when all that you had access to was basic threading. In particular, they added the `java.util.concurrent` package in Java 5 to make many of the common tasks in parallel programming easier to do.

9.2.1 Executors and Executor Services

As we discussed, the process of creating new threads is expensive. It also is not abstract, meaning it cannot be easily varied for different platforms. The `java.util.concurrent` package addresses this with the `Executor` and `ExecutorService` types. The `Executor` type is a very simple abstract type with a single method.

```
def execute(command: Runnable): Unit
```

When this method is called, the `run` method of the `command` is called. What you do not know at this level is how that will be done. It is possible that a new thread will be created and it will run in that thread. However, it is also possible that it could be run in a thread that had been created earlier or even run in the current thread. Note that this last option implies that the current thread might not continue until after the `command` has finished.

The abstract type `ExecutorService` extends `Executor` and adds significantly more functionality to it. The following methods are significant for us. Look in the Java API for a full list.

- `shutdown(): Unit` - Tells the service to start an orderly shutdown allowing current tasks to finish. New tasks will not be accepted.

- `submit[T](task: Callable[T]): Future[T]` - Submits the specified task for execution and returns a `java.util.concurrent.Future[T]` object representing the computation. Note that the Java version of `Future` is not the same as the Scala version of `Future`.

- `submit(task: Runnable): Future[_]` - Submits the specified task for execution and returns a `java.util.concurrent.Future[_]` object representing the computation.

There are two new types presented in these methods: `Callable[T]` and `Future[T]`.

9.2.2 Callable and Java Futures

With normal threads, the computation can only be submitted as a `Runnable`. Unfortunately, the `run` method in `Runnable` does not return a value. This is a significant limitation as we often want to get results from the threads. The `Callable[T]` type contains one method that has a return value.

```
def call[T](): T
```

The submitted task might take a while to finish, so the `submit` method cannot return the `T` object as that would require blocking. Instead, it returns a `java.util.concurrent.Future[T]` object that can be used to check on the computation (use `isDone(): Boolean`) and get the result when it is complete (use `get(): T`). Note that get on a Java Future is a blocking call. This is one of the shortcomings of the Java Future. The only way to retrieve the value requires a blocking call. Java 8 introduced a new subtype of java.util.concurrent.Future called the CompletableFuture that has methods like thenAccept and thenApply, which allow you to schedule further tasks with the results.

Both Executor and ExecutorService are abstract. The standard way to get a concrete instance is to call one of several methods on java.util.concurrent.Executors. The methods you are likely to use are newCachedThreadPool(): ExecutorService and newFixedThreadPool(nThreads: Int): ExecutorService. A cached thread pool will make as many threads as there are active tasks. Once a task has completed, the thread is stored in a wait state so that it can be reused for a later task. A fixed thread pool only has the specified number of threads. If a new task is submitted when all the threads are filled, it will wait until one of the active threads has completed.[4]

To demonstrate these things in action, we will look at a version of factorial that uses BigInt and is multithreaded. Recall that a single-threaded version of this can be written in the following way.

```scala
def fact(n:BigInt) = (BigInt(1) to n).product
```

It can be done in parallel using parallel collections by just adding a call to par, but we want to explore other alternatives that focus on the Java libraries. If you call this for sufficiently large values of n, you will find that it can take a while to complete the calculation.[5] That makes it a reasonable function for us to consider parallelizing. How should we go about doing that using the facilities of the Java libraries? We want a function that breaks the multiplications up across a number of threads using an executor service. So the question that so often comes up when creating parallel programs is, how do we break up the work?

One approach to this is to break all the numbers into groups. If n were divisible by nThreads, this could be viewed as n! = 1 * ... * n/nThreads * (n/nThreads+1) * ... * 2*n/nThreads * (2*n/nThreads+1)*... When n is not divisible by nThreads, we need to make some of the groups bigger. Here is code that does that.

```scala
def parallelFactorial(n: BigInt, es: ExecutorService, nThreads: Int): BigInt = {
  val block = n / nThreads
  val rem = n % nThreads
  var i = BigInt(1)
  val futures = Array.tabulate(nThreads)(j => es.submit(new Callable[BigInt] {
    def call(): BigInt = {
      val start = i
      val end = start + block + (if (BigInt(j) < rem) 1 else 0)
      i = end
      (start until end).product
    }
  }))
  futures.map(_.get).product
}
```

This function starts with calculations of the base size for each block as well as how many blocks need to be one bigger. It then creates a variable i that helps us keep track of where in the range we are. A call to Array.tabulate is used to submit the proper number of tasks to the ExecutorService. Each task uses i as the start value and adds an appropriate offset

---

[4]You must be careful when using a fixed thread pool if you add new tasks to the pool inside of other tasks. It is possible for the second task to be stuck waiting for the first to finish. If this happens and the first task has code to get the result of the second task, you will get deadlock.

[5]Depending on your computer, sufficiently large will be at least several tens of thousands.

to get the end value. It then updates i and uses the standard **product** method to multiply the values in that chunk. The last thing this code does is map the various **Future[BigInt]** objects with **get**. This will block until they are done and give us a collection of the results. Another call to **product** produces the final result.

You can put this in an application to test it. When you do this, you probably want to start off with an import, then make your **ExecutorService**.

```
import java.util.concurrent.Executors
val es = Executors.newCachedThreadPool()
```

Now you can call the function and compare it to the non-parallel version above. Ideally you would specify the number of threads to be equal to the number of cores in your computer, but you can play with this to see how it impacts execution time. Either at the end of the script or before you exit the REPL, you need to call **es.shutdown()**.

This approach to breaking up the work is not the only option we have. One potential problem with this approach is that the different threads we created really do not have equal workloads. The way the BigInt works, dealing with large numbers is slower than dealing with small numbers. Given the way the work was broken up here, the first thread winds up with a much smaller value than the last one and, as such, has a smaller work load. An alternate approach would be to have each thread work with numbers spread out through the whole range. Specifically, the first thread does 1*(nThreads+1)*(2*nThreads+1)*... while the second thread does 2*(nThreads+2)*(2*nThreads+2)*..., etc. The methods of **Range** types in Scala actually makes this version a lot shorter.

```
def parallelFactorial(n: BigInt, es: ExecutorService, nThreads: Int): BigInt = {
  val futures = Array.tabulate(nThreads)(j => es.submit(new Callable[BigInt] {
    def call(): BigInt = {
      (BigInt(j + 1) to n by nThreads).product
    }
  }))
  futures.map(_.get).product
}
```

The by method lets us get a range of values with the proper spacing. As long as it starts at the proper value, everything else will work the way we need.

For comparison, here are two possible versions using the Scala libraries that were discussed in the previous chapter. Note that the one that uses Scala **Futures** gives back a **Future[BigInt]** instead of just a regular **BigInt** as the others do. If your program were using **Futures** extensively, this would be the desired approach so that you could minimize blocking.

```
def parallelFactorial(n: BigInt, nThreads: Int): Future[BigInt] = {
  val futures = List.tabulate(nThreads)(j => Future {
    (BigInt(j + 1) to n by nThreads).product
  })
  Future.sequence(futures).map(_.product)
}
```

```
def parallelFactorial(n:BigInt) = (BigInt(1) to n).par.product
```

If you play with the five versions of factorial that are presented here, one sequential and four parallel, you should find that the parallel versions are faster than the sequential version. If you have access to a machine with enough cores, you probably also noticed that there was a point of diminishing returns. Going from one thread to two shows a significant boost, perhaps even more than a factor of two, and you might have seen similar gains going up to four threads. However, if you have enough cores on your machine, at some point, adding additional threads becomes less and less beneficial, even before you reach your core count. There are a number of things going on here, but one of the big ones is what is known as Amdahl's law [1].

The basic idea of Amdahl's law is that the benefit you get from improving the speed of a section of your code depends on how much of the code you can make this improvement to. In the case of parallel processing, not all of the program will be done in parallel. Some sections are still done sequentially. If the fraction of your code that is done in parallel is represented by P, and the fraction that remains sequential is $(1 - P)$, then Amdahl's law states that the maximum speed-up you can get with N processors is

$$\frac{1}{(1 - P) + \frac{P}{N}}.$$

As N goes to infinity, this converges to

$$\frac{1}{1 - P}.$$

So no matter how many processors you have, you are limited in the speed boost you can see by how much of the code you can actually do in parallel.

How much work are our functions doing sequentially? In the second version it is pretty clear that the only sequential work is submitting the tasks and calculating the final product. For large values of n, submitting the tasks should be fairly negligible. Unfortunately, the final product will not be. Remember that operations with `BigInt` take longer for bigger numbers. The biggest numbers of the calculation will be the ones at the end that come up during this sequential section. If we really wanted to make this as fast as possible, we would need to make that last product happen in parallel as well. That is easy to do if you use the `par` method, but challenging otherwise. That task is left as an exercise for the reader.

Java 7 also introduced a new type of `ExecutorService` called a `ForkJoinPool` that is the ideal for many tasks these days. They produce their own subtype of `java.util.concurrent.Future` that has the ability to do work steeling. So if a task finished with its work, and some other task could be broken up, some of the remaining work can be handed off to one of the existing threads. Scala parallel collections are built on top of this type of functionality.

9.2.3   Parallel Data Structures

There are a number of different data structures provided in `java.util.concurrent` that help in performing common tasks. Instead of simply telling you what these do, we will consider scenarios where you want to do different things and then present the data structures that can help you to do those things. These have not been explicitly rewritten in a Scala style, so even if you write most of your code using Scala libraries, these can still be very helpful.

9.2.3.1   Shared Barriers

Earlier we had the example of a large number of people working to dig a trench that would then be filled with water. This is the type of task where you have many workers, represented by threads on the computer, working on a task and there are certain critical points where no one can proceed until everyone has gotten to that critical point.

Consider a computer program that has to deal with collisions between moving objects in a scene. Collisions need to be handled at the exact time they occur for this to be accurate. In the program there are other things that need to be computed at certain intervals. We will call this interval the time step and it has a length of Δt. Collisions are processed in a more continuous way through the time step. So in any given step you would look for collisions that happen given the initial configuration, process collisions, with updates that happen as a result of earlier collisions, then do whatever processing is required at the end of the step.

To make things go fast, you want to break this work across as many threads as you can. You break the particles up into groups and have one thread run through and find all collisions during that time step given the initial trajectories of the particles. You cannot start processing collisions until all the threads are done identifying the first round because it is possible that the last one found could be the first one you need to process. So all the threads have to finish finding collisions before the first one can be processed. There are schemes for processing collisions in parallel as well that make sure that you do not ever work on two collisions at the same time if they are too close. All collisions have to be processed and all threads have to get to the same point before the other processing can be done.

There are three different types in `java.util.concurrent` that can help give you this type of behavior. The first one is a type called `CountDownLatch`. You instantiate a `CountDownLatch` with `new CountDownLatch(count)`, where `count` is an integer value. The two main methods of this type are `countDown`, which decrements the count by one, and `await`, which will block until the count reaches zero. If the count is already at zero, a call to `await` does nothing.

The second type that can provide this type of behavior is the `CyclicBarrier`. While the `CountDownLatch` is only good for one use, the `CyclicBarrier` can be cycled many times. You make a `CyclicBarrier` with `new CyclicBarrier(parties)`, where parties is an integer number specifying how many threads should be involved in the barrier. When each thread has finished doing its task, it makes a call to the `await` method on the `CyclicBarrier`. Once all "parties" have called `await`, they are all allowed to proceed forward.

The third type is the `Phaser` type. This type is more complex than the other two. It is also newer, having been added in Java 1.7. Interested readers should consult the API for usage of this type.

9.2.3.2 The Exchange

Imagine a situation where two parties work mostly independently, but they perform calculations that require some information from another party. You do not want to try reading the information at any old time as the value might not contain a valid, final result most of the time. Instead, you need to arrange a point when the parties pause and come together for an exchange. The image of two spies subtly exchanging briefcases at a park bench comes to mind.

A programming example might be a large simulation of a solid body like an airplane. These are often done with finite-element models. Each thread would have a chunk of the plane to work with and it would calculate the stresses on the elements in that chunk as they change over time. The stresses on one chunk will depend on what is happening with all the elements in adjacent chunks that are connected to elements in the first one. In order to do a full calculation, the threads need to meet up to exchange information about the boundary elements. This type of thing can be done with an `Exchanger`.

You instantiate an `Exchanger` with `new Exchanger[A]()`, where `A` is the type of the information you want to exchange. When a thread is ready to exchange, it calls `exchange(x: A): A` on the `Exchanger`. This method will block for the first thread until the second thread also calls it. At that point, both calls return and each thread gets the value that was passed in by the other one.

9.2.3.3 Assembly Line

When you picture humans working in parallel on a task, one way to break things up is the assembly line. Here you have one group of people who do some part of the work and get products to a certain point of completion. Then they pass their results on to others in a different group who do the next step. This continues until the final result is produced. This arrangement has a critical junction at the point where the product moves from one station to the next. Generally, one person in line cannot start working on something until the person before them is done. If there are not any products in the right state, then that person needs to wait until one shows up.

There are two different types in `java.util.concurrent` that will allow you to achieve this behavior, `BlockingQueue[A]` and `BlockingDeque[A]`. The two main methods on both of these types are `put` and `take`. The `put` method will add a new value in. Some implementations of these types will have limited space. If there is not space to add the new value, the call will block until space is freed up. The `take` method will remove a value. If no value is available, then the call will block until one becomes available.

Both of these types are abstract. In the case of `BlockingDeque` there is one implementation provided called `LinkedBlockingDeque`. For the `BlockingQueue` type there are a number of different implementations.

- `ArrayBlockingQueue` works like a normal array-based queue with a fixed size. It blocks if you try to remove something when nothing is there or if you try to add something when it is already full.

- `DelayQueue` works like an unbounded queue, but items cannot be taken out until after some delay beyond which they are added. Attempts to remove before an element is ready will block.

- `LinkedBlockingQueue` works like a queue with unbounded size that blocks if you try to dequeue when nothing is there.[6]

- `LinkedTransferQueue` works like an unbounded queue that blocks with a thread tries to dequeue and nothing is there. It also has the threads that enqueue block until another thread dequeues their element.[7]

- `PriorityBlockingQueue` works as an unbounded priority queue that blocks on a dequeue when nothing is there. The priority queue ADT is discussed in chapters 13 and 18.

- `SynchronousQueue` blocks on a dequeue when nothing is there, but only allows an element to be enqueued if another thread is waiting for it. If a thread is not waiting for the next item, attempts to add return false and do not add the element.

---

[6]We will talk about how you can implement the queue's ADT using linked lists the way this `class` does in chapter 12.

[7]Think of an assembly line, but when you are done working on your part, you do not put it down someplace and move to the next one, instead you carry it to the place where someone else will take it from you. You have to stay there until someone comes to get it.

9.2.3.4 Ticketed Passengers

Some types of work are limited by resources. If you are building a house the number of people involved in driving nails will be limited by the number of hammers that you have. If you are transporting people on a plane, there are a limited number of seats. On a computer, there might be limits to how many threads you want involved in activities that deal with hardware, like reading from the network. There might be some types of processing that are particularly memory intensive for which you need to limit the number of threads executing these processes at any given time. For these types of situations, the `Semaphore` provides a way to limit how many threads can get to a certain region of code at a given time. Unlike the monitors described with `synchronized`, the `Semaphore` is more flexible than just allowing one thread in at a time.

You can instantiate a `Semaphore` with `new Semaphore(permits)`, where `permits` is an `Int` specifying how many of the resource you have. The main methods for the `Semaphore` are `acquire` and `release`. The `acquire` method will get one of the permits. If none are available, it will block until a different thread calls `release` to make one available. There is also a `tryAcquire` method that returns a `Boolean` that does not block. If a permit was available, it takes it and returns `true`. Otherwise it simply returns `false`.

9.2.3.5 Other Threadsafe Types

There are various other types in `java.util.concurrent` for working in applications with multiple threads. These types all begin with `Concurrent`. The primary one is the `ConcurrentHashMap`, which can be instantiated like this.

```
val cmmap = new ConcurrentHashMap[String,Data]
```

This type is a bit more of a pain to work with because you have to explicitly call `get` and `set`, and it lacks the methods you are used to from Scala. You can get a version with a Scala-like interface by calling `asScala`. For this to work, you need to include `import scala.collection.JavaConverters._` in your code.

9.2.4 Atomic (`java.util.concurrent.atomic`)

Another critical term that you will see mentioned in discussions of parallel processing is the term "atomic". Atomic operations are typically safe in multithreaded code. What does it mean for an operation to be atomic? The Greek root of the word atom means indivisible. The name was given to the particles because it was believed at the time that they were the smallest pieces of matter and could not be split. While that turns out to not really be true of atoms in nature, it is true of atomic operations on a computer. In this alternate sense, when we say it cannot be split, we mean that it cannot be interrupted. Once an atomic operation begins, nothing else can take over before it finishes. It helps for operations to be atomic in multithreading because otherwise, two threads can interact in ways that you are not expecting that messes up code as part of a race condition.

This is why we cannot use simple `Boolean` variables for locking and why we use things like monitors with `synchronized`. Consider the following code where we try to do our own synchronization without using `synchronized`.

```scala
var isSafe = true
var cnt = 0
for (i <- (1 to 1000000000).par) {
  if(isSafe) {
    isSafe = false
```

```
6        cnt += 1
7        isSafe = true
8      }
9    }
10   println(cnt)
```

This code fails to produce the correct answer, and it does so rather slowly compared to the sequential code. The reason it gets the wrong answer is that the operations on lines 4 and 5 are not atomic. As a result, two or more threads can both look at `isSafe` and see true before either one sets `isSafe` to false.

Synchronization is not the only operation that benefits from being atomic. One of the two packages under `java.util.concurrent` is `java.util.concurrent.atomic`. This package contains a number of different classes that provide you with ways of storing and manipulating data in an atomic manner. These `classes` do what they do without locking or standard synchronization. Instead, they use an approach called lock-free, which is based on a single atomic operation called get-and-set that is supported in most computer hardware.

Back in section 8.6 we looked at a number of different programs that did counting across multiple threads. One of the solutions that we considered in that section synchronized the increments. This was needed because the seemingly simple statement `cnt += 1` is not atomic; it can be interrupted leading to race conditions. Indeed, without synchronization, using this statement with a single mutable variable inevitably produced incorrect results.

The problem with using synchronization in that situation was that there is a lot of overhead in checking and locking monitors. The `atomic` package provides us with a slightly lighter-weight solution. One of the classes in `java.util.concurrent.atomic` is `AtomicInteger`. Instances of this class store a single integer value. The advantage of using them is that they provide a number of different methods for dealing with that value that operate atomically without doing full synchronization. Here is code that does the counting using `AtomicInteger`.

```
package multithreading2

import java.util.concurrent.atomic.AtomicInteger

object CountAtomic extends App {
  var cnt = new AtomicInteger(0)
  for (i <- (1 to 1000000000).par) {
    cnt.incrementAndGet()
  }
  println(cnt.get)
}
```

In limited testing, this code ran about twice as fast as a version with full synchronization. That is still significantly slower than a sequential version or better-written parallel version that uses one counter per thread and combines the results at the end, but there are some situations where you truly need a single mutable value. In those situations, types like `AtomicInteger` are the ideal solution.

9.2.5   Locks (`java.util.concurrent.locks`)

The other package under `java.util.concurrent` is `java.util.concurrent.locks`. Locks were added to address a shortcoming in using `synchronized`. When you use `synchronized`, you are able to lock a monitor and restrict access from other threads for one block of code, which has often been one method in our examples. However, it is not uncommon to have code where you want to restrict access beginning with one method call and not release it back until some other method is called. There is not a concise way to do this with `synchronized`, but there is with a `Lock`. `Lock` is an abstract type with the methods `lock`, `unlock`, and `tryLock`. It is much like a `Semaphore` with only one permit. The main concrete subclass is called `ReentrantLock`, which you can make by calling `new ReentrantLock`.

9.3 End of Chapter Material

9.3.1 Summary of Concepts

- Through the Java libraries, Scala programs have access to basic thread operations. You can easily create instances of `java.lang.Thread` and start work running in the various threads.

  - A way to control threads to help them work together is with `wait` and `notifyAll`. The `wait` method causes a thread to pause. It will stay paused until a notification wakes it up.

  - The `Thread` class has other methods like `sleep` and `yield` that can be quite useful.

  - The `dumpStack` method will print out the current stack trace.

  - The `join` method is a useful blocking call that will stop the current thread until another thread has completed.

  - You can also set the priority of a thread using the `setPriority` and `getPriority` methods.

- There are many tasks that are common to a lot of parallel programs. For this reason, a concurrency library was added to Java in the `java.util.concurrent` package.

  - `Executors` and `ExecutorServices` abstract away the task of picking a method for running a piece of code. There are some useful methods in `Executors` that will provide `ExecutorService` objects, which behave in generally useful ways. Many of these allow for the reuse of threads as there is significant overhead in creating `Thread` objects.

  - The `Callable` type allows you to run a task on an `ExecutorService` that gives you back a value. When such a task is submitted, you are given a `Future` object that will provide the result.

  - The concurrency library includes a number of parallel data structures as well.

    * The `CyclicBarrier` and `CountDownLatch` classes make it easy to have multiple threads pause at a point in the code until all threads have gotten there.

    * An `Exchanger` can be used to pass a value between threads at some particular point.

    * The `BlockingQueue` is a data structure that lets one group of threads exchange objects with another group of threads and blocks in situations when there is not space to put something or nothing is available.

* A `Semaphore` provides a way to control how many threads are working in a particular part of the code at any given time.

− Java 7 also introduced a new type of `ExecutorService` called a `ForkJoinPool`, which is ideal for many tasks these days. They produce their own subtype of `java.util.concurrent.Future` that has the ability to do work stealing.

− The `java.util.concurrent.atomic` package provides `classes` for basic types that have operations which cannot be interrupted. These often provide a simpler and more efficient approach to preventing race conditions than synchronization.

− The `java.util.concurrent.locks` package defines locks that can be used for synchronization-type behaviors that can span beyond a single method call or code block.

9.3.2   Exercises

1. In the discussion of the parallel factorial and Amdahl's law we concluded that the primary limitation in the code we had written for scaling was the final product. Using techniques from `java.util.concurrent`, write code to make that operation happen in parallel.

2. A good way to make sure you understand parallelism is to work on implementing a number of different simple problems using each of the different techniques that were presented in this chapter. The problems you should solve are listed below. For each of them, try to write a solution using the following methods.

   • Normal `java.lang.Threads` or calls to `Future` with `synchronized` and `wait/notifyAll`.

   • Elements from `java.util.concurrent` such as `ExecutorServices`, thread-safe data structures, and atomic values.

   • Scala parallel collections. (Note that this is not flexible enough to handle all problems so only use it for those that work.)

   For each solution that you write, you should do timing tests to see how the different solutions compare for speed using the highest core-count machine you have access to. For speed comparisons, you should consider writing a sequential version as well. For most of these to be of interest for parallel, the inputs will need to be fairly large.

   (a) Find the minimum value in a sequence of `Doubles`.

   (b) Given a `Seq[Int]`, find the value for which the following function returns the largest value.

```scala
def goldbach(num:Int):Int = {
  var ret = 0
  var n = num
  while (n > 1) {
    n = if (n % 2 == 1) 3 * n - 1 else n / 2
    ret += 1
  }
  ret
}
```

   (c) Matrix addition using two matrices stored as `Array[Array[Double]]`.

(d) Matrix multiplication using two matrices stored as `Array[Array[Double]]`.

(e) Find all occurrences of a substring in a given `String`.

9.3.3 Projects

As you would expect, all of the projects for this chapter have you adding multithreading to your project. How you do this depends a lot on the project. There are two aspects of this that are worth noting in general. First, you cannot generally multithread drawing to the screen. That is because the graphics is being done by a single device. Often, drawing to different images in parallel will not help either because drawing to images often involves the graphics pipeline on your graphics card so it is a bottleneck, even when you have separate threads doing separate work on separate images.

The other thing to note is not to overdo multithreading. Spawning many threads with little work for each to do will only slow your program down. Spawning too many can even consume enough resources to bring your machine to its knees. It is easy to spawn threads with things like `Actor.actor`, but you need to think about what you are doing when you use it. Overuse of threading can also lead to race conditions that are extremely difficult to debug. Appropriate usage of the thread-safe collections from `java.util.conurrent` can help dramatically with that problem.

1. If you chose not to use the actor model for the MUD, you should be using `Futures` to make things happen in parallel while the program is waiting for the user to type things. Use elements of `java.util.concurrent` to help make this safe.

2. Networking, like reading from and writing to a disk, can be a significant bottle neck. Unlike the disk though, it is possible that the bottleneck is not related to your machine. For this reason, there can be significant speed benefits to having multiple threads reading from the network all at once. If you are working on the web spider, this is exactly what you should do for this project. You can also set the stage for multiple threads processing data as well, in case that becomes a bottleneck.

   To do this, you can basically make two pools of threads, one that reads data and another that processes the data. Over time you can balance the number of threads in each to get maximum performance. Data needs to be passed between these threads in a thread-safe manner. One or more `java.util.concurrent.BlockingQueue`s would probably work well. As the data-collecting threads pull down the data and get them into a form with minimal processing, they put their results on a queue. The analysis threads pull data off the queue and do the proper analysis on it. In the final version of this, the analysis will inevitably find more data that needs to be collected. That will be put on a different queue that the reading threads will pull from.

   You can stick with pulled data from files for this project, though it is unlikely the threading will help in that situation or that you have a large enough data set. In the next two chapters we will introduce concepts that will make it easier to pull the data from the web or other network sources.

3. There is no specific task associated with multithreading using features of the Java languages for the second games project.

4. There is no specific task associated with multithreading using features of the Java languages for the math worksheet project.

5. There is no specific task associated with multithreading using features of the Java languages for the Photoshop® project.

6. There is no specific task associated with multithreading using features of the Java languages for the simulation workbench project.

7. There is no specific task associated with multithreading using features of the Java languages for the stock project.

8. There is no specific task associated with multithreading using features of the Java languages for the movies project.

9. There is no specific task associated with multithreading using features of the Java languages for the L-systems project.

Chapter 10

Stream I/O and XML

In chapter 1 we saw how to use `scala.io.Source` and `java.util.Scanner` to read from text files. We also saw how we could use `java.io.PrintWriter` to write text files. In this chapter we will explore the overall structure of the `java.io` package and see how streams can be used to represent a general approach to input and output. This will set us up in the following chapter to use streams for network communication. We will also explore the use of XML in Scala as another format for storing or communicating data.

10.1 The `java.io` Package

The original library for doing input and output in Java was the `java.io` package. While there is now a `java.nio` package (non-blocking io) that can provide higher performance, the original library is still used as a foundation for many other parts of the Java libraries, which makes it worth knowing and understanding.

The majority of the classes in the `java.io` package are subtypes of one of four different abstract types: `InputStream`, `OutputStream`, `Reader`, and `Writer`. The `InputStream` and `OutputStream` have basic operations that work with `Bytes`. The `Reader` and `Writer` have almost the same set of methods, but they work with `Chars`. Here are the signatures of the methods in `InputStream`.

```
available():Int
close():Unit
mark(readlimit:Int):Unit
markSupported():Boolean
read():Int // This method is abstract
read(b:Array[Byte]):Int
read(b:Array[Byte], off:Int, len:Int):Int
reset():Unit
skip(long n):Long
```

The basic `read()` method is abstract. The other read methods have default implementations that depend on this one, though they can be overridden to provide more efficient implementations as well. The `read()` method returns the value of the next `Byte` in the stream giving an `Int` in the range of 0–255 if the read succeeds. If the end of the stream has been reached, it returns -1. The methods for `Reader` are very similar. One major difference is that the read methods take an `Array[Char]` instead of an `Array[Byte]`. We will be focusing on the streams in this chapter. Refer to the Java API for full details on `Reader` and `Writer`.

The term STREAM comes from the way that data is accessed. The data is a flow and you can see what is there, but you cannot really jump around to random places. Imagine water flowing through a stream. You can see the water as it goes by or you can look away and skip it. However, once it has passed, you do not get the option to repeat seeing that same water. In other words, you must access the data sequentially. That is the idea of a data stream. Note that the `InputStream` has methods called `markSupported`, `mark`, and `reset`. For certain streams, these methods will allow you to effectively cheat a bit and jump backward in a stream. However, not all streams will support this behavior and even in those that do, you are constrained to only being able to return to the point that was previously marked.

The `OutputStream` has a slightly shorter set of methods. Like the `InputStream`, it works with `Bytes`. In this case, writing data to the stream is like throwing something into flowing water. It gets whisked away and you do not have the option of making alterations. Here are the methods for `OutputStream`.

```
close():Unit
flush():Unit
write(b:Array[Byte]):Unit
write(b:Array[Byte], off:Int, len:Int):Unit
write(b:Int):Unit // This method is abstract
```

Once again, there is one critical method, in this case it is `write(b:Int)`, that is abstract. While it takes an `Int`, only the lowest eight bits (the lowest byte) will be written out. The other 24 bits are ignored.

10.2   Streams for Files

The `InputStream` and `OutputStream` do not actually include code to read from/write to anything. This is why they are abstract. Their sole purpose is to serve as base classes for inheritance. Some of the subtypes are more specific about what source of data they are attached to. The most obvious things to attach streams to are files. The `FileInputStream` and `FileOutputStream` are subtypes that implement the **read/write** methods so that they work with files. You instantiate these types using **new** and passing in the file you want either as a string or as an instance of `java.io.File`. Here is a simple example that illustrates using a `FileInputStream` to read a file. It prints each byte as a numeric value.

```scala
package iostreams

import java.io.FileInputStream

object ReadBytes extends App {
  val fis = new FileInputStream(args(0))
  var byte = fis.read()
  while (byte >= 0) {
    print(byte+" ")
    byte = fis.read()
  }
  println()
  fis.close()
}
```

Running this program produces the following output.[1]

```
$ scala -cp bin iostreams.ReadBytes src/iostreams/ReadBytes.scala
112 97 99 107 97 103 101 32 105 111 115 116 114 101 97 109 115 10 10 105 109 112
    111 114 116 32 106 97 118 97 46 105 111 46 70 105 108 101 73 110 112 117 116
    83 116 114 101 97 109 10 10 111 98 106 101 99 116 32 82 101 97 100 66 121 116
    101 115 32 101 120 116 101 110 100 115 32 65 112 112 32 123 10 32 32 118 97
    108 32 102 105 115 32 61 32 110 101 119 32 70 105 108 101 73 110 112 117 116
    83 116 114 101 97 109 40 97 114 103 115 40 48 41 41 10 32 32 118 97 114 32 98
    121 116 101 32 61 32 102 105 115 46 114 101 97 100 40 41 10 32 32 119 104 105
    108 101 32 40 98 121 116 101 32 62 61 32 48 41 32 123 10 32 32 32 32 112 114
    105 110 116 40 98 121 116 101 43 34 32 34 41 10 32 32 32 32 98 121 116 101 32
    61 32 102 105 115 46 114 101 97 100 40 41 10 32 32 125 10 32 32 112 114 105
    110 116 108 110 40 41 10 32 32 102 105 115 46 99 108 111 115 101 40 41 10 125
```

A slight modification will produce a script that can be used to copy files.

```scala
package iostreams

import java.io.FileInputStream
import java.io.FileOutputStream

object ByteCopy extends App {
  val fis = new FileInputStream(args(0))
  val fos = new FileOutputStream(args(1))
  var byte = fis.read()
  while (byte >= 0) {
    fos.write(byte)
    byte = fis.read()
  }
  fis.close()
  fos.close()
```

---

[1]Note that this code is in an Eclipse project, so the compiled .class files are in the **bin** directory and the source files are in the **src** directory.

```
}
```

This script takes two arguments, just like the Linux `cp` command. It uses the first as the file name for the `FileInputStream` and the second for the name of the `FileOutputStream`. Instead of printing the bytes as numbers to standard output, they are written to the output stream. When everything is done, both files are closed.

Wrapping an `InputStream` in a `Source`

One advantage of the `scala.io.Source` type is that it was a Scala collection. This means that all the methods we have gotten comfortable with when using other collections will work on it as well. If you have an `InputStream` that is reading text you can take advantage of this benefit. Instead of calling `Source.fromFile(name:String)`, you can call `Source.fromInputStream(is:InputStream)` to get a `Source` that will read from the specified `InputStream`. Much of the code we will use with streams will be reading binary data instead of text data, but you should know that this option exists.[2]

10.3   Exceptions

If you run a program and provide a name for an input file that is not valid, you will get something that looks like the following.

```
java.io.FileNotFoundException: notthere.txt (No such file or directory)
     at java.io.FileInputStream.open0(Native Method)
     at java.io.FileInputStream.open(FileInputStream.java:195)
     at java.io.FileInputStream.<init>(FileInputStream.java:138)
     at java.io.FileInputStream.<init>(FileInputStream.java:93)
     at
        iostreams.ReadBytes$.delayedEndpoint$iostreams$ReadBytes$1(ReadBytes.scala:6)
     at iostreams.ReadBytes$delayedInit$body.apply(ReadBytes.scala:5)
     ...
```

You have probably seen quite a few outputs like this while working with Scala. This is a stack trace and it was printed out because something bad happened that caused the program to throw an exception.

Exceptions are meant to occur in exceptional circumstances, such as when there is an error. In the example above, you can see that it was a `java.io.FileNotFoundException` that occurred. This should make sense given that we tried to open a file that was not there. Real applications need to be able to deal with this. For example, if a user specified a file that did not exist, the program should tell the user that and perhaps ask for an alternate file. It should not crash. In order to prevent the crash, we use a `try/catch/finally` as was discussed in section 1.5.3.

Methods can throw any subtype of the `java.lang.Throwable` class. The two immediate subtypes of this are `java.lang.Error` and `java.lang.Exception`. Typically an `Error` represents something that you will not be able to recover from. An `Exception` is something that you should consider dealing with. If something goes wrong in your own code, you can throw your own exception using the `throw` keyword. When you do this, you should typically throw a subtype of Exception that matches what has happened. If such a subtype exists in the libraries, use it. If not, create your own subtype. For example if something was passed into a method that does not match the requirements of that method, you might have the code **throw new IllegalArgumentException("The value of grade must be between 0 and 100.")**.[3]

Like in the example, most subtypes of **Exception** can be passed a string argument. One of the main advantages of exceptions over other ways of dealing with errors is that exceptions can be informative. The ideal exception gives the programmer all the information they need to fix the error. All exceptions give you a stack trace so that you can see what line caused the exception and how that was reached. Often you need additional information to know what went wrong. In the case of the **FileNotFoundException**, the name of the file is critical information. If you try to access a value that is out of bounds on an **Array**, knowing what index was being accessed and how big the **Array** really was are critical pieces of information. So when you throw an exception or make a new subtype of **Exception**, you should make sure that it has all the information that will be needed to fix it. There are good odds that the first person to see that exception message will be you, and you will thank yourself for good error messages that make debugging easier.

10.3.1 try-catch-finally Revisited

Any time you open a file that was provided to you by the user, you know that something might go wrong and it might throw an exception. To stop that exception from crashing the thread that it is running in, you tell Scala that you are going to "try" to execute some code. If an exception is thrown that you want to deal with, you should "catch" it. The **try** keyword is followed by the block of code that you want to try to run. After the block of code you put the **catch** keyword followed by a block of cases that include the handlers for different exceptions that might arise. In the case of opening a file and printing the contents, code might look like this.

```scala
package iostreams

import java.io.IOException
import java.io.FileInputStream
import java.io.FileNotFoundException

object ReadBytesTry extends App {
  try {
    val fis = new FileInputStream(args(0))
    var byte = fis.read()
    while (byte >= 0) {
      print(byte+" ")
      byte = fis.read()
    }
    println()
    fis.close()
  } catch {
    case ex: FileNotFoundException =>
      println("The file "+args(0)+" could not be opened.")
    case ex: IOException =>
      println("There was a problem working with the file.")
```

---

[3]In Scala you would probably use a **require** call for this instead of an explicit **throw**, but this shows you the syntax of throwing.

```
  }
}
```

This has two different `cases` in the `catch`. The first deals with the exception we saw before, where the file could not be opened. This can happen even if the file exists if you do not have permission to read that file. The second `case` catches the more general `IOException`. This is a supertype of `FileNotFoundException` and represents any type of error that can occur with input and output. In this case, it would imply that something goes wrong with the reading of the file. Other than hardware errors, it is hard to picture anything going wrong with this particular code because it makes no assumptions about the structure of the data in the file. It simply reads bytes until it runs out of data to read.

This code has a problem with it that would be even more significant if we were doing more interesting things with the reading. The problem arises if the file opens, but an exception occurs while it is being read. Imagine that an `IOException` is thrown by the `fis.read()` call in the `while` loop because something happens to the file. The exception causes the `try` block to stop and control jumps down to the proper `case` in the `catch`. After the code in that `case` is executed, the program continues with whatever follows the `catch` block. This means that the code in the `try` block after the location where the exception occurs is never executed. This is a problem, because that code included the closing of the file. Failure to close files can be a significant problem for large, long running programs as any given program is only allowed to have so many files open at a given time.

What we need to fix this problem is a way where we can specify a block of code that will run after a `try` block regardless of what happens in the `try` block. This is done with a `finally` block. You can put a `finally` block after a `catch` block or directly after a `try` block. To do our file-reading problem correctly we could use the following code.

```scala
package iostreams

import java.io.IOException
import java.io.FileInputStream
import java.io.FileNotFoundException

object ReadBytesTryFinally extends App {
  try {
    val fis = new FileInputStream(args(0))
    try {
      var byte = fis.read()
      while (byte >= 0) {
        print(byte+" ")
        byte = fis.read()
      }
      println()
    } catch {
      case ex: IOException =>
        println("There was a problem working with the file.")
    } finally {
      fis.close()
    }
  } catch {
    case ex: FileNotFoundException =>
      println("The file "+args(0)+" could not be opened.")
  }
}
```

This has one `try` block that contains everything, and inside of that is another `try` block that reads the file with a `finally` for closing the file. This is a bit verbose. We will see shortly how we can set things up so that we do not have to repeat this.

As with nearly every other construct in Scala, the `try-catch-finally` produces a value and can be used as an expression. If the `try` block executes without throwing an exception, then the value will be that of the `try` block. If an exception is thrown and caught, the value of the whole expression is the value of the `catch` for that `case`. Any value for `finally` is ignored as `finally` is typically used for clean-up, not part of the normal computation.

There is also the possibility that the `try` block executes and throws an exception that is not caught. In this situation the value of the full `try` expression is `Nothing`. You might remember this type from figure 4.4. It is at the very bottom of the type diagram. The `Nothing` type is a subtype of everything else. There are no instances of `Nothing`. It is mostly used by the type inference system. When trying to figure out the type of a complex expression with multiple possible return values, Scala picks the lowest type in the hierarchy that is a supertype of all the possibilities. Since a `try` can throw an exception that is not caught, it is almost always possible for it to produce a `Nothing`. The fact that `Nothing` is at the base of the type hierarchy means that it does not impact the inferred type.

If you use a `try-catch-finally` as an expression, you have to be careful of the types of the last expressions in each of the different cases of the `catch`. In particular, you probably do not want the last thing in any of the cases to be a print statement. Calls like `println` have a return type of `Unit`. Unless all the other options are `Unit`, the inferred type will almost certainly be `AnyVal` or `Any`. Those are types that do not provide you with significant functionality. The `catch` `case`s of a `try` that is used as an expression probably should not print anything. If it should be an expression, then it should communicate results with a result value instead of through side effects.

10.3.2 Effect of Exceptions

So far we have treated an uncaught exception as causing the thread to terminate. This is not really the case. When an exception is thrown in a method/function, it causes execution of the current block to stop and control jumps either to the `catch` of an enclosing `try` block, or out of that method/function all together. If there is no enclosing `try`, or if the `catch` on that `try` does not have a `case` for the exception that was thrown, the exception will pop up the stack to the method that called the current function. If the call point in that function is in a `try` block, the `case`s for that `try` block are checked against the exception. This process continues up the call stack until either a `case` is found that matches the exception or the top of the stack is reached and the thread terminates.

The fact that exceptions propagate up the call stack means that you often do not catch them at the place where they originate. Instead, you should catch exceptions at a place in the code where you know how to deal with them. Often this is several calls above the one that caused the exception to occur.

10.3.3 Loan Pattern

The need to double nest `try` expressions with a `finally` to properly handle closing a file is enough of a pain that it would prevent most programmers from putting in the effort to do it. This same problem existed in Java so they altered the language with an enhanced `try-catch` in Java 7. The syntax of Scala lets you streamline this code without altering the language by making it easy to implement the *loan pattern* [10].[4] Here is code that does the reading we have been doing in this way.

```scala
package iostreams

import java.io.IOException
import java.io.FileInputStream
import java.io.FileNotFoundException

object LoanPattern extends App {
  def useFileInputStream[A](fileName: String)(body: FileInputStream => A): A = {
    val fis = new FileInputStream(fileName)
    try {
      body(fis)
    } finally {
      fis.close()
    }
  }

  try {
    useFileInputStream(args(0))(fis => {
      var byte = fis.read()
      while (byte >= 0) {
        print(byte+" ")
        byte = fis.read()
      }
      println()
    })
  } catch {
    case ex: FileNotFoundException =>
      println("The file "+args(0)+" could not be opened.")
    case ex: IOException =>
      println("There was a problem working with the file.")
  }
}
```

Here the `try` with the `finally` has been moved into a separate function called `useFileInputStream`. This function takes a file name in one argument list and a function in a second. The function is used in the application code with a `try` that is only responsible for handling the errors. There are still two `try` statements, but the one in the `useFileInputStream` only has to be written once. The rest of the code can call that function and safely include only a single `try-catch` without a `finally`.

While it is not used in this example, the `useFileInputStream` function takes a type argument that allows it to be used as an expression. The type is inferred from the code in the `body` function.

---

[4]The loan pattern loans a resource to a function of code.

10.4 Decorating Streams

The `InputStream` and `OutputStream` types only allow reading/writing of single `Bytes` or `Arrays` of `Bytes`. This is technically sufficient to do anything you want as all the data on a computer is stored as a sequence of bytes when you get down to a certain level. Being sufficient though does not mean that it is ideal. What do we do if we are reading a stream that is supposed to contain a mixture of different data that includes things like `Int` or `Double` and not just `Byte`. You can build an `Int` from four `Bytes`, but you do not want to have to write the code to do so yourself, especially not every time you need that functionality. This limitation of the basic streams, and the file streams, is addressed by decorating streams.[5]

Types like `FileInputStream` know what they are pulling data from, but they are not very flexible or efficient. This shortcoming is addressed by having stream types that are constructed by wrapping them around other streams. This wrapping of one stream around another to provide additional functionality is the decorating.

10.4.1 Buffering

One of the shortcomings of `FileInputStream` and `FileOutputStream` deals with performance. Accessing files is a slow process, it is one of the slower things that you can do on a computer. Reading one byte at a time is particularly slow. It is much better to read a large block of data than to read all the bytes, one at a time. This is the reason why `BufferedInputStream` and `BufferedOutputStream` were created. These types have the same methods as the standard `InputStream` and `OutputStream`. The only difference is how the `read/write` method is implemented.

Both types keep a buffer in memory. In the `BufferedInputBuffer` if that buffer has unread `Bytes` in it, the read pulls from the buffer. When the buffer runs out of data, a full buffer is read at once. For the `BufferedOutputStream`, the `Bytes` that are written go into the buffer until it is full, then all of them are dumped out at once. If you need to force things to go out before the buffer is full, you can call the `flush` method.

The buffered types do not actually read from or write to any particular source. In order to use one of the buffered types you need to give it a stream to pull data from or push data to. The syntax for doing that could look like this.

```
val bis = new BufferedInputStream(new FileInputStream(fileName))
```

The fact that disk access has really high latency means that you should always wrap file streams with buffered streams. The speed boost of doing so for large files can be remarkable.

Latency and Bandwidth

To understand why it is better to read/write a big block of data than to read separate bytes, you need to know something about how we characterize communication speeds. When we talk about how fast data can move inside of a computer or between computers, there are two values that are significant, latency and bandwidth. Most of the time you hear people refer to bandwidth. This is how much data can be moved in

---

[5]This terminology comes about because the `java.io` library uses the *decorator pattern*. [6]

a particular period of time. Bandwidths often fall in the range from a few Mb/s to a few Gb/s[6].

The bandwidth value is only relevant in the middle of a block of communication. Every time a new communication starts, there is a pause called latency. The amount of time it takes to read a block of N bits is roughly given by $time = latency + N/bandwidth$. If you read your data in one-byte/eight-bit increments, the second term is small and you spend almost all your time waiting for the latency. Reading larger blocks minimizes the latency overhead and gives you an actual speed closer to the full bandwidth.

10.4.2   Binary Data

The other drawback of the standard `InputStream` and `OutputStream` is that they only work with bytes. Pretty much all applications will need to work with data in some more complex format. We have seen how to do this using flat text files. There are a number of drawbacks to plain text files though. They are not very informative, they are slow, and they can take a lot more space than is needed to store the data in them. It turns out that you cannot easily solve all of those problems at once. In this section we will discuss storing data in binary format, which can help address the speed and space issues.

The first way to read/write data in binary format is with the `DataInputStream` and `DataOutputStream`. These types get wrapped around other streams like the `BufferedInputStream` and `BufferedOutputStream`. What they give us is a set of additional methods. For the `DataInputStream`, these methods include the following.

```
readBoolean():Boolean
readByte():Byte
readChar():Char
readDouble():Double
readFloat():Float
readInt():Int
readLong():Long
readShort():Short
readUTF():String
```

Each of these reads the specified data value from the stream that is wrapped. The `DataOutputStream` has matching methods for writing data. It also has a few extras that can write strings in other ways.

```
writeBoolean(v:Boolean):Unit
writeByte(v:Int):Unit
writeBytes(s:String):Unit
writeChar(v:Int):Unit
writeChars(s:String):Unit
writeDouble(v:Double):Unit
writeFloat(v:Float):Unit
writeInt(v:Int):Unit
writeLong(v:Long):Unit
writeShort(v:Int):Unit
writeUTF(str:String):Unit
```

The combination of these methods gives you the ability the write data using the more basic types to a file and read it back in without having to go all the way down to `Bytes`.

If you are working with files for your binary data, you still really need to have buffering for performance reasons. The beauty of the way the `java.io` streams library works is that you can decorate streams however you want. In this situation you want to wrap the data stream around the buffered stream, which is wrapped around the file stream. Code for that looks like the following.

```
val dis = new DataInputStream(new BufferedInputStream(new FileInputStream(file)))
```

The order is significant here, mainly because the methods we want to be able to call are part of the `DataInputStream`. The general rule is that the outermost type needs to have the methods that you want to call. The ones between that and the actual source stream should implement the basic methods in an altered fashion, such as buffering. Those should be stacked in an order that makes sense for the application.

The challenge in working with binary data files is that they cannot be easily edited with any standard programs. To understand this, consider the following code.

```scala
package iostreams

import java.io.FileInputStream
import java.io.FileOutputStream
import java.io.BufferedInputStream
import java.io.BufferedOutputStream
import java.io.DataInputStream
import java.io.DataOutputStream

object ReadWriteBinary {
  def withDOS[A](fileName: String)(body: DataOutputStream => A): A = {
    val dos = new DataOutputStream(new BufferedOutputStream(new
        FileOutputStream(fileName)))
    try {
      body(dos)
    } finally {
      dos.close()
    }
  }

  def withDIS[A](fileName: String)(body: DataInputStream => A): A = {
    val dis = new DataInputStream(new BufferedInputStream(new
        FileInputStream(fileName)))
    try {
      body(dis)
    } finally {
      dis.close()
    }
  }

  def writeDoubleArray(fileName: String, data: Array[Double]): Unit = {
    withDOS(fileName)(dos => {
      dos.writeInt(data.size)
      data.foreach(x => dos.writeDouble(x))
    })
  }

  def readDoubleArray(fileName: String): Array[Double] = {
    withDIS(fileName)(dis => {
```

```
    Array.fill(dis.readInt)(dis.readDouble)
  })
 }
}
```

This code contains two functions that can be used to generally work with `DataOutputStreams` and `DataInputStreams`. In an application, these would be joined with a number of other such utility methods. The other two methods use those first two and write out or read back in an `Array` of `Doubles`. If you run a REPL with your project in the path, you can test it with the following.

```
scala> iostreams.ReadWriteBinary.writeDoubleArray("data.bin",
    Array.fill(10)(math.random))

scala> iostreams.ReadWriteBinary.readDoubleArray("data.bin")
res0: Array[Double] = Array(0.3269452122685005, 0.029280830398166646,
    0.3791670999357305, 0.18297861033677165, 0.34052948432594365,
    0.8869272299029871, 0.7410153573200774, 0.8323511762671919,
    0.35609195645359537, 0.4658746077421544)
```

After you have done this, you should go look at the contents of the "data.txt" file. What you will find using `less`, `cat`, or `vi` is that the contents look like random garbage characters. If you were to edit the file using `vi`, then try to read it in with the `readDoubleArray`, the data will almost certainly be messed up if it manages to read at all.

The problem with looking at a file like "data.bin" is that normal characters only account for a fairly small fraction of the possible values that each byte can take. Binary data tends to use all possible values, including many that do not print well. There are command-line tools like `hexdump` and `xxd` that can be used to view binary files. The following shows the output of `xxd` on a Linux system.

```
> xxd data.bin
0000000: 0000 000a 3fd4 ecab 9c91 ba2a 3f9d fbcb  ....?......*?...
0000010: 43d6 e280 3fd8 4446 157c 5c2e 3fc7 6bd7  C...?.DF.|\.?.k.
0000020: d5a1 c980 3fd5 cb3c 2da0 3c2c 3fec 61b5  ....?..<-.<,?.a.
0000030: 36cb afbe 3fe7 b665 d6b0 c067 3fea a29e  6...?..e...g?...
0000040: ef1b 59df 3fd6 ca35 ead5 8eec 3fdd d0e3  ..Y.?..5....?...
0000050: bb12 853a                                ...:
```

The first column shows the position in the file. The next eight columns show hexadecimal values for the contents of the file. There are two characters for each byte because $256 = 16^2$. So each line shows 16 bytes. The last column shows the ASCII characters with any non-printable characters appearing as a dot. The `xxd` tool also does a reverse encoding. You can have it output to a file, edit the hex section of the file, and run the reverse encoding to get back a binary file. Doing so is a somewhat delicate operation because it is easy to mess up.

Big Endian vs. Little Endian

The data in a file produced by a `DataOutputStream` probably does not exactly match what was in the memory of your computer because Java libraries write the data out in a platform-independent way. When computer makers started laying out bytes in memory for larger groups like `Int` or `Double`, different computer makers picked different orders. For the x86 chips, Intel put the least significant byte first. This order is called

Little Endian. Most other chip makers used the opposite ordering, called Big Endian. (The terms Big Endian and Little Endian are a reference to Gulliver's Travels where the Lilliputians were fighting a bitter war over which end of an egg one should crack when eating a soft boiled egg.)

Looking closely at the hex dump above, you can see that Java writes files out in Big Endian order. You can tell this because we know that the first thing in the file is the `Int` value 10. An `Int` is stored in four bytes so the hex is the eight characters "0000000a". If the file were written using Little Endian this would be "0a000000". The reason Java uses this format, even though your computer inevitably does not (as you are most likely using an x86-based machine to run Scala), is inevitably related to the fact that when Sun® created Java, they had their own SPARC® architecture which was Big Endian.

10.5 Serialization

When you are saving data to a file, you often want to write out whole objects. The process of converting an object to some format that you can write out and read back in later is called serialization. Java and a number of other modern platforms have built-in serialization methods. The Java serialization method uses a binary output format. You can also write your own code to serialize objects in other formats. XML happens to be a useful serialization format that we will explore later in this chapter. We will look at each of these separately.

10.5.1 Binary Serialization

The Java platform has a rather powerful form of serialization that is built into the system. Other systems have different ways of supporting this, and there are several 3rd party libraries for Scala that provide binary serialization. For simplicity, we will use the default on the JVM.

To make a type that can be serialized in Scala, have it inherit from `Serializable`.[7] The reason you would want to have things that are serializable is that they can be used with `ObjectInputStream` and `ObjectOutputStream`. These types have the same methods as `DataInputStream` and `DataOutputStream` plus the additional methods `readObject():AnyRef` and `writeObject(obj:AnyRef):Unit` respectively.

To help illustrate this, consider the following little application that includes a serializable class and some code to either write an instance of it to a file or read an instance in from a file then print it.

```
package iostreams

import java.io.FileInputStream
import java.io.FileOutputStream
import java.io.ObjectInputStream
import java.io.BufferedInputStream
```

---

[7]`case classes` automatically inherit from `Serializable`.

```scala
import java.io.ObjectOutputStream
import java.io.BufferedOutputStream

class Student(val name: String, val grades: Array[Int]) extends Serializable

object SerializationTest extends App {
  args(0) match {
    case "-r" =>
      val ois = new ObjectInputStream(new BufferedInputStream(new
          FileInputStream(args(1))))
      ois.readObject() match {
        case s: Student => println(s.name+" "+s.grades.mkString(", "))
        case _ => println("Unidentified type.")
      }
      ois.close()
    case "-w" =>
      val oos = new ObjectOutputStream(new BufferedOutputStream(new
          FileOutputStream(args(1))))
      val s = new Student(args(2), args.drop(3).map(_.toInt))
      oos.writeObject(s)
      oos.close()
    case _ =>
      println("Usage: -r filename | -w filename name g1 g2 ...")
  }
}
```

From the command line you can invoke this first step using the command below.[8]

```
scala -cp bin iostreams.SerializationTest -w obj.bin John 98 78 88 93 100 83
```

After running this there will be a file called "obj.bin". You can look at it with **cat** or **xxd**. You will see that this is clearly a binary file, but there are some parts that are human-readable strings. One of these is the name itself, but there are others that give type names like **Student** and **java/lang/String**. These have to be in the file because when you *deserialize* the file, it has to know what types of objects to create. You can verify that this process works by running the program with the arguments **-r obj.bin**. This will read back in the file you just created and print the **Student** object that was read.

One critical aspect to note about the code for the **-r** option is that it includes a **match**. If you leave out the **match** and assign the result of **ois.readObject()** to a variable, the variable will have type **AnyRef**. That is because **readObject** has a return type of **AnyRef**.[9] You will not be able to get **name** or **grades** from an **AnyRef** because that type does not have those. The **match** allows you to check the type of the object that was read in and do the print statement we want, if it is a **Student**, or print an error message if it is not.[10]

When an object is serialized, some indication of its type is written out, followed by a serialization of its contents. This only works if all the contents are serializable. If you try running this code here, you will find that it throws and exception.

---

[8]This example execution assumes the Eclipse style of having the compiled files in a directory called **bin**. You can do this inside of an IDE by setting the arguments to use when the program is run. In Eclipse you need to open the Run settings and change the arguments. Using command line arguments is easier in the terminal. If you want to test this in Eclipse you might consider changing from using **args** to standard input and output.

[9]In the Java API you will see the type **Object**. **AnyRef** is the Scala equivalent to **java.lang.Object**.

[10]This same type of operation can be done with the **isInstanceOf[A]** and **asInstanceOf[A]** methods. However, the use of those methods is strongly frowned upon in Scala. Using a **match** is the appropriate Scala style for determining the type of an object and getting a reference to that object of the proper type.

```
class OtherData(val id: String, val course: String)
class Student(val name: String, val grades: Array[Int], val od: OtherData)
    extends Serializable

val oos = new ObjectOutputStream(new FileOutputStream("fail.bin"))
val s = new Student("John", Array(98, 90), new OtherData("0123", "CS2"))
oos.writeObject(s)
oos.close()
```

The details of the exception are shown here.

```
Exception in thread "main" java.io.NotSerializableException: iostreams.OtherData
        at java.io.ObjectOutputStream.writeObject0(ObjectOutputStream.java:1184)
```

The problem with this code is that od: OtherData is a member of Student, but it is not serializable. So when the serialization process gets to the od member in a Student, it fails. In this case it is simple enough to fix that by making it so that OtherData extends Serializable. In other situations you will not have that type of control because the type that you are dealing with might have been written by someone else and not be Serializable.

One way to deal with information that is not serializable is to simply not write it out. There are other times when this approach is valid as well. For example, the DrawRectangle type keeps the propPanel variable so that it does not make a new GUI component every time the user looks at the settings. However, if you save off a DrawRectangle, there is no reason to save all the information associated with the GUI component. That could be easily re-created from the other information. In order to do this in Scala, we use an ANNOTATION.

An annotation is specified by a normal name that is preceded by an @. Annotations provide meta-information about a program. That is information that is used by higher-level tools and is not really part of the normal program code. There are a number of standard annotations that are part of the Scala compiler. You can identify many of them in the API because they start with lowercase letters, while all standard types begin with capital letters. The two annotations associated with serialization are @transient and @SerialVersionUID. If you have a member in a class that you do not want to have serialized, simply annotate it with @transient. Here we have code where this has been done with od.

```
class OtherData(val id: String, val course: String)
class Student(val name: String, val grades: Array[Int],
              @transient val od: OtherData) extends Serializable

val oos = new ObjectOutputStream(new FileOutputStream("pass.bin"))
val s = new Student("John", Array(98, 90), new OtherData("0123", "CS2"))
oos.writeObject(s)
oos.close()

val ois = new ObjectInputStream(new FileInputStream("pass.bin"))
ois.readObject() match {
  case s2: Student => println(s2.name+" "+s2.grades+" "+s2.od)
  case _ => println("Unknown type read.")
}
ois.close()
```

This code also reads the object back in and prints the different fields. This is done to illustrate what happens when you read a serialized object that has a transient field. That field is not part of the serialization, so when the object is read back in, that field is given a default value. For any subtype of **AnyRef**, the default value is **null**. Running this code will show you that the last value printed is indeed **null**.

The fact that the default serialization in Java provides **nulls** for deserialized **transient** values is an unhappy fact of life for Scala developers. Good style in Scala is to avoid the use of **null** as much as possible. If something might not have a value, you are strongly encouraged to use the **Option** type. Avoidance of **null** was a design goal of the Scala language because the most common runtime error in Java is, by far, the NullPointerException. It is so common that it is often just referred to as an NPE. Following good Scala style can pretty much eliminate the possibility of having NPEs. Unfortunately, many Java libraries that we want to use from within Scala still make extensive use of **null**, so we cannot avoid it completely.

In addition to opening the door to **NullPointerExceptions**, anything that is transient must be a **var** instead of a **val**. This is due to the fact that when you deserialize such an object, that member will have a default value that almost certainly needs to be changed. The other side of this is that places that use any transient members likely need to have conditional code that will initialize the values if they are not set properly. This is similar to what we have been doing with **propPanel** in the **Drawable** types, but we had been following proper Scala style using **Option** and starting with the value of **None**. When we add code to save our drawings later in this chapter, we will have to regress somewhat and use **null** to indicate no value instead of having proper use of **Option**. If the **transient** field is accessed a lot, you should consider having a **private** local var that is only accessed through a method. Here is an alternate version of **Student** which follows that rule.

```scala
class Student(val name: String,
              val grades: Array[Int],
              @transient private var _od: OtherData) extends Serializable {
  assert(_od != null)
  def od: OtherData = {
    if (_od == null) {
      _od = new OtherData("012345", "Default")
    }
    _od
  }
}
```

You can substitute this version in the code above and you will no longer get **null** in the print out. This is because now the reference to **s2.od** is a call to the method that will create a default value if the member **_od** has not been assigned. Note that if it were reasonable for the field to not have a value, the method could return an **Option** type so that outside code could handle the lack of a value more safely. The **apply** method on **Option** was designed for this situation. If you call **Option(v)** and v is **null**, you will get back a **None**, otherwise you will get back a **Some(v)**.

The other annotation used for serialization is **@SerialVersionUID**. It is used to attach a version number to a type. This is done so that when you load an object back in, the version number of the saved object can be compared to the version of the current compiled code to make sure they are the same. Java will automatically generate these for you, but they change nearly every time the code for the type changes. This will cause saved files to break, even if they would still work fine. To prevent this, you might consider putting this annotation, followed by an argument of a numeric ID, before the type declaration.

Custom Serialization

Default serialization is wonderful in many situations. However, it can be very inefficient for some types. In addition, there might be situations where you really do need to store information for a member even though it is not serializable. For this reason, it is possible to override the default and use your own custom serialization.

To override the default implementation, you implement the private methods `writeObjects(oos:ObjectOutputStream)` and `readObjects(ois:ObjectInputStream)`. The first thing you will do in these methods is call `oos.defaultWriteObject()` and `ois.defaultReadObject()` respectively. This will write out or read in all of the information that should be part of the default serialization. Even if there are no fields to be serialized, you still need to call this. After that you write to or read from the stream. Here is an example of `Student` using this approach.

```scala
class Student(val name: String,
              val grades: Array[Int],
              @transient private var _od: OtherData) extends Serializable {
  def od = _od
  private def writeObject(oos: ObjectOutputStream): Unit = {
    oos.defaultWriteObject()
    oos.writeUTF(od.id)
    oos.writeUTF(od.course)
  }
  private def readObject(ois: ObjectInputStream): Unit = {
    ois.defaultReadObject()
    _od = new OtherData(ois.readUTF, ois.readUTF)
  }
}
```

The `_od` member is still a `private var`. It needs to be `private` because it is a `var`. It needs to be a `var` because of the line in `readObject` that makes an assignment to it. Outside code gets access to the value using the `od` method. Note that the method is simpler in this case because it does not have to check the value of `_od`.

This might seem like a silly example, but it is not hard to find a real use for custom serialization. It happens that the `scalafx.scene.paint.Color` class is not serializable, but we use it in both `DrawRectangle` and `DrawText`. If we want to serialize those, which we will want before the end of this chapter, we need to be able to serialize the value of the color. For example, we could change the declaration to `@transient private var _color: Color` and insert the following code.

```scala
private def writeObject(oos: ObjectOutputStream): Unit = {
  oos.defaultWriteObject()
  oos.writeDouble(_color.red)
  oos.writeDouble(_color.green)
  oos.writeDouble(_color.blue)
  oos.writeDouble(_color.opacity)
}
private def readObject(ois: ObjectInputStream): Unit = {
  ois.defaultReadObject()
  _color = Color(ois.readDouble(), ois.readDouble(), ois.readDouble(),
      ois.readDouble())
}
```

This code writes the red, green, blue, and opacity values, then makes a `Color` from reading them back in.

If you ever have a hard time deciding what values you need in order to save it as `writeObject`, you should write some using `readObject` first. Whatever values you need for constructing the instance in `readObject` will need to be saved off in `writeObject`.

10.5.2   XML Serialization

Java's default serialization comes with the standard advantages and limitations of a binary format. In addition, it only works for programs running under the Java Virtual Machine (JVM). That includes not only Java and Scala, but a number of other languages as well. It does not include all languages. C and C++ stand out as languages that do not have a JVM implementation. There are also some situations where you do not need the benefits of binary format and you would rather have the portability and readability of some other format.

Plain text is another option. You could simply write all the information out to text files and then have code that reads it back in. You can put whatever you want into text, and you could read it back in pretty much any language that you wanted. However, if you use plain text, you will have to write all the code to parse the plain text file in every language that you use, and formatting some types of data in plain text can be challenging. In addition, plain text is not very self-documenting. This can cause issues when you move the files around. In addition, while it is editable with basic text editors, plain text formatting for a lot of data winds up being brittle and hard to edit. It is nice for the text to follow some standard format that many tools know how to read, and that allows you to attach some meaning to what is in the files.

There are two standard text formats in broad usage today for storing and communicating information. In chapter 8 we saw that the web API for `MetaWeather.com` responded with data using the JSON format. The other format that is broadly in use is XML, short for eXtensible Mark-up Language. At the time that Scala was created, most text data was using the XML format, so the creator of Scala, Martin Odersky, decided to include support for it at the language level and in the libraries.[11] This makes it very easy for us to include XML in our programs. In particular, we want to see how we can serialize out objects to XML and deserialize them back into our programs.[12]

10.5.2.1   Nature of XML

Before we discuss serialization and deserialization, you have to know the basics of XML. If you have ever looked at the HTML behind a web page, then XML will look very familiar. XML is written in text, so you can use a plain text editor to edit it, but lots of programming editors know about XML and will help with the process. An XML file is a combination of markup and content. The markup is bracketed by either < and > characters or & and ; characters. More specifically, the < and > characters are used to create tags and the & and ;

---

[11]It should be noted that Martin now considers this a significant mistake and it is likely that language level support for XML will go away in a future version of Scala. Thanks to the macro capabilities of modern Scala, it is likely that there will be library support that rivals the current language-level support at that time.

[12]It is also possible to serialize to and deserialize from JSON. However, the lack of language support makes this harder without bringing in additional 3rd-party libraries.

characters are used to specify text that is special in XML, like <, which cannot be entered directly.

The tags have text inside of them. They come in three varieties. Note that the square brackets are used to denote something that is optional; they are not part of the XML syntax.

- *<tagName [attributes]>* denotes an open tag.

- *</tagName>* denotes a close tag.

- *<tagName [attributes]/>* denotes an empty element tag.

Each open tag must have a matching close tag, and the combination defines an element. The empty element format is just a shorthand for an open tag followed by a close tag with nothing between them. The *tagName* is a single word that generally specifies what type of information the element contains. The *attributes*, which are optional, are of the form *name="value"* and allow you to attach additional data to an element. Note that the double quotes around the values are required in XML, even if the data is numeric. You can have as many attributes as you want.

The contents of an element can be whatever combination of plain text and other markup that you wish. So you can nest elements inside of one another. Indeed, a big part of what makes XML useful is the ability to nest information. The last thing to note is that the entire XML document has to be contained in a single top-level element.

To help make this clearer, consider the following XML document that might be a reasonable version of saving off a `Drawing`.

```
<drawing>
  <drawable type="transform" transType="translate" value1="0" value2="0">
    <drawable type="rectangle" x="10" y="0" width="200" height="400">
      <color red="0.0" green="0.0" blue="1.0" opacity="1.0"/>
    </drawable>
    <drawable type="text" x="100" y="100">
      <color red="0.0" green="0.0" blue="0.0" opacity="1.0"/>
      <text>This is the text that is drawn.</text>
    </drawable>
  </drawable>
</drawing>
```

You can see that the whole thing is in an element with the name "drawing". The only thing inside of the "drawing" element is a single "drawable" with the type value of "transform". That element also has other attributes that would normally be associated with a `DrawTransform`. There are two other elements with the name "drawable" inside of the transform. These have types "rectangle" and "text". Most of the values for the rectangle and the text are specified as attributes. However, the colors for both and the display string for the text are nested in subelements.

This leads to a very interesting question. How do you decide if information should be in the form of an attribute, a subelement, or just content text? There is no hard and fast rule for this, and different projects choose to do things in different ways. The rule of thumb that we will follow is that if there is only one value for the information and it is fairly short, so it will never need more than half a line of text, then we use an attribute. That is part of why there are so many attributes in this sample. If there a multiples of a value, such as the multiple `Drawables` that are the children of a `DrawTransform`, or the value is either complex enough that it needs parsing or longer than about half a line, it will go inside of the element, most likely as a subelement. Text contents are only used when we get to an element like `<text>` in the above example, which has only a single value that does not need parsing, but which might be long. So the simple version is, short and single valued is an attribute while anything else is probably a subelement.

There are lots of other ways that you could have represented the data in the example. This is just one possibility, but it follows the rough rule that was just stated about how to decide between attributes and contents. Other formats could be chosen that follow slightly different approaches.

10.5.2.2   XML in Scala Code

As we said earlier, there is language-level support for XML in Scala as well as support in the libraries. The language support primarily comes in the form of XML literals, which allow you to put XML directly in your Scala programs, so you could enter the above XML directly into a Scala program just like any other literal expression. When you do this, the type is `scala.xml.Elem`.[13] You can do more as well. You can put Scala code in your XML to fill in attribute values or contents of an element. To do this, you put the expression with the value you want in curly braces. The following code gives a short example of this.

```
val name = "Ginny"
val age = 16
val description = "Red hair and second hand textbooks."
val xmlRecord = <person name={name} age={age.toString}>{description}</person>
```

Note the call to `toString` in the age attribute. This is required as the values put in attributes must be strings. This restriction does not apply to the content of elements. The code in the curly braces here is all fairly simple, but it can be arbitrary Scala expressions that might include embedded XML literals. To help show what that might look like, consider the following code. Note that the outer element is plural, "costs", and the inner elements are singular, "cost".

```
val costs = List(8,5,6,9)
val costXML = <costs>{costs.map(c => <cost>{c}</cost>)}</costs>
```

After executing this code, the value in `costXML` is the following.

```
<costs><cost>8</cost><cost>5</cost><cost>6</cost><cost>9</cost></costs>
```

The standard library includes types for dealing with XML in the `scala.xml` package. There are four `classes`/`objects` in this package that are of significance to us. They are `Elem`, `Node`, `NodeSeq`, and `XML`. The first three are related through inheritance. `Elem extends Node` and `Node extends NodeSeq`. The `Node` type represents any of the various types of things that can go into an XML document, including an element, so it is natural that `Elem` would inherit from `Node`. It might seem odd though that `Node` would inherit from `NodeSeq`, but having this be the case makes things work very nicely for searching for various data inside of an XML document.

The `XML` singleton `object` contains a variety of helpful methods, including `load`, `loadFile`, `loadString`, `save`, and `write`. If you put the previous sample XML for a drawing into a file called "drawing.xml", you can load that in with the line `XML.loadFile("drawing.xml")`. Similarly, you could save the `costXML` variable that we built out to file with `XML.save("cost.xml", costXML)`.

---

[13]Like the parsers used in chapter 8, the XML section of the Scala standard library was pulled into a separate module in Scala 2.11. So it is possible that your IDE will not use it by default. If you get errors when you try to put XML in your code, add scala_xml JAR file from your Scala install into the build path.

You now know enough to do XML serialization. In order to be able to deserialize the XML, we need one more piece of information, which is the ability to search for information inside of XML, and specifically, inside of a `NodeSeq` as that is the most general type used in the XML libraries. There are two operators defined on `NodeSeq` that we use to do this. They are \ and \\.[14] These take an argument that is a `String` that specifies what they are looking for. You can search for elements or attributes. If you are searching for an attribute, you need to put a "@" at the beginning of the string.

The following REPL session, loads in our drawing XML from earlier and does three different searches.

```
scala> val drawXML = XML.load("drawing.xml")
drawXML: scala.xml.Elem =
<drawing>
  <drawable value2="0" value1="0" transType="translate" type="transform">
    <drawable height="400" width="200" y="0" x="10" type="rectangle">
      <color opacity="1.0" blue="1.0" green="0.0" red="0.0"/>
    </drawable>
    <drawable y="100" x="100" type="text">
      <color opacity="1.0" blue="0.0" green="0.0" red="0.0"/>
      <text>This is the text that is drawn.</text>
    </drawable>
  </drawable>
</drawing>

scala> drawXML \ "drawable"
res1: scala.xml.NodeSeq =
NodeSeq(<drawable value2="0" value1="0" transType="translate" type="transform">
    <drawable height="400" width="200" y="0" x="10" type="rectangle">
      <color opacity="1.0" blue="1.0" green="0.0" red="0.0"/>
    </drawable>
    <drawable y="100" x="100" type="text">
      <color opacity="1.0" blue="0.0" green="0.0" red="0.0"/>
      <text>This is the text that is drawn.</text>
    </drawable>
  </drawable>)

scala> drawXML \ "@x"
res2: scala.xml.NodeSeq = NodeSeq()

scala> drawXML \\ "@x"
res3: scala.xml.NodeSeq = NodeSeq(10, 100)
```

You can see that the type of all the search results is listed as `NodeSeq`. That is safe, because it is a common supertype for anything that might be found, including multiple `Nodes`. The first search for `"drawable"` gives back a `NodeSeq` with a single, somewhat large element inside of it. The second and third searches look for the attribute x using the string `"@x"`. The first one finds nothing and the second one finds the values 10 and 100.

The difference between the second and third searches is that the second one uses \ and the third on uses \\ to illustrate the difference between the two. The \ only searches at the top level of the first operand. The \\ does a deep search that finds all occurrences, regardless of how many levels down they are nested.

---

[14]These are modeled off of operators that were used in the XML handling standard, XPath. XPath used / and // instead. Scala could not use // because that is how you made a single line comment, so they decided to use backslashes.

What if we wanted a search that finds the `Nodes` for both the rectangle and the text? Using `\\` on `"drawable"` gives back too much as it also gives the transform that contains the rectangle and the text. Instead, we can utilize the fact that the result of a search is a `NodeSeq` that we can search on and do the following.

```scala
scala> drawXML \ "drawable" \ "drawable"
res4: scala.xml.NodeSeq =
NodeSeq(<drawable type="rectangle" x="10" y="0" width="200" height="400">
        <color red="0.0" green="0.0" blue="1.0" opacity="1.0"/>
      </drawable>, <drawable type="text" x="100" y="100">
        <color red="0.0" green="0.0" blue="0.0" opacity="1.0"/>
        <text>This is the text that is drawn.</text>
      </drawable>)
```

What if we wanted to find the `x` attribute, but only of the rectangle? There are several ways that we could do that. If we know that the rectangle element is the first one in the transform, we could use indexing to pull out the first element as shown here.

```scala
scala> (drawXML \ "drawable" \ "drawable")(0) \ "@x"
res5: scala.xml.NodeSeq = 10
```

We could also use the fact that the `NodeSeq` is an immutable `Seq[Node]`. This means that we can call all the methods we would use on other types of sequences on the results of a search. So we could make more general code with a `filter` that finds all the rectangles and gets their `x` values with code like this.

```scala
scala> (drawXML \ "drawable" \ "drawable").filter(n => (n \
    "@type").text=="rectangle") \ "@x"
res6: scala.xml.NodeSeq = 10
```

Note that to get the `String` value of an attribute on a search, we call the `text` method. This method will work on elements and produces all the text contents, minus mark-up, inside of that element.

What if we want to get the `x`, `y`, and `type` attributes of both the rectangle and the text and store them as 3-tuples? We can do this by using `map` instead of `filter`.

```scala
scala> (drawXML \ "drawable" \ "drawable").map(n => ((n \ "@x").text, (n \
    "@y").text, (n \ "@type").text))
res7: scala.collection.immutable.Seq[(String, String, String)] =
    List((10,0,rectangle), (100,100,text))
```

With this search, we can clearly see that we have a rectangle at $(10, 0)$ and a text at $(100, 100)$.

Most of the time when we are using this, we pull the values out of the XML in a systematic way, and use it to construct instance objects of whatever type we want. This is what we will do for the deserialization process.

10.5.2.3 Serializing to XML

There is not a built-in method of converting objects to XML or getting them back, but you can add this type of functionality to your own code fairly easily. To do this, we will follow three fairly simple rules with each class we want to be able to serialize.

- Put a `toXML: scala.xml.Node` method in the class that returns an XML element with all the information you want to save for those objects.

- Include a companion object with an `apply(node: scala.xml.Node): Type` method that deserializes the XML node and returns an object of the type you have created.

- Make all the fields that are serialized be arguments to the class so that we can pass in their values on construction.

If you do this consistently across your types, when one type includes a reference to another, you can simply call `toXML` on that type.[15] You can use the normal methods in `scala.xml.XML` to read the XML from a file or write it back out to a file.

Compressed Streams

If you have spent much time doing things on a computer or downloading files from the Internet, odds are good that at some point you have come across a ZIP file. Compressed files in Windows® use this format. It allows you to combine many files into one, and for some files, particularly those that store text data, it makes them much smaller. If you have large text files, note that large XML files would qualify, zipping them can save a significant amount of space.

When your program creates those large files it would be more efficient to have them written out directly to the ZIP format and then read back in from that as well. Fortunately, there are some wrapper streams in `java.util.zip` which can do exactly that for you. The `ZippedInputStream` and `ZippedOutputStream` can be used to decorate other streams in exactly the same fashion as a `BufferedInputStream` or `BufferedOutputStream`. These streams significantly alter the contents of what is written or read back in so that everything uses the ZIP format.

The Zip format has the ability to store multiple files inside of a single zip file. There is a type called `ZipEntry` that represents a single file. If you are working with Zip files, you have to make some extra calls to position the stream in a particular entry.

10.6 Saving Drawings (Project Integration)

Now it is time to integrate these different concepts into our drawing project. We will do this by giving our program the ability to save drawings. For completeness, there will be three different save options: default serialized binary, XML, and zipped XML. These three options will allow us to cover nearly everything that was introduced in the chapter in a manner that is more significant than the small samples shown with each topic.

There are three separate pieces to adding this code. We need to have GUI elements for the user to interact with. Those GUI elements need to call code that deals with the files. Then changes have to be made to the `Drawing` and `Drawables` so that they can be serialized and deserialized using both the default method and XML. We will start with the GUI code for two reasons. First, it is fairly easy to add menu items. Second, in order to test the other parts we need to have the ability to make the program call that code, and that is what the GUI elements do. In `DrawingMain` we change our single `saveItem` to the following three menu items and add them all to the file menu.

---

[15]The only shortcoming of this approach is that if there are two or more references to any given object, it will be duplicated. The Java serialization code caches objects so this does not happen.

```scala
val saveBinaryItem = new MenuItem("Save Binary")
val saveXMLItem = new MenuItem("Save XML")
val saveZipItem = new MenuItem("Save Zip")
...
fileMenu.items = List(newItem, openItem, saveBinaryItem, saveXMLItem,
    saveZipItem, closeItem, new SeparatorMenuItem, exitItem)
```

We also still have the openItem that was in the menus before. We need those four items to have action handlers that do the correct things.

```scala
openItem.onAction = (ae:ActionEvent) => open()
saveBinaryItem.onAction = (ae:ActionEvent) => saveBinary()
saveXMLItem.onAction = (ae:ActionEvent) => saveXML()
saveZipItem.onAction = (ae:ActionEvent) => saveZip()
```

Now to implement those four methods. You can start by making empty stubs so that your code compiles. The saveBinary is the simplest. It should let the user select a file to write to, open a FileInputStream, wrap it in some way, and then write the currently selected drawing out to the stream. The saveXML method is going to have the user select a file, then use scala.xml.XML.save to write a Node to disk. The node needs to come from the currently selected drawing and given what was said above, this will probably be done by adding toXML methods to Drawing and other places. The saveZip method will start similar to the saveXML, but instead of using XML.save, we will use XML.write and the power of the java.io library to get it to write out to a ZIP file. All the save options are also going to name the tab the name of the file it is saved as. The load method needs to allow the user to select a file, then identify the file type by the extension and execute the appropriate deserialization code.

There is certain functionality that is the same for all three of the save options. They all have to open a file chooser dialog and let the user select a file, then they need to use that file to save things. They should also all do proper error checking. They also all need to set the name on the tab and they have to know the drawing for the currently open tab. To prevent code duplication, we will start off by writing some helper methods to consolidate this shared code.

```scala
private def withOutputStream[A, B <: OutputStream](os: B)(body: B => A): A = {
  try {
    body(os)
  } finally {
    os.close()
  }
}

private def withSaveFile(body: File => Unit): Unit = {
  val chooser = new FileChooser
  val file = chooser.showSaveDialog(stage)
  if (file != null) {
    try {
      body(file)
    } catch {
      case ex: FileNotFoundException => ex.printStackTrace()
      case ex: IOException => ex.printStackTrace()
    }
  }
}
```

```
private def setTabAndGetDrawing(name: String): Option[Drawing] = {
  val current = tabPane.selectionModel.value.selectedIndex.value
  if (current >= 0) {
    val (drawing, treeView) = drawings(current)
    tabPane.tabs(current).text = name
    Some(drawing)
  } else None
}
```

The `withOutputStream` method is just a version of the Loan pattern we saw previously. One added feature is that there is a second type parameter B that has a type bound. It has to be a subtype of `OutputStream`. In the section on the Loan pattern, we just used a `FileOutputStream`. That will not work here. Each of the output types is going to use a different type of `OutputStream`. The second type parameter allows that to happen.

The `withSaveFile` method handles the file chooser. It takes in a function that will accept the `java.io.File` that the user selects as an input. This function is called once the user has told us what file they want to save to. The error handling currently just prints stack traces, but you could extend this to pop up dialogs giving the user information that something bad happened. The `setTabAndGetDrawing` method does pretty much what the name says. It finds what tab is currently open and sets the text at the top of the tab to the name of the selected file. It returns an `Option[Drawing]` that will have the proper drawing if a tab is selected or `None` if for some reason there is not a currently active tab.

We can now implement the `saveBinary` method using those three helper methods in the following way.

```
private def saveBinary(): Unit = {
  withSaveFile(file => {
    val oos = new ObjectOutputStream(new BufferedOutputStream(new
        FileOutputStream(file)))
    withOutputStream(oos)(strm => {
      setTabAndGetDrawing(file.getName()).foreach(drawing =>
          strm.writeObject(drawing))
    })
  })
}
```

As you can see, most of the logic is in the helper methods. Other than calling those methods in the proper sequence, all that this method has to do is set up the `ObjectOutputStream` and call `writeObject`.

This code compiles, but it does not yet run. In order to get it to run, we have to make it so that the `Drawing` and `Drawable extend Serializable`. Then we have to do something to deal with any members that are not `Serializable`. We will show the details of that later so we can discuss those additions along with the ones needed for the XML serialization.

We can reuse the `withSaveFile` and `setTabAndGetDrawing` methods in the `saveXML` method using the following code.

```
private def saveXML(): Unit = {
  withSaveFile(file => {
    setTabAndGetDrawing(file.getName()).foreach(drawing =>
        xml.XML.save(file.getAbsolutePath(), drawing.toXML))
  })
}
```

In order to get this code to compile, you have to add a `toXML : xml.Node` method to the `Drawing` class. At this point, that function can just return `<drawing></drawing>`. Not only is that sufficient to get it to compile, you can now test this option and make sure that you get an output XML file with that one element in it.

The `saveZip` method winds up being a little longer. It can reuse `withSaveFile` and `setTabAndGetDrawing`, but we have to play some tricks to get the XML into a Zip file.

```scala
private def saveZip(): Unit = {
  withSaveFile(file => {
    setTabAndGetDrawing(file.getName()).foreach(drawing => {
      val zos = new ZipOutputStream(new BufferedOutputStream(new
          FileOutputStream(file)))
      zos.putNextEntry(new ZipEntry(file.getName().dropRight(3)+"xml"))
      val sw = new OutputStreamWriter(zos)
      try {
        setTabAndGetDrawing(file.getName()).foreach(drawing => xml.XML.write(sw,
            drawing.toXML, "", false, null))
      } finally {
        sw.close
      }
    })
  })
}
```

On line 4 we create a ZipOutputStream. Line 5 creates an entry in the Zip file. Remember that a Zip file can hold multiple files, so this is how you tell it that you want to start including a new file in the Zip file. In our case, we only have one file that we are compressing. We assume that the filename that user gave ended in ".zip" and we replace the "zip" with "xml". On line 8 we call the `XML.write` method. This method take a `java.io.Writer` instead of a `java.io.OutputStream`. This makes sense given that the `Writer` is designed to work with character data, and XML is a plain text format. However, that does mean an extra step for us. Line 6 needs to decorate our `ZipOutputStream` in an `OutputStreamWriter`. This type simply wraps an `OutputStream` inside of a `Writer` so that we can use it in places that require a `Writer`, such as with `XML.write`. The fact that the `Writer` type is not a subtype of `OutputStream` means that we cannot take advantage of `withOutputStream` for this method.

The **open** method is a bit longer because it can handle any of the three types of files based on the extension.

```scala
private def open(): Unit = {
  val chooser = new FileChooser()
  val file = chooser.showOpenDialog(stage)
  if (file != null) {
    if (file.getName().endsWith(".bin")) {
      val ois = new ObjectInputStream(new BufferedInputStream(new
          FileInputStream(file)))
      withInputStream(ois)(strm => {
        deserializeDrawingFromStream(strm, file.getName())
      })
    } else if (file.getName().endsWith(".zip")) {
      val zis = new ZipInputStream(new BufferedInputStream(new
          FileInputStream(file)))
      zis.getNextEntry
      val nd = Drawing(xml.XML.load(zis))
```

```
      val (tree, tab) = makeDrawingTab(nd, file.getName)
      drawings = drawings :+ (nd, tree)
      tabPane += tab
      nd.draw()
    } else if (file.getName().endsWith(".xml")) {
      val nd = Drawing(xml.XML.loadFile(file))
      val (tree, tab) = makeDrawingTab(nd, file.getName)
      drawings = drawings :+ (nd, tree)
      tabPane += tab
      nd.draw()
    }
  }
}
```

Similar helper files are used here as those that were used for saving. For the XML and Zip options to work, a companion object must be added for `Drawing` that has an `apply` method that takes an `xml.Node` and returns a `Drawing`. For now this can simply return a new `Drawing`.

A complete listing of the revised `DrawingMain` is shown here.

```scala
 1  package iostreams.drawing
 2
 3  import scalafx.Includes._
 4  import scalafx.application.JFXApp
 5  import scalafx.application.Platform
 6  import scalafx.geometry.Orientation
 7  import scalafx.scene.Scene
 8  import scalafx.scene.canvas.Canvas
 9  import scalafx.scene.control.Menu
10  import scalafx.scene.control.MenuBar
11  import scalafx.scene.control.MenuItem
12  import scalafx.scene.control.ScrollPane
13  import scalafx.scene.control.SeparatorMenuItem
14  import scalafx.scene.control.Slider
15  import scalafx.scene.control.SplitPane
16  import scalafx.scene.control.Tab
17  import scalafx.scene.control.TabPane
18  import scalafx.scene.control.TextArea
19  import scalafx.scene.control.TextField
20  import scalafx.scene.control.TreeView
21  import scalafx.scene.layout.BorderPane
22  import scalafx.scene.paint.Color
23  import scalafx.event.ActionEvent
24  import scalafx.scene.control.TreeItem
25  import scalafx.scene.control.SelectionMode
26  import scalafx.scene.control.Label
27  import scalafx.scene.layout.Priority
28  import scalafx.scene.control.ComboBox
29  import scalafx.scene.control.Dialog
30  import scalafx.scene.control.ChoiceDialog
31  import scala.concurrent.Future
32  import scala.concurrent.ExecutionContext.Implicits.global
33  import scalafx.stage.WindowEvent
34  import akka.actor.ActorSystem
35  import java.io.IOException
```

```scala
36  import java.io.FileNotFoundException
37  import java.io.File
38  import scalafx.stage.FileChooser
39  import java.io.FileOutputStream
40  import java.io.ObjectOutputStream
41  import java.io.BufferedOutputStream
42  import java.io.OutputStream
43  import java.util.zip.ZipOutputStream
44  import java.util.zip.ZipEntry
45  import java.io.OutputStreamWriter
46  import java.io.FileInputStream
47  import java.io.ObjectInputStream
48  import java.io.BufferedInputStream
49  import java.util.zip.ZipInputStream
50  import java.io.InputStream
51
52  object DrawingMain extends JFXApp {
53    private var drawings = List[(Drawing, TreeView[Drawable])]()
54    val system = ActorSystem("DrawingSystem")
55    val tabPane = new TabPane
56
57    private val creators = Map[String, Drawing => Drawable](
58      "Rectangle" -> (drawing => DrawRectangle(drawing, 0, 0, 300, 300, Color.Blue)),
59      "Transform" -> (drawing => DrawTransform(drawing)),
60      "Text" -> (drawing => DrawText(drawing, 100, 100, "Text", Color.Black)),
61      "Maze" -> (drawing => DrawMaze(drawing)),
62      "Mandelbrot" -> (drawing => DrawMandelbrot(drawing)),
63      "Julia Set" -> (drawing => DrawJulia(drawing)))
64
65    stage = new JFXApp.PrimaryStage {
66      title = "Drawing Program"
67      scene = new Scene(1000, 700) {
68        // Menus
69        val menuBar = new MenuBar
70        val fileMenu = new Menu("File")
71        val newItem = new MenuItem("New")
72        val openItem = new MenuItem("Open")
73        val saveBinaryItem = new MenuItem("Save Binary")
74        val saveXMLItem = new MenuItem("Save XML")
75        val saveZipItem = new MenuItem("Save Zip")
76        val closeItem = new MenuItem("Close")
77        val exitItem = new MenuItem("Exit")
78        fileMenu.items = List(newItem, openItem, saveBinaryItem, saveXMLItem,
79            saveZipItem, closeItem, new SeparatorMenuItem, exitItem)
79        val editMenu = new Menu("Edit")
80        val addItem = new MenuItem("Add Drawable")
81        val copyItem = new MenuItem("Copy")
82        val cutItem = new MenuItem("Cut")
83        val pasteItem = new MenuItem("Paste")
84        editMenu.items = List(addItem, copyItem, cutItem, pasteItem)
85        menuBar.menus = List(fileMenu, editMenu)
86
87        // Tabs
88        val drawing = Drawing()
89        val (tree, tab) = makeDrawingTab(drawing, "Untitled")
```

```
90      drawings = drawings :+ (drawing, tree)
91      tabPane += tab
92
93      // Menu Actions
94      newItem.onAction = (ae: ActionEvent) => {
95        val drawing = Drawing()
96        val (tree, tab) = makeDrawingTab(drawing, "Untitled")
97        drawings = drawings :+ (drawing, tree)
98        tabPane += tab
99        drawing.draw()
100     }
101     openItem.onAction = (ae: ActionEvent) => open()
102     saveBinaryItem.onAction = (ae: ActionEvent) => saveBinary()
103     saveXMLItem.onAction = (ae: ActionEvent) => saveXML()
104     saveZipItem.onAction = (ae: ActionEvent) => saveZip()
105     closeItem.onAction = (ae: ActionEvent) => {
106       val current = tabPane.selectionModel.value.selectedIndex.value
107       if (current >= 0) {
108         drawings = drawings.patch(current, Nil, 1)
109         tabPane.tabs.remove(current)
110       }
111     }
112     addItem.onAction = (ae: ActionEvent) => {
113       val current = tabPane.selectionModel.value.selectedIndex.value
114       if (current >= 0) {
115         val (drawing, treeView) = drawings(current)
116         val dtypes = creators.keys.toSeq
117         val dialog = new ChoiceDialog(dtypes(0), dtypes)
118         val d = dialog.showAndWait() match {
119           case Some(s) =>
120             val d = creators(s)(drawing)
121             val treeSelect = treeView.selectionModel.value.getSelectedItem
122             def treeAdd(item: TreeItem[Drawable]): Unit = item.getValue match {
123               case dt: DrawTransform =>
124                 dt.addChild(d)
125                 item.children += new TreeItem(d)
126                 drawing.draw()
127               case d =>
128                 treeAdd(item.getParent)
129             }
130             if (treeSelect != null) treeAdd(treeSelect)
131           case None =>
132         }
133       }
134     }
135     exitItem.onAction = (ae: ActionEvent) => {
136       // TODO - Add saving before closing down.
137       sys.exit()
138     }
139
140     // Top Level Setup
141     val rootPane = new BorderPane
142     rootPane.top = menuBar
143     rootPane.center = tabPane
144     root = rootPane
```

```
145
146            drawing.draw()
147            onCloseRequest = (we: WindowEvent) => system.terminate()
148          }
149        }
150
151        private def makeDrawingTab(drawing: Drawing, tabName: String):
               (TreeView[Drawable], Tab) = {
152          val canvas = new Canvas(2000, 2000)
153          val gc = canvas.graphicsContext2D
154          drawing.graphicsContext = gc
155
156          // left side
157          val drawingTree = new TreeView[Drawable]
158          drawingTree.selectionModel.value.selectionMode = SelectionMode.Single
159          drawingTree.root = drawing.makeTree
160          val treeScroll = new ScrollPane
161          drawingTree.prefWidth <== treeScroll.width
162          drawingTree.prefHeight <== treeScroll.height
163          treeScroll.content = drawingTree
164          val propertyPane = new ScrollPane
165          val leftSplit = new SplitPane
166          leftSplit.orientation = Orientation.Vertical
167          leftSplit.items ++= List(treeScroll, propertyPane)
168
169          // right side
170          val slider = new Slider(0, 1000, 0)
171          val rightTopBorder = new BorderPane
172          val canvasScroll = new ScrollPane
173          canvasScroll.content = canvas
174          rightTopBorder.top = slider
175          rightTopBorder.center = canvasScroll
176          val commandField = new TextField
177          val commandArea = new TextArea
178          commandArea.editable = false
179          commandField.onAction = (ae: ActionEvent) => {
180            val text = commandField.text.value
181            if (text.nonEmpty) {
182              commandField.text = ""
183              val future = Future { Commands(text, drawing) }
184              future.foreach(result => Platform.runLater {
185                commandArea.text = (commandArea.text+"> "+text+"\n"+result+"\n").value
186              })
187            }
188          }
189          val commandScroll = new ScrollPane
190          commandArea.prefWidth <== commandScroll.width
191          commandArea.prefHeight <== commandScroll.height
192          commandScroll.content = commandArea
193          val rightBottomBorder = new BorderPane
194          rightBottomBorder.top = commandField
195          rightBottomBorder.center = commandScroll
196          val rightSplit = new SplitPane
197          rightSplit.orientation = Orientation.Vertical
198          rightSplit.items ++= List(rightTopBorder, rightBottomBorder)
```

```scala
199      rightSplit.dividerPositions = 0.7
200
201      // Top Level
202      val topSplit = new SplitPane
203      topSplit.items ++= List(leftSplit, rightSplit)
204      topSplit.dividerPositions = 0.3
205
206      // Tree Selection
207      drawingTree.selectionModel.value.selectedItem.onChange {
208        val selected = drawingTree.selectionModel.value.selectedItem.value
209        if (selected != null) {
210          propertyPane.content = selected.getValue.propertiesPanel()
211        } else {
212          propertyPane.content = new Label("Nothing selected.")
213        }
214      }
215
216      val tab = new Tab
217      tab.text = tabName
218      tab.content = topSplit
219      tab.closable = false
220      (drawingTree, tab)
221    }
222
223    def labeledTextField(labelText: String, initialText: String, action: String =>
             Unit): BorderPane = {
224      val bp = new BorderPane
225      val label = Label(labelText)
226      val field = new TextField
227      field.text = initialText
228      bp.left = label
229      bp.center = field
230      field.onAction = (ae: ActionEvent) => action(field.text.value)
231      field.focused.onChange(if (!field.focused.value) action(field.text.value))
232      bp
233    }
234
235    private def open(): Unit = {
236      val chooser = new FileChooser()
237      val file = chooser.showOpenDialog(stage)
238      if (file != null) {
239        if (file.getName().endsWith(".bin")) {
240          val ois = new ObjectInputStream(new BufferedInputStream(new
                 FileInputStream(file)))
241          withInputStream(ois)(strm => {
242            deserializeDrawingFromStream(strm, file.getName())
243          })
244        } else if (file.getName().endsWith(".zip")) {
245          val zis = new ZipInputStream(new BufferedInputStream(new
                 FileInputStream(file)))
246          zis.getNextEntry
247          val nd = Drawing(xml.XML.load(zis))
248          val (tree, tab) = makeDrawingTab(nd, file.getName)
249          drawings = drawings :+ (nd, tree)
250          tabPane += tab
```

```scala
251          nd.draw()
252        } else if (file.getName().endsWith(".xml")) {
253          val nd = Drawing(xml.XML.loadFile(file))
254          val (tree, tab) = makeDrawingTab(nd, file.getName)
255          drawings = drawings :+ (nd, tree)
256          tabPane += tab
257          nd.draw()
258        }
259      }
260    }
261
262    private def withInputStream[A, B <: InputStream](is: B)(body: B => A): A = {
263      try {
264        body(is)
265      } finally {
266        is.close()
267      }
268    }
269
270    private def deserializeDrawingFromStream(ois: ObjectInputStream, name: String) {
271      val obj = ois.readObject()
272      obj match {
273        case nd: Drawing =>
274          val (tree, tab) = makeDrawingTab(nd, name)
275          drawings = drawings :+ (nd, tree)
276          tabPane += tab
277          nd.draw()
278        case _ =>
279      }
280    }
281
282    private def saveBinary(): Unit = {
283      withSaveFile(file => {
284        val oos = new ObjectOutputStream(new BufferedOutputStream(new
               FileOutputStream(file)))
285        withOutputStream(oos)(strm => {
286          setTabAndGetDrawing(file.getName()).foreach(drawing =>
                 strm.writeObject(drawing))
287        })
288      })
289    }
290
291    private def saveXML(): Unit = {
292      withSaveFile(file => {
293        setTabAndGetDrawing(file.getName()).foreach(drawing =>
               xml.XML.save(file.getAbsolutePath(), drawing.toXML))
294      })
295    }
296
297    private def saveZip(): Unit = {
298      withSaveFile(file => {
299        setTabAndGetDrawing(file.getName()).foreach(drawing => {
300          val zos = new ZipOutputStream(new BufferedOutputStream(new
               FileOutputStream(file)))
301          zos.putNextEntry(new ZipEntry(file.getName().dropRight(3)+"xml"))
```

```
302        val sw = new OutputStreamWriter(zos)
303        try {
304          setTabAndGetDrawing(file.getName()).foreach(drawing => xml.XML.write(sw,
                 drawing.toXML, "", false, null))
305        } finally {
306          sw.close
307        }
308      })
309    })
310  }
311
312  private def withOutputStream[A, B <: OutputStream](os: B)(body: B => A): A = {
313    try {
314      body(os)
315    } finally {
316      os.close()
317    }
318  }
319
320  private def withSaveFile(body: File => Unit): Unit = {
321    val chooser = new FileChooser
322    val file = chooser.showSaveDialog(stage)
323    if (file != null) {
324      try {
325        body(file)
326      } catch {
327        case ex: FileNotFoundException => ex.printStackTrace()
328        case ex: IOException => ex.printStackTrace()
329      }
330    }
331  }
332
333  private def setTabAndGetDrawing(name: String): Option[Drawing] = {
334    val current = tabPane.selectionModel.value.selectedIndex.value
335    if (current >= 0) {
336      val (drawing, treeView) = drawings(current)
337      tabPane.tabs(current).text = name
338      Some(drawing)
339    } else None
340  }
341
342 }
```

The withInputStream (lines 262–268) and withOutputStream (lines 312–316) methods are curried to facilitate the type inference. Without the currying, it would be necessary to put a type on the strm argument for the functions that are passed in.

The error handling in this code is minimal, at best. Having a print of a stack trace in the code can be very helpful for programmers. However, it tells the user nothing. For a GUI-based application like this one, it is likely that the user will not even see a console where the stack trace would print out. This code should pop up a window letting the user know what has happened. It has been left out here to not bloat the code and is left as an exercise for the reader.

If you try to run this code at this point, none of the options will really work. To make them work, we have to alter Drawing and the Drawable hierarchy to support default and

XML serialization. We will start with `Drawing` because that is what we are actually serializing. For the default serialization, this means that the `Drawing` class needs to extend
`Serializable` and that members that are not saved should be made `transient`. In addition, code needs to be altered so that `transient` fields do not cause problems when they
are deserialized with a value of `null`.

For the XML-based serialization, that means implementing the `toXML` method to put
any values we want saved into an XML element and implementing a deserializing `apply`
method in a companion object. To make this work, values that are saved should also be
moved up to be arguments of the class. After doing this to `Drawing`, we get code that looks
like the following.

```scala
package iostreams.drawing

import collection.mutable
import scalafx.scene.canvas.GraphicsContext
import scalafx.scene.control.TreeItem
import scalafx.scene.paint.Color

class Drawing(private[drawing] val vars:mutable.Map[String, Double]) extends
    Serializable {
  private var root = DrawTransform(this)
  @transient private var gc: GraphicsContext = null

  def graphicsContext_=(g: GraphicsContext): Unit = {
    gc = g
  }

  def graphicsContext = gc

  def draw(): Unit = {
    gc.fill = Color.White
    gc.fillRect(0, 0, 2000, 2000)
    root.draw(gc)
  }

  def makeTree: TreeItem[Drawable] = {
    def helper(d: Drawable): TreeItem[Drawable] = d match {
      case dt: DrawTransform =>
        val item = new TreeItem(d)
        item.children = dt.children.map(c => helper(c))
        item
      case _ => new TreeItem(d)
    }
    helper(root)
  }

  def toXML: xml.Node = {
    <drawing>
      {root.toXML}
      {for((k,v) <- vars) yield <var key={k} value={v.toString()}/>}
    </drawing>
  }
}

object Drawing {
```

```scala
44    def apply(): Drawing = {
45      new Drawing(mutable.Map())
46    }
47
48    def apply(n: xml.Node): Drawing = {
49      val vars = mutable.Map((n \ "var").map(vnode => (vnode \ "@key").text -> (vnode
            \ "@value").text.toDouble):_*)
50      val drawing = new Drawing(vars)
51      drawing.root = DrawTransform(n \ "drawable", drawing)
52      drawing
53    }
54
55    def colorToXML(color: Color): xml.Node = {
56      <color red={color.red.toString()} green={color.green.toString()}
            blue={color.blue.toString()} opacity={color.opacity.toString()}/>
57    }
58
59    def xmlToColor(n: xml.NodeSeq): Color = {
60      Color((n \ "@red").text.toDouble,(n \ "@green").text.toDouble,(n \
            "@blue").text.toDouble,(n \ "@opacity").text.toDouble)
61    }
62  }
```

The primary alterations are at the end. In order for this code to compile, you must add a **toXML** method and a companion object with an **apply** method to **DrawTransform**. We do not pass in the **GraphicsContext**. Instead, we added a method that allows us to set it. If you look back that the listing of **DrawingMain**, you can see that this is called on line 154 when we are building the tab for that drawing. This is needed by both approaches to serialization. The **GraphicsContext** is not **Serializable**, so we have to be able to set it after loading a drawing in from binary file. We also cannot write one out to XML, so we have to build the **Drawing** without it, then set it when we are putting together the GUI.

In addition to the **apply** methods that are used to create instances of **Drawing**, there are two other methods in the companion object. These are intended to reduce code duplication. There are currently two subtypes of **Drawable** that store **Colors** and it is easy to see how we could add several more. To prevent us from rewriting the code to convert these to and from XML, methods for doing that are placed here. If there were other elements that were shared across many **Drawable** subtypes, methods to convert them to and from XML could go here as well.

You might wonder why this code refers to **xml.Node** instead of using an **import** and just calling it **Node**. The reason lies in the fact that the **propertiesPanel** returns a **scalafx.scene.Node**. We do not want the name **Node** to be ambiguous. We could use a remaining **import** to give the **xml.Node** another name. Perhaps something like **import xml.Node => XNode**. However, we feel that using **xml.Node** is not long enough to be onerous, and it is very clear to any reader what it represents.

The fact that **DrawTransform** is part of an inheritance hierarchy, combined with the fact that default serialization is done through inheritance means that we will want to start at the top. So before looking at **DrawTransform**, we will look at **Drawable**. Here is the code for that **class**.

```scala
1   package iostreams.drawing
2
3   import scalafx.scene.canvas.GraphicsContext
4   import scalafx.scene.Node
```

```
 5
 6  abstract class Drawable(val drawing: Drawing) extends Serializable {
 7    def draw(gc: GraphicsContext): Unit
 8    def propertiesPanel(): Node
 9    def toXML: xml.Node
10  }
11
12  object Drawable {
13    def makeDrawable(n: xml.Node, d: Drawing): Drawable = {
14      (n \ "@type").text match {
15        case "rectangle" => DrawRectangle(n, d)
16        case "text" => DrawText(n, d)
17        case "transform" => DrawTransform(n, d)
18        case "maze" => DrawMaze(n, d)
19        case "mandelbrot" => DrawMandelbrot(n, d)
20        case "julia" => DrawJulia(n, d)
21      }
22    }
23  }
```

We have made it inherit from `Serializable` and put in an abstract `toXML` method. The former now makes all subtypes serializable. The latter will force us to put in concrete `toXML` methods in the subtypes. For the subtypes then, we need to make the proper values transient, write `toXML`, and add a companion object with the appropriate `apply` method.

The interesting part of `Drawable` is in the companion object. You might recall from the sample XML earlier that we didn't make separate tags for each of the different `Drawable` types. Instead, there is a tag called `drawable` and the `type` attribute specified what type of `Drawable` it was. This was very intentional, and you will see how that benefits us in `DrawTransform`. The first hint comes in the `makeDrawable` method on lines 13–21 here though. The first argument is a `xml.Node` that we know has the tag `drawable`. It does a match on the value of the `type` attribute so that it can make the correct subtype of `Drawable`.

For the subtypes, we will start with `DrawRectangle` because it is fairly simple. Here is code for the revision to that class.

```
 1  package iostreams.drawing
 2
 3  import scalafx.Includes._
 4  import scalafx.scene.canvas.GraphicsContext
 5  import scalafx.scene.paint.Color
 6  import scalafx.scene.Node
 7  import scalafx.scene.layout.VBox
 8  import scalafx.scene.control.TextField
 9  import scalafx.scene.control.ColorPicker
10  import scalafx.event.ActionEvent
11  import scalafx.scene.layout.Priority
12  import java.io.ObjectInputStream
13  import java.io.ObjectOutputStream
14
15  class DrawRectangle(
16      d: Drawing,
17      private var _x: Double,
18      private var _y: Double,
19      private var _width: Double,
```

```scala
20      private var _height: Double,
21      @transient private var _color: Color) extends Drawable(d) {
22
23    @transient private var propPanel: Node = null
24
25    override def toString = "Rectangle"
26
27    def draw(gc: GraphicsContext): Unit = {
28      gc.fill = _color
29      gc.fillRect(_x, _y, _width, _height)
30    }
31
32    def propertiesPanel(): Node = {
33      if (propPanel == null) {
34        val panel = new VBox
35        val xField = DrawingMain.labeledTextField("x", _x.toString, s => { _x =
                s.toDouble; drawing.draw() })
36        val yField = DrawingMain.labeledTextField("y", _y.toString, s => { _y =
                s.toDouble; drawing.draw() })
37        val widthField = DrawingMain.labeledTextField("width", _width.toString, s =>
                { _width = s.toDouble; drawing.draw() })
38        val heightField = DrawingMain.labeledTextField("height", _height.toString, s
                => { _height = s.toDouble; drawing.draw() })
39        val colorPicker = new ColorPicker(_color)
40        colorPicker.onAction = (ae: ActionEvent) => {
41          _color = colorPicker.value.value
42          drawing.draw()
43        }
44        panel.children = List(xField, yField, widthField, heightField, colorPicker)
45        propPanel = panel
46      }
47      propPanel
48    }
49
50    def toXML: xml.Node = {
51      <drawable type="rectangle" x={ _x.toString() } y={ _y.toString() } width={
                _width.toString() } height={ _height.toString() }>
52        { Drawing.colorToXML(_color) }
53      </drawable>
54    }
55
56    private def writeObject(oos: ObjectOutputStream): Unit = {
57      oos.defaultWriteObject()
58      oos.writeDouble(_color.red)
59      oos.writeDouble(_color.green)
60      oos.writeDouble(_color.blue)
61      oos.writeDouble(_color.opacity)
62    }
63    private def readObject(ois: ObjectInputStream): Unit = {
64      ois.defaultReadObject()
65      _color = Color(ois.readDouble(), ois.readDouble(), ois.readDouble(),
                ois.readDouble())
66    }
67  }
68
```

```scala
69  object DrawRectangle {
70    def apply(d: Drawing, x: Double, y: Double, width: Double, height: Double, color:
          Color) = {
71      new DrawRectangle(d, x, y, width, height, color)
72    }
73
74    def apply(n: xml.Node, d: Drawing) = {
75      val x = (n \ "@x").text.toDouble
76      val y = (n \ "@y").text.toDouble
77      val width = (n \ "@width").text.toDouble
78      val height = (n \ "@height").text.toDouble
79      val color = Drawing.xmlToColor(n \ "color")
80      new DrawRectangle(d, x, y, width, height, color)
81    }
82  }
```

We have to make both _color and propPanel **transient** as neither can be serialized. Unfortunately, this means that we change the **propPanel** from the nice type safe **Option[Node]** that we had been using back to a plain **Node** that starts as **null**. We do this because when we deserialize one of these objects, the value will be **null**, even if we do not want it to be, so the **Option** winds up being overhead that does not help us at that point. The change alters the code in **propertiesPanel** a bit.

Next we look at the **DrawTransform** type.

```scala
1   package iostreams.drawing
2
3   import collection.mutable
4   import scalafx.Includes._
5   import scalafx.scene.Node
6   import scalafx.scene.canvas.GraphicsContext
7   import scalafx.scene.layout.VBox
8   import scalafx.scene.control.ComboBox
9   import scalafx.event.ActionEvent
10  import scalafx.scene.transform.Rotate
11  import scalafx.scene.transform.Translate
12  import scalafx.scene.transform.Scale
13  import scalafx.scene.transform.Shear
14
15  class DrawTransform(
16      d: Drawing,
17      private val _children: mutable.Buffer[Drawable],
18      private var transformType: DrawTransform.Value = DrawTransform.Translate,
19      private var value1: Double = 0.0,
20      private var value2: Double = 0.0) extends Drawable(d) {
21
22    @transient private var propPanel: Node = null
23
24    def children = _children
25
26    override def toString = "Transform"
27
28    def addChild(d: Drawable): Unit = {
29      _children += d
30    }
31
```

```scala
32    def removeChild(d: Drawable): Unit = {
33      _children -= d
34    }
35
36    def draw(gc: GraphicsContext): Unit = {
37      gc.save()
38      transformType match {
39        case DrawTransform.Rotate => gc.rotate(value1)
40        case DrawTransform.Translate => gc.translate(value1, value2)
41        case DrawTransform.Scale => gc.scale(value1, value2)
42        case DrawTransform.Shear => gc.transform(1.0, value1, value2, 1.0, 0.0, 0.0)
43      }
44      _children.foreach(_.draw(gc))
45      gc.restore()
46    }
47
48    def propertiesPanel(): Node = {
49      if (propPanel == null) {
50        val panel = new VBox
51        val combo = new ComboBox(DrawTransform.values.toSeq)
52        combo.onAction = (ae: ActionEvent) => {
53          transformType = combo.selectionModel.value.selectedItem.value
54          drawing.draw()
55        }
56        combo.selectionModel.value.select(transformType)
57        val v1Field = DrawingMain.labeledTextField("x/theta", value1.toString, s => {
               value1 = s.toDouble; drawing.draw() })
58        val v2Field = DrawingMain.labeledTextField("y", value2.toString, s => {
               value2 = s.toDouble; drawing.draw() })
59        panel.children = List(combo, v1Field, v2Field)
60        propPanel = panel
61      }
62      propPanel
63    }
64
65    def toXML: xml.Node = {
66      <drawable type="transform" value1={ value1.toString() } value2={
             value2.toString() } transType={ transformType.toString() }>
67        { _children.map(c => c.toXML) }
68      </drawable>
69    }
70  }
71
72  object DrawTransform extends Enumeration {
73    val Rotate, Scale, Shear, Translate = Value
74
75    def apply(d: Drawing): DrawTransform = {
76      new DrawTransform(d, mutable.Buffer[Drawable]())
77    }
78
79    def apply(n: xml.NodeSeq, d: Drawing): DrawTransform = {
80      val children = (n \ "drawable").map(dnode => Drawable.makeDrawable(dnode,
             d)).toBuffer
81      val transType = (n \ "@transType").text
82      val value1 = (n \ "@value1").text.toDouble
```

```
83    val value2 = (n \ "@value2").text.toDouble
84    new DrawTransform(d, children, DrawTransform.values.find(_.toString ==
          transType).get, value1, value2)
85  }
86 }
```

The most interesting feature of this code is the XML serialization and deserialization. Line 80 makes use of the `Drawable.makeDrawable` that we looked at earlier. Hopefully this line makes it clear why we do not want separate tags for `rectangle`, `text`, `transform`, etc. We want a collection of `Drawable` type, and the searching mechanism in XML does this naturally if all the elements in question have the same tag.

We finish with the `DrawJulia class` because it was probably the most challenging to modify so that it could be safely serialized and deserialized in both manners we are working with.[16]

```
1  package iostreams.drawing
2
3  import scalafx.Includes._
4  import scalafx.scene.Node
5  import scalafx.scene.canvas.GraphicsContext
6  import scalafx.scene.layout.VBox
7  import scalafx.scene.paint.Color
8  import scalafx.scene.image.WritableImage
9  import scalafx.scene.image.PixelWriter
10 import scala.concurrent.Future
11 import scala.concurrent.ExecutionContext.Implicits.global
12 import scalafx.application.Platform
13 import akka.actor.Props
14 import akka.routing.BalancingPool
15 import akka.actor.Actor
16 import akka.actor.ActorRef
17
18 class DrawJulia(
19     d: Drawing,
20     private var rmin: Double,
21     private var rmax: Double,
22     private var imin: Double,
23     private var imax: Double,
24     private var creal: Double,
25     private var cimag: Double,
26     private var width: Int,
27     private var height: Int,
28     private var maxCount: Int) extends Drawable(d) {
29   @transient private var img = new WritableImage(width, height)
30   @transient private var juliaActor: ActorRef = null
31   @transient private var router: ActorRef = null
32   @transient private var propPanel: Node = null
33
34   makeActors()
35   juliaActor ! MakeImage
36
37   override def toString() = "Julia Set"
```

---

[16]The Zip save is really an XML serialization that is being passed through an extra layer of processing to compress the results.

```scala
38
39   def draw(gc: GraphicsContext): Unit = {
40     if (img != null) gc.drawImage(img, 0, 0)
41     else {
42       if (juliaActor == null) makeActors()
43       juliaActor ! MakeImage
44     }
45   }
46
47   def propertiesPanel(): Node = {
48     def checkChangeMade[A](originalValue: A, newValue: => A): A = {
49       try {
50         val nv = newValue
51         if (originalValue == nv) originalValue else {
52           if (juliaActor == null) makeActors()
53           juliaActor ! MakeImage
54           nv
55         }
56       } catch {
57         case nfe: NumberFormatException => originalValue
58       }
59     }
60     if (propPanel == null) {
61       val panel = new VBox
62       panel.children = List(
63         DrawingMain.labeledTextField("C-real", creal.toString(), s => { creal =
              checkChangeMade(creal, s.toDouble) }),
64         DrawingMain.labeledTextField("C-imaginary", cimag.toString(), s => { cimag
              = checkChangeMade(cimag, s.toDouble) }),
65         DrawingMain.labeledTextField("Real Min", rmin.toString(), s => { rmin =
              checkChangeMade(rmin, s.toDouble) }),
66         DrawingMain.labeledTextField("Real Max", rmax.toString(), s => { rmax =
              checkChangeMade(rmax, s.toDouble) }),
67         DrawingMain.labeledTextField("Imaginary Min", imin.toString(), s => { imin
              = checkChangeMade(imin, s.toDouble) }),
68         DrawingMain.labeledTextField("Imaginary Max", imax.toString(), s => { imax
              = checkChangeMade(imax, s.toDouble) }),
69         DrawingMain.labeledTextField("Max Count", maxCount.toString(), s => {
              maxCount = checkChangeMade(maxCount, s.toInt) }),
70         DrawingMain.labeledTextField("Width", width.toString(), s => { width =
              checkChangeMade(width, s.toInt) }),
71         DrawingMain.labeledTextField("Height", height.toString(), s => { height =
              checkChangeMade(height, s.toInt) }))
72       propPanel = panel
73     }
74     propPanel
75   }
76
77   def toXML: xml.Node = {
78     <drawable type="julia" rmin={ rmin.toString() } rmax={ rmax.toString() } imin={
         imin.toString() } imax={ imax.toString() } creal={ creal.toString() }
         cimag={ cimag.toString() } width={ width.toString() } height={
         height.toString() } maxCount={ maxCount.toString() }/>
79   }
80
```

```scala
81    private def makeActors(): Unit = {
82      juliaActor = DrawingMain.system.actorOf(Props(new JuliaActor), "JuliaActor")
83      router = DrawingMain.system.actorOf(BalancingPool(5).props(Props(new
          LineActor)), "poolRouter")
84    }
85
86    private def juliaIter(zr: Double, zi: Double, cr: Double, ci: Double) = (zr * zr
        - zi * zi + cr, 2 * zr * zi + ci)
87
88    private def juliaCount(z1r: Double, z1i: Double): Int = {
89      var ret = 0
90      var (zr, zi) = (z1r, z1i)
91      while (ret < maxCount && zr * zr + zi * zi < 4) {
92        val (tr, ti) = juliaIter(zr, zi, creal, cimag)
93        zr = tr
94        zi = ti
95        ret += 1
96      }
97      ret
98    }
99
100   case object MakeImage
101   case class Line(row: Int, imaginary: Double)
102   case class LineResult(row: Int, colors: Array[Color])
103
104   class JuliaActor extends Actor {
105     private var image: WritableImage = null
106     private var writer: PixelWriter = null
107     private var lineCount = 0
108     def receive = {
109       case MakeImage =>
110         image = new WritableImage(width, height)
111         writer = image.pixelWriter
112         lineCount = 0
113         for (i <- 0 until height) {
114           val imag = imax - i * (imax - imin) / width
115           router ! Line(i, imag)
116         }
117       case LineResult(r, colors) =>
118         lineCount += 1
119         for (i <- colors.indices) {
120           writer.setColor(i, r, colors(i))
121         }
122         if (lineCount >= height) {
123           Platform.runLater {
124             img = image
125             drawing.draw()
126           }
127         }
128     }
129   }
130
131   class LineActor extends Actor {
132     def receive = {
133       case Line(r, imag) =>
```

```
134       sender ! LineResult(r, Array.tabulate(width)(i => {
135         val real = rmin + i * (rmax - rmin) / width
136         val cnt = juliaCount(real, imag)
137         if (cnt == maxCount) Color.Black else
138           Color(1.0, 0.0, 0.0, math.log(cnt.toDouble) / math.log(maxCount))
139       }))
140     }
141   }
142 }
143
144 object DrawJulia {
145   def apply(d: Drawing) = new DrawJulia(d, -1, 1, -1, 1, -0.5, 0.6, 600, 600, 100)
146
147   def apply(n: xml.Node, d: Drawing) = {
148     val rmin = (n \ "@rmin").text.toDouble
149     val rmax = (n \ "@rmax").text.toDouble
150     val imin = (n \ "@imin").text.toDouble
151     val imax = (n \ "@imax").text.toDouble
152     val creal = (n \ "@creal").text.toDouble
153     val cimag = (n \ "@cimag").text.toDouble
154     val width = (n \ "@width").text.toInt
155     val height = (n \ "@height").text.toInt
156     val maxCount = (n \ "@maxCount").text.toInt
157     new DrawJulia(d, rmin, rmax, imin, imax, creal, cimag, width, height, maxCount)
158   }
159 }
```

There are two main things that make `DrawJulia` more challenging than the other `Drawable` types. The first is that when we first wrote it, we did not pass in a lot of the values, instead opting to initialize them in the **class** during instantiation. That made the code back in `DrawingMain` a bit easier to write initially, but it led to more work at this stage because we really want our XML deserialization to be able to pass in those values.

The more significant challenge comes from the fact that we have four **transient** members in this **class**, including the two actors used for the parallel calculations. When the default deserialization occurs, those values are going to be **null**. As such, we included a **makeActors** method that is called if the `juliaActor` is **null** before any of the calls that tell that actor to make a new image.

We are not going to go through the other three subtypes of `Drawable` that we have created to this point. You should feel free to look at them in the GitHub repository for this book. You can find those files at **https://github.com/MarkCLewis/OOAbstractDataStructScala/tree/master/src/iostreams/drawing**.

With these changes in place, you now have the ability to make whatever drawings you wish and save them off in one of three styles, then load them back in. In practice you would likely choose to support one form of serialization as few applications would need to support both. We included three styles of saving the files so that you could see how they could be implemented.

10.7   End of Chapter Material

10.7.1   Summary of Concepts

- The `java.io` package contains quite a few useful types. There are four extended type hierarchies rooted in `InputStream`, `OutputStream`, `Reader`, and `Writer`. The first two have operations for reading and writing bytes. The second two operate on characters. We previously saw how to use `scala.io.Source` and `java.io.PrintWriter` for text access, so this chapter focuses mainly on the two stream hierarchies.

- The `InputStream` and `OutputStream` classes are both abstract. Their subtypes come in two general forms, those that actually specify a source or sink for reading or writing, and those that modify the manner or functionality of reading and writing. The `FileInputStream` and `FileOutputStream` are of the former types. They provide stream behaviors attached to files.

- Exceptions are a way of dealing with unexpected conditions or other errors in code. When something goes wrong, an exception can be `thrown`. It is then up to the code that called the function where the error occurred to figure out what to do about it.

  - Exception handling is done through the `try-catch` expression. Code that might fail is put in a `try` block. The `catch` partial function has cases for the different exceptions the coder knows how to handle at that point.

  - When an exception is thrown, it immediately begins to pop up the call stack until it comes to a `catch` that can handle it. If none exists, it will go all the way to the top of the call stack and crash that thread.

  - The loan pattern is a handy approach to use in Scala to make it easy to write code that deals properly with exceptions.

- There are other subtypes of the primary stream types that do not inherently attach to a source or sink. These `class`es alter behavior of the basic methods or add additional methods to provide new functionality. You use them to "decorate" an existing stream.

  - One of the most common forms of decoration is buffering. This makes sure that slow operations like disk access are done in large chunks.

  - The `DataInputStream` and `DataOutputStream` provide additional methods that allow you to read or write data in binary. This is generally faster and more compact than text, but it loses the convenience of human readability.

  - The `ObjectInputStream` and `ObjectOutputStream` go a step beyond providing basic binary data reading and writing capabilities, they allow you to serialize whole objects assuming that the objects adhere to certain rules.

10.7.2   Exercises

1. Make the save for a drawing be a single option, and give the dialog different file types and extensions.

2. Add code to what was written in the chapter where you pop up error messages if something goes wrong.

3. The code shown for saving drawings has a significant bug that occurs if you try to save when there are no open drawings. Figure out what this is, and edit the code to fix it.

4. Make a `class` called `Matrix` that has inside of it an `Array[Array[Double]]`. Write code so that your class can be serialized and deserialized using Java serialization, XML, and custom binary formats.

5. Make a binary file with 1,000,000 integer values written with a `DataOutputStream` using `writeInt`. Try reading that back in with and without a `BufferedInputStream` between the `DataInputStream` and the `FileInputStream`. Time the results to see how much it matters on your hardware.

6. Write a binary file of integers with ten values in it. Make the values all different, perhaps random. Using `xxd`, attempt to increment every value by one and put the result in a different file. Write Scala code to read back in that file and see if what you did worked.

7. If you did the last exercise and you are looking for a significant challenge, try to do the same thing with `Doubles` instead of `Ints`.

8. Write code to save some XML data out to a file. The challenge here is that it should be written to a Zip file, not as plain text. Use an unzip utility to see if what you did worked. Also try to read it back in using Scala.

10.7.3   Projects

The projects for this chapter involve saving and loading data through serializing objects. You can decide whether you want to take the route of default binary serialization, XML-based serialization, or a custom binary serialization. Which is best depends a lot on the project you are doing.

Something to note in general about the default binary serialization is that if you go this route, you almost certainly want to label your serializable classes as `@SerialVersionUID` so that the saved files do not break quite so often. They will still break, just not as often. The extra control you have over the XML gives it something of an advantage in this regard.

1. For the MUD, you could choose at this point to update the file that saves all of your rooms and other information to something other than a plain text file. XML is actually a rather good choice for this.

   The main thing that will need to be saved in a MUD is the players. Over time they will collect different items and should probably gain in abilities. At this point, you can only have one player playing the game at a time. However, you can add the ability for a player to log in and log back out. That way the game can shift from one player to another. When the player logs out, their information should be saved. You can decide the format and method you want to use for saving the players.

2. If you are building the web spider, you inevitably want to store the data that you are interested in locally in a format that you can easily work with. You might not want to store images locally, but at least keep a local copy of the URL. For text data that you process, storing it in the processed form will make it much easier to work with.

   So for this project you need to start building an internal representation of the data you care about, then work on serializing that out. Using the default binary serialization is probably ideal for numeric or tabular data. Things that are really text and images could work with XML. Choose the format that matches what you are interested in.

3. Like the MUD, a networked, multi-player game will inevitably need a way to store off information about how various players are doing. Implement that for this project along with menu options for saving.

4. Even games that are not networked likely have a state that needs to be remembered. If you are working on that type of project, add menu options for saving and loading games.

5. If you are working on the math worksheet, that needs to be saved. You can pick between default serialization and XML for that.

6. Obviously, a Photoshop® type application needs to be able to save what you are drawing. This is one option where the default serialization is probably the easiest route. You will have to do some customized serialization code for images, but images do not store well in XML either.

7. Storing simulation data in plain text can make it easy to load in with simple plotting tools. However, large simulations need the space efficiency and speed of binary. For this, the ideal is neither XML, nor default serialization. Instead, you want to write custom binary serialization. For example, if you have N particles with x, y, z, v_x, v_y, and v_z coordinates, as well as a mass, the efficient binary serialization starts with writing N as a binary `Int`, then following it with $7N$ binary `Doubles` for the data values of the particles.

8. Keeping all of the user stock portfolio information is not easily done with plain text files. Using binary serialization or XML would be much better. Choose the format that matches what you are interested in and implement the necessary changes in your project.

9. Although using a regular text file for keeping a listing of all the movies your company has available may be sufficient, keeping all of the user data listings, and most especially the rental history of each of the movies, would benefit more from using binary serialization or XML. Choose one of these and implement the necessary changes in this project.

10. If you have been working with L-systems, you can certainly make it so that users can build their own systems of productions in your GUI and then include the ability to save and load those using either binary or XML serialization.

Chapter 11

Networking

In the 1990s Sun Microsystems®, a server hardware company and the creator of Java, had an advertising campaign with the slogan, "The network is the computer". Given the date, their proclamation had remarkable foresight. Today there are few things that you do on a computer that do not use network access and many computing devices are built specifically for use with a network connection of some form. You might often find that your devices become far less useful to you when they are not connected to the Internet.

At a basic level, networking is just having computers talk to one another. This allows them to share information. This includes having programs on one computer react to things that happen on another computer. So far we have written programs that only run locally and only deal with data they can read as files.[1] In this chapter we will branch out beyond that and learn how to make programs that can talk to other computers.

11.1 TCP and UDP

When computers talk via a network they send bits either over a wire/fiber-optic cable or through wireless signals. In order for other computers to understand those bits and for the bits to get to the computer they are supposed to arrive at, there have to be agreed-upon ways of interpreting things. This agreement for how to interpret the binary data is what we call a network PROTOCOL. There are many different protocols that have been developed and there are different ones for different levels of communication. For moving information across a network, the Java libraries provide support for Transmission Control Protocol (TCP) and User Datagram Protocol (UDP) in the `java.net` package.

Both of these protocols break the information you send up into packets. Each packet contains a certain amount of your data as well as protocol data with information such as

---

[1] It is quite possible that the some of the work you have done has happened on a networked file system. In that case, you have been using a network connection, even if you did not realize it.

where that data should go. Most network communication uses TCP, and we will follow along with that. The reason is that TCP is "safe". You can think of TCP as being something like certified mail. When a TCP packet is sent out, it goes through the network until it reaches its destination. The receiver sends back a verification that it got the packet. If something is lost, TCP will try to resend it. If it fails too many times, the program sending the data will be notified with an error message. In Scala, that notification will come in the form of an exception.

UDP is more like standard mail and lacks the safety mechanisms of TCP. You put your information in a UDP packet and send it out and just hope it arrives. If you are on a good network, most of the packets will. However, UDP will not actually tell you if they do or do not. It is your responsibility as a programmer to write the code in such a way that packets can be dropped and the program can deal with it. The advantage of UDP being "unsafe" is that it can be much faster. The system does not ever sit around waiting for verification that data got across, it just sends things out. This makes it the preferred protocol for things like games where the network speed can significantly impact user experience. It also places an extra development cost on game companies because effort does have to go into making the programs deal with dropped packets.

11.2   Sockets

Communication between machines is done through SOCKETS. The program on one machine, often called the "server", will wait for connections from the outside. Programs running on other computers, called "clients", then make a connection to the one that is waiting. In order to support multiple different types of connections, each computer has many "ports" that sockets can connect to. The server specifies which port it will listen to and clients should connect to the appropriate port for the server they want to talk to.

11.2.1   TCP Sockets

The libraries for using sockets can be found in `java.io.net`. For TCP connections, the program on the server side creates a `ServerSocket` and gives it a port number to use. A call to `accept` on the `ServerSocket` will block until some other computer makes a connection to this port on the server computer. The `accept` method returns an instance of `Socket`. If you wanted to accept a single connection and do something with it, the code might look like this.

```scala
package networking

import java.net.ServerSocket

object DoServer extends App {
  val ss = new ServerSocket(8000)
  val serverSock = ss.accept()
  // Stuff to communicate across serverSock
}
```

The counterpart to this code on the client side just uses the `Socket class` to make a connection to a remote computer. You pass in the machine to connect to and a port number like this.

```
package networking

import java.net.Socket

object DoClient extends App {
  val clientSock = new Socket("localhost", 8000)
  // Stuff to communicate across clientSock
}
```

The port numbers on the client and server need to agree for them to connect. In this example, the client would be run on the same machine as the server. The name "localhost" is used to represent the local machine. You could also provide a remote machine by name or IP address.

Details of Port Numbers

Port numbers are 16-bit unsigned values, so they can range from 0 to 65535. The first 1024 (from 0 to 1023) of these can only be used by programs run with administrator privileges. Most of these are assigned to well-known system-level programs. For example, ssh uses port 22 by default while the SMTP mail service uses 25.

The ports from 1024 to 49151 are registered ports that have assigned uses based on applications to the IANA (Internet Assigned Numbers Authority). On most systems these are open for users, though you cannot use a port that is already in use by some other application. The code in this chapter uses port 8000. Technically, this port has been registered for the Intel® Remote Desktop Management Interface (iRDMI). However, unless you happen to be running that on your system, it should work just fine for your programs.

The ports from 49152 to 65535 cannot be registered and are used for custom applications or temporary purposes. You should feel free to have your application use a port in that range to make sure that you do not conflict with anything on whatever system the application might be run on.

11.2.2 UDP Sockets

With UDP, the role of client and server are less obvious. There is a single DatagramSocket class that is used for both sending and receiving data. It does this through a DatagramPacket. The two simplest ways to build a DatagramPacket are the following.

```
new DatagramPacket(buf: Array[Byte], len: Int): Unit // Receiving packet
new DatagramPacket(buf: Array[Byte], len: Int, address: InetAddress, port: Int):
    Unit // Sending packet
```

In the first case, you are building a packet to receive data. The data goes into the specified array and the specified number of bytes are supposed to be read. The length of the array does not have to match len, but for simple usage it probably should.

The second is building a packet to send data and takes additional arguments that tell where the data should be sent. The InetAddress type is also in java.net. There are several different methods in InetAddress that can be used to get an instance of the type. The two you are most likely to use are the following.

```
getByAddress(addr: Array[Byte]): InetAddress
```

getByName(host: String): InetAddress

You are probably most familiar with using names to represent computers and will likely use the second option.

The DatagramSocket has methods called **send** and **receive** that each take a DatagramPacket and do the appropriate action with them. The **receive** method will block until a message comes in on the proper port. The real challenge of working with UDP is that you have to pack whatever data you want to send into the arrays of bytes in the packets yourself. For this reason, we will not be making much use of UDP in this book. The following example is as far as we will go.

```scala
package networking

import java.net.DatagramPacket
import java.net.DatagramSocket
import java.net.InetAddress
import scala.collection.mutable

object Datagram extends App {
  def packDouble(d: Double, v: mutable.IndexedSeq[Byte]):Unit = {
    var dlong = java.lang.Double.doubleToLongBits(d)
    for (i <- v.indices) {
      v(i) = dlong.toByte
      dlong >>= 8
    }
  }

  def unpackDouble(v: mutable.IndexedSeq[Byte]): Double = {
    val dlong = v.foldRight(0L)((b, dl) => (dl << 8) | (b & 0xff))
    java.lang.Double.longBitsToDouble(dlong)
  }

  def packArray(ds: Array[Double]): Array[Byte] = {
    val ret = new Array[Byte](8 * ds.length)
    for (i <- ds.indices) packDouble(ds(i), ret.view(8 * i, 8 * (i + 1)))
    ret
  }

  if (args.length > 1) {
    val socket = new DatagramSocket()
    val data = packArray(args.map(_.toDouble))
    val packet = new DatagramPacket(data, data.length,
        InetAddress.getByName("localhost"), 8000)
    socket.send(packet)
  } else if (args.length == 1) {
    val socket = new DatagramSocket(8000)
    val num = args(0).toInt
    val data = new Array[Byte](8 * num)
    val packet = new DatagramPacket(data, data.length)
    socket.receive(packet)
    for (i <- 0 until 8 * num by 8) println(unpackDouble(data.view(i, i + 8)))
  } else println("Specify one int for number to read or multiple doubles to send.")
}
```

This code has the ability to send an **Array** of **Doubles** using UDP. Note that in order to pack the **Doubles** into an **Array** of bytes, we use some calls in the Java library such as doubleToLongBits as well as bitwise operations like **>>**, **<<**, **|**, and **&**.

To run this code you first start the receiving end, as the **receive** method will block. If we want to send three numbers you could do that with **scala DatagramPoint 3**. This will just sit there until we run the sender with a call like this, **scala DatagramPoint 2.3 7.8 1.8775**. After this call the other instance will print out the three values on different lines.

Clearly this is code that you would only go through the effort to write if you really need the speed of UDP. Instead of hand packing the data, you can also use the java.io.ByteArrayInputStream wrapped in either a java.io.DataInputStream or a java.io.ObjectInputStream. If you use this approach, you can get the data out with a java.io.ByteArrayOutputStream wrapped in the appropriate data or object output stream.

11.2.3   Streams from Sockets

For UDP we saw that data has to be packed into the **DatagramPackets** to be sent to other machines and then unpacked when they arrive. We did not discuss how data is sent using the plain **Socket** class for TCP. To do this, we use streams. The **Socket** class has the following methods.

```
getInputStream(): InputStream
getOutputStream(): OutputStream
```

These methods give you the input stream and the output stream associated with the socket. If you call **getInputStream** on a socket, it will return an implementation of **InputStream** that connects to the socket so that data sent through the socket from the other machine will be read. You can wrap these streams in the same manner you would have wrapped streams attached to files. This is part of the flexibility of the stream library. You can write code that uses an **InputStream** or an **OutputStream** and it can work regardless of the type of stream that is passed to it. So the same code can be used to work with sockets or files.

To see how this works, we will create a simple chat room with a server and the ability to include multiple clients. To understand what we really want, we should do some analysis. Figure 11.1 shows a use case diagram for what we want to do. There are two types of users of the system. One will start up the chat room as a server on a particular computer. Other users will be clients of the server and will connect into the room and have options for chatting and leaving. When a person connects, they have to give a handle they will go by so that others will know who has chatted certain messages and so that private messages can be sent. For this to work, only a single user at any given time can have a particular handle.

We will start with the client code because it is simpler. It only needs to make a socket to the server, get the user's handle, ask the user for chat lines/statements that are sent to the server, and display chat text coming from the server. Here is code that does that with a console interface.

```scala
package networking.chat

import io.StdIn.readLine
import java.io.PrintStream
import java.net.Socket
import java.util.Scanner
import scala.annotation.tailrec
```

Chat Room

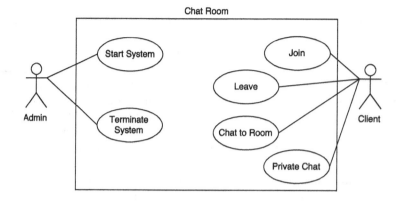

FIGURE 11.1: This figure shows the different use cases for the socket/stream-based chat program that we want to write.

```scala
 8  import scala.concurrent.Future
 9  import scala.concurrent.ExecutionContext.Implicits.global
10
11  object ChatClient extends App {
12    if (args.isEmpty) {
13      println("Usage: scala ChatClient host [port]")
14    } else {
15      val port = if (args.length > 1) args(1).toInt else 8000
16      val sock = new Socket(args(0), port)
17      val sc = new Scanner(sock.getInputStream())
18      val ps = new PrintStream(sock.getOutputStream())
19      println("What is your handle for the chat room?")
20      val name = readLine()
21      ps.println(name)
22      val response = sc.nextLine()
23      if (response != ":quit") {
24        Future { incoming(sc) }
25        println("Welcome! Begin chatting.")
26        outgoing(ps)
27      } else {
28        println("The server has rejected you.")
29      }
30      sock.close()
31    }
32
33    @tailrec private def incoming(sc: Scanner) {
34      val line = sc.nextLine()
35      println(line)
36      incoming(sc)
37    }
38
39    @tailrec private def outgoing(ps: PrintStream) {
40      print("> ")
41      val input = readLine().trim
42      ps.println(input)
43      if (input != ":quit") outgoing(ps)
```

```
44     }
45   }
```

The most interesting part of this code is how the `InputStream` of the `Socket` is wrapped in a `Scanner` and the `OutputStream` is wrapped in a `PrintStream`. Our chat program is using basic text to communicate. These types have helpful methods for reading and writing text so that we do not have to go down to the level of working with arrays of bytes.

Typing in ":quit" as a chat line will terminate the chat. Note that the incoming messages are handled in a separate thread using `Future`. The reason for doing this is that both incoming and outgoing message handling contain blocking methods in the form of `nextLine` and `readLine`. To handle both in a single thread we would have to write more elaborate code that only reads when there are things to be read to prevent blocking. We will do this in the server.

The `incoming` and `outgoing` methods include something new that we have not seen before, an annotation called `@tailrec`. This annotation tells the Scala compiler that a particular recursive method must be optimized for tail recursion. If it cannot, the compiler should generate an error. What does this mean? Tail recursion is a term used to describe a recursive function that contains only one recursive call and does not do significant work after the call returns. These types of recursive functions can often be transformed into `while` loops. In this way you can get the efficiency of a `while` loop while getting the programming benefits of recursion. The `@tailrec` annotation will force an error if the Scala compiler cannot convert the function to a loop. This is important, because you might recall that recursive calls consume memory on the stack, and if you recurse too deeply, you run out of memory and the program crashes from a stack overflow. That does not happen with tail recursive calls. Note that Scala will make recursive methods tail recursive if it can, even if you do not use the annotation. Using the annotation is just a safety to prevent you from accidentally making a non-tail recursive method when you need it to be tail recursive.

The code for the server is a bit more complex because it has more work to do. It needs to have a `ServerSocket` that is accepting new connections, code that takes in chat messages from the sockets, and code that lets the person who runs the server enter a command to stop it. Like the client, some of these things involve blocking calls so multithreading can be helpful. Here is one way of writing the code.

```
1    package networking.chat
2
3    import java.io.InputStream
4    import java.io.PrintStream
5    import java.net.ServerSocket
6    import java.net.Socket
7    import java.util.concurrent.ConcurrentHashMap
8    import scala.collection.JavaConverters._
9    import scala.concurrent.Future
10   import scala.concurrent.ExecutionContext.Implicits.global
11   import io.StdIn.readLine
12
13   object ChatServer extends App {
14     private case class User(name: String, sock: Socket, is: InputStream, ps:
             PrintStream)
15     private val users = new ConcurrentHashMap[String, User]().asScala
16
17     val port = if (args.isEmpty) 8000 else args(0).toInt
18     Future { startServer(port) }
19     while (true) {
```

```scala
20        for ((name, user) <- users) {
21          doChat(user)
22        }
23        Thread.sleep(100)
24      }
25
26      private def startServer(port: Int): Unit = {
27        val ss = new ServerSocket(port)
28        while (true) {
29          val sock = ss.accept()
30          addUser(sock)
31        }
32      }
33
34      private def nonbockingReadLine(is: InputStream): Option[String] = {
35        if (is.available() > 0) {
36          val buf = Array.fill(is.available())(0.toByte)
37          is.read(buf)
38          Some(new String(buf).trim)
39        } else None
40      }
41
42      private def addUser(sock: Socket): Unit = {
43        val is = sock.getInputStream()
44        val ps = new PrintStream(sock.getOutputStream())
45        Future {
46          while (is.available() == 0) Thread.sleep(10)
47          val name = nonbockingReadLine(is).get
48          if (users.contains(name)) {
49            ps.println(":quit")
50            sock.close()
51          } else {
52            ps.println(":accept")
53            val user = User(name, sock, is, ps)
54            users += name -> user
55          }
56        }
57      }
58
59      private def doChat(user: User): Unit = {
60        nonbockingReadLine(user.is) match {
61          case Some(input) =>
62            if (input != ":quit") {
63              val index = input.indexOf("<-")
64              if (index > 0) {
65                val oname = input.take(index).trim
66                if (users.contains(oname)) {
67                  users(oname).ps.println(user.name+" -> "+input.drop(index + 2).trim)
68                } else {
69                  user.ps.println(oname+" is not a valid handle.")
70                }
71              } else {
72                val output = user.name+" : "+input
73                for ((name, ouser) <- users; if name != user.name) {
74                  ouser.ps.println(output)
```

```
75            }
76          }
77        } else {
78          user.sock.close()
79          users.remove(user.name)
80        }
81      case None =>
82    }
83  }
84 }
```

The body of the application, on lines 17–24 gets a port number, starts the server in a separate thread, then enters an infinite loop that checks the users to see if they have said anything.

The function that creates the `ServerSocket` goes into its own infinite loop, where it accepts connections and calls `addUser`. Lines 34–40 declare a method called `nonblockingReadLine` that takes an `InputStream`, produces an `Option[String]`, and then checks to see if the stream has any data available. If it does, it reads all the data into an `Array[Byte]` and returns `Some` of a `String` that is built from the information read. If no data is available, it returns `None`. As the name implies, whether there is data available or not, this method does not block. It will return quickly and either give back a value or say it did not get anything.

The `addUser` method gets streams from the socket and wraps the `OutputStream` in a `PrintStream`. It then starts a `Future` so that it can wait for the user to type in their name. Once data is available, it reads it from the `InputStream` using our `nonblockingReadLine` method. Note that the server defined a `case class` called `User` and keeps a thread-safe mutable map associating names with users. This map allows us to easily ensure that there are never two users with the same handle. It also makes it easier to do private messages. The name that is read in is checked against the map. If the map already has that as a key, it tells the client to terminate before closing the socket. Otherwise, it tells the client that it was accepted, and adds a new `User` to the map.

The `doChat` method uses the `nonblockingReadLine` to check if a user has sent anything to the server. If they have not, it does nothing but return so that the next user can be processed. If it gets input from the user, it reads that input and checks to see if they have disconnected. If they have not, it either sends out a private message to the correct user or runs through all users sending everyone a message. If they quit, that user name is removed from the map and there is no recursive call. The way that private messages are handled is this method looks for `<-` in the text and assumes that what is before that symbol is a user handle and what is after it is the message to send.

Enter these files and check that they work. If you use "localhost" as the machine name you can have conversations with yourself on a single computer. After playing with this for a while you might notice some areas in which it is lacking. First, it is not very pretty. It is plain text in the console. Chat messages that come in will print in the middle of what you are typing in a way that can make it not only a bit ugly, but can also make it hard to use as you cannot keep track of what you are writing.

It is interesting to note that because of the plain text interface, you can use the chat server with the `telnet` program instead of our chat client. It does not look quite as nice, but it is functional. Note that the version of `telnet` that comes with Windows behaves oddly in that it sends individual characters instead of full lines. If you have a Windows machine, you can see what impact this has. A fun challenge is trying to rewrite the server so that it can handle this.

Network Security

The topics of security and information assurance in computing are extremely broad in scope and of critical importance in our modern, computer-driven world. Programs that deal with confidential information need to take efforts to protect that information so that outside entities cannot get to it. Programs that run critical systems need to take precautions to make sure that they cannot be taken down by individuals without the authorization to do so.

The chat program we just wrote does virtually nothing related to security. Adding the proper error handling would be a major step in the direction to make sure that only the person running the server can take it down. However, every message coming into and out of the server is sent pretty much as plain text that could be easily read at any number of points between the client and the server if sent across a network. Granted, this was not intended to be a highly secure system. It does nothing to even verify who is connecting to it. This does not mean that it would never be used to discuss things that people might want to keep private. Someone might feel that if they set up the server on a machine just temporarily, then no one they do not invite would have a chance to come in a snoop on what is being said. The problem with this assumption is that another party could be watching communication without even knowing about the server.

The standard way to prevent someone from listening to network communications is to encrypt messages. The person could still see the message going by, but because it is encrypted they cannot read it. The simplest way to do this with socket communication in Scala is to use a `javax.net.ssl.SSLServerSocket` on the server and connect with `javax.net.ssl.SSLSocket`s. These make sockets that encrypt all communications using the Secure Socket Layer (SSL) protocol. This is the same protocol used by `ssh` and `scp`.

11.3   URLs

As was mentioned at the beginning of the chapter, you have likely been using networked programs for years. Most notable of these would be your web browser. Web browsers were originally written as little more than a program to pull information down from servers and display that information on your screen. Many web applications now have some 2-way communication, but there is still a lot on the web that can be accessed by simply pulling down information.

When you use a web browser to access information you have to tell it where you want to pull the information from and what information you want to pull down. This is done with a Uniform Resource Locator or URL. Most of the URLs you enter into a browser begin with "http". This tells the browser that it is supposed to communicate with the server using the Hypertext Transport Protocol (HTTP). This is a standard for requesting information across sockets using fairly simple text commands. By default, HTTP servers sit on port 80 though other port numbers can be specified as part of the URL. You could pull down information from an HTTP server by opening a socket and sending the appropriate commands. However, because HTTP is so ubiquitous, there is a `java.net.URL` class that will help to do this for you.

There are a number of ways to build a URL where you can specify different parts of the URL, but the simplest to describe just takes a `String` with the same text that you would enter into a browser giving a complete specification of what information you want. For example, you could use the following line of code to get a URL for the main page of Scala.

```
import java.net.URL
val scalaURL = new URL("http://www.scala-lang.org/")
```

We saw back in chapter 8 that we can read the contents of a URL using `Source.fromURL`. The `URL` class can be used to do this as well. It has a method called `openStream` that will return an `InputStream` that can be used to read the contents of the URL. We could use this in code like the following to read the complete contents of that URL and convert it to a `String`.

```
import java.io._
import collection.mutable
val urlis = new BufferedInputStream(scalaURL.openStream())
val buffer = mutable.Buffer[Byte]()
var res = urlis.read()
while (res>=0) {
  buffer += res.toByte
  res = urlis.read()
}
val contents = new String(buffer.toArray)
```

This code opens the stream and wraps it in a `BufferedInputStream` to improve performance. It then creates a `Buffer[Byte]` to hold the values that are read, then enters a loop where bytes are read one at a time and added to the buffer. This process ends when the value read is less than zero, indicating that the end of the stream has been reached. After the end has been reached, the buffer is converted to an array and a new `String` is build from that array.

The use of a `while` loop here is called for because we do not know how much data we are loading in advance and other approaches would require that. Instead of a `while` loop this could easily be converted to use a recursive function as well. Just make sure to specify `@tailrec` so that you know what you write will not overflow the stack. Knowing the length of the content would allow this alternate version, which is significantly simpler. The challenge is knowing what goes in place of the question marks.

```
import java.io._
val urlis = new BufferedInputStream(scalaURL.openStream())
val buffer = new Array[Byte](???) // Needs the size of the contents in bytes.
urlis.read(buffer)
val contents = new String(buffer)
```

It is tempting to call `available` on the stream. Unfortunately, this is not a good solution. That method returns the number of bytes that can be read before blocking. That number is often different from the total size. If the stream was just opened it will be small. Waiting for the full contents to become available is not an appropriate solution.

One way to do this that will work for some content is using a `java.net.URLConnection`. The `URL` class has a method called `openConnection: URLConnection` that can return one of these objects. The `URLConnection` has a `getContentLength: Int` method that will tell you the size of the contents of that URL. There is also a `getContentLengthLong: Long` method that can give you the size for content that exceeds the maximum size of an `Int`. A call to `getInputStream: IntputStream` on the `URLConnection` will provide an input stream that you can read from.

```scala
import java.io._
val connection = scalaURL.openConnection()
val urlis = new BufferedInputStream(connection.getInputStream())
val buffer = new Array[Byte](connection.getContentLength)
urlis.read(buffer)
val contents = new String(buffer)
```

Unfortunately, the `getContentLength` method does not work with all types of content. If the length is not available, a value of -1 is returned, which will cause this code to crash. For this reason, a `while` loop is probably the most robust solution.

If the URL pointed to something other than a text file, the data that was read would have to be handled in a different way. For example, it might point to an image. The `URLConnection` has a method called `getContentType: String` that can help you determine what type of data the URL references. You would then have to write code to load in the data and process it appropriately. There is also a method called `getContent: AnyRef` that is part of both `URL` and `URLConnection`. This method can return various things depending on the type of the contents. For example, `getContent` will return an image object if that is what the URL refers to. The fact that it returns `AnyRef` means that you should do a match on the value and handle it appropriately depending on what it is. Unfortunately, the types you get for images are not associated with JavaFX or ScalaFX. If you know a URL refers to an image, there is a constructor for `scalafx.scene.image.Image` that takes a URL and reads it.

11.4   Remote Method Invocation (RMI)

In the next section we will be adding networking to the project in a way that will let users collaborate on a drawing while working on different computers. In order for this to work, they need to be able to share a drawing, chat with one another, and modify elements of that shared drawing. We could also consider adding other features to the sharing such as a sketch pad where they can doodle ideas. Now picture the code to implement these features using sockets.

The challenge that such an implementation will face is that there are many different types of messages being sent between the two users and those different messages can have an impact on very different parts of the code on either end. The code to handle this type of situation using sockets gets messy. It might be tempting to use a different port for each type of message and keep many different sockets open, but such a solution is sloppy and does not scale. The operating system puts limits on the number of resources a single application can use. Instead, you have to send additional information with each message that identifies what type of message it is, and what objects it should be directed toward.

Dealing with that extra bookkeeping is challenging and time consuming to code. It is also difficult to debug. For these reasons, and the fact that this is something that happens in many applications, the Java libraries provide a simplified alternative called Remote Method Invocation (RMI). As the name implies, RMI lets you make method calls on objects that are located on other machines. It automatically handles the bookkeeping so that information is sent to the right place and the correct code is invoked on the opposite side. The downsides of RMI are that it does have a bit of a learning/setup curve, it only works between languages running on the JVM , and it runs slower than well-done specialized code for the same task. The rest of this section will talk about the setup and help you climb the learning curve. Our application is running on the JVM so that is not a problem for us. Our application is not highly dependent on performance either, so the costs involved will not be significant and will be more than make up for with the convenience that RMI provides.

There are a number of steps that are required to get an application to use RMI. They are listed here.

1. Set up abstract types that specify the methods that can be called remotely. This is done by making completely abstract **traits** for any types that you want to have remote access to. Being completely abstract means the only things in the **trait** should be declarations using **def** that do not specify any code to run.[2] You should annotate these **traits** with **@remote**. The **@remote** annotation does two things. It makes the **trait** extend **java.rmi.Remote** and it has each method need to say it can throw a **java.rmi.RemoteException**. More on that below.

2. Write a class that extends **java.rmi.server.UnicastRemoteObject** with the **trait** you just created. You have to implement all the methods from the **trait**. Provide other methods and data members as needed to make a complete implementation for your program.[3]

3. Create a server application that binds the server object to an RMI registry. You do this by calling **bind** or **rebind** on **java.rmi.Naming**. Those methods take two arguments. The first is the name that the registry will know the object by, and the second is the object to bind. The **bind** method will throw an exception if that name is already in use. The **rebind** method will unbind the old object and bind this new one in its place.

4. Create a client application that uses the **lookup** method of **java.rmi.Naming** to get hold of the server object. The **lookup** method takes a single argument that is a URL for the server and object to look up. The protocol for the URL should be "rmi" and the name should be separated from the server by a slash. An example would be "rmi://computer.college.edu/ServerObjectName". The name used here must match what the server used when it was bound to the registry.

5. Lastly you need to run **rmiregistry** then start up the server and clients. The **rmiregistry** program is part of the Java JDK installation. You can run **rmiregistry** manually from the command line, but you have to worry about the classpath so that it can find your files if you do that. If you are only going to have one program using the registry on a given computer, you can execute **java.rmi.registry.LocateRegistry.createRegistry(port: Int)** at the top of your server as one of the first things it does.

This might seem like a daunting list, but it really is only five steps and once you have done

---

[2]Java has a construct called an interface that is purely abstract. That is what RMI actually requires. Scala **traits** with no data that have only abstract methods get compiled to Java interfaces. Java 8 added default method implementations to interfaces, so starting with Scala 2.12, it is likely that **traits** that are not completely abstract will still compile to interfaces. You should still make your remote **traits** completely abstract.

[3]There are other servers you can use instead of **UnicastRemoteObject**. That one is specified here because it is the most straightforward.

that, it becomes much easier to extend your networked application. We will go back through the steps shortly as part of an example.

Before the example, there is one other detail of RMI that needs to be covered, the passing of arguments. As we have discussed, Scala has two normal passing conventions, pass-by-reference and pass-by-name. The pass-by-reference behavior is the default. With RMI there are two different ways to pass things: pass-by-value and pass-by-reference. Which happens depends on the type of the object being passed. Pass-by-value implies that the object is copied and the other machine gets its own copy that is independent of the first one. This happens for serializable data and includes the standard types you are used to working with in Scala. Objects with a type that implements `Remote` are passed as a remote reference. Instead of getting a copy of the object, the remote machine gets an object that implements the remote interface and which has the ability to communicate calls across the network. Methods called on these objects go back to the original version of the object on the other machine. The fact that there is network communication involved means that they are much slower than calls on local objects.

If you try to pass a value that is not serializable or remote, you will get an exception. The reason for this should be fairly clear to you; RMI is utilizing object streams to perform communication. Only objects that can be serialized will go through those. So the discussion of how to make objects serializable in section 10.5 is relevant when using RMI as well.

To help you see how this all goes together, we will write another chat application that uses RMI instead of making direct use of sockets. We will run through the steps from above and point out what code is written for each one. Before we can write it though, we need to figure out what we are writing. This chat application is going to be a bit different from the last one. In particular, it will have a GUI for the client. The GUI will include not only standard chat elements, but a list of everyone logged onto the server. The idea is that users can send private messages by selecting one or more users from the list when they do their chats. Only the selected users will get their messages.

Starting with step 1 from the RMI steps, we lay out the types we want to have remote access to and put in the methods that we need to be able to call remotely. We start with the server.

```scala
package networking.chat

@remote trait RemoteServer {
  def connect(client: RemoteClient): String
  def disconnect(client: RemoteClient): Unit
  def getClients: Seq[RemoteClient]
  def publicMessage(client: RemoteClient, text: String): Unit
}
```

The fundamental structure is that we have a `trait` that extends `java.rmi.Remote`. Inside of that `trait` there are four methods that we want to be able to call on a server remotely. These allow the client to connect, disconnect, and get a list of the current clients. There is also a method called `publicMessage` that should be called whenever a client sends out a message publicly. This allows the server to keep a brief history so that someone new logging in can see what has been discussed recently.

A second trait is set up for the methods that we want to be able to call remotely on the client. It looks like the following.

```scala
package networking.chat

@remote trait RemoteClient {
  def name: String
```

```
5    def message(sender: RemoteClient, text: String): Unit
6    def clientUpdate(clients: Seq[RemoteClient]): Unit
7  }
```

There are only three methods on the client. These get the name of the client, deliver a message to this client, and tell the client to update its list of clients logged into the server.

Looking at these two interfaces you might feel like there is something missing: methods on the server for sending out public or private chats. In the last chat, when one person wrote a chat message, that was sent to the server as just a **String**, and the server figured out what to do with it and then sent it to the various chat participants. This uses the standard client-server network application structure where all communication goes through the server. This worked well for sockets and made our life easier. However, it is not very efficient as the server becomes a significant bottle neck. What is more, if you really wanted your messages to be private, you lack some security because they all have to go through the server for processing. Basically, the server sees everything and does a lot of work sending out all the messages.

RMI makes it easy for us to deviate from the client-server structure and use more of a peer-to-peer approach. The clients will not just have a remote reference to the server, they will have remote references to all the other clients. This means that each client can call the **message** method on the other clients it wants messages to go to. This reduces pressure on the server and makes it less of a bottleneck. If we did not want to keep a history of what was being said on the server, chat traffic would not be sent to the server at all.

This next piece of code includes steps 2, 3, and 5 for the server. Step 2 is completed by the **extends** and **with** clauses along with the methods implemented on lines 17–33. Step 3 is done on line 12 with the call to **rebind**. Step 5 is completed on line 11 with the creation of a registry. This call uses the port number 1099, which is the default port for RMI. If you use any other value, you will have to specify the port number in the URLs that you use to connect in **Naming.lookup**, so life is easier if you use 1099.

```
1   package networking.chat
2
3   import java.rmi.Naming
4   import java.rmi.RemoteException
5   import java.rmi.server.UnicastRemoteObject
6   import javax.management.remote.rmi.RMIServerImpl
7   import scala.collection.mutable
8   import java.rmi.registry.LocateRegistry
9
10  object RMIServer extends UnicastRemoteObject with App with RemoteServer {
11    LocateRegistry.createRegistry(1099)
12    Naming.rebind("ChatServer", this)
13
14    private val clients = mutable.Buffer[RemoteClient]()
15    private var history = mutable.ListBuffer("Server Started\n")
16
17    def connect(client: RemoteClient): String = {
18      clients += client
19      sendUpdate
20      history.mkString("\n")+"\n"
21    }
22
23    def disconnect(client: RemoteClient) {
24      clients -= client
```

```
25      sendUpdate
26    }
27
28    def getClients: Seq[RemoteClient] = clients
29
30    def publicMessage(client: RemoteClient, text: String) {
31      history += client.name+" : "+text
32      if (history.length > 10) history.remove(0)
33    }
34
35    private def sendUpdate: Unit = {
36      val deadClients = clients.filter(c =>
37        try {
38          c.name
39          false
40        } catch {
41          case ex: RemoteException => true
42        })
43      clients --= deadClients
44      clients.foreach(_.clientUpdate(clients))
45    }
46  }
```

The class defines two fields for storing the clients and history, then implements the four methods from the interface with rather short methods. You can see that the **connect** method returns the history. Both **connect** and **disconnect** make use of the one private method that is also included, **sendUpdate**. This method sends out a message to all the clients letting them know that the client list has been modified. This method is a bit longer because it first runs through all the clients and calls their **name** method. That value is not used for anything, but it serves as a check that the client is still there. This is part of a filter operation that is checking for dead clients. The clients that cannot be reached are removed from the list before that list is sent out to all the active clients.

This next piece of code shows steps 2 and 4 for the client. This code is a bit longer because it includes support for the GUI. Even that does not make it all that long.

```
1   package networking.chat
2
3   import scalafx.Includes._
4   import java.rmi.server.UnicastRemoteObject
5   import java.rmi.RemoteException
6   import java.rmi.Naming
7   import scalafx.application.JFXApp
8   import scalafx.scene.control.ListView
9   import scalafx.scene.control.TextArea
10  import scalafx.scene.control.TextField
11  import scalafx.scene.control.TextInputDialog
12  import scalafx.application.JFXApp.PrimaryStage
13  import scalafx.scene.Scene
14  import scalafx.event.ActionEvent
15  import scalafx.scene.control.ScrollPane
16  import scalafx.stage.WindowEvent
17  import scalafx.scene.layout.BorderPane
18  import scalafx.collections.ObservableBuffer
19  import scalafx.application.Platform
20
```

```scala
21  object RMIClient extends UnicastRemoteObject with JFXApp with RemoteClient {
22    val mDialog = new TextInputDialog("localhost")
23    mDialog.title = "Server Machine"
24    mDialog.contentText = "What server do you want to connect to?"
25    mDialog.headerText = "Server Name"
26    val (myName, server) = mDialog.showAndWait() match {
27      case Some(machineName) =>
28        Naming.lookup("rmi://"+machineName+"/ChatServer") match {
29          case server: RemoteServer =>
30            val nDialog = new TextInputDialog("")
31            nDialog.title = "Chat Name"
32            nDialog.contentText = "What name do you want to go by?"
33            nDialog.headerText = "User Name"
34            val name = nDialog.showAndWait()
35            if (name.nonEmpty) (name.get, server)
36            else sys.exit(0)
37          case _ =>
38            println("That machine does not have a registered server.")
39            sys.exit(0)
40        }
41      case None => sys.exit(0)
42    }
43
44    private val chatText = new TextArea(server.connect(this))
45    chatText.editable = false
46    private var clients = server.getClients
47    private val userList = new ListView(clients.map(_.name))
48    private val chatField = new TextField()
49    chatField.onAction = (ae: ActionEvent) => {
50      if (chatField.text.value.trim.nonEmpty) {
51        val recipients = if (userList.selectionModel.value.selectedItems.isEmpty) {
52          server.publicMessage(this, chatField.text.value)
53          clients
54        } else {
55          userList.selectionModel.value.selectedIndices.map(i => clients(i)).toSeq
56        }
57        recipients.foreach(r => try {
58          r.message(this, chatField.text.value)
59        } catch {
60          case ex: RemoteException => chatText.text = chatText.text.value+"Couldn't
                send to one recipient."
61        })
62        chatField.text = ""
63      }
64    }
65
66    stage = new PrimaryStage {
67      title = "RMI Based Chat"
68      scene = new Scene(600, 600) {
69        val borderPane = new BorderPane
70        val scrollList = new ScrollPane
71        scrollList.content = userList
72        borderPane.left = scrollList
73        val nestedBorder = new BorderPane
74        val scrollChat = new ScrollPane
```

```scala
75        scrollChat.content = chatText
76        nestedBorder.center = scrollChat
77        nestedBorder.bottom = chatField
78        borderPane.center = nestedBorder
79        root = borderPane
80
81        userList.prefHeight <== borderPane.height
82        chatText.prefWidth <== nestedBorder.width
83        chatText.prefHeight <== nestedBorder.height - chatField.height
84      }
85      onCloseRequest = (we: WindowEvent) => {
86        server.disconnect(RMIClient.this)
87        sys.exit(0)
88      }
89    }
90
91    def name: String = myName
92
93    def message(sender: RemoteClient, text: String): Unit = Platform.runLater {
94      chatText.text = chatText.text.value + sender.name+" : "+text+"\n"
95    }
96
97    def clientUpdate(cls: Seq[RemoteClient]): Unit = Platform.runLater {
98      clients = cls
99      if (userList != null) userList.items = ObservableBuffer(cls.map(c =>
100        try {
101          c.name
102        } catch {
103          case ex: RemoteException => "Error"
104        }))
105    }
106  }
```

Lines 22–42 are responsible for connecting to the server and getting the user's name. This is done with some dialog boxes that have text input. If the user chooses to not answer either of the dialog boxes, the application stops.

The GUI includes a **TextArea** to display chat text on line 44, a **TextField** for the user to enter chat text into on line 48, and a **ListView** to show the names of the clients so that private messages can be sent on line 47. The text field has event handling code to send out messages when the user hits the Enter key on the field. This is where the peer-to-peer aspect of the code comes into play as each client can directly call **message** on the other clients.

Lines 66–89 are responsible for actually setting up the GUI. It has two nested **BorderPane**s. Both the list and the main area are embedded in **ScrollPane**s. The sizes of the **ListView** and the **TextArea** are bound to elements above them so that the GUI looks reasonable. There is also event code so that when the window is closed, this client will disconnect from the server. The GUI as a whole looks like figure 11.2.

Line 86 does something that we have not seen before. It includes the expression **RMIClient.this** in a call to the server. You should recall that **this** is the name we use as the reference to the current instance object. There is a complication with that line though. Recall that code like **new PrimaryStage** ... where a **new** expression is followed by curly braces creates a new anonymous **class** that is a subtype of **PrimaryStage**. Line 86 is nested inside of that block of code. So if you just use **this**, you are referring to the current instance of the subtype of **PrimaryStage**, but we need an instance of **RemoteClient**. You can refer

FIGURE 11.2: This is a screen shot of the RMI chat client with two users connected.

to the current instance of a **class** or **object** of a binding scope by putting the name of the **class** or **object** before **this** separated by a dot. So **RMIClient.this** refers to the bounding **RMIClient**.[4]

The implementations of the remote methods are quite short for the client as well. Only the updating of the client list has any length and that is due to the fact that the call to **name** is a remote call that needs some error handling so that the code does not die on the first client that cannot be reached. Make sure that you note the use of **Platform.runLater** on lines 93 and 97. RMI calls execute in their own threads. As such, it is essential that you put any code that modifies a GUI inside of **Platform.runLater** as is done here. If you fail to do so, your code will exhibit random failures when threads happen to conflict.

There is one aspect of this application that could be seen as a weakness. Calls to the client **name** method are made more often than they probably should be. The fact that **name** is a method in a remote **trait** means that every call to this will incur network overhead. In this case, "frequent" really just means when chats go through or when client lists are updated. Given that those things are driven by human interactions, this likely is not a problem. Frequently updating names likely makes it harder to "spoof" the system and act like someone else. In general, you do need to be cognizant of when you are making remote calls because they do have a significant overhead. There might be times when you pass things by value to allow a client to do fast processing, then pass a modified copy back just to avoid the overhead of many remote calls.

To really understand the benefits of RMI, consider what it takes to add features. Perhaps you want to make menu options for things like sharing images or having a shared

---

[4]We cheated a little here. The fact that **RMIClient** is a singleton **object** means that you can remove the `.this` and the code still compiles. However, this syntax would be required if **RMIClient** were a **class**, and you need to know this syntax as you are likely to run into situations where you need it, so we left it in to motivate this discussion.

sketchpad. Using socketing, those additions would make the code significantly more complex. With RMI, you would just add one or two methods into the `RemoteClient trait` and the `RMIClient object`. You implement those methods and put in code to call them and you are done. Setting up the RMI is a bit more hassle, but once in place, you can add new features to the RMI-based application just about as easily as you would a non-networked application.

11.5   Collaborative Drawing (Project Integration)

Now it is time to put things together into the drawing program. We will do this in two ways. To demonstrate socketing, we will add a simple feature that allows a copy of a drawing to be sent from one user to another. This will be simple to implement because we already have code that sends a drawing to a stream that writes to a file. The other aspect of networking that we want to add is more significant and we will use RMI. This will be a collaboration panel where multiple users can post drawings for others to pull down, chat with one another, and even draw basic sketches. Note that the only reason to include both RMI and the socketing is pedagogical. We could easily implement all of this using RMI, but we want to demonstrate both approaches in this one application.

We will add a third menu with the title "Collaborate" onto the menu bar for the frame in `DrawingMain`. This menu will have four options. The first two will work with the basic sending of drawings over the network. The second two will work with the more sophisticated RMI-based collaboration scheme. We will add all four to the menu, but only implement the first two initially. This can be done by adding the following code to `DrawingMain` in the menu's setup and menu handlers.

```scala
val collaborateMenu = new Menu("Collaborate")
val acceptItem = new MenuItem("Accept Drawings")
val sendItem = new MenuItem("Send Drawing")
val joinItem = new MenuItem("Join Collaboration")
val shareItem = new MenuItem("Share Drawing")
shareItem.disable = true
collaborateMenu.items = List(acceptItem, sendItem, joinItem, shareItem)
menuBar.menus = List(fileMenu, editMenu, collaborateMenu)
...
acceptItem.onAction = (ae: ActionEvent) => {
  startServer()
  acceptItem.disable = true
}
sendItem.onAction = (ae: ActionEvent) => {
  sendDrawing()
}
```

The "Accept Drawings" option calls `startServer` which is intended to set up a `ServerSocket` on the current machine that will sit there until the program is stopped and accept connections from other machines that want to share drawings. The "SendDrawing" option calls `sendDrawing` which allows us to send the drawing in the tab that is currently selected to another machine. For this code to compile, we need to implement the `startServer` and `sendDrawing` methods.

These two methods can be put directly in `DrawingMain` and do not require too much additional work because we have already made it so that the `Drawing` type can be serialized. If this had not been done for saving through streams earlier, we would have to add that functionality now. A possible implementation of the `startServer` method is shown here.

```
private def startServer() {
  Future {
    val ss = new ServerSocket(8080)
    while (true) {
      val sock = ss.accept
      val ois = new ObjectInputStream(new
        BufferedInputStream(sock.getInputStream()))
      val sender = ois.readUTF()
      val title = ois.readUTF()
      ois.readObject() match {
        case drawing: Drawing => Platform.runLater {
          val response = showConfirmationDialog("Accept "+title+" from
            "+sender+"?", "Accept Drawing?")
          if (response) {
            addDrawing(drawing, title)
          }
        }
        case _ =>
          println("Got the wrong type.")
      }
      ois.close()
      sock.close()
    }
  }
}
```

This spawns a thread in a `Future`, makes a new `ServerSocket`, and then enters into an infinite loop of accepting connections. This code needs to go into a separate thread because the loop will never exit. Recall from chapter 8 that all events in a GUI are handled in the same thread. If you do not put the server code in a separate thread, the GUI will become completely unresponsive because this method never ends.

That infinite loop accepts connections, opens the stream with wrappings to buffer and read objects, then it reads two strings and an object. The strings tell us who the sender is and the title of the drawing. The last read is for the drawing itself. A `match` statement is used to make sure that the object that is read, is really a `Drawing`. If it is, the user is asked if they want to accept that drawing from that sender. If they answer yes, it is added as a new tab using the `addDrawing` method shown here.

```
def addDrawing(nd: Drawing, name: String): Unit = {
  val (tree, tab) = makeDrawingTab(nd, name)
  drawings = drawings :+ (nd, tree)
  tabPane += tab
  nd.draw()
}
```

This code is duplicated in several places that can be refactored to use this code.[5] Note that the code interacting with the GUI is in a call to `Platform.runLater`. This is required because this code is in a `Future`.

---

[5] A full discussion of refactoring is found in chapter 14. Here we just replace the places where we had been doing this code to a call to a method.

The line asking the user if they are willing at accept the drawing uses the following method that utilizes a `scalafx.scene.control.Alert` to bring up a dialog box for the user.

```scala
def showConfirmationDialog(message: String, title: String): Boolean = {
  val alert = new Alert(AlertType.Confirmation)
  alert.title = title
  alert.contentText = message
  alert.showAndWait() == Some(ButtonType.OK)
}
```

Note that this method is public as it is potentially useful for other GUI code outside of `DrawingMain`. Once everything else is done, the stream and the socket are closed.

The other side of this communication is sending a drawing to a different machine. This method is invoked when we select the "Collaborate → Send Drawing" menu option.

```scala
private def sendDrawing() {
  currentTabDrawing.foreach { drawing =>
    val host = showInputDialog("What machine do you want to send to?", "Machine
      Name")
    if (host.nonEmpty) {
      val sock = new Socket(host.get, 8080)
      val oos = new ObjectOutputStream(new
        BufferedOutputStream(sock.getOutputStream()))
      val name = showInputDialog("Who do you want to say is sending?", "Name")
      if (name.nonEmpty) {
        val title = showInputDialog("What is the title of this drawing?", "Title")
        if (title.nonEmpty) {
          oos.writeUTF(name.get)
          oos.writeUTF(title.get)
          oos.writeObject(drawing)
          oos.close()
          sock.close()
        }
      }
    }
  }
}
```

This method asks for the machine we want to send to, then opens a socket to that machine and wraps the `OutputStream` of the socket for sending objects. Once those are open, it asks for the name of the sender and the title of the drawing. Those are sent through the socket, followed by the drawing in the currently selected tab. Once that is sent, everything is closed.

This code uses two additional short methods, one for getting simple text input from the user in a dialog box and the other to give us the drawing in the current tab. Those are shown here.

```scala
def showInputDialog(message: String, title: String): Option[String] = {
  val dialog = new TextInputDialog
  dialog.title = title
  dialog.contentText = message
  dialog.showAndWait()
}

private def currentTabDrawing: Option[Drawing] = {
```

```
    val current = tabPane.selectionModel.value.selectedIndex.value
    if (current >= 0) {
      val (drawing, _) = drawings(current)
      Some(drawing)
    } else None
  }
```

Neither `startServer` nor `sendDrawing` includes proper error checking. That helps to significantly reduce the length of the code, but it is something that production network code really cannot ignore as networking is a task with the ability to produce many types of exceptions.

This basic socketing approach is technically sufficient to let us share a drawing, but if you use it for a while you will notice that it really does not feel very collaborative. There is not any real interaction between the different users. Putting in real interaction requires that there are many types of different messages sent between the machines. This is the type of thing that leans toward the strengths of RMI.

The ideal feature to add would be a true collaboration on a drawing where two or more users could view and edit a single drawing at the same time. This path was considered for this chapter, but eventually dismissed because the required code was too involved. It is made more challenging do to the fact that we are already using standard serialization for saving drawings and RMI features interact with serialization. In addition, users on both sides of the network connection need to interact with elements through a GUI. ScalaFX GUI elements are not serializable, and the event handling is complex to do properly. In the end it was decided to go with this simpler form of collaboration that demonstrates real-time interactions without including many other details that would obscure the educational objectives.

Instead, we are going to give the ability to pop up a separate window that two users can interact with as a shared "sketchpad", chat system, and drawing repository. The sketchpad functionality will be similar to what you think of as a simple paint program with the ability to draw lines, rectangles, and ellipses in different colors. There will also be a text field and text area for chatting. Lastly, we want to have the ability for users to take drawings from the main drawing program and make them available for others.

The use of RMI means that we need to follow the basic steps of RMI. That begins with determining what remote interfaces we want to have and what methods should go into them. Our application here is still simple enough that we can make due with one server type and one client type. In this case, the server type will be fairly simple. Both the remote `trait` and the implementation are put in the same file to simplify the code layout.

```
1   package networking.drawing
2
3   import java.rmi.RemoteException
4   import java.rmi.server.UnicastRemoteObject
5
6   import scala.collection.mutable
7
8   @remote trait RemoteCollaborationServer {
9     def joinCollaboration(col: RemoteCollaborator): (Array[RemoteCollaborator],
          Array[(String, Drawing)])
10    def addDrawing(title: String, drawing: Drawing): Unit
11  }
12
13  class CollaborationServer extends UnicastRemoteObject with
        RemoteCollaborationServer {
```

```
14    private val collaborators = mutable.Buffer[RemoteCollaborator]()
15    private val drawings = mutable.Buffer[(String, Drawing)]()
16
17    def joinCollaboration(col: RemoteCollaborator): (Array[RemoteCollaborator],
          Array[(String, Drawing)]) = {
18      collaborators += col
19      (collaborators.toArray, drawings.toArray)
20    }
21
22    def addDrawing(title: String, drawing: Drawing): Unit = {
23      drawings += title -> drawing
24      for (c <- collaborators) {
25        try {
26          c.addDrawing(title, drawing)
27        } catch {
28          case ex: RemoteException =>
29        }
30      }
31    }
32  }
```

The **trait** has only two methods, one for joining and one for sharing a drawing. The implementation includes those two methods, which are both fairly short, along with two data members for storing who is currently participating in the collaboration and the drawings that have been shared, along with their titles. The `joinCollaboration` method returns all of the collaborators as well as the drawings that have been shared. It is interesting to point out that these things are passed differently. The drawings and their titles are all serializable but not remote, so they are passed by value and the new collaborator so that it gets a copy of them. On the other hand, the collaborators are remote, so the array that goes across will be filled with remote references that allow the new user to communicate with everyone else.

It is worth noting that we have used a slightly different approach to arranging files for these types. Instead of having only a single **class** or **trait** in the file, this file has both the remote **trait** and the implementing **class**.[6] This was done because the two are intimately related and it cuts down on additional **imports** of things like `java.rmi`. The downside of doing this is that someone looking for `RemoteCollaborationServer` might take a long time to find it as there is no file with that name. This is very much a style issue and something that different instructors and employers might request different rules for.[7]

The majority of the code for implementing the collaboration feature goes into the client code. This includes a remote interface for the `Collaborator` as well as an implementation. Here again, the two are put into a single file. This time the file is longer because the implementation is longer. If both the **trait** and the implementation were long we would certainly split them into two files to avoid long files. However, moving the **trait** out does not significantly reduce the length of this file. The code gives us the ability to produce a window that looks like figure 11.3.

```
1  package networking.drawing
2
3  import java.rmi.RemoteException
```

---

[6]This is something that you can do in Scala, but not in Java. In Java, you can only have one public type per file and it must have the same name as the file.

[7]Modern IDEs can make the issue of finding code less important. For example, in Eclipse, if you put your cursor on something and press F3, it will jump you to the declaration. You can get similar functionality holding Ctrl and clicking on some part of code.

```scala
4   import java.rmi.server.UnicastRemoteObject
5
6   import scala.collection.mutable
7
8   import scalafx.Includes._
9   import scalafx.application.Platform
10  import scalafx.event.ActionEvent
11  import scalafx.geometry.Orientation
12  import scalafx.scene.Scene
13  import scalafx.scene.canvas.Canvas
14  import scalafx.scene.control.Button
15  import scalafx.scene.control.ColorPicker
16  import scalafx.scene.control.Label
17  import scalafx.scene.control.RadioButton
18  import scalafx.scene.control.ScrollPane
19  import scalafx.scene.control.SplitPane
20  import scalafx.scene.control.TextArea
21  import scalafx.scene.control.TextField
22  import scalafx.scene.control.ToggleGroup
23  import scalafx.scene.input.MouseEvent
24  import scalafx.scene.layout.BorderPane
25  import scalafx.scene.layout.FlowPane
26  import scalafx.scene.layout.VBox
27  import scalafx.scene.paint.Color
28  import scalafx.scene.text.Font
29  import scalafx.stage.Stage
30
31  @remote trait RemoteCollaborator {
32    def post(text: String): Unit
33    def requestSketch: Seq[Sketchable]
34    def updateSketch(who: RemoteCollaborator, sketch: Seq[Sketchable]): Unit
35    def addDrawing(title: String, drawing: Drawing): Unit
36  }
37
38  /**
39   * This implementation will run on the clients and bring up a window that will show
40   * thumb nails of the shared drawings, a simple sketch pad, and chat window.
41   */
42  class Collaborator(server: RemoteCollaborationServer) extends UnicastRemoteObject
        with RemoteCollaborator {
43    // Set up sketch and drawing variable
44    private val sketch = mutable.Buffer[Sketchable]()
45
46    private val (sketches, drawings) = {
47      val (cols, draws) = server.joinCollaboration(this)
48      mutable.Map(cols.map(c => {
49        try {
50          Some(c -> c.requestSketch)
51        } catch {
52          case ex: RemoteException => None
53        }
54      }).filter(_.nonEmpty).map(_.get): _*) -> mutable.Buffer(draws: _*)
55    }
56
57    // Set up chatting variables
```

```scala
58    private val nameField = new TextField
59    private val chatField = new TextField
60    private val chatArea = new TextArea
61    chatArea.editable = false
62    chatField.onAction = (ae: ActionEvent) => {
63      if (chatField.text.value.nonEmpty) {
64        val text = nameField.text.value+": "+chatField.text.value
65        chatField.text = ""
66        foreachCollaborator { _.post(text) }
67      }
68    }
69
70    // Set up the shared drawing view
71    val sharedVBox = new VBox
72    sharedVBox.prefWidth = 200
73    for ((n, d) <- drawings) drawShared(n, d)
74
75    def drawShared(name: String, drawing: Drawing): Unit = {
76      val sharedCanvas = new Canvas(200, 200)
77      val sharedGC = sharedCanvas.graphicsContext2D
78      sharedGC.fill = Color.White
79      sharedGC.fillRect(0, 0, sharedCanvas.width.value, sharedCanvas.height.value)
80      sharedGC.scale(0.15, 0.15)
81      sharedGC.fill = Color.Black
82      drawing.drawTo(sharedGC)
83      sharedVBox.children += sharedCanvas
84      sharedVBox.children += new Label(name)
85      sharedCanvas.onMouseClicked = (me: MouseEvent) => {
86        if (me.clickCount == 2) DrawingMain.addDrawing(drawing, name)
87      }
88    }
89
90    // Set up the main GUI and sketching controls
91    val colorPicker = new ColorPicker(Color.Black)
92    private var sketchCreator: MouseEvent => Sketchable = (me: MouseEvent) => {
93      val c = colorPicker.value.value
94      new SketchPath(me.x, me.y, c.red, c.green, c.blue, c.opacity)
95    }
96    val sketchCanvas = new Canvas(2000, 2000)
97    val sketchGC = sketchCanvas.graphicsContext2D
98    sketchCanvas.onMousePressed = (me: MouseEvent) => {
99      sketch += sketchCreator(me)
100   }
101   sketchCanvas.onMouseDragged = (me: MouseEvent) => {
102     sketch.last.mouseDragged(me.x, me.y)
103     redrawSketch()
104   }
105   sketchCanvas.onMouseReleased = (me: MouseEvent) => {
106     sketch.last.mouseReleased(me.x, me.y)
107     redrawSketch()
108     for ((c, _) <- sketches) {
109       c.updateSketch(this, sketch)
110     }
111   }
112
```

```
113  sketchUpdated()
114  redrawSketch()
115
116  val stage = new Stage {
117    title = "Collabortive Sketch"
118    scene = new Scene(800, 600) {
119      val chatBorder = new BorderPane
120      chatBorder.top = chatField
121      val chatScroll = new ScrollPane
122      chatScroll.content = chatArea
123      chatBorder.center = chatScroll
124      val sketchBorder = new BorderPane
125      val flowPane = new FlowPane
126      val buttonGroup = new ToggleGroup
127      val freeformButton = new RadioButton("Freeform")
128      freeformButton.selected = true
129      freeformButton.onAction = (ae: ActionEvent) => {
130        if (freeformButton.selected.value) sketchCreator = (me: MouseEvent) => {
131          val c = colorPicker.value.value
132          new SketchPath(me.x, me.y, c.red, c.green, c.blue, c.opacity)
133        }
134      }
135      val lineButton = new RadioButton("Line")
136      lineButton.onAction = (ae: ActionEvent) => {
137        if (lineButton.selected.value) sketchCreator = (me: MouseEvent) => {
138          val c = colorPicker.value.value
139          new SketchLine(me.x, me.y, c.red, c.green, c.blue, c.opacity)
140        }
141      }
142      val rectButton = new RadioButton("Rectangle")
143      rectButton.onAction = (ae: ActionEvent) => {
144        if (rectButton.selected.value) sketchCreator = (me: MouseEvent) => {
145          val c = colorPicker.value.value
146          new SketchRect(me.x, me.y, c.red, c.green, c.blue, c.opacity)
147        }
148      }
149      val ellipseButton = new RadioButton("Ellipse")
150      ellipseButton.onAction = (ae: ActionEvent) => {
151        if (ellipseButton.selected.value) sketchCreator = (me: MouseEvent) => {
152          val c = colorPicker.value.value
153          new SketchEllipse(me.x, me.y, c.red, c.green, c.blue, c.opacity)
154        }
155      }
156      val clearButton = new Button("Clear")
157      clearButton.onAction = (ae: ActionEvent) => {
158        sketch.clear()
159        redrawSketch()
160        for ((c, _) <- sketches) {
161          c.updateSketch(Collaborator.this, sketch)
162        }
163      }
164      flowPane.children = List(freeformButton, lineButton, rectButton,
             ellipseButton, clearButton, colorPicker)
165      buttonGroup.toggles = List(freeformButton, lineButton, rectButton,
             ellipseButton)
```

```scala
166        sketchBorder.top = flowPane
167        val sketchScroll = new ScrollPane
168        sketchScroll.content = sketchCanvas
169        sketchBorder.center = sketchScroll
170        val collabSplit = new SplitPane
171        collabSplit.orientation = Orientation.Vertical
172        collabSplit.items ++= List(sketchBorder, chatBorder)
173        collabSplit.dividerPositions = 0.7
174        val nameBorder = new BorderPane
175        nameBorder.left = new Label("Name")
176        nameBorder.center = nameField
177        val topBorder = new BorderPane
178        topBorder.top = nameBorder
179        topBorder.center = collabSplit
180        topBorder.left = sharedVBox
181        root = topBorder
182      }
183    }
184
185    // Local helper methods
186    private def foreachCollaborator(f: RemoteCollaborator => Unit) {
187      for (c <- sketches.keys) try {
188        f(c)
189      } catch {
190        case ex: RemoteException =>
191      }
192    }
193
194    private def redrawSketch(): Unit = {
195      sketchGC.fill = Color.White
196      sketchGC.fillRect(0, 0, sketchCanvas.width.value, sketchCanvas.height.value)
197      for ((_, s) <- sketches; sk <- s) {
198        sk.draw(sketchGC)
199      }
200      for (sk <- sketch) {
201        sk.draw(sketchGC)
202      }
203    }
204
205    private def sketchUpdated() {
206      foreachCollaborator(c => {
207        c.updateSketch(this, sketch)
208      })
209    }
210
211    // Remote methods
212    def post(text: String) {
213      chatArea.text = chatArea.text.value + text+"\n"
214    }
215
216    def requestSketch: Seq[Sketchable] = sketch
217
218    def updateSketch(who: RemoteCollaborator, sketch: Seq[Sketchable]): Unit = {
219      sketches(who) = sketch
220      redrawSketch()
```

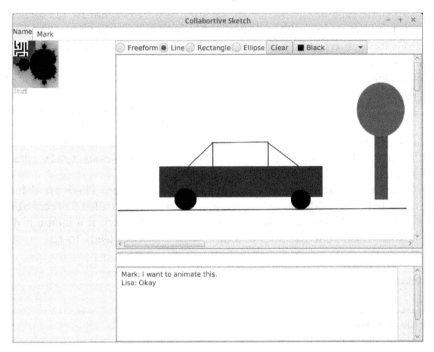

FIGURE 11.3: This is a screen shot of the collaboration window. On the left is a column showing shared drawings. On the bottom is a chat area. The main portion is a shared sketch region where different collaborators can draw using some simple types.

```
221     }
222
223     def addDrawing(title: String, drawing: Drawing) {
224       drawings += title -> drawing
225       Platform.runLater { drawShared(title, drawing) }
226     }
227   }
```

The **class** starts with the creation of a **Buffer** of a type called **Sketchable** that represents the things we can put in a sketch. We will look at that code in a bit. After that, is code that joins the collaboration, then uses the return from that to get the sketches of all the other collaborators as well as all current shared drawings.

Lines 58–68 set up some variables used for chatting. There is a **nameField** that goes at the top of the GUI for the user to enter their name. Below that are the **chatField** and the **chatArea**. There is a handler on the **chatField** that runs through all the collaborators sending them the proper chat message.

Lines 71–89 make the GUI elements for the shared drawings and draw thumbnails of these shared drawings to canvases that are stacked vertically in a **VBox** with **Labels** between them. Note the use of the **scale** method. This is what makes the drawings come out smaller. To allow the drawings to potentially draw to a different **Canvas** without causing problems, we added a **drawTo** method that is called on line 82. It does the same thing as **draw**, but using the **GraphicsContext** that is passed in.

Lines 91–183 set up the overall GUI and handle the sketching. There is a **Canvas** that the sketch is drawn to that has handlers added for the mouse being pressed, dragged, and released. When the mouse is pressed, a new instance of **Sketchable** has to be created and added to the current sketch. To abstract the creation of the `Sketchable`, lines 92–95 create a variable called `sketchCreator`. The default makes an instance of a type called `SketchPath`. Radio buttons for each type of thing that can be drawn are created on lines 129–155. Each one, when selected, changes the value of `sketchCreator` so that it makes the proper type. There is also a `Button` that clears the sketch of the current user and a `ColorPicker` so that users can draw in different colors.

Lines 186–209 define some methods that are used locally for executing code on each collaborator and for drawing out the sketches. Note that each collaborator has their own sketch. Those are drawn in order and the current user's sketch is draw on top of the others.

The file ends with the methods that are declared in the remote `trait`, which are all reasonably short and straightforward.

The collaboration code makes multiple uses of `Sketchable`s. These are defined in the following code. It begins with the definition of a `sealed trait` called `Sketchable` that has code for drawing to a `Canvas` as well as methods for handling when a mouse is dragged or released. It is assumed that the mouse being pressed is what leads to the creation of an instance.

```scala
 1  package networking.drawing
 2
 3  import scala.collection.mutable
 4  import scalafx.scene.canvas.GraphicsContext
 5  import scalafx.scene.paint.Color
 6
 7  sealed trait Sketchable extends Serializable {
 8    def draw(gc: GraphicsContext): Unit
 9    def mouseDragged(x: Double, y: Double): Unit
10    def mouseReleased(x: Double, y: Double): Unit
11  }
12
13  class SketchPath(val sx: Double, val sy: Double, val red: Double, val green:
        Double, val blue: Double, val opacity: Double) extends Sketchable {
14    private val points = mutable.Buffer[(Double, Double)]()
15    def draw(gc: GraphicsContext): Unit = {
16      gc.stroke = Color(red, green, blue, opacity)
17      gc.beginPath()
18      gc.moveTo(sx, sy)
19      for ((x, y) <- points) gc.lineTo(x, y)
20      gc.strokePath()
21    }
22    def mouseDragged(x: Double, y: Double): Unit = {
23      points += (x -> y)
24    }
25    def mouseReleased(x: Double, y: Double): Unit = {
26      points += (x -> y)
27    }
28  }
29
30  class SketchLine(val sx: Double, val sy: Double, val red: Double, val green:
        Double, val blue: Double, val opacity: Double) extends Sketchable {
31    private var ex, ey = 0.0
32    def draw(gc: GraphicsContext): Unit = {
33      gc.stroke = Color(red, green, blue, opacity)
34      gc.strokeLine(sx, sy, ex, ey)
35    }
```

```scala
36    def mouseDragged(x: Double, y: Double): Unit = {
37      ex = x
38      ey = y
39    }
40    def mouseReleased(x: Double, y: Double): Unit = {
41      ex = x
42      ey = y
43    }
44  }
45
46  class SketchRect(val sx: Double, val sy: Double, val red: Double, val green:
        Double, val blue: Double, val opacity: Double) extends Sketchable {
47    private var (ex, ey) = (sx, sy)
48    def draw(gc: GraphicsContext): Unit = {
49      gc.fill = Color(red, green, blue, opacity)
50      gc.fillRect(sx min ex, sy min ey, (sx - ex).abs, (sy - ey).abs)
51    }
52    def mouseDragged(x: Double, y: Double): Unit = {
53      ex = x
54      ey = y
55    }
56    def mouseReleased(x: Double, y: Double): Unit = {
57      ex = x
58      ey = y
59    }
60  }
61
62  class SketchEllipse(val sx: Double, val sy: Double, val red: Double, val green:
        Double, val blue: Double, val opacity: Double) extends Sketchable {
63    private var (ex, ey) = (sx, sy)
64    def draw(gc: GraphicsContext): Unit = {
65      gc.fill = Color(red, green, blue, opacity)
66      gc.fillOval(sx min ex, sy min ey, (sx - ex).abs, (sy - ey).abs)
67    }
68    def mouseDragged(x: Double, y: Double): Unit = {
69      ex = x
70      ey = y
71    }
72    def mouseReleased(x: Double, y: Double): Unit = {
73      ex = x
74      ey = y
75    }
76  }
```

There are four subtypes of **Sketchable** declared in the file for paths, lines, rectangles, and ellipses. Each one takes six arguments at creation that specify the location of the initial mouse pressed event that leads to their creation as well as the components of the current color for drawing. The point of this inheritance hierarchy is to provide a serializable type that abstracts the drawing process in terms of both user interactions with the mouse and rendering to the screen.

In order to make the collaboration window useful, we have to implement the handlers for the last two menu items. Code for that is shown here.

```scala
joinItem.onAction = (ae: ActionEvent) => {
  if (server == null) {
```

```scala
    val host = showInputDialog("What server host do you want to use?",
        "Server Name")
    if (host.nonEmpty && host.get.trim.nonEmpty) {
      try {
        Naming.lookup("rmi://"+host.get+"/Collaborate") match {
          case svr: RemoteCollaborationServer =>
            server = svr
            client = new Collaborator(server)
          case _ => throw new NotBoundException("Wrong type found.")
        }
      } catch {
        case (_: NotBoundException) | (_: RemoteException) =>
          val s = new CollaborationServer()
          Naming.bind("Collaborate", s)
          server = s
          client = new Collaborator(server)
      }
    }
  }
  if (client != null) {
    client.stage.show()
    shareItem.disable = false
  }
}
shareItem.onAction = (ae: ActionEvent) => {
  val title = showInputDialog("What do you want to call this drawing?",
      "Drawing Title")
  if (title.nonEmpty) {
    currentTabDrawing.foreach(drawing => server.addDrawing(title.get,
        drawing))
  }
}
```

The handler for `joinItem` asks what server the user wants to connect to and tries to make a connection there. If it does not find a server, cannot get an object from the server, or gets the wrong type of object from the server, it creates a `CollaborationServer` locally and connects to that. At the end, if there is a client, it calls `client.stage.show()` to open the window for the collaboration client. It also activates the `shareItem` menu item so that users can use it.

When the user selects the `shareItem`, they are asked for a name to give the drawing and then the current drawing, if there is a tab up, is sent to the server for distribution.

This section contains a fair bit of code that really requires some exploration. We have not shown the changes to `Drawing`. In addition, there were a number of tweaks made to `DrawMandelbrot` and `DrawJulia` to support having the multithreading code render to an alternate `Canvas`. You can look at that code and the other code from that chapter at the GitHub repository at `https://github.com/MarkCLewis/OOAbstractDataStructScala/tree/master/src/networking`.

11.6 End of Chapter Material

11.6.1 Summary of Concepts

- Networking has become an essential aspect of computing. When computers talk to one another over a network they have to go about it in an agreed-upon manner known as a protocol.

  - The most commonly used networking protocol in the Transmission Control Protocol, TCP. A TCP connection guarantees transmission or throws an exception so that the program knows it has failed. This makes it slower, but safe.

  - Applications needing higher speed that are willing to live with the occasional loss of a packet typically use the User Datagram Protocol, UDP. This protocol throws out packets of information and does not care if they get where they are going. As a result, some will be lost. The program will have to have mechanisms in place for dealing with packets that do not get through.

- Computers communicate through sockets. There are different socket types in the `java.net` package.

  - For TCP connections, you want to create a `ServerSocket` on one machine that will accept connections on a particular port. Other machines can create `Sockets` that go to the port on the machine with the `ServerSocket`.

  - The `Socket` class has the ability to give you an `InputStream` or an `OutputStream`. These can be used just like any other streams to send or receive information.

  - A machine that wants to receive connections over UDP can open a `DatagramSocket`. The information has to be packed up by a client in a `DatagramPacket`.

- The `java.net.URL` class provides a simple way to read data information from web servers.

- Java includes an approach to network communication called Remote Method Invocation, RMI. Using RMI, you can make method calls to objects that reside on other machines using the standard syntax of a method call. There are a few steps you have to take to get an application set up with RMI.

11.6.2 Exercises

1. Set up a server socket and have it accept one connection. Have it open both the input and output streams and run through a loop where it parrots back to output what was written in input with minor modifications of your choosing. If the input is "quit", have it close the socket and terminate. Run this on a machine, then use `telnet` to connect to that port and see what happens.

2. Write a simple chat server that works using `telnet` as the client.

3. Fix the error handling in the first chat program example in the chapter.

4. Clean up the allowed names code in the first chat program example in the chapter.

5. Take the chat we have and add some extra commands. For example, `:who` should display everyone in the room. Other options might include private interactions other than normal chat. You might look in `scala.Console` for some ideas about other settings.

6. Make a chat using sockets that has a GUI client with a list of all people in the chat room that is kept updated at all times.

7. Write a text-based chat client and server with RMI as the mode of communicating between the two sides.

8. Write a GUI-based chat client and server with RMI as the mode of communication.

9. Using the serialization concepts from the last chapter, add functionality to the GUI-based chats that you have written to send other types of information as serialized objects.

11.6.3   Projects

For this chapter, the obvious addition to the projects is to put in networking. For those projects that are based on networking, it is easy to see how this will happen. For the other projects we are going to force it in a bit, but only in ways that are potentially useful to you.

1. If you are working on a MUD, it is customary for players to be able to access the game through `telent`. If you want to take that approach, then you simply add a `ServerSocket` to your map game and set it up so that many people can connect and give commands. You can also choose to create a more advanced viewer, but keep in mind that this is supposed to be text based.

   How you do this will depend a bit on whether you opted to use actors for your parallelism or not. You will have a separate thread handling the `ServerSocket` with calls to `accept`. The new connections, once the user is logged in, will need to be handed off to the main game thread or have appropriate messages sent to manager actors. It is suggested that you use one of the types in `java.util.concurrent` for doing this. The main game code need not change too much for this addition. Instead of reading from `readLine`, you will be pulling from the `InputStream` of the users `Socket`. You can also support multiple players so that on each iteration, the program checks for input from all users and executes them in order. Keeping the main processing sequential is one way to prevent race conditions. Using actors with a non-blocking read will also prevent race conditions.

2. If you are writing the web-spider, it is finally time to add real functionality. Use the `URL class` and give your program the ability to download pages. You could add a very simple search for "`href=`" so that you can find links to other pages. Use the ability of the `URL class` to build relative URLs to make this easier. Have your program pull down a few hundred pages and collect the data that you want from them. If you had the code set up as described in the previous chapter, this should be as simple as making the threads that are acquiring data pull from a URL instead of from files.

   If you do not already have a graphical display of information, you should add that in. That way you can display to the user what pages have been downloaded, then let them look at the data you have collected in some manner. Remember that you need to have a `Set` or `Map` to make sure that you do not visit the same page multiple times.

3. For the multiplayer graphical game, you have to make it so that two or more instances of the game can be run on separate computers that talk to one another. Exactly how you do this depends on the style of game that you are creating. If your game allows many users to play in a networked fashion, you should probably use a client-server model where there is a separate application running as the server and then various clients connect.

In the client-server model, much of the game's logic will likely be processed on the server. The clients become graphical displays and the medium for user interactions. The commands given to the clients are likely sent to the server with little interpretation.

If your game is only a two-player networked game, you might skip having a server and instead have a single application where the user can set their instance up as something of a local server for a second instance to connect to. In this situation, the game logic will happen primarily on that "main" instance and the second client is primarily just sending and displaying information.

One thing that you will probably realize fairly quickly when you work on this is that distributed parallel applications, those running on multiple machines across a network, can have their own types of race conditions as the different clients might not always be seeing the same world.

4. A game that is not supposed to involve networking is probably the most difficult project to tie networking into. One reason that you might want to include networking is to be able to keep track of things going on in the game outside of the normal display or provide other types of basic interaction. This is largely useful for debugging.

Large and complex software often needs to have the ability to do "logging." This is where different steps in the code write out what they are doing to a file or something similar. When things go wrong, you can consult the log files to see what happened. Instead of logging to file, you can make your game log to the network. Have your game create a `ServerSocket` that accepts connections. All logging should go to all the connections. If there are no connections, the logging is basically skipped.

While logging to the network lacks some of the advantages of logging to a file, like an easy way to go back over things, it provides the flexibility that you can use commands written to the network connection. These commands could be to query information for use in debugging, or they could be like a superuser mode to set up situations that would be hard to create when you are testing your program.

5. The math worksheet program could involve networking in at least two different ways. One would be the addition of collaborative features, like those added to the drawing program in this chapter, so that users can send a complete or partial worksheet from one computer to another.

It is also possible to add logging features, like those described in project 4. The logging features for a math worksheet are most likely going to be for debugging purposes to inspect what is happening in the background.

You can decide whether you want to implement one or both of these options into your code. Pick what interests you most and fits with your objectives for the project.

6. For the Photoshop® project you could add networking in the form of collaboration, logging, and/or using a URL to pull down images from the web. The collaboration could be much like what was shown in the chapter, though it should be altered to fit the flow of the program you are developing. Logging would be most useful for

debugging as described in project 4. Both of those options are fairly significant, and you only need to pick one.

The addition of loading images from URLs across the web is very straightforward and really is something that should be in this type of application. Whichever of the other two networking options you choose, you should probably add this possibility on top of it.

7. The simulation workbench provides a very different feature to add networking to the application. The fact that simulations can be highly processor intensive means that there can be benefits to breaking the work load not only across threads, but also across multiple machines. This form of distributed computing requires networking.

   You should make a second application for a helper node that you can run on other computers. That helper application will make a network connection to the main application and wait to be given work. The simplest way to divide the work is to have separate simulations on each machine with information passed back occasionally for the purposes of displaying data to the user and perhaps saving. If the number of particles in a simulation is really large, it can make sense to distribute that one simulation across multiple machines. This is a more challenging problem as the different machines have to communicate the state of the particles they are controlling for force calculations.

   It is left to the student to decide how involved you wish to get with distributing your simulations.

8. Because the stock portfolio management system is a business product, and you will want many users to be able to use the system, you should probably use a client-server model where there is a separate application running as the server and then various clients connect.

   In the client-server model, much of the business management logic, such as loading stock data from other websites, will likely be processed on the server. The clients become graphical displays and the medium for user interactions. The commands given to the clients are likely sent to the server with little interpretation.

9. The movie project is very similar to the stock portfolio management system in that you should use a client-server model. Much of the movie management processes, such as updating movie information from the web, will be done on the server. The clients become graphical displays and the medium for user interactions. The commands given to the clients are likely sent to the server with little interpretation.

10. We do not have a clear use for networking for the L-system project. Perhaps you can think of something interesting to do with it.

Chapter 12

Linked Lists

So far in this book we have been very happy to use the collections libraries without fully understanding how they work. That is fine for much of what we have done and in a first year of CS study, but real developers need to have knowledge of what is really happening so that they can make informed decisions about what collection types to use in different situations. This chapter will begin to reveal how we can build our own structures for storing data by linking objects that store small pieces of data together.

12.1 The List/Seq ADT

The topic of this chapter is the linked list. This is an approach to storing data in a way that can be applied to some different ADTs. Later in the chapter we will see how a linked list can be used as the storage for stacks and queues. Before that, we want to introduce a new ADT that is a step up on the ladder of capabilities from the stack and queue, the list ADT. This ADT provides the ability to store a sequence of data elements that are indexed by integers with functionality for random retrieval, insertion, and removal. In a language-independent way, one might expect these abilities to be implemented in methods like the following.

```
def get(index: Int): A
def set(index: Int, data: A)
def insert(index: Int, data: A)
def remove(index: Int): A
```

Here, A is the type of data being stored in the list. The `insert` and `remove` methods shift things around so that all elements after the specified index are moved back one index on an insert or forward one index for a remove.

This set of methods is most appropriate for a mutable list implementation. It is also possible to make immutable list implementations. These would return new collections when operations such as inserting or removing are performed. They also typically avoid doing general insertions or removals as those operations can be very inefficient.

In the Scala collections, the behaviors of this ADT are represented without `insert` and `remove` functionality by the `Seq trait` and in a more complete sense by the `Buffer trait`. They have slightly different methods than what was listed above. We will see these in later sections where we will build implementations that inherit from those `traits`.

12.2 Nature of Arrays

To understand linked lists, it is helpful to have something to compare them to. Arrays are the simplest type of collection. An array is a block of memory that is big enough to hold everything that needs to be stored. Memory in the computer is laid out with numeric addresses and each element of an array has to be the same size. This means that the location of any element of an array can be found with a few mathematical operations. So if you know where to find the first element of an array, you can find any other with $O(1)$ work.

The list ADT can be implemented using an array for the storage. This is what an `ArrayBuffer` does. When you build a list with an array, the `get` and `set` methods can happen in $O(1)$ operations because they simply do a direct access into the array. This makes them very fast, and if you are going to be calling these a lot on the list, then an array-based list could be ideal. On the other hand, the `insert` and `remove` methods in an array-based list have to copy $O(n)$ elements in order to either make room for a new element or fill in a hole left when an element is removed. More specifically, all the elements after the specified index have to be moved. This is not only $O(n)$ operations, the writing to memory part of the copy tends to be particularly slow on most modern hardware so it is $O(n)$ in a fairly costly operation. If you are always adding to or removing from the end of the list, then the access is much like the stack we wrote previously and everything is $O(1)$ or close to it. If your program will spend a lot of time doing insertions and removes at random locations, or worse, doing them near the beginning of the list, then the array-based version probably is not ideal.

Like the array-based version of the stack and the queue, the list should allow for the array to be larger than the current list with the true size stored in an `Int`. In addition, the array should be grown in multiplicative steps instead of additive ones so that the amortized cost is $O(1)$. The actual construction of such a list is left as an exercise for the reader. Figure 12.1 gives you a illustration of what is going on, and should help you do this exercise.

12.3 Nature of Linked Lists

In some ways, the linked list can be viewed as an alternative to the array. The linked list is strong in the areas where the array is weak, but weak in the areas where the array is strong.

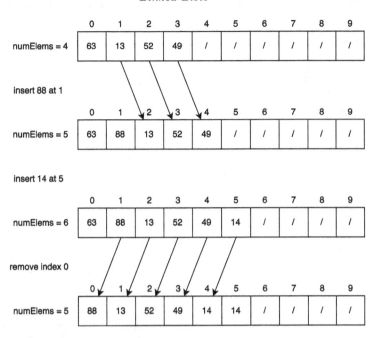

FIGURE 12.1: This figure shows what happens when you do an insert and a remove from an array-based implementation of the list ADT. Note how operations near the beginning have to do many copies, but operations near the end do not.

Instead of having a single large block of memory such that the computer can efficiently access any element if it simply knows the location of the first one, linked lists are composed of nodes that store individual elements which are linked together by references/pointers.[1] The organization in memory is effectively random so the only nodes you can directly reach from one node are the ones it links to. Unlike the array, there is no math formula you can do to find other elements in $O(1)$ operation. There are many different types of linked data structures that have this type of form. A linked list has a sequential structure, and each node will link to the node after it and possibly the node before it. When there is only a link to the next node it is called a singly linked list. When there is a link to both the previous and the next it is called a doubly linked list.

Figure 12.2 shows a graphical representation of a number of different types of linked lists. In this figure, each node is represented by a set of boxes. For a singly linked list, the first box represents a reference to the data stored in that node while the second is a reference to the next node. Arrows show where the references link to. No arrows are shown for the data as that is typically independent of the functionality. In a sense, lists do not care about what they contain. This type of graphical representation can be extremely helpful for understanding the code used to manipulate linked lists.

From the figure, you can see that both singly and doubly linked lists can be set up in regular or circular configurations. In the regular configuration, the end (first or last element) of the list is represented by a **null** reference. In the circular configuration these point back to the other end of the list.

The program must always keep track of at least one element of a linked list. For a

---

[1]Whether the link is a reference or a pointer depends on the language and is mainly a difference in terminology. Scala does not technically have "pointers", but the references in Scala do "point" to locations in memory. The distinction between the two is not significant for this text.

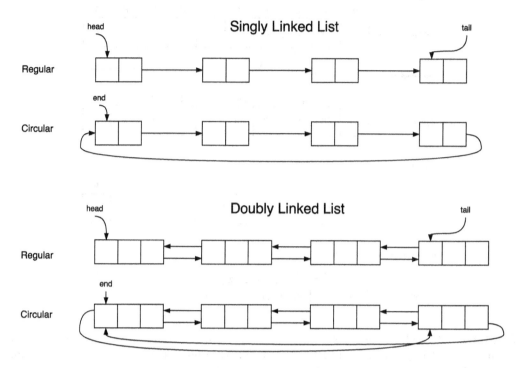

FIGURE 12.2: This figure shows the basic structure of some different forms of linked lists.

regular, singly linked list, this must be the head. All elements of the list are reachable from the head, but because references are one way, if you lose your reference to the head, and you are currently referencing a node, then any nodes that had been in front of the current one referenced will be lost. The figure shows another reference at the other end of the list to a tail. Storing the tail is not generally required, but it can make certain operations significantly faster. In particular, adding to the end of a mutable list is $O(n)$ if you do not keep track of the tail, but can be done in $O(1)$ if you do keep track of the tail. For this reason, we will keep track of the tail in our mutable, singly linked list despite the fact that it will add complexity to the code.

12.4 Mutable Singly Linked List

The first form of linked list that we will implement is the mutable, singly linked list. We will include only the four methods that were listed above. Even though this minimal implementation is not very useful, it is instructive to start here so that the details of other methods do not get in the way of your understanding.

The **class** below shows what such a linked list might look like. The **get** and **set** methods have been changed to **apply** and **update** to take advantage of the Scala syntax rules for these special methods, but the logic inside of them has not changed.

```
1  package linkedlist.adt
2
3  /**
```

```scala
 4    * This is a very simple implementation of a basic linked list.
 5    * @tparam A the type stored in the list.
 6    */
 7   class SinglyLinkedList[A] {
 8     private class Node(var data: A, var next: Node)
 9     private var head: Node = null
10
11     /**
12      * This methods gets the value at a specified index. If the index is beyond the
13      * length of the list a NullPointerException will be thrown.
14      * @param index the index to get.
15      * @return the value at that index.
16      */
17     def apply(index: Int): A = {
18       assert(index >= 0)
19       var rover = head
20       for (i <- 0 until index) rover = rover.next
21       rover.data
22     }
23
24     /**
25      * Sets the value at a particular index in the list. If the index is beyond the
26      * length of the list a NullPointerException will be thrown.
27      * @param index the index to set the value at.
28      * @param data the value to store at that index.
29      */
30     def update(index: Int, data: A): Unit = {
31       assert(index >= 0)
32       var rover = head
33       for (i <- 0 until index) rover = rover.next
34       rover.data = data
35     }
36
37     /**
38      * Inserts a new value at a particular index in the list. If the index is beyond
39      * the length of the list a NullPointerException will be thrown.
40      * @param index the index to set the value at.
41      * @param data the value to insert at that index.
42      */
43     def insert(index: Int, data: A): Unit = {
44       assert(index >= 0)
45       if (index == 0) {
46         head = new Node(data, head)
47       } else {
48         var rover = head
49         for (i <- 0 until index - 1) rover = rover.next
50         rover.next = new Node(data, rover.next)
51       }
52     }
53
54     /**
55      * Removes a particular index from the list. If the index is beyond the length
56      * of the list a NullPointerException will be thrown.
57      * @param index the index to remove.
58      * @return the data value that had been stored at that index.
```

```
59      */
60    def remove(index: Int): A = {
61      assert(index >= 0)
62      if (index == 0) {
63        val ret = head.data
64        head = head.next
65        ret
66      } else {
67        var rover = head
68        for (i <- 0 until index - 1) rover = rover.next
69        val ret = rover.next.data
70        rover.next = rover.next.next
71        ret
72      }
73    }
74  }
```

The **class** takes a type parameter named **A** which represents the type of data the list contains. This **class** contains a small private **class** called **Node** that stores a reference to the data and a reference to the next node. The linked list **class** also keeps track of the head of the list in a variable called **head**.

The four methods in the **class** are all fairly simple and they contain very similar structures. Let us look first at the **apply** method. It begins with an **assert** to ensure that the index is not less than zero. The comments also say that if an index that is beyond the length of the list is provided, a **NullPointerException** will be thrown. While this is sufficient for our current needs, this is not ideal behavior. It would be far better if asking for an index that is out of bounds threw an appropriate exception with information on what was requested and how many there were. This shortcoming will be corrected in the more complete implementation that comes later.

All of the methods in this class declare a variable called **rover** that is used to run through the list. With an array, we move through the elements by counting with an index. Nodes do not have to be consecutive in memory, so we must run through the elements by following references from one node to another. This is what the assignment **rover = rover.next** inside of the **for** loop does.

The **update** method is very similar to the **apply** method in that they both run through a **for** loop moving **rover** forward to find the correct element. The **apply** method gives back the value in the **Node** it stops on. The **update** method sets the value of the data in that **Node**.

To help you understand what happens with **insert** and **remove**, consider what is shown in figure 12.3. In the code, the **insert** and **remove** methods have an element that is not present in the **apply** and **update**, a special case for dealing with the beginning of the list. Both **insert** and **remove** alter references in the list and have to include a special case for when that reference is the first element. The normal case alters a **next** reference, while the special case alters **head**.

The reader should also notice that **insert** and **remove** have **for** loops that only run $n - 1$ times. This is because in a singly linked list, adding or removing a node requires altering the **next** reference of the node before it. If you want to remove the node at index 3, you must have a reference to the node at index 2 and change its **next** reference to point to the node currently at index 4. This is what happens on line 70 where it says **rover.next = rover.next.next**. If **rover** were to go all the way to the node we wanted to remove, we would have no access to the reference that needs to be modified because it is not possible to move backward through a singly linked list.

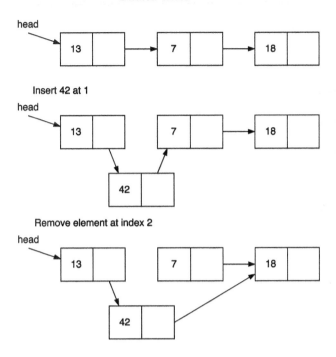

FIGURE 12.3: This figure shows what happens with insert and remove on a singly linked list.

Question: How could you remove duplicate code from this simple implementation of a linked list?

To convince ourselves that this code works, we could implement the following tests that do some minimal checking.

```scala
package test.linkedlist.adt

import org.junit._
import org.junit.Assert._
import linkedlist.adt.SinglyLinkedList

class TestSinglyLinkedList {
  var list: SinglyLinkedList[Int] = null

  @Before def setup: Unit = {
    list = new SinglyLinkedList[Int]
  }

  @Test def appendOne: Unit = {
    list.insert(0, 10)
    assertEquals(10, list(0))
  }

  @Test def appendTwo: Unit = {
    list.insert(0, 10)
    list.insert(0, 20)
```

```
22        assertEquals(20, list(0))
23        assertEquals(10, list(1))
24      }
25
26      @Test def appendTwoB: Unit = {
27        list.insert(0, 10)
28        list.insert(1, 20)
29        assertEquals(10, list(0))
30        assertEquals(20, list(1))
31      }
32
33      @Test def appendTwoUpdate: Unit = {
34        list.insert(0, 10)
35        list.insert(0, 20)
36        list(0) = 5
37        list(1) = 8
38        assertEquals(5, list(0))
39        assertEquals(8, list(1))
40      }
41
42      @Test def appendRemove: Unit = {
43        list.insert(0, 5)
44        list.insert(1, 10)
45        list.insert(2, 15)
46        list.insert(3, 20)
47        assertEquals(5, list(0))
48        assertEquals(10, list(1))
49        assertEquals(15, list(2))
50        assertEquals(20, list(3))
51        list.remove(2)
52        assertEquals(5, list(0))
53        assertEquals(10, list(1))
54        assertEquals(20, list(2))
55        list.remove(0)
56        assertEquals(10, list(0))
57        assertEquals(20, list(1))
58      }
59    }
```

As you should be coming to expect in this text, the tests shown here are not really sufficient to give us confidence in the code. One test that has been left out, which is of particular significance for collections like this, is a test that adds to the list, empties it completely using **remove**, then begins adding again. Such boundary cases are particularly common sources of errors in collections.

12.4.1   Implementing `mutable.Buffer`

The previous example of a linked list was extremely minimal and lacked a lot of the functionality that is really needed to make a collection useful. One way for us to access significantly more functionality without all the additional effort is to build on top of the foundation provided by the `collection` packages in the Scala libraries. In otherwords, we will look at inheriting from them. For a mutable linked list, the obvious choice for a foundation type is the `mutable.Buffer` type.

If you look in the API you will find that `mutable.Buffer` lists the following nine methods as being abstract.

```scala
def +=(elem: A): Buffer.this.type
def +=:(elem: A): Buffer.this.type
def apply(n: Int): A
def clear(): Unit
def insertAll(n: Int, elems: Traversable[A]): Unit
def iterator: Iterator[A]
def length: Int
def remove(n: Int): A
def update(n: Int, newelem: A): Unit
```

This means that if we want to extend the `mutable.Buffer` type, we must implement these nine methods. The reward for doing this is that the large number of other methods that are part of `mutable.Buffer` will automatically work for us. This happens because those other methods are implemented in terms of the nine abstract methods. So for the cost of nine you get many, many others.

The following code shows an implementation of just such a class. This implementation keeps track of both a head and a tail for the list. They are called `hd` and `tl` so that they do not conflict with the `head` and `tail` members that are part of the general Scala sequence. The class also keeps track of an `Int` named `numElems` so that calls to `length` and bounds checking can both be done efficiently.

```scala
package linkedlist.adt

import collection.mutable

class MutableSLList[A] extends mutable.Buffer[A] {
  private var hd: Node = null
  private var tl: Node = null
  private var numElems = 0
  private class Node(var data: A, var next: Node)

  def +=(elem: A) = {
    if (tl == null) {
      tl = new Node(elem, null)
      hd = tl
    } else {
      tl.next = new Node(elem, null)
      tl = tl.next
    }
    numElems += 1
    this
  }

  def +=:(elem: A) = {
    hd = new Node(elem, hd)
    if (tl == null) {
      tl = hd
    }
    numElems += 1
    this
  }

```

```scala
32    def apply(n: Int): A = {
33      if (n < 0 || n >= numElems) throw new IndexOutOfBoundsException(n+" of
            "+numElems)
34      var rover = hd
35      for (i <- 0 until n) rover = rover.next
36      rover.data
37    }
38
39    def clear(): Unit = {
40      hd = null
41      tl = null
42      numElems = 0
43    }
44
45    def insertAll(n: Int, elems: Traversable[A]): Unit = {
46      if (n < 0 || n >= numElems + 1) throw new IndexOutOfBoundsException(n+" of
            "+numElems)
47      if (elems.nonEmpty) {
48        val first = new Node(elems.head, null)
49        var last = first
50        numElems += 1
51        for (e <- elems.tail) {
52          last.next = new Node(e, null)
53          last = last.next
54          numElems += 1
55        }
56        if (n == 0) {
57          last.next = hd
58          hd = first
59          if (tl == null) tl = last
60        } else {
61          var rover = hd
62          for (i <- 1 until n) rover = rover.next
63          last.next = rover.next
64          rover.next = first
65          if (last.next == null) tl = last
66        }
67      }
68    }
69
70    def iterator = new Iterator[A] {
71      var rover = hd
72      def hasNext = rover != null
73      def next: A = {
74        val ret = rover.data
75        rover = rover.next
76        ret
77      }
78    }
79
80    def length: Int = numElems
81
82    def remove(n: Int): A = {
83      if (n < 0 || n >= numElems) throw new IndexOutOfBoundsException(n+" of
            "+numElems)
```

```
84    numElems -= 1
85    if (n == 0) {
86      val ret = hd.data
87      hd = hd.next
88      if (hd == null) tl = null
89      ret
90    } else {
91      var rover = hd
92      for (i <- 1 until n) rover = rover.next
93      val ret = rover.next.data
94      rover.next = rover.next.next
95      if (rover.next == null) tl = rover
96      ret
97    }
98  }
99
100   def update(n: Int, newelem: A): Unit = {
101     if (n < 0 || n >= numElems) throw new IndexOutOfBoundsException(n+" of
          "+numElems)
102     var rover = hd
103     for (i <- 0 until n) rover = rover.next
104     rover.data = newelem
105   }
106
107   override def toString = mkString("MutableSLList(", ", ", ")")
108 }
```

Comments on public methods have been omitted as they would mirror those from the main API.

As with the previous list, `insert` and `remove` have special cases in them for handling the end cases. These are situations where either the head or the tail of the list is modified. There is more special case code here, because a reference to the tail is included. The primary motivation for including this reference is that the `+=` method for appending to the list is $O(1)$ when we have a tail and we should expect this to be called fairly regularly. Without a tail reference, this call would be $O(n)$.

The other method that is interesting is the `insertAll` method. The way in which it works might seem a bit odd. It starts by running through the `Traversable` of elements to add, building a separate linked list. While it does this, it keeps track of the first and last elements. When it is done, it inserts that linked list into the original linked list. This example shows a great strength of linked lists: you can insert whole lists into other lists in $O(1)$ operations, assuming that you have a reference to where you want the inserted list to go.

12.4.2 Iterators

This code also includes an implementation for the `iterator` method. This method is supposed to give us back an `Iterator` that can be used to run through the list efficiently. Using a loop with an index and `apply` to go through a linked list is not efficient. To be more specific, running through a linked list in that way requires $O(n^2)$ operations. This is because `apply` runs through the entire list up to the requested index. So the first request has to take 0 steps, but the next takes 1 and the third takes 2. This continues up to the last call which has to run through $n - 1$ links. The sum of these is $n(n - 1)/2$ which is $O(n^2)$. Using an `Iterator` provides a way for any outside code to walk through the elements of the list in $O(n)$ operations. An `Iterator` has methods called `next` and `hasNext` that let us walk through the elements. These are the only things we have to write to create our own `Iterator`. In the implementation of our list, they look much like the code we use for the loops that are internal to the class. We have a `rover` that begins at the head. The `hasNext` method simply checks if `rover` has gotten to the `null` at the end of the list. The `next` method moves `rover` to be `rover.next` and returns the data `rover` had been referencing.

Fitting the Scala Collection Libraries

If you play a bit in the REPL with the code that is shown here for the mutable singly linked list, you will find that while it works correctly, there are a few behaviors that are less than ideal. In particular, if you call methods like `map` or `filter`, the resulting collection will be an `ArrayBuffer`. This is not the behavior you have come to expect from Scala collections where these methods tend to return the same type they operate on. This happens because we did not write those methods, and we are relying on the implementations in the `Buffer` supertype to provide them. The `Buffer` type defaults to giving back an `ArrayBuffer` so that is what we see here.

If the solution to this required rewriting all of the methods that return collections, it would greatly reduce the benefit of extending the `Buffer trait`. The Scala collections library has been designed so that this is not required, but getting these methods to return a different type is still a bit challenging. To make your collection really fit in with the main collections library, there are a few things that need to be done. Instead of just extending the `Buffer trait`, we also extend the `BufferLike trait`. The `BufferLike trait` is one of a number of `traits` that end in `Like` that can be used to add new collections types onto the libraries.

The `BufferLike trait` takes two parameters. One is the type of the values in the collection and the other is the collection type. In this case, `MutableSLList[A]`. That second type parameter tells it what should be returned for calls that create a new collection, such as `filter` and `map`. Beyond this, things get a bit more complex. We need to specify a way for these types to be built. For our code, this leads to inheriting from `GenericTraversableTemplate`, making a companion object that is a `SeqFactory`, and making our class inherit from `Builder` as well. Putting it all together, the changes look like this.

```scala
import collection.mutable
import collection.generic._

class MutableSLList[A] extends mutable.Buffer[A]
    with GenericTraversableTemplate[A, MutableSLList]
    with mutable.BufferLike[A, MutableSLList[A]]
    with mutable.Builder[A, MutableSLList[A]] {

  ...

  override def companion: GenericCompanion[MutableSLList] = MutableSLList

  def result() = this

  ...
}
```

```
object MutableSLList extends SeqFactory[MutableSLList] {
  implicit def canBuildFrom[A]: CanBuildFrom[MutableSLList[_], A,
      MutableSLList[A]] =
    new GenericCanBuildFrom[A]
  def newBuilder[A]: mutable.Builder[A, MutableSLList[A]] =
    new MutableSLList[A]
}
```

This begins to pull in aspects of the Scala language that are mainly important for library writers. As the goal of this book is not to turn you into a proficient Scala library writer, we will not be putting this type of functionality into future collections as it obfuscates the concepts that you should be learning. Instead, we will add our own versions of `foreach`, `map`, and `filter` to a later collection because you should understand how those work.

12.5 Mutable Doubly Linked List

Now we want to consider a doubly linked list. The obvious advantage of a doubly linked list over a singly linked list is that it can be traversed in either direction. This comes with the cost of an extra reference per node and extra code to handle it. However, as we saw in the last section, the main source of additional code was not handling the next references, but dealing with the special cases for head and tail references. Fortunately, there is a way we can remove that special case code when we are using a doubly linked list so that handling the previous reference is more than compensated for. The way we do this is through the introduction of a special node called a SENTINEL.

A sentinel is an extra node that does not store data, but which functions as both the beginning and end of the list. To make this possible, the list needs to be circular. If we call the sentinel `end` then `end.next` is the head of the list and `end.prev` is the tail of the list. This alteration saves a significant amount of code because now the head and tail are just `next` and `prev` references, so adding and removing at those locations is no different from doing so at other locations.

Code for a doubly linked list that implements the `Buffer` trait is shown here. The declarations at the top of the class are very similar to those in the singly linked list, except that there is only one node reference called `end` and initially the `next` and `prev` of that node are set to point back to it. There are two other points to note about the top part of this class. The `end` node has to be given a data value. This causes a bit of a problem, because there is not any single value that can be used as the default for the type parameter A. To get around this, we create a **private var** with a default value for type A that we call `default` on line 6. This is done using a special syntax where the value is represented as an underscore. Note that we have to specify the type of the variable in order for this to be meaningful. When the sentinel is created, we pass it `default` as the value to store. There are other solutions to this that use inheritance to make a sentinel that does not store any value, but they require more code and potentially obfuscate the purpose of this example.

```
package linkedlist.adt

import collection.mutable
```

```scala
 4   import scala.collection.generic.CanBuildFrom
 5
 6   class MutableDLList[A] extends mutable.Buffer[A] {
 7     private var default: A = _
 8     private var numElems = 0
 9     private class Node(var data: A, var prev: Node, var next: Node)
10     private val end = new Node(default, null, null)
11     end.next = end
12     end.prev = end
13
14     def +=(elem: A) = {
15       val newNode = new Node(elem, end.prev, end)
16       end.prev.next = newNode
17       end.prev = newNode
18       numElems += 1
19       this
20     }
21
22     def +=:(elem: A) = {
23       val newNode = new Node(elem, end, end.next)
24       end.next.prev = newNode
25       end.next = newNode
26       numElems += 1
27       this
28     }
29
30     def apply(n: Int): A = {
31       if (n < 0 || n >= numElems) throw new IndexOutOfBoundsException(n+" of
               "+numElems)
32       var rover = end.next
33       for (i <- 0 until n) rover = rover.next
34       rover.data
35     }
36
37     def clear(): Unit = {
38       end.prev = end
39       end.next = end
40       numElems = 0
41     }
42
43     def insertAll(n: Int, elems: Traversable[A]): Unit = {
44       if (n < 0 || n >= numElems + 1) throw new IndexOutOfBoundsException(n+" of
               "+numElems)
45       if (elems.nonEmpty) {
46         var rover = end.next
47         for (i <- 0 until n) rover = rover.next
48         for (e <- elems) {
49           val newNode = new Node(e, rover.prev, rover)
50           rover.prev.next = newNode
51           rover.prev = newNode
52           numElems += 1
53         }
54       }
55     }
56
```

```scala
57    def iterator = new Iterator[A] {
58      var rover = end.next
59      def hasNext = rover != end
60      def next: A = {
61        val ret = rover.data
62        rover = rover.next
63        ret
64      }
65    }
66
67    def length: Int = numElems
68
69    def remove(n: Int): A = {
70      if (n < 0 || n >= numElems) throw new IndexOutOfBoundsException(n+" of
                "+numElems)
71      numElems -= 1
72      var rover = end.next
73      for (i <- 0 until n) rover = rover.next
74      val ret = rover.data
75      rover.prev.next = rover.next
76      rover.next.prev = rover.prev
77      ret
78    }
79
80    def update(n: Int, newelem: A) {
81      if (n < 0 || n >= numElems) throw new IndexOutOfBoundsException(n+" of
                "+numElems)
82      var rover = end.next
83      for (i <- 0 until n) rover = rover.next
84      rover.data = newelem
85    }
86
87    override def foreach[U](f: A => U): Unit = {
88      var rover = end.next
89      while(rover != end) {
90        f(rover.data)
91        rover = rover.next
92      }
93    }
94
95    override def filter(pred: A => Boolean): MutableDLList[A] = {
96      val ret = new MutableDLList[A]()
97      var rover = end.next
98      while(rover != end) {
99        if (pred(rover.data)) ret += rover.data
100       rover = rover.next
101     }
102     ret
103   }
104
105   def myMap[B](f: A => B): MutableDLList[B] = {
106     val ret = new MutableDLList[B]()
107     var rover = end.next
108     while(rover != end) {
109       ret += f(rover.data)
```

```
110      rover = rover.next
111    }
112    ret
113  }
114
115  override def toString = mkString("MutableDLList(", ", ", ")")
116 }
```

The other methods should be fairly straightforward after you have seen the singly linked list. The biggest change is that there are two pointers that have to be assigned for each node-node link and that all the special case code is gone. On the whole, this version is shorter, less complex, and less error prone due to the removal of the special case code. The reader should draw pictures to see the effect of code like `rover.prev.next = rover.next` on line 74. This particular line in `remove` takes advantage of the fact that doubly linked lists can be traversed forward and backward so a reference to the node that we are removing is sufficient. We do not need to have a direct reference to the prior node as was required for the singly linked list.

Another aspect that needs to be noted about working with linked lists, especially the doubly linked lists, is that the order of assignment for the links can be very important. Consider lines 49 and 50 in `insertAll`. If you flip the order of those two lines, the code breaks. The reason for this is that line 50 changes `rover.prev` and line 49 uses `rover.prev`. So the order of how you change things can be very significant. We said before that it helps to draw pictures for your linked lists. When you do this, make sure that you change the arrows in the same order that you are doing it in the code. That will help you figure out what is going wrong if you mess something up.

As you go through the code for the doubly linked list, consult figure 12.4 and imagine how the arrows would move around for each line of the code. Note that an expression like `new Node(e, rover.prev, rover)` not only makes a new node, it also sets that node to point to two other nodes.

At the end of this `class` on lines 87–113 we added implementations of `foreach`, `filter`, and a custom version of `map`. The goal here is to show you how simple these methods can be. They all have a loop that runs through the collection using a rover. They all take functions that are applied to each of the elements. How they differ is in what they do with the result of those function applications. The `foreach` method simply calls the function and does nothing with the result. The `filter` method checks if the result is true, and if so it adds that element to a new list that is returned. The `map` adds all the results of the function calls to a new list. Note that we call our method `myMap` because the actual type signature of the `map` method is more complex than what we want to deal with here. The complexity in the libraries deals with the details of the result type and the fact that not all types are efficient to build. Fortunately, the `MutableDLList class` is efficient to build using `+=`, so we can have a simple implementation.

Sentinels through Inheritance

Making a `private var` to store a default value that will then be used as a data value in the sentinel, is less than ideal from a stylistic standpoint. There is an alternative. Using inheritance, we can make a special node type to represent the sentinel, which does not have data in it. Alternate code that uses that approach is shown here.

```
class MutableDLList2[A] extends mutable.Buffer[A] {
```

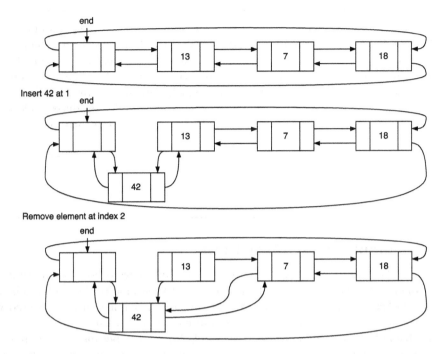

FIGURE 12.4: This figure shows what happens with insert and remove on a doubly linked list.

```
private var numElems = 0
private trait Node {
  var data: A
  var prev: Node
  var next: Node
}
private class DNode(var data: A, var prev: Node, var next: Node) extends Node
private class ENode extends Node {
  def data: A = throw new IllegalArgumentException("Requested value of
      sentinel node.")
  def data_=(d: A) = throw new IllegalArgumentException("Value assigned to
      sentinel node.")
  var next: Node = this
  var prev: Node = this
}
private val end = new ENode
```

Using this approach, the Node type is now a **trait** with subtypes called DNode and ENode. This Node defines abstract **vars** for the different fields we had in Node earlier. The DNode type inherits from this and otherwise looks just like our original Node class. Where this approach differs is in the ENode for the sentinel. This class uses regular **var** declarations for **next** and **prev**, but it handles **data** quite differently. Instead of declaring a **var**, there are two **def** declarations for getting and setting **data**. Both of these throw exceptions instead of returning a value. This works because a **var** declaration actually compiles to these getter and setter methods and the **throw** expression has the type

Nothing, which is a subtype of every other type in Scala. The three points in the code that did **new Node** before would be updated to **new DNode** to get this to compile.

The choice between these approaches is mostly one of style, but it is possible that the use of inheritance here could produce slightly slower code when it is compiled. Inherited methods use virtual function calls, which have more overhead than calling a **final** method that nothing overrides. Whether the compiler produces code that is slower in this case depends on the compiler. Testing this on your own platform is left as an exercise for the reader.

12.6   Immutable Singly Linked List

The two linked list implementations we have looked at were both for mutable linked lists. We could instantiate instances of these lists and then modify the contents in various ways. This worked well with the **Buffer trait**, but the first linked list we worked with in Scala was the **List** class, which represents an immutable linked list.

Immutable linked lists only make sense as singly linked lists because you are not allowed to change data or links. Once a list is created, it must remain completely intact and unchanged. Adding to the front of the list, an operation that we have done with : :, appends a new node without altering the ones after it in any way. Unlike the mutable linked lists, where the list wraps around the nodes and hides them from view, in an immutable linked list, the nodes are the list. There is nothing wrapped around them.

The code below gives an implementation of an immutable, singly linked list. This uses a design similar to that for the **List class** itself. Instead of being a single **class**, it is a small hierarchy with the primary **class** we normally use to refer to the list being abstract and two subtypes to represent empty and non-empty nodes.

```scala
package linkedlist.adt

import collection.immutable.LinearSeq

sealed abstract class ImmutableSLList[+A] extends LinearSeq[A] {
  def ::[B >: A](elem: B): ImmutableSLList[B] = new Cons(elem, this)

  override def iterator = new Iterator[A] {
    var rover: LinearSeq[A] = ImmutableSLList.this
    def hasNext = !rover.isEmpty
    def next: A = {
      val ret = rover.head
      rover = rover.tail
      ret
    }
  }
}

final class Cons[A](hd: A, tl: ImmutableSLList[A]) extends ImmutableSLList[A] {
  def length = 1 + tl.length
  def apply(index: Int): A = if (index == 0) hd else tl.apply(index - 1)
  override def isEmpty = false
```

```
23    override def head = hd
24    override def tail = tl
25  }
26
27  object MyNil extends ImmutableSLList[Nothing] {
28    def length = 0
29    def apply(index: Int) = throw new IllegalArgumentException("Nil has no contents.")
30    override def isEmpty = true
31    override def head = throw new IllegalArgumentException("Nil has no head.")
32    override def tail = throw new IllegalArgumentException("Nil has no tail.")
33  }
34
35  object ImmutableSLList {
36    def apply[A](data: A*): ImmutableSLList[A] = {
37      var ret: ImmutableSLList[A] = MyNil
38      for (d <- data.reverse) {
39        ret = new Cons(d, ret)
40      }
41      ret
42    }
43  }
```

There are a few elements of this code that we have not seen before that could use some explanation. Starting at the top, the main **class** is labeled as **sealed** and **abstract**. We have seen **abstract** before, but **sealed** is new. What this keyword implies is that all the subtypes of this **class** have to appear in the current file. This can be used to provide better type checking for patterns in **match**. In this case it provides us with the security of knowing that other programmers cannot produce undesired subtypes that behave in ways different from that desired for our list. Often we can enforce that by using **final** to make sure that nothing inherits from a class. In this case that is not an option because we need to have some inheritance, but there are restrictions we want to place on the subtype. In this situation **sealed** provides us with safety and the ability to use inheritance.

Next there is the type parameter for **ImmutableSLList**. It is +A instead of just A. The plus sign indicates that the list is COVARIANT in the type parameter. This is a topic that is covered in more detail in the on-line appendices.[2] For now it is sufficient to say that this means that if B is a subtype of A, then **ImmutableSLList[B]** is a subtype of **ImmutableSLList[A]**. Without the plus sign, the type parameters are INVARIANT and types with different type parameters produce types that are effectively unrelated. The **Array** and **Buffer** types are invariant. So you cannot use a **Buffer[Rectangle]** in code that accepts a **Buffer[Shape]**. On the other hand, List, like the **ImmutableSLList** shown here, is covariant, so List[Rectangle] is a subtype of List[Shape] and code that accepts a List[Shape] will accept a List[Rectangle].

The first method in the **class** is called :: to mimic that used in the normal **List** class. The odd aspect of this is the bounds on the type parameter. This is the first usage we have seen for the >: type bounds. This indicates that we can use any type, B, as long as it is the same as A or a supertype of A. This notation was mentioned in subsection 4.3.3, but we did not have a use for it until now. To see the effect that this can have, consider the following code using the standard **List** in the REPL.

```
scala> val nums = List(1,2,3)
nums: List[Int] = List(1, 2, 3)
```

---

[2]The appendices are available at http://www.programmingusingscala.net.

```
scala> "hi" :: nums
res0: List[Any] = List(hi, 1, 2, 3)
```

Note the type of res0. It is a List[Any]. Our list will behave in this same way. The new element, "hi" is not an Int, so Scala has to figure out a type B to use. The only type that is a supertype of both String and Int is Any, so that is what is used. If there were a common supertype lower in the inheritance hierarchy, that would be used instead.

In the iterator method, the declaration of rover needs an explicit type because rover.tail is a method of the supertype. It is also interesting to note that the rover is initialized to be the current node. This is a side effect of the fact that in this implementation, the nodes are not hidden in a list, they truly represent a list. In fact, each node represents a list either as the empty list or the head of some list.

Following the definition of the main class are the two subtypes. The first represents a non-empty node. It is called Cons here and it keeps track of the data element in that node as well as a node for the rest of the list. To complete the LinearSeq, the class has to implement length and apply. To make the implementation more efficient for standard usage, we also overload isEmpty, head, and tail. These same methods are put in the object MyNil, which represents an empty list. This is an object because we only need one of them to exist. It extends ImmutableSLList[Nothing] so that it can be a subtype of any list type. This is where the covariance of ImmutableSLList with respect to A becomes important. Without that, a separate empty list object would be needed for every type of list.

The code ends with a companion object that has an apply method. This makes it possible to create our new list type using the same syntax that we are accustomed to with List.

Some test code works well to demonstrate the usage of our list. You can see that it is very much like the normal List type in how it functions.

```
1   package test.linkedlist.adt
2
3   import org.junit._
4   import org.junit.Assert._
5   import linkedlist.adt.MyNil
6   import linkedlist.adt.ImmutableSLList
7
8   class TestImmutableSLList {
9     @Test def nilIsEmpty: Unit = {
10      assertTrue(MyNil.isEmpty)
11    }
12
13    @Test def buildCons: Unit = {
14      val lst = 1 :: 2 :: 3 :: MyNil
15      assertEquals(3, lst.length)
16      assertEquals(1, lst.head)
17      assertEquals(2, lst(1))
18      assertEquals(3, lst(2))
19    }
20
21    @Test def buildApply: Unit = {
22      val lst = ImmutableSLList(1, 2, 3)
23      assertEquals(3, lst.length)
24      assertEquals(1, lst.head)
25      assertEquals(2, lst(1))
26      assertEquals(3, lst(2))
27    }
```

28 `}`

One thing that our code will not do is pattern matching on `::` as that requires more advanced techniques.

The actual `scala.collection.immutable.List` code is more complex than what we have written here, but mainly to introduce further efficiencies. This class is completely immutable in that no node can ever change after it is created. This forces us to build them by appending to the head if we want to have any efficiency at all. This appears in our code at the end when the input to the `apply` method is reversed on line 38 in the `for` loop. This is one type of overhead which the true library version manages to get around.

12.7 Linked List-Based Stacks and Queues

Recall that the primary characteristic of an ADT is that it tells us what different methods do, not how they are done. The linked list is a specific implementation and, as we have seen, it can be used to implement the list ADT. It can also be used as the storage mechanism for other ADTs. In particular, it works well for the only other ADTs that have been considered so far, the stack and the queue. As with the array-based implementation, we will require that all the operations on these collections be $O(1)$ because they are fundamentally simple.

12.7.1 Linked List-Based Stack

To implement a stack using a linked list, we could put an instance of one of the lists that we wrote in a previous section in this chapter as a member and perform operations on that. Such an approach would have value from a software engineering standpoint where code reuse can significantly boost productivity. However, the focus of this chapter is on the nature and structure of linked lists. For that purpose, there is value to building things from scratch and including a node type and code to manipulate it directly in this class. Such an approach also helps us to have greater control over the efficiency of what we write.

The operations on a stack just need to be able to add and remove from the same location in a collection. On the remove, it needs to get fast access to the next most recently added element. That combination of operations is efficiently implemented by a singly linked list. The push adds a new element at the head of the list and the pop element removes the head.

An implementation of a stack using a singly linked list is shown here. The `Node` type is made using a `case class` because the values in it do not change and the `case class` provides an easy way to make an immutable node.

```
1   package linkedlist.adt
2
3   class ListStack[A] extends Stack[A] {
4     private case class Node(data: A, next: Node)
5     private var top: Node = null
6
7     def push(obj: A): Unit = {
8       top = Node(obj, top)
9     }
10
11    def pop(): A = {
12      assert(!isEmpty, "Pop called on an empty stack.")
```

```scala
13      val ret = top.data
14      top = top.next
15      ret
16    }
17
18    def peek: A = top.data
19
20    def isEmpty: Boolean = top == null
21  }
```

The `top` member is a `var` that serves as the head of the list. The implementation of the methods are straightforward. As you can see, this is a very short `class`, even though we are re-implementing a singly linked list here. The simplicity of the implementation is possible because of the simplicity of the stack ADT. No loops or complex logic is required because all access, both adding and removing, happens at the top.

Test code for this class is identical to that for `ArrayStack` with the word `Array` changed to `List`. Only one of those substitutions is functional instead of cosmetic, the instantiation of the stack in the `@Before` method.

12.7.2   Linked List-Based Queue

The same procedure can be used to create a queue that uses a linked list for the data storage. For the stack we only needed to keep one end of the list that we could add to and remove from. In the case of a queue, we need to keep track of both ends of the list. One of them will be the front and the other will be the back. Which is the head and which is the tail is determined by the functionality that we need. New elements should be added at the back and removed at the front. We just saw that the head can do both of these operations efficiently. What about the tail?

In the case of a singly linked list, we can only add efficiently at the tail, we cannot easily remove from there. This goes back to that problem that in order to remove from a singly linked list, you need to know the node in front of the one you are removing and we cannot walk backward up the list. For this reason, the back of the queue must be the tail of the list if we want to maintain $O(1)$ performance.

Code for a list-based queue is shown here. There are a few more special cases that have to be dealt with which make this code longer than that for the stack. In particular, there are adjustments to the back that are required when an element is added to an empty queue or a remove takes out the last element from the queue.

```scala
1   package linkedlist.adt
2
3   class ListQueue[A] extends Queue[A] {
4     private class Node(val data: A, var next: Node)
5     private var front: Node = null
6     private var back: Node = null
7
8     def enqueue(obj: A): Unit = {
9       if (front == null) {
10        front = new Node(obj, null)
11        back = front
12      } else {
13        back.next = new Node(obj, null)
14        back = back.next
15      }
```

```
16    }
17
18    def dequeue(): A = {
19      assert(!isEmpty, "Dequeue called on an empty queue.")
20      val ret = front.data
21      front = front.next
22      if (front == null) back = null
23      ret
24    }
25
26    def peek: A = front.data
27
28    def isEmpty: Boolean = front == null
29  }
```

The GitHub repository, https://github.com/MarkCLewis/OOAbstractDataStructScala, includes basic test code for all the classes shown in this chapter, including those for which tests are not shown in the text.

12.8 Project Integration

There is no specific code presented to integrate the linked lists with the drawing project. However, it is not hard to see how it could be used in that context. In particular, the DrawTransform keeps a Buffer of the children. It is a fairly straightforward proposition to change that to use a subtype of Buffer that we have written here.

12.9 End of Chapter Material

12.9.1 Summary of Concepts

- A slightly more complex ADT than the stack or queue is the List/Seq ADT. This ADT provides methods for random access of elements.

- The Arrays that we have been working with are stored as single blocks of memory that can hold the required number of references. This means when you know where the block is, you know where every element is. However, inserting and removing is slow.

- Linked lists are an alternative approach to storing values in a sequential manner. Instead of having a large block, there are nodes that link to one or two other nodes. The list keeps track of at least one node and other nodes can be reached from that one. This makes random access slow, but inserting and removing can be done with $O(1)$ memory moves.

- In a singly linked list, each node keeps a reference to the next node in the list and the list must keep track of the first node. It can also keep track of the last node. These boundaries become special cases that make the code more complex.

- A doubly linked list has nodes that keep references to both the next element and the previous element. They are best coded with a sentinel node that stores no data and is effectively the first and last node in the list. The sentinel keeps the code for a doubly linked list short, without special cases.

- While random access is slow for linked lists, iterators allow efficient linear access.

- A singly linked list can also be written to be immutable. This is what the `List` type does. They are generally written where the node itself represents a list and there is a special `object` for an empty list.

- The fact that an ADT does not specify how things are done can be illustrated by reimplementing the stack and queue ADTs using linked lists as the mechanism for storing data.

12.9.2 Exercises

1. Write an array-based implementation of the list ADT. You can choose to implement just the four methods listed in the first section of this chapter or implement the `mutable.Buffer trait`.

2. Compare speed of your array-based list and a linked list using the following set of manipulations.

    - Add at the end, randomly read elements, randomly set elements, remove from the end.
    - Add at random locations, randomly read elements, randomly set elements, remove from random locations.
    - Add at the beginning, randomly read elements, randomly set elements, remove from the beginning.
    - Add at random locations, read sequential elements with an iterator, remove from random locations.

3. Compare speed of doubly linked lists using the two different node types using the same operation as above.

4. Test `MutableDLList2` on your own platform.

5. Early in the chapter you saw code written to implement a singly linked list. How could you remove duplicate code from this simple implementation of a linked list?

6. Add more tests to the linked list test code. Make sure you include the case where the list is completely emptied using remove and then refilled.

7. Compare the speed of the linked list-based stack to the array-based stack for different sizes and usage modes.

8. Compare the speed of the linked list-based queue to the array-based queue for different sizes and usage modes.

9. Alter the code for a linked list so that you pull out the node walking using `rover` into a single function and call that function in places where that behavior is needed.

12.9.3 Projects

For this chapter there are not separate descriptions for each of the projects. In all of them, you should already be using some form of `Seq` or `Buffer` for storing significant data. For this project, you need to convert your code to use a linked-list class of your construction instead of the Scala library classes you have been using.

Chapter 13

Priority Queues

The queue ADT that we looked at previously can be thought of as modeling a service line. All the entities that are added in are considered equal, and they are taken off in the order they arrive. Not every system works in such an equitable way. In real life there are things like VIP lines. Some systems have a serious reason to treat entities differently. An example of such a system is an emergency room. If you show up to an ER with a paper cut, you can expect to wait for a very long time. This is because the order in which people are served in a ER is based much more on the seriousness of the person's condition than on the time of arrival. Situations where people set up appointments are another example. Arriving long before your appointment time generally does not lead to you being seen significantly earlier.

These types of systems call for a new ADT, the priority queue. A priority queue has the same methods as a normal queue, but there is a different behavior for the dequeue and peek methods in that they give the highest-priority object on the queue.

13.1 Two Approaches

We will consider two different approaches to writing a priority queue in this chapter, both use a linked list for storage. The difference between the two is whether extra work is done when new items are added to the queue or if that work is put off until the items are removed. We will discuss both and compare their efficiency, but only code the one that is better. Unlike the normal queue, it is not possible to get $O(1)$ performance for all the methods in a priority queue. For the implementations we will talk about here, some methods will be $O(n)$. We will discuss another implementation that can improve on that method in chapter 18.

All implementations of the priority queue will extend the following trait.

```
package priorityqueues.adt
```

```
/**
 * This trait defines a mutable Priority Queue ADT.
```

```
 * @tparam A the type of data stored
 */
trait PriorityQueue[A] {
  /**
   * Add an item to the priority queue.
   * @param obj the item to add
   */
  def enqueue(obj: A): Unit

  /**
   * Remove the item that has the highest priority or, in the case of a tie,
   * has been on the longest.
   * @return the item that was removed
   */
  def dequeue(): A

  /**
   * Return the item that has the highest priority or, in the case of a tie,
   * has been on the longest without removing it.
   * @return the most recently added item
   */
  def peek: A

  /**
   * Tells whether this priority queue is empty.
   * @return true if there are no items on the priority queue, otherwise false.
   */
  def isEmpty: Boolean
}
```

13.1.1 Searching by Priority

One way to create a priority queue using a linked list is to add elements to the tail end just like a regular queue, but then alter the code in **dequeue** and **peek** so that they run through the elements to find the first one with the highest-priority. In the case of **dequeue**, that one is then removed.

We can evaluate the quality of this approach by looking at the amount of work done by each of the methods. The **enqueue** and **isEmpty** methods will be $O(1)$. However, both **dequeue** and **peek** have to do a search through all the elements to find the one with the highest-priority. This requires $O(n)$ time. Looking closer than just order, this search is particularly costly because it is impossible to know what we are looking for early on so every element must be considered for each search.

13.1.2 Sorted Linked List

A second approach is to do the additional work when elements are added, and keep the list in sorted order so that we always know where the highest-priority item is and we can remove it efficiently. Using this approach, the **dequeue**, **peek**, and **isEmpty** methods are all $O(1)$. The **enqueue** method, however, must perform a search to find the correct insertion locations, and is therefore $O(n)$.

Both of these approaches contain at least one method which is $O(n)$. As most applications will have roughly equal numbers of calls to **enqueue** and **dequeue**, and the number of calls will be at least $O(n)$ as well, we can expect either of these implementations to have an overall performance of $O(n^2)$. This second approach is better for most applications for a few reasons. The first is that calls to **peek** are $O(1)$. While there is a logical reason why calls to **enqueue** and **dequeue** should roughly balance, there could easily be more calls to **peek** than either of the other two. In addition, placing items into a sorted list does not require walking the full list. This procedure is like what happens in the inner loop of an insertion sort. You only have to walk through the list until you find the right place to put the new element. For many applications, new elements will have a naturally lower priority than many previous elements. This means that the code will do comparisons to fewer than half the elements, on average, for each **enqueue**, compared to comparisons of all elements for **dequeue** and **peek** in the other approach.

Here is a listing of a sample implementation of a priority queue using a sorted linked list.

```scala
package scalabook.adt

import reflect.ClassTag

class SortedListPriorityQueue[A : ClassTag](comp: (A,A)=>Int) extends
    PriorityQueue[A] {
  private var default: A = _
  private class Node(var data: A, var prev: Node, var next: Node)
  private val end = new Node(default, null, null)
  end.next = end
  end.prev = end

  def enqueue(obj:A): Unit = {
    var rover = end.prev
    while (rover != end && comp(obj,rover.data) > 0) rover = rover.prev
    rover.next.prev = new Node(obj, rover, rover.next)
    rover.next = rover.next.prev
  }

  def dequeue(): A = {
    val ret = end.next.data
    end.next = end.next.next
    end.next.prev = end
    ret
  }

  def peek: A = end.next.data

  def isEmpty: Boolean = end.next==end
}
```

Note that a comparison function is passed in at creation so that any type can be used. This implementation uses a doubly linked list with a sentinel to reduce special cases and keep the code compact. The highest-priority items are kept at the head of the list, **end.next**, and this is where **dequeue** pulls from. The **enqueue** method runs from the tail of the list backward until it gets to the sentinel or finds a node with higher-priority data. Note that if each new item had a lower priority than the one before it, this version of **enqueue** would be $O(1)$.

The test code for this **class** can look just like that for the regular queue with a few minor changes. In addition to changing the type, there is a comparison function required for the priority queue at creation. Here we can use the `compareTo` method that is part of `Int` and most other built-in types in the library that have a natural ordering. This makes higher values have a higher priority. If you were in a situation where lower values should have the higher priority, then you could use `b.compareTo(a)` or `-a.compareTo(b)` instead.

```scala
var queue: PriorityQueue[Int] = null

@Before def initQueue: Unit = {
  queue = new SortedListPriorityQueue[Int]((a,b) => a.compareTo(b))
}

@Test def enqueue100Dequeue100: Unit = {
  val nums = Array.fill(100)(util.Random.nextInt)
  nums.foreach(queue.enqueue)
  nums.sorted.reverse.foreach(assertEquals(_, queue.dequeue))
}
```

Using the priority queue also requires that the code checking the dequeue order needs to be altered. The last method shown here was one that uses a queue of random values. The checks can be done against a reversed, sorted version of the array because the comparison made higher values have higher priority and they will come off first.

13.1.3   Problems with Arrays

For both the regular stack and the regular queue, it was possible to write an implementation using an array instead of a linked list utilizing a similar amount of code and having very similar performance. This is not the case for the priority queue using either of these approaches, because they involve either insertion or deletion from random locations in the array, which requires doing $O(n)$ copies to move elements around.[1]

In one sense, the array-based implementation is still $O(n)$ in the same methods. However, the type of operations they are $O(n)$ in has changed. The linked list approaches are $O(n)$ in comparisons and memory reads, but they are only $O(1)$ in memory writes. Standard array-based implementations add $O(n)$ write overhead on top of the comparisons and memory reads.

This overhead is completely unavoidable for the version that keeps the array sorted all the time as it is not possible to maintain a sorted array without doing inserts that shift elements around. The sort and remove approach can get around this if we use a slightly different approach to filling the hole left by an element that has been removed.[2] The approach is that, when the highest-priority item is found, it can be removed and the last item in the array can be put into its place instead of copying all the items down. Figure 13.1 shows how this looks in memory.

---

[1]In chapter 18, we will see that an array is exactly what we want to use for a priority queue implementation that is more efficient than either of the options considered here.

[2]So far, all the implementations we have discussed would use a standard queue ordering for elements with the same priority. This more efficient approach to filling the hole will break that. This is not as bad as it might sound because many applications will have very few, if any, ties, or if they do, the order will not matter.

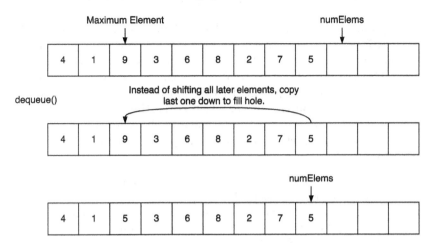

FIGURE 13.1: This figure shows how we can do remove from an unsorted array-based priority queue in $O(1)$ operations after we have done an $O(n)$ search to find the highest-priority element.

13.2 Project Integration: Discrete Event Simulation

One area that makes significant use of priority queues is the field of DISCRETE EVENT SIMULATION. This is a style of simulation in which all changes to the system are modeled as events that happen at specific times and the simulation jumps instantly from one event to the next. The way this is normally done is to have a priority queue that events are placed onto. Calls to dequeue pull events off in order by the time they occur.[3] Most events have the ability to schedule one or more other events. The simulation continues until either a certain amount of time has passed, or until the queue becomes empty.

A very standard example of a discrete event simulation is a line of people at a teller. The events in this system are people arriving and getting in line and tellers finishing with one person and taking another. Simulations of this type allowed businesses to determine how many people need to be working during different hours based on measurements of how often customers arrive. A minor variation on these can be used to set schedules for traffic lights as well.

13.2.1 Cell Splitting

A very basic example of a discrete event simulation is a simple model of cells dividing in a dish. Part of the simplicity of this simulation is that there is only one type of event, a cell dividing. The dish starts off with a single cell. After a certain period of time, that cell divides into two cells. Each of those cells later divides into two more. This process continues until we stop the simulation. We ignore cell death to keep things simpler.

The delay between when a cell is first "born" and when it splits is not a fixed value. Instead, it has a certain randomness. It is possible to spend a whole semester covering random numbers, how to pull random numbers from various statistical distributions, and

---

[3]It is worth noting that a priority queue ordered by time will definitely give smaller values higher priority because events that occur earlier in time have smaller time values than those that occur later. So we will have to make sure that we flip the sort order in these examples.

how to find appropriate distributions for different data sets. As these details do not really concern us, we will say that each cell splits somewhere between a minimum and a maximum number of time units after the split that created it with an equal probability of any value in that range. This uses a uniform distribution, which is what we get from `math.random`.

The events, in this case, are splitting times, and can be represented with a `Double`. We begin by placing one value on the queue, which is the time the first cells splits. We also need to start a counter that keeps track of how many cells there are. It begins at one and it gets incremented by one for each event as one cell splits into two. Basic code for this looks like the following.

```scala
private def runSimulation(): mutable.Buffer[Double] = {
  val splits = mutable.Buffer[Double]()
  val pq = new SortedListPriorityQueue[Double]((a, b) => b.compareTo(a))
  pq.enqueue(_minSplitTime + math.random * (_maxSplitTime - _minSplitTime))
  while (splits.length + 1 < _maximumPopulation) {
    val time = pq.dequeue()
    splits += time
    pq.enqueue(time + _minSplitTime + math.random * (_maxSplitTime -
        _minSplitTime))
    pq.enqueue(time + _minSplitTime + math.random * (_maxSplitTime -
        _minSplitTime))
  }
  splits
}
```

This code will run the simulation, but it does not really do anything useful for us. It returns the times at which different splits happened. We can produce something that is a bit more functional by putting this inside of a `class` that is `Drawable` and having it plot the population as a function of time. Code that does that is shown here.

```scala
package priorityqueues.drawing

import scala.collection.mutable
import scalafx.scene.Node
import scalafx.scene.canvas.GraphicsContext
import scalafx.scene.paint.Color
import scalafx.scene.layout.VBox
import priorityqueues.adt.SortedListPriorityQueue

class DrawCellSim(
    d: Drawing,
    private var _width: Double,
    private var _height: Double,
    private var _maximumPopulation: Int,
    private var _minSplitTime: Double,
    private var _maxSplitTime: Double,
    private var _numSims: Int) extends Drawable(d) {

  @transient private var propPanel: Node = null
  @transient private var allSplits = Vector[mutable.Buffer[Double]]()

  runAllSims()

  override def toString = "Cell Simulation"

```

```scala
26   def draw(gc: GraphicsContext): Unit = {
27     if (allSplits == null) runAllSims()
28     gc.stroke = Color.Black
29     val maxTime = allSplits.map(_.last).max
30     for (s <- allSplits) {
31       var (lastX, lastY) = (0.0, _height - _height / _maximumPopulation)
32       for (i <- s.indices) {
33         val (x, y) = (s(i) * _width / maxTime, _height - (i + 1) * _height /
              _maximumPopulation)
34         gc.strokeLine(lastX, lastY, x, y)
35         lastX = x
36         lastY = y
37       }
38     }
39   }
40
41   def propertiesPanel(): Node = {
42     if (propPanel == null) {
43       val panel = new VBox
44       val widthField = DrawingMain.labeledTextField("Width", _width.toString, s =>
              { _width = s.toDouble; runAllSims() })
45       val heightField = DrawingMain.labeledTextField("Height", _height.toString, s
              => { _height = s.toDouble; runAllSims() })
46       val maxPopField = DrawingMain.labeledTextField("Maximum Pop",
              _maximumPopulation.toString, s => { _maximumPopulation = s.toInt;
              runAllSims() })
47       val minSplitField = DrawingMain.labeledTextField("Min Split Time",
              _minSplitTime.toString, s => { _minSplitTime = s.toDouble; runAllSims() })
48       val maxSplitField = DrawingMain.labeledTextField("Max Split Time",
              _maxSplitTime.toString, s => { _maxSplitTime = s.toDouble; runAllSims() })
49       val numSimsField = DrawingMain.labeledTextField("Num Sims",
              _numSims.toString, s => { _numSims = s.toInt; runAllSims() })
50       panel.children = List(widthField, heightField, maxPopField, minSplitField,
              maxSplitField, numSimsField)
51       propPanel = panel
52     }
53     propPanel
54   }
55
56   def toXML: xml.Node = {
57     <drawable type="cellSim" width={ _width.toString() } height={
          _height.toString() } maxPop={ _maximumPopulation.toString() } minSplit={
          _minSplitTime.toString() } maxSplit={ _maxSplitTime.toString() } numSims={
          _numSims.toString() }>
58     </drawable>
59   }
60
61   private def runAllSims(): Unit = {
62     allSplits = Vector.fill(_numSims)(runSimulation())
63     drawing.draw()
64   }
65
66   private def runSimulation(): mutable.Buffer[Double] = {
67     val splits = mutable.Buffer[Double]()
68     val pq = new SortedListPriorityQueue[Double]((a, b) => b.compareTo(a))
```

```scala
69    pq.enqueue(_minSplitTime + math.random * (_maxSplitTime - _minSplitTime))
70    while (splits.length + 1 < _maximumPopulation) {
71      val time = pq.dequeue()
72      splits += time
73      pq.enqueue(time + _minSplitTime + math.random * (_maxSplitTime -
          _minSplitTime))
74      pq.enqueue(time + _minSplitTime + math.random * (_maxSplitTime -
          _minSplitTime))
75    }
76    splits
77  }
78  }
79
80  object DrawCellSim {
81    def apply(d: Drawing, width: Double, height: Double, maximumPopulation: Int,
          minSplitTime: Double, maxSplitTime: Double, numSims: Int): DrawCellSim = {
82      new DrawCellSim(d, width, height, maximumPopulation, minSplitTime,
          maxSplitTime, numSims)
83    }
84
85    def apply(n: xml.Node, d: Drawing): DrawCellSim = {
86      val width = (n \ "@width").text.toDouble
87      val height = (n \ "@height").text.toDouble
88      val maximumPopulation = (n \ "@maxPop").text.toInt
89      val minSplitTime = (n \ "@minSplit").text.toDouble
90      val maxSplitTime = (n \ "@maxSplit").text.toDouble
91      val numSims = (n \ "@numSims").text.toInt
92      new DrawCellSim(d, width, height, maximumPopulation, minSplitTime,
          maxSplitTime, numSims)
93    }
94  }
```

This lets the user set a number of different parameters for running the simulation and plots lines for many different simulations. The plots are scaled to fit the range in the specified size. No attempt is made to show scale values as that is a significantly more challenging problem. Figure 13.2 shows an example using this class in a drawing.

13.2.2   Collision Handling

Another example where discrete event handling makes sense is in realistic hard-sphere collision handling. The events in this system are times when balls run into one another or into other obstacles such as walls. They have to be handled in order because one collision can alter or completely prevent a later one. It is possible to do this constantly for the whole simulation, though there are reasons to bring all the balls up to a particular time every so often. For example, if you are going to render the simulation or save off the configuration to disk, it helps if all the balls have a location for the same time instead of having them at the last position for which they had an event.

Using events for collisions is more accurate and potentially significantly faster than letting balls move forward for a whole time step, then checking for overlap. The latter approach requires small time steps and it potentially misses details. The obvious example of this is if two small, fast-moving balls should collide in the middle of a time step. Unless the actual times of collisions are found, they can pass through one another and have no

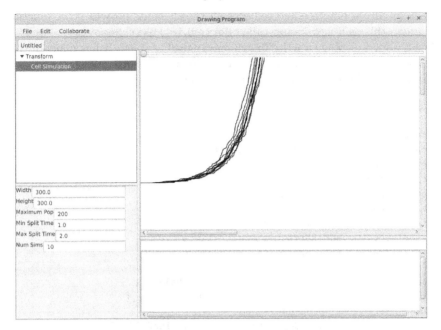

FIGURE 13.2: This screen shot shows the output of the cell simulation `Drawable`. The properties that are used are visible on the left side of the window.

overlap at the end of the step. Similar types of errors can occur when one ball should collide with two or more other balls during a step.

In a purely event-driven simulation, the "time steps" can also be handled as events. So when these events are reached, all the ball positions are updated and any special handling, such as rendering the balls or the application of non-collisional forces like gravity, can be performed. This type of handling can also be performed by breaking out of the event handling loop occasionally. We are going to write a `DrawBouncingBalls class` that will use discrete events to draw 2D balls bouncing around in a box. We are also going to add an `AnimationTimer` to the `DrawingMain` and give `Drawables` the ability to update so that they can be animated on the timer. Our previous `Drawables`, with the exception of `DrawTransform`, will not do anything for advance. All that `DrawTransform` will do is tell its children to advance.

There is also a modification to the priority queue that has to be made. Unlike the other examples that have been mentioned, this example requires that we be able to remove events from the queue without processing those events. Each time we process an event that modifies the velocity of a ball, all future events involving that ball must be removed. The most general way for us to add that functionality is with a method like the following.

```scala
def removeMatches(f: A => Boolean): Unit = {
  var rover = end.next
  while (rover != end) {
    if (f(rover.data)) {
      rover.prev.next = rover.next
      rover.next.prev = rover.prev
    }
    rover = rover.next
  }
}
```

This takes a predicate function and removes all the elements for which the function is true. We have added this as a new `class` called `SortedListPQWithRemove`.

With this added in, we can write `DrawBouncingBalls` itself. It has many of the features that you should expect to see in a `Drawable` at this point, but it also has a fair bit of code that deals with the discrete event simulation aspect.

```scala
package priorityqueues.drawing

import scalafx.Includes._
import scalafx.scene.canvas.GraphicsContext
import scalafx.scene.paint.Color
import scalafx.scene.Node
import scalafx.scene.layout.VBox
import scalafx.event.ActionEvent
import scalafx.scene.control.Button
import priorityqueues.adt.SortedListPQWithRemove
import priorityqueues.adt.SortedListPQWithRemove

class DrawBouncingBalls(d: Drawing, private var balls:
    Vector[DrawBouncingBalls.Ball]) extends Drawable(d) {
  @transient private var propPanel: Node = null
  @transient private var workCopy: Array[DrawBouncingBalls.Ball] = null
  @transient private var pq: SortedListPQWithRemove[CollEvent] = null

  override def advance(dt: Double): Unit = {
    if(workCopy == null) {
      workCopy = new Array[DrawBouncingBalls.Ball](balls.length)
      pq = new SortedListPQWithRemove((e1, e2) => e2.time.compareTo(e2.time))
    }
    for (i <- balls.indices) workCopy(i) = balls(i).copy(vy = balls(i).vy + 0.01,
        time = 0.0)
    for (i <- balls.indices) {
      findEventsFor(i, i + 1 until balls.length, 0.0, dt)
    }
    while (!pq.isEmpty) {
      val event = pq.dequeue()
      event.handle(dt)
    }
    balls = Vector(workCopy.map(_.advanceTo(dt)): _*)
  }

  def draw(gc: GraphicsContext): Unit = {
    gc.fill = Color.Black
    gc.fillRect(0, 0, 400, 400)
    gc.fill = Color.Green
    for (DrawBouncingBalls.Ball(x, y, _, _, s, _) <- balls) {
      gc.fillOval((x - s) * 400, (y - s) * 400, s * 800, s * 800)
    }
  }

  def propertiesPanel(): Node = {
    if (propPanel == null) {
      val panel = new VBox
      propPanel = panel
    }
```

```scala
48     propPanel
49   }
50
51   override def toString = "Bouncing Balls"
52
53   def toXML: xml.Node =
54     <drawable type="bouncingBalls">
55       { balls.map(_.toXML) }
56     </drawable>
57
58   private def collisionTime(b1: DrawBouncingBalls.Ball, b2:
           DrawBouncingBalls.Ball): Double = {
59     val (sx1, sy1) = (b1.x - b1.vx * b1.time, b1.y - b1.vy * b1.time)
60     val (sx2, sy2) = (b2.x - b2.vx * b2.time, b2.y - b2.vy * b2.time)
61     val radSum = b1.size + b2.size
62     val (dx, dy) = (sx1 - sx2, sy1 - sy2)
63     val (dvx, dvy) = (b1.vx - b2.vx, b1.vy - b2.vy)
64     val c = dx * dx + dy * dy - radSum * radSum
65     val b = 2 * (dx * dvx + dy * dvy)
66     val a = dvx * dvx + dvy * dvy
67     val root = b * b - 4 * a * c
68     if (root < 0) {
69       -1.0
70     } else {
71       (-b - math.sqrt(root)) / (2 * a)
72     }
73   }
74
75   private def findEventsFor(i: Int, against: Seq[Int], curTime: Double, dt:Double) {
76     for (j <- against) {
77       val t = collisionTime(workCopy(i), workCopy(j))
78       if (t >= curTime && t < dt) {
79         pq.enqueue(new BallBallColl(t, i, j))
80       }
81     }
82     for ((tfunc, bfunc) <- DrawBouncingBalls.wallInfo) {
83       val t = tfunc(workCopy(i))
84       if (t < dt) { // t >= curTime &&
85         pq.enqueue(new BallWallColl(t, i, bfunc))
86       }
87     }
88   }
89
90   private trait CollEvent {
91     def time: Double
92     def handle(dt: Double): Unit
93   }
94
95   private class BallBallColl(val time: Double, val b1: Int, val b2: Int) extends
           CollEvent {
96     def handle(dt: Double) {
97       val ball1 = workCopy(b1).advanceTo(time)
98       val ball2 = workCopy(b2).advanceTo(time)
99       val m1 = ball1.size * ball1.size * ball1.size
100      val m2 = ball2.size * ball2.size * ball2.size
```

```scala
101        val cmvx = (ball1.vx * m1 + ball2.vx * m2) / (m1 + m2)
102        val cmvy = (ball1.vy * m1 + ball2.vy * m2) / (m1 + m2)
103        val dx = ball1.x - ball2.x
104        val dy = ball1.y - ball2.y
105        val dist = math.sqrt(dx * dx + dy * dy)
106        if (dist > 1.01 * (ball1.size + ball2.size)) {
107          println("Warning: collision with big separation. "+b1+" "+b2+" "+dist)
108        }
109        if (dist < 0.99 * (ball1.size + ball2.size)) {
110          println("Warning: collision with little separation. "+b1+" "+b2+" "+dist)
111        }
112        val vx1 = ball1.vx - cmvx
113        val vy1 = ball1.vy - cmvy
114        val nx = dx / dist
115        val ny = dy / dist
116        val mag = nx * vx1 + ny * vy1
117        workCopy(b1) = ball1.copy(vx = ball1.vx - 2.0 * mag * nx, vy = ball1.vy - 2.0
                  * mag * ny)
118        workCopy(b2) = ball2.copy(vx = ball2.vx + 2.0 * mag * nx * m1 / m2, vy =
                  ball2.vy + 2.0 * mag * ny * m1 / m2)
119        pq.removeMatches(_ match {
120          case bbc: BallBallColl => bbc.b1 == b1 || bbc.b2 == b1 || bbc.b1 == b2 ||
                  bbc.b2 == b2
121          case bwc: BallWallColl => bwc.b == b1 || bwc.b == b2
122          case _ => false
123        })
124        val others = workCopy.indices.filter(b => b != b1 && b != b2)
125        findEventsFor(b1, others, time, dt)
126        findEventsFor(b2, others, time, dt)
127      }
128    }
129
130    private class BallWallColl(val time: Double, val b: Int, val newDir: (Double,
            Double, Double, Double) => (Double, Double)) extends CollEvent {
131      def handle(dt: Double) {
132        val ball = workCopy(b)
133        val nx = ball.x + (time - ball.time) * ball.vx
134        val ny = ball.y + (time - ball.time) * ball.vy
135        val (nvx, nvy) = newDir(ball.x, ball.y, ball.vx, ball.vy)
136        workCopy(b) = DrawBouncingBalls.Ball(nx, ny, nvx, nvy, ball.size, time)
137        pq.removeMatches(_ match {
138          case bbc: BallBallColl => bbc.b1 == b || bbc.b2 == b
139          case bwc: BallWallColl => bwc.b == b
140          case _ => false
141        })
142        findEventsFor(b, workCopy.indices.filter(_ != b), time, dt)
143      }
144    }
145  }
146
147  object DrawBouncingBalls {
148    def apply(d: Drawing, minSize: Double = 0.01, maxSize: Double = 0.05) = {
149      new DrawBouncingBalls(d, Vector.fill(20) {
150        val size = minSize + math.random * (maxSize - minSize)
151        Ball(size + math.random * (1 - 2 * size), size + math.random * (1 - 2 * size),
```

```
152         (math.random - 0.5) * 0.02, (math.random - 0.5) * 0.02, size, 0.0)
153     })
154   }
155
156   def apply(data: xml.Node, d: Drawing) = {
157     new DrawBouncingBalls(d, Vector((data \ "ball").map(bxml => {
158       val x = (bxml \ "@x").text.toDouble
159       val y = (bxml \ "@y").text.toDouble
160       val vx = (bxml \ "@vx").text.toDouble
161       val vy = (bxml \ "@vy").text.toDouble
162       val size = (bxml \ "@size").text.toDouble
163       Ball(x, y, vx, vy, size, 0.0)
164     }): _*))
165   }
166
167   case class Ball(x: Double, y: Double, vx: Double, vy: Double, size: Double, time:
          Double) {
168     def toXML = <ball x={ x.toString } y={ y.toString } vx={ vx.toString } vy={
          vy.toString } size={ size.toString }/>
169
170     def advanceTo(t: Double) = {
171       val dt = t - time
172       copy(x = x + vx * dt, y = y + vy * dt, time = t)
173     }
174   }
175
176   private val wallInfo = Seq[(Ball => Double, (Double, Double, Double, Double) =>
          (Double, Double))](
177     (b => if (b.vx < 0) b.time - (b.x - b.size) / b.vx else 1000, (x, y, vx, vy) =>
          (vx.abs, vy)),
178     (b => if (b.vx > 0) b.time + (1 - b.x - b.size) / b.vx else 1000, (x, y, vx,
          vy) => (-vx.abs, vy)),
179     (b => if (b.vy < 0) b.time - (b.y - b.size) / b.vy else 1000, (x, y, vx, vy) =>
          (vx, vy.abs)),
180     (b => if (b.vy > 0) b.time + (1 - b.y - b.size) / b.vy else 1000, (x, y, vx,
          vy) => (vx, -vy.abs)))
181 }
```

To understand this code, it probably helps to begin looking down near the bottom at lines 167–174, where we define a **case class** called **Ball** that stores a position, velocity, size, and time. It has methods that can write it out as XML or advance it a certain amount of time.

The running of the simulation happens in the **advance(dt: Double)** method on lines 18–32. You can see that this method has the **override** keyword. Because most of the **Drawables** do not need to do anything here, a default implementation was put in **Drawable** and we only **override** it in the subtypes that need an implementation. The first thing that **advance** does is check if the **transient** variables used by the simulation are **null**. If so, they are created. This has to be done for the purposes of serialization. The **workCopy** variable is initialized to have a copy of the balls with gravity applied. You can see this as adding 0.01 to the *y*-velocity.[4] Then for each ball, all events that happen during the next time unit are found. Each ball only checks for collisions against the other balls after it is in the array. The

---

[4]Remember that graphics coordinates grow down the screen, so adding to vy makes the balls fall on screen.

condition of only checking against balls later in the array prevents us from finding collisions twice. Lines 27–30 make it so that as long as the priority queue is not empty, we pull off the next event and handle it. When there are no more events, the balls are copied over to a `Vector` that will be used for the purposes of drawing. The code for drawing, making the properties, and producing XML are fairly straightforward.

After the `toXML` method there are a number of methods and `classes` that are used for finding collisions and defining/handling discrete events. First is a method that takes two `Ball` objects and returns a collision time for the two. This method assumes the balls move on straight lines, so the distance between them squared is a quadratic in time. We simply have to solve for the point in time when they get to a distance equal to the sum of the radii. Each ball keeps track of the "time" it is at during a step so that has to be accounted for. If there is no collision, this method returns -1.0. Otherwise, it returns the lesser root as that is when the balls hit. The other root would be when they separate if they moved through one another. If you are interested in the details of the math, here is a derivation of the coefficients in the quadratic equation.

$$d^2(t) = ((x_1 + vx_1 * t) - (x_2 + vx_2 * t))^2 + ((y_1 + vy_1 * t) - (y_2 + vy_2 * t))^2$$
$$= (\Delta x + \Delta vx * t)^2 + (\Delta y + \Delta vy * t)^2$$
$$= \Delta x^2 + \Delta y^2 + 2(\Delta x \Delta vx + \Delta y \Delta vy)t + (\Delta vx^2 + \Delta vy^2)t^2$$

When ball motion is more complex than a straight line, more complex code has to be involved to find the collision time as normally there is not a closed form solution like there is in this case.

After the method that finds the time of a collision is a method that finds all the events for a particular ball against some set of other balls and the walls. The way the code has been set up, the structure for handling the balls and the walls is very similar. We loop over the objects we want to check against, calculate an event time, and if that event time falls into the valid range for the current step, we add it to the queue. We will come back to how the wall code is written to work in this way.

This simulation has two types of collisions that are subtypes of the `CollEvent trait`. They are `BallBallColl` and `BallWallColl`. Each of these knows a time when the event happens and contains code for handling the event. The wall `class` is simpler and is probably worth looking at to see what is going on. First, the position of the ball at the time of the event is found. Then we call a function that was passed in at the construction of the event which takes the position and velocity of the ball and returns a new velocity. That new velocity and position are used to build a new ball object which is stored in the array in place of the old one. Then future events involving that ball are removed and new events are found.

The code for ball-to-ball collisions is doing the same thing, but for two balls, and it involves more math for handling that type of collision. It is not significant that you understand this math unless you want to. It starts by calculating relative masses of the balls, then finding the center of mass velocity of the two balls.[5] This code checks the distance against the sum of the radii and prints an error message if it is not close enough to the expected value. This code can be very helpful for making sure that everything is working correctly, and while it is not technically needed in the final code, it can be helpful to keep in there to prevent problems with future modifications. The modification in velocity for each ball is calculated as a multiple of the projection of the velocity along the line separating the balls. In this code, the constant 2.0 is used so we have perfect conservation of energy. Collisions that dissipate energy would use a value between 1.0 and 2.0. Unfortunately, the gravity

---

[5]Collisions are much easier to calculate in the center of mass frame. In this moving frame of motion, it is as if each ball ran into a stationary wall.

integration for this code does not conserve energy, so it is helpful to have the collisions damp the system a bit.[6]

While the math in this code might be a bit confusing, the most interesting piece of the code is probably the way in which collisions with the walls are handled. The event-finding code uses something called `wallInfo` and functions from that are passed into the `BallWallColl` objects. The magic is contained in the declaration of `wallInfo` at the bottom of the code. This is a sequence of tuples where each element has two functions. These functions represent a wall. The first one tells when a given ball would collide with that wall and the second one gives the modified velocity of the ball after it collides with the wall.

The beauty of this approach is that it makes it fairly simple to add additional barriers. For example, the following tuple defines a barrier that runs diagonally across the middle of the cell that bounces balls toward the top left corner.

```
(b => if(b.vx>0 || b.vy>0) b.time+(1-b.x-b.y)/(b.vx+b.vy) else -1,
    (x,y,vx,vy) => (-vy.abs,-vx.abs))
```

This is not technically a wall because it only considers the center of the ball, not the edges. It also allows balls to pass up through it, just not down. However, it demonstrates the flexibility of the approach. With a bit more math, it would be possible to throw in barriers like a round wall that is a cup at the bottom. As soon as the functions are added to the `wallInfo` sequence, it automatically works with the rest of the code.

You can get the full implementation for the GitHub repository at `https://github.com/MarkCLewis/OOAbstractDataStructScala/tree/master/src/priorityqueues`. If you add a bouncing balls drawable, you will see 20 balls of random sizes added in, just like before. You can click the button to start the timer going to see it run. If you watch for a while, you will see that the large balls bunch up near the bottom and the small ones occasionally get shot higher up into the box. This type of behavior is called equipartition of energy and is a standard feature of physical systems where bodies of different masses interact. The fact that we see it in this system is a good indication that our code is behaving properly. Not everything is perfect though. Occasionally balls will pass through one another. Numerical errors makes this pretty much impossible to avoid unless you do something to force balls apart when they overlap.

13.3   End of Chapter Material

13.3.1   Summary of Concepts

- A priority queue has the same methods as a queue, but the `dequeue` method functions differently.

  - Instead of always taking the object that has been present the longest, it takes the object with the highest-priority.

  - Tied priority can go to the one that has been present longest, though not all implementations will do this.

  - We can create a reasonably efficient priority queue using a sorted linked list.

---

[6]This code applies a full downward acceleration to each ball at the beginning of the step. If the ball reverses vertical direction part way through the step, the influence of gravity should have been reversed. That does not happen here, so balls can gain energy with each bounce off the bottom.

- Discrete event simulations use priority queues.

  - Events are prioritized by the time they occur.
  - We looked at a cell-splitting simulation and collision handling.
  - Collision handling required an extra method to remove certain events.

13.3.2　Exercises

1. Code the search-based priority queue and do speed tests to compare it to the version that maintains a sorted linked list. For the speed tests, insert elements with random priorities and dequeue them. Vary the number of elements used across an exponential range to measure how that impacts performance.

2. Code `Array`-based implementations of a priority queue using both a sorted `Array` and an unsorted `Array` that you search. Do speed tests to compare them to each other and to the linked list based implementations. For the search-based method with an `Array`, do both shifting elements down and swapping the last element into the hole left by the one that is removed.

3. Implement priority queues using `ArrayBuffer` and `ListBuffer` to see how efficient they are relative to our "custom" implementations.

4. Implement some extra wall types for the collision simulation. Consider trying walls that are not lines or writing a method that creates wall segments.

5. Look up some different distributions for random variables and try the cell-splitting simulation using those.

13.3.3　Projects

1. There are a few ways that you could use a priority queue in your MUD to good effect. All of them involve having a single priority queue that actions can be added to so that they will run at a later time. This will allow things like spells to have delayed effects or to wear off, or for weapons to have different speeds for attacks. It can also be used by computer-controlled characters and to implement random spawning of characters. The advantage of a priority queue is that you do not have to have checks in every single character/room/item for whether something should happen. Instead, all those things could place events on the priority queue with a delay, and the main thread only pulls off things as needed.

   The ideal here uses a sorted list or `Array` for the priority queue so that the `peek` operation is fast. The idea is that most "ticks" there will not be any events on the priority queue to process. In that case, the code only does a `peek`, sees that the time of the first element is after the current time, then goes on to whatever else needs to be done.

   If your MUD is using an actor system, put the priority queue in a top-level actor that all the other actors can schedule events with. The events can just send messages back when the event has occurred so that the original sender can respond correctly.

2. If you are working on the web spider project, you might have noticed that many websites are extremely large. Even if you limit your spider to a single domain, say

only pages coming from domains ending with cnn.com, espn.go.com, or google.com, there can still be an extremely large number of pages. If you let the processing proceed in the order that links are found, you are likely to never get to a lot of the data that you are interested in because of resource limitations. One way to deal with this is to make the queue of pages to be read work as a priority queue.

If you know the types of pages you are looking for and have an idea of what paths they are at on the site you are running through, you can give those pages a higher priority. That way, you will visit those pages first. As long as the program keeps finding pages that have a structure you have coded to be interesting, it will work on those pages and not go to other pages that are less likely to have interesting information.

3. If you are working on either a networked or non-networked graphical game, you can use a priority queue in the main loop of the game in much the same way as is described above for the MUD. It is possible that players will be able to do things that should not take effect for a while. It is also possible that you will have many units in the game that do not need to update every single tick. In either of these cases, you can make a priority queue of events/commands that are only executed when it comes time for them to happen.

4. The math worksheet does not provide any obvious places to include a priority queue. Two possibilities would be to give the ability for the user to build simple discrete event simulations in a worksheet or to have elements of a worksheet prioritized for processing. The latter might be useful if you either want things that are currently on screen to be calculated before things off screen or if you decide that processing of graphical elements should be lower priority than processing numeric elements.

5. The Photoshop® project also does not automatically lend itself to including a priority queue. If you can think of an application, you could write one. Prioritizing draw elements could work, but you have to make sure you do not break the user control over what covers what in a painter's algorithm.

6. This chapter gave a reasonable introduction to the topic of discrete event simulation. Add some capabilities to your simulation workbench that support general discrete event simulations as well as collisional simulations. Make a few specific instances of discrete event simulations that users can create and run. Allow them to adjust certain significant parameters and track significant aspects of the status of the simulation.

7. We do not have a clear use for priority queues for the stock portfolio management project. Perhaps you can think of something interesting to do with it.

8. The movie system can definitely use a priority queue to manage the movie requests from the users. Remember, each user should have a list of the movies they would like to watch that are ranked in order of preference to indicate which one they would like to watch first, which they would like to watch second, etc. Use a priority queue on these movie requests so that when a movie becomes available it will be allocated to the user. Remember that users will add, remove, and change movies in their watch request list.

9. We do not have a good use of priority queues for the L-systems project.

Chapter 14

Refactoring

As you have worked with longer pieces of code, there are a few things that you should have realized about working on bigger programs. For example, the first time that you write a piece of code, you probably do not do it in an optimal way. Also, when you first lay out how to put something together, you likely miss details that have to be added. What is more, sticking pieces of code onto other code in a haphazard way often leads to really brittle code that is hard to work with. If you were writing software in a professional environment, there would be another rule that could be added on here—the customer often does not really know what he/she wants.

When you put all of these together, it leads to the general rule that code needs to be dynamic and adaptable. You have to be willing and able to modify how things are done over time. It is also ideal for this to be done in a good way so that the results do not just pile mistaken approach upon mistaken approach in a crudely rigged solution to a final goal.

In this chapter, we will look at one of the primary tools in the toolkit of the software developer for effectively managing code over time as old approaches are replaced by newer ones and as features get piled up. This tool is called refactoring and it is the alteration of existing code without changing functionality. At first the idea of changing code without altering the functionality might seem odd, but there are good reasons for why it should be done.

14.1 Smells

Not all code is created equal. While there are effectively an infinite number of ways to build correct solutions to a problem, you have hopefully learned that many solutions that are technically correct, are less than ideal. In some situations this is obvious. In others, it depends on the application or even who you ask. Code is very much an art and, as such, beauty is often in the eye of the beholder. Even when code is not quite ideal, that does not mean it is worth the effort to rewrite to a better form. The technical term for code that has a property that makes it less than ideal is that it *smells*.

The term smell carries significant imagery. Smells can be strong or mild. Even if something smells bad, if it is really mild, then you probably will not go to any lengths to correct it unless there is a good reason to. In addition, not all programmers will be equally sensitive to different smells. In his book *Refactoring: Improving the Design of Existing Code*, Martin Fowler, along with Kent Beck, list 22 different smells that can afflict code [5]. We will list them all here with brief explanations along with possible examples of where they have potentially popped up in code we have written.

Alternative Classes with Different Interfaces - When you have methods in different classes that do the same thing, but call it different names. This makes code very hard to work with. You should pick a name that fits and rename everything to match that.

Comments - Some comments are definitely good. Too many, especially when they do not really add extra information, can be smelly. Comments that are out of date with the code are also horribly smelly. Programmers will read comments to help them figure out what is happening in the code. If the comments have not been updated with the code, this can lead to significant confusion and waste a lot of time. You are better off with no comments than bad comments, but a total lack of comments also makes code much harder to read, which makes that approach smelly as well.

Data Class - These are classes that do nothing but store data and have no real functionality. Littering your code with individual files that contain these makes it very hard for people to navigate the code. You might consider having these nested inside of objects that have a meaningful relationship to the class.

Data Clumps - Pieces of data that are almost always found together and should be united into a class. A simple example of this is information for a data point. Instead of keeping x, y, and z separate and passing them as three separate arguments into many different methods, use a class that can store the three values. One benefit of this is that calling code explicitly knows the data is a point, not three unrelated `Doubles`.

Divergent Change - When you change the same piece of code (often a class) at different times for different reasons. This can imply that code is taking on overly broad responsibilities. Try to break it into smaller pieces that are more focused in their tasks.

Duplicated Code - The name says it all here. We have been talking about ways to avoid this smell since very early in the book. The reader should also be well aware by this point of the pitfalls of duplicate code, such as duplication of errors and the need to modify all copies for certain changes. Multiple approaches have been introduced to reduce code duplication, from functions to collections and loops, then finally to abstraction and polymorphism, you should strive to write any particular piece of logic as few times as possible.

Feature Envy - This is when a method in one class makes lots of calls to another class, especially if it does so just to get data for doing a calculation. This likely implies that the method is in the wrong class and should be moved closer to the data it is using.

Inappropriate Intimacy - An object is supposed to encapsulate an idea. If you have two classes that are constantly referring to elements of one another, it implies that things have not been broken up appropriately. Consider moving things around so that each class stands more on its own.

Incomplete Library Class - This smell occurs when a library that you are using lacks functionality that you need. The significance of it being in a library is that you typically cannot modify library code. In Scala, you can often get around this using implicit conversions.

Large Class - Classes that get too large become hard to manage. You want to have the functionality of any given class be specific to a particular task. If a class gets too large it is an indication that it should probably be broken into parts.

Lazy Class - If you have a class that does not do much of anything, then it is lazy and you should consider getting rid of it. At the very least, put classes like this inside of an appropriate other class or object so they do not take up their own file.

Long Method - Long methods/functions are error prone. If a method starts to get too long it should be broken up into pieces.

Long Parameter List - Methods that have extremely long parameter lists are hard to use.[1] Often, long parameter lists go along with the "Data Clumps" smell. The parameter list can be shortened by passing in a smaller number of objects containing the needed values.

Message Chains - This smell occurs when one class goes through many objects to get to what it needs. For example, `val stu = school.teacher.course.student`. The problem with this is that whatever class you write this in now depends on the exact name and general contract for the methods `teacher`, `course`, and `student` in three different classes. If any of those things change, this code will break. That is bad.

Middle Man - This smell arises when too many methods in one class do nothing but forward calls to another class. The one doing the forwarding is the "middle man." If this happens too much, and is not providing other benefits, you should consider going straight to the object that does the work.

Parallel Inheritance Hierarchies - This is a special case of "Shotgun Surgery." In this case, you have two inheritance hierarchies that mirror one another. When a class is added to one, it must be added to the other as well.

Primitive Obsession - In Java and many other languages, types like `Int`, `Double`, `Char`, and `Boolean` are called PRIMITIVE types and they are treated differently than objects.[2] Scala does not maintain this distinction, but this smell can still apply. Code has this smell when you use basic or built-in types too much instead of creating your own types. An example of this would be using tuples too much for grouping data instead of writing case classes. This makes for smelly code because references to `_1` and `_2` tell you nothing about the meaning of the value. Similarly, an `(Int, Int)` could be a month/year combination or coordinates on a planar grid. Using types specific to the task not only makes code easier to read, it allows the type checker to help you determine when you have made an error by producing a syntax error on a type mismatch.

Refused Bequest - This smell occurs when a class inherits from another class or trait, but does not "want" some of the methods in that supertype. This is typically a sign of a

---

[1] Having default values and named parameters in Scala does minimize this problem a bit.

[2] Primitive types are built into the hardware of the machine so they execute much more quickly and take up less space than an object. When you use an `Int` in Scala, the compiler will try to compile it to a Java primitive `int` for better performance.

poor design choice. The motivation behind inheritance should be to become a subtype of the supertype. That means that any method calls that could be made on an instance of the supertype should work on the subtype. If this is not the case, then inheritance is probably the wrong approach. The `java.util.Iterator` type was designed with this flaw. It has a `remove` method that is listed as optional. Implementations that do not support removing are supposed to throw an exception. That means that something that really should be a syntax error becomes a runtime error.

Shotgun Surgery - This is when a particular change in your code has a tendency to force you to make edits in several different places. This is smelly because it is too easy to forget one of those edits, which leads to bugs. Adding a new `Drawable` has something of a shotgun surgery feel to it. When we come up with a new `Drawable` type, we also have to edit the `makeDrawable` method in the `Drawable` companion object and add a creator in `DrawingMain`.

Speculative Generality - We often want to have a lot of flexibility in our code, but code that is far more general can be artificially hard to understand, use, and maintain. This happens when a code writer expects to need certain flexibility that the final usage does not take advantage of.

Switch Statements - Scala does not have switch statements. The closest approximation is the `match`, which is far more powerful than a normal switch statement. This smell can still occur in Scala code if matches are used in particular ways. This smell occurs when you have multiple `match` statements/expressions that have exactly the same cases and have the property that whenever a case has to be added to one, it must be added to all. This turns the adding of a case into a shotgun surgery. Generally, this should be dealt with by making a `trait` that has one method for each of the different `match` occurrences and a different concrete subtype for each case. Each `match` in the code then becomes a call to the appropriate method of an object that keeps track of the state.

Temporary Field - This is when you have a field of a class that is only needed at certain times. This makes code hard to understand because the field will have a value at other times, but that value might not really have any meaning for the object.

The goal of refactoring, and what you should be thinking about when considering smells in code, is that you want to keep your code readable and maintainable as well as keeping it correct and sufficiently optimized for your needs. Some of these smells almost compete with one another, pushing you in opposite directions. Before you make changes, consider the motivation. Will the change improve the code and make it easier to work with? Would the side effects of a cure for a particular smell be worse than the original smell? Just because you see something in your code that seems like it matches one of these smells does not mean the code is really smelly. That judgment requires thinking about how the code behaves and how the programmer interacts with it.

14.2   Refactorings

Fowler's book on refactoring lists 72 different refactoring methods. Only a few will be discussed here. Interested readers can turn to the original book for a full description, but do not worry too much about doing so. Many of the refactoring methods presented in the book involve concepts that are beyond the scope of this book or what the reader is expected to understand at this point in his/her computer science education.

14.2.1 Built-in Refactoring Methods

Many of the refactoring methods are algorithmic in nature. For this reason, IDEs, such as Eclipse, have a number of different refactoring functions written within it. Using built-in refactoring tools can make the process much faster and safer. They are also set up so that they have no impact on the behavior of the code. They change the structure without changing the functionality.

The following are some of the options that are available under the refactoring menu when you right-click on code in Eclipse under Scala. There are others, and the list goes with new releases of the tool.

- Rename - This can be applied to any programmer-given name such as variable names, method names, or class names. Just as comments that no longer fit the code are smelly, using names in your programs that are hard to understand or, worse yet, are misleading, should have you detecting a rank stench. Of course, you can change the name of something on your own, but that means finding all the places it is used and changing them as well. For something like the name of a public method in a large project, this can mean tracking down invocations in many different files. This refactoring option does all that for you. So if you find that a name you chose early in the development of a project no longer fits, select this option and change it to something that does.

- Inline Local - This will remove a local `val` and change all references to it to the value it is initialized to. This should only be done with immutable values as this operation can change behavior if the object is mutated in some of the usages. You would use this if you put something in a variable expecting it would be long or used many times and then come to find out neither is true.

- Organize Imports - This very handy option, which has the keystroke Ctrl-Alt-O, will automatically add imports for any types that are lacking them. It also organizes the import statements at the top of the file. It should be noted that this uses Java-style imports that go at the top of the file and have one class per line. If you need an import to be local, you should add that yourself. The fact that it uses a longer format is not that important as you do not have to type it.

- Extract Local - This is the inverse of Inline Local. It can be used if there is an expression in a method that either appears multiple times, or which would make more sense if a name were associated with it. This is sometimes also called "Introduce Explaining Variable".

- Extract Method - This remarkably convenient refactoring option should be used to break up large methods or to extract code into a method when you find that you need to use it more than you originally believed. Simply highlight the code you want in the method and select this option. It will figure out the parameters that are needed and you just pick a name.

The built-in refactoring tools in IDEs like Eclipse can make certain types of changes remarkably easy. The most important aspect of this is that they reduce the barrier so that you really have no excuse not to use them. If you leave a poorly named variable or method in your code, or if you decide to live with a method that is 200 lines long, you are simply being lazy.

14.2.2   Introduce Null Object

Not all refactoring methods are quite so easy to perform. Some will require a bit of effort on your part, but when they are called for, they are definitely worth it. To see this, we can consider one of the exercises from chapter 12 where you were asked to write a doubly linked list without using a sentinel. If you actually went through that exercise you found that the code wound up being a nest of conditionals set up to deal with the boundary cases where a **prev** or **next** reference was **null**. This type of special case coding to deal with **null** references if fairly common. So common, in fact, that there is a refactoring method specifically for dealing with it called, "Introduce Null Object." As the name implies, the goal is to introduce an object that is used in place of **null** references. This object should have methods on it that are basically stubs with limited functionality. You want to do this if you find that your code has many lines that look something like this.

```
if (ref == null) ...
```

The sentinel played the role of a Null Object in the doubly linked list. An even better example is **Nil** for the immutable singly linked list. You could theoretically make an immutable singly linked list without making a **Nil** object. Instead, the lists would be terminated with **null** references as they were for our mutable singly linked list. Such code tends to be far messier to write and to work with, making it harder to debug and to maintain. We will see another usage of a Null Object in chapter 16 when we make our immutable binary search tree.

14.2.3   Add and Remove Parameter

The code we wrote back in chapter 12 used the Null Objects initially; we did not have to refactor for it. Describing such a refactoring would be challenging. To make it clearer how this might work, consider the "Add Parameter" refactoring. The idea here is that you have written a method that takes a certain list of parameters. As an example, consider a method that logs employee pay. This might be part of some larger application used by a company. The current version looks like this.

```
def logHours(employeeID:ID, hoursWorked:Int, wagesPaid:Int):Unit = {
  // Stuff to do logging
}
```

It works and everything is fine until it is determined that the logs need to include additional information to comply with some new government regulation. For example maybe this is a large, multi-national corporation and they also have to keep records of what location the employee was working at for those hours. Clearly this is going to require changes to the **logHours** method. The idea of refactoring is that you break those changes into two parts. The first part is to perform the Add Parameter refactoring. That gives you this code.

```
def logHours(employeeID:ID, hoursWorked:Int, wagesPaid:Int,
    location:OfficeLocation):Unit = {
  // Same stuff as before
}
```

Note that the code in the method is unchanged. Adding this parameter forces changes at the call sites, but it does not actually change the behavior of the code. You should make this change, get the code to compile again, then run all existing tests and verify that things work. Once you have confirmed that the code is still functional, you can proceed to the second step, where you actually alter the functionality of the method.

```scala
def logHours(employeeID:ID, hoursWorked:Int, wagesPaid:Int,
    location:OfficeLocation):Unit = {
  // Modified stuff to include the location.
}
```

The whole goal of refactoring is to formalize your approach to change so that when code needs to change, you are more willing to do it and more confident that it will work. Changing existing code can be a very scary thing to do. Often students will wind up with large segments of their code commented out, but not deleted, because they are afraid to really commit to changes. Combining the techniques of refactoring and automated unit testing can help to give you the courage to make needed changes. Code should be dynamic and flexible. That only happens if programmers are willing to change things.

The counterpart to Add Parameter is "Remove Parameter". The motivation for this refactoring would be that alterations to a method or the code around it have made it so that parameters, which at one point provided needed information, are no longer being used. Perhaps alterations in the system have made it so that the value of the wages paid should be calculated using the employee information and how many hours were worked instead of being passed in. In that case, you might have code that still has **wagesPaid** as a parameter, but that parameter is never used. That is smelly as it is not only overhead for the code, it is likely to confuse people looking at it. Such a parameter should be removed to make the code easier to understand and work with.

14.2.4 Cures for Switch Statements

As was mentioned above, Scala does not have a switch statement in the traditional sense. However, the smell that is generally associated with switch statements can be produced in any conditional execution including not only **match**, but also **if**. Running with the example from above, consider that we need to calculate a number of different values associated with employees and the nature of the calculation depends on the type of employee under consideration. This could be done with code that looks something like this.

```scala
val monthlyPay = employeeType match {
  case Wage => ...
  case Salary => ...
  case Temp => ...
}
```

This same pattern might be repeated in different parts of the code for calculating benefits and leave time as well. The values **Wage**, **Salary**, and **Temp** could be **vals** defined in some object, or enumerations. This could also appear in code using **if** expressions.

```scala
val monthlyPay = if (employeeType == Wage) ...
  else if (employeeType == Salary) ...
  else ...
}
```

The key to this being a smell is that the pattern above is repeated in multiple locations. If a new category of employee were added, all of those would need to be modified. Missing one or more occurrences will lead to difficult-to-track errors.

There are two refactoring methods that can be employed to remove this smell. They are "Replace Conditional with Polymorphism" and "Replace Type Code with Subclasses". This would appear in the code as an abstract supertype that has methods for each of the different types of actions/calculations that are required. That might look like the following.

```scala
trait EmployeeType {
  def monthlyPay(hours: Int, wage: Int): Int
  def logBenefits(employee: Employee, hours: Int): Unit
  def leaveTime(hours: Int): Double
}
```

Using this, the `match` or `if` constructs shown above would be replaced by this simple line.

```scala
val monthlyPay = employeeType.monthlyPay(hours,wage)
```

This works because the `employeeType` will be a reference to an appropriate subtype of `EmployeeType`. That subtype will have the appropriate code in it for that type. The code might look like the following with proper implementations of all the methods. Then the parts of the code that give a value to `employeeType` would provide the appropriate subtype.

```scala
class WageEmployee extends EmployeeType {
  def monthlyPay(hours: Int, wage: Int): Int = ...
  def logBenefits(employee: Employee, hours: Int): Unit = ...
  def leaveTime(hours: Int): Double = ...
}

class SalaryEmployee extends EmployeeType {
  def monthlyPay(hours: Int, wage: Int): Int = ...
  def logBenefits(employee: Employee, hours: Int): Unit = ...
  def leaveTime(hours: Int): Double = ...
}

class TempEmployee extends EmployeeType {
  def monthlyPay(hours: Int, wage: Int): Int = ...
  def logBenefits(employee: Employee, hours: Int): Unit = ...
  def leaveTime(hours: Int): Double = ...
}
```

Note that if each of these had no state, members that differ between different instances, they could be declared as objects instead of classes.

There are two main benefits to this change that improve upon the switch statement smell. The first is that this new structure makes it significantly easier to add new types of employees. If it is decided that any of these categories needs to be resolved into different classifications, that can be done easily. It also localizes code from any given type of employee and makes it much easier to find the code when changes need to be made.

14.2.5 Consolidate Conditional Expression

In the calculation of how much you should pay wage employees, it is likely that you will have to include expressions for normal overtime and perhaps even double overtime. These might look something like `hours >= 40 && hours < 60` and `hours >= 60`. The first problem that you notice with these expressions is that they include "magic numbers" in the form of 40 and 60. As a general rule, you do not want to have these magic numbers spread through your code. The other potential problem is that these expressions do not have obvious meaning and it is possible that the requirements might need to be changed in the future. All of these can be fixed, to one degree or another, by pulling the Boolean expression out into a method like this.

```
def overtime(hours: Int): Boolean = hours >= 40 && hours < 60
```

Then your condition in other parts of the code becomes `overtime(hours)`. This is easy to read, highly informative, and makes it easy to change the nature of the condition. However, if the expression is not used multiple times, this approach is overkill and only adds to the complexity of the code.

14.2.6 Convert Procedural Design to Objects

Object-orientation introduces additional concepts that are not essential for solving small problems. The approach where data and functionality are viewed as being independent is called the procedural approach. Bigger programs gain a lot from being built in a proper object-oriented manner. There is also a benefit to consistent design across a project. For this reason, if you have aspects of a large program that contain occasional segments that are more procedural, you should refactor them to use an object-oriented approach. This is done by grouping the functionality with the data that it should operate on and putting the two together into objects. This can be done in an incremental way by extracting methods from long procedures and moving them into the appropriate `class`, `trait`, or `object`.

14.2.7 Encapsulate Collection

Picture a class that has a data member that is a mutable collection. This might be something like an `Array` or a `Buffer`. For example, in our company you have a type for a `Department` that has a collection filled with the `Employee` type. Now imagine that the class has a method like the following.

```
def employees:mutable.Buffer[Employee] = _employees
```

Note that this method is public and gives any outside code access to the local employee collection. This is a risky thing to do with a mutable collection. The reason being that any code can call this method and then start making whatever changes it desires to the employees in that department without the `Department` class having any control over it. That is not something one would expect to allow in real life and it is best avoided in code as well. Even if it seems like a reasonable thing to do when you write it, even if the collection is not that critical to the working of the class, giving up that much control is likely to cause problems down the line.

The alternative is to return an immutable copy from this method and then write other methods for adding or removing elements as needed. This could be implemented in the simplest terms with the following code.

```
def employees:Seq[Employee] = _employees.toSeq
def addEmployee(e:Employee) = _employee += e
def removeEmployee(index:Int) = _employee.remove(index,1)
```

With just this code, there is not much of a benefit. However, if at some point you decide that you need to check conditions on new employees or to log when employees are removed, that will be easy to do. Using the original implementation, that would have been nearly impossible.

The general rule of thumb here is that mutable data should not be publicly viewable and any changes to mutable data should go through methods that have a chance to limit access or do something in response to it. While the ability to define special member assignment methods in Scala makes it reasonable to have public **var** members, care should still be taken in regard to how mutable members are shared with outside code.

14.2.8   Push Down or Pull Up Field or Method

Imagine that in the drawing program we had decided to put the methods for handling children, which are required in **DrawTransform**, up in **Drawable** because we needed more general access to them. All the types other than **DrawTransform** would have implementations of those methods that basically said there were no children. This would cause significant code duplication. In that situation, we could remove the code duplication by creating another type called **DrawLeaf** to be a subtype of **Drawable** and a supertype of all the **Drawable** types that do not have children. After that class was created, the common methods and fields should have been pulled up into that class. This action would have utilized the "Pull Up Field" and "Pull Up Method" refactoring.

These refactorings provide a simple way of dealing with any duplication of code that you find between different types inheriting from a common subtype. If all the subtypes have the same implementation, it is clear that the shared method implementation should be moved up to the common supertype. If only some share those methods, you should consider making a new intermediate type and moving the methods into that type.

There are also times when you want to do the opposite. You originally feel that a method will be implemented the same way for most types in a hierarchy, only to realize later that most subtypes wind up overriding that method. This starts to have the smell of a refused bequest. In that situation, you should perform "Push Down Field" or "Push Down Method" to minimize the odor. If there are some common implementations, introduce intermediate types to hold that code. If a large number of the subtypes are going to override a method, you should strongly consider having it be abstract in the highest-level type that contains the method. The advantage of this is that because you know many implementations are going to be different, you force anyone who inherits from your type to do something with that method or member. This forces them to think about it and reduces the odds that they will accidentally inherit an inappropriate implementation that leads to challenging to find bugs.

14.2.9   Substitute Algorithm

The last refactoring to be considered in this chapter is "Substitute Algorithm." There are many different reasons why your first choice of an algorithm for some part of a problem might not wind up being ideal. A simple example that we have dealt with was a sort algorithm for a small number of data values. By default, the built-in sort from the libraries was selected. It was easy to use, and it was a well-tuned, general purpose merge sort algorithm. One of the language library writers had certainly worked hard to make it efficient. However, when the full application was completed, performance turned out to be quite poor. Profiling indicated that an inordinate amount of time was being spent in the sort. A simple insertion sort written with the specific comparison required for that code was put in place instead. This switch of algorithm did not alter the final results of the program at all. However, it led to a remarkable improvement in speed.

When you first solve a problem, you are likely to use whatever approach is easiest to write correctly. That is exactly what you should do, because most of the time that will work best and be the easiest to maintain. There are situations where your initial guesses about the nature of the data will be incorrect and different approaches turn out to be significantly faster. Optimizing for speed is one of the last things you should do in developing a program, and even then it should only be done when you know that it is required. A well-known quote from Donald Knuth is that "We should forget about small efficiencies, say about 97% of the time: premature optimization is the root of all evil" [7]. When you do come to realize that an early choice of algorithm is slowing things down in an unacceptable way, do not be afraid to refactor the code and put in a different algorithm that is better suited for the specific task at hand.

14.3 End of Chapter Material

14.3.1 Summary of Concepts

- Smells

  - Something in the code that makes it less than ideal.
  - Sometimes related to a specific piece of code, but often deals with structure and includes many distributed elements.
  - They often build up over time as code is modified.
  - Make code harder to work with and maintain.

- Refactoring

  - The act of changing the structure of code or how it works without changing the behavior.
  - Can be done to remove smells.
  - Also done to help change the structure of code to allow new features to be implemented.
  - Many IDEs include certain options for automatic refactoring.

14.3.2 Exercises

1. Test each of the different built-in refactoring options in Eclipse.

2. The website for the book has some examples of smelly code. Go through each one, figure out why it is smelly, and refactor it.

14.3.3 Projects

Instead of giving individual descriptions for each of the projects, this chapter has more of a general description. Go through the code you have written and find things that smell. When you find them, refactor them so that they do not smell, or at least smell less.[3]

---

[3]We understand that novice programmers are notorious for writing smelly code. In part that is because novice programmers often do not have a good feel for when something is smelly, they are just challenged to make things work.

Going forward, there will likely be projects that will force you to change some of the design aspects of your project. When those things come up, make sure that you refactor the code first, then add the new features. Strive to not mix the steps. Do the refactoring and verify that the code works as it did before, then add to it.

Chapter 15

Recursion

At this point you should definitely have some basic familiarity with recursion, either from previous work or from consulting the on-line appendices[1] and earlier examples in this book. This chapter will look at the concept of recursion to help solidify it in your mind.[2] The next three chapters will utilize recursion, making this an ideal time to provide a refresher and integrate it into our project.

15.1 Refresher

A recursive function is nothing more than a function that calls itself. This means that it is defined in terms of itself. This seemingly circular type of definition works because all recursive functions need some form of base case where they do not call themselves. The other cases should progress toward the base case. Recursive functions that only call themselves once provide iteration and are typically replaced with loops in most programming languages.[3]

The real power of recursion comes out when a function calls itself more than once. This allows a recursive function to test multiple alternatives. The memory of the stack is essential to this, as what really happens is that one call is executed, and after it has finished, the stack keeps track of where the code was and control resumes there so that other recursive calls can be executed. Functions like this can be converted into loops, but it typically requires significantly more effort. Certain problems that would be quite challenging to solve using

---

[1]The appendices are available at http://www.programmingusingscala.net.

[2]Recursion is a topic many students struggle with, but which is very powerful, so there is value in returning to it several times.

[3]It is worth noting that even if a function only calls itself once, it can still benefit from being recursive if it does some work as it pops back up the stack. We will see examples of this in chapter 16.

normal loops, can be dealt with very easily by means of recursive code. We are going to explore some of those problems in this chapter.

15.2   Project Integration: A Maze

Back in chapter 7 we looked at the problem of completing a maze. In particular, we wanted to find the length of the shortest path through the maze. We used a queue to create a breadth-first solution to this problem. The breadth-first approach visits all locations that are n steps away before going to the locations that are $n+1$ steps away. This is very efficient for finding the shortest path through a maze, but it does not lend itself to other related problems like finding the longest path[4] or counting the number of distinct paths.

In this section, we are going to consider how we could solve this problem using recursion. Using recursion will give us a depth-first search through the maze. This approach will not be as efficient for shortest path, but it will be more flexible for solving other types of problems. We will do this by adding code to our **DrawMaze class**. In addition to adding these other search algorithms, we will add functionality to edit the maze in the drawing and to view the algorithms as they are working.

15.2.1   The Recursive Approach

The **depthFirsthortestPath** method uses a basic recursive approach to this problem. Given a particular location, it is supposed to tell you how many steps it would take to get to the exit on the shortest path. Note that we are not building an algorithm here that returns the actual path the way we did for the breadth-first search. This code can be extended to do so, but we leave that as an exercise for the reader.

As with most recursive methods, we need to start off by checking for some base cases. One base case is that we have reached the end. It takes zero steps to get from the end of the maze to the end of the maze, so this case produces zero. The other base cases are for invalid locations. The obvious invalid locations are locations that are out of bounds or that are the positions of walls. There is a third type of invalid location, a place that you already visited on this path. As with the breadth-first approach, we have to somehow keep track of the places that we have visited, otherwise the computer will happily go back and forth between two adjacent positions. In the breadth-first algorithm, we did this with a set. We could pass a set through the recursive calls, but for our purposes it also works to mutate the array that stores the maze. All of these base cases need to return a value that cannot possibly be the length of the shortest path through the maze. That could be a negative value, or it could be a positive value that is much larger than the number of squares in the maze.

When none of those base cases apply, we get to the recursive case. The way we define this function recursively is to assume that we somehow know the distance to the exit from our four neighboring locations. If we knew those four values, then the distance from the current location would be the smallest of those four plus one for the one step it takes to get from where we are to one of the neighboring locations. We can easily get the smallest of several values in Scala using **min** as in the expression **a min b min c min d**. Note that this logic works better if our base case for an invalid location produces a really large value

---

[4]We only consider paths that do not go through the same location more than once because allowing that immediately means the longest path is infinitely long.

instead of a negative one. The way that we get the value for the neighboring locations is through recursive calls. This is part of the magic of recursion—we assume that the recursive function works and use it as if it does. Assuming we set things up properly, it will work and everything falls together.

The other things that we have to do in the recursive case are put a marker in the maze saying we have been to the current location and then pick it back up after we have done the recursive calls.

Code that implements this approach might look like the following.

```
private def depthFirstShortestPath(x: Int, y: Int): Int = {
  if (x == endX && y == endY) 0
  else if (x < 0 || x >= maze.length || y < 0 || y >= maze(x).length ||
      maze(x)(y) < 0) {
    1000000000
  } else {
    maze(x)(y) = -2
    val ret = 1 + (depthFirstShortestPath(x + 1, y) min
      depthFirstShortestPath(x - 1, y) min
      depthFirstShortestPath(x, y + 1) min
      depthFirstShortestPath(x, y - 1))
    maze(x)(y) = 0
    ret
  }
}
```

The base cases for invalid locations return 1000000000 as a number that cannot be the solution and will not ever be the minimum. It might be tempting to return `Int.MaxValue`, but that is not a good choice because `Int.MaxValue+1` is `Int.MinValue`, which would definitely be the minimum value for lines 7–10.

Line 6, which says `maze(x)(y) = -2`, is dropping a "breadcrumb" on the maze so that the program recognizes that this location has been visited and does not run in loops. You can think of this as moving the recursion toward a base case. Negative values in squares are bases cases and this line makes another square negative before calling the recursion. That can only happen a finite number of times before the recursion runs out of empty squares. This action is reversed on line 11 when the four recursive calls are done so that a square can be revisited on a future path that gets to it in a different way. In the **DrawMaze class** that is shown below, there are a few extra lines added to this that will animate the process.

If you compare this code to that for the breadth-first search, you will see that the recursive code is shorter and in many ways, less complex. Then again, it does not return the path, only the length, so it would have to be extended to be more comparable, but even then, it will still be simpler. As it happens, you can write a depth-first traversal in the same style that we wrote the breadth-first traversal, and the only difference is that instead of using a queue, you use a stack.[5] The primary reason why the recursion can be simpler is that recursion gets the use of a stack implicitly. The program keeps something referred to as the CALL STACK, and every time a method/function is called, it gets a chunk of memory, referred to as a STACK FRAME, to store local variables and other things, like what part of the code is executing, that we do not have direct access to. When the method/function is done and control returns to where it had been called from, the stack frame for that method/function pops off the call stack and is free for use with a subsequent call.

---

[5]It turns out that handling the "breadcrumbs" or the `visited Set` when you switch to a stack is also more challenging because the depth-first approach must be allowed to visit the same location more than once to find an optimal solution.

The `depthFirstShortestPath` method makes heavy use of the call stack implicitly. That is a big part of why it can be simpler than code that explicitly uses a stack or a queue.

15.2.2   Graphical Editing

Previously, we hard coded the maze and just worked with what was there. Ideally, we would like to be able to edit the maze in the drawing program so you can add and remove walls as well as moving the start and end location. It would be possible to put this type of editing into the properties panel, but that is a bit silly considering that the maze is already being drawn in the drawing. There is no reason we should not be able to make it so that the user can interact with the maze through mouse activity on the maze in the drawing.

There are some challenges associated with implementing this type of feature. These arise from three different sources. The most obvious challenge is that the `Canvas` the user will be clicking on is in the `DrawingMain`, not in the `DrawMaze` or other `Drawable` types that we might want to have performing the reaction. Another challenge is that the elements in the drawing can be transformed so that the location of the click on the panel is very indirectly linked to the location of the click relative to the drawn object. The most obscure challenge, and perhaps the one that is hardest to figure out how to solve, is that we do not want to implement changes that break the serialization. All three of these problems can be partially addressed by adding the following `trait` and making some changes to `DrawingMain`.

```
package recursion.drawing

import scalafx.scene.input.MouseEvent

trait Clickable {
  def mouseEvent(me: MouseEvent):Unit
}
```

The idea is that this `trait` should be inherited by any `Drawable` that we want the user to be able to interact with using the mouse. That `class` will then implement this method and check for the type of event so it can take the appropriate action.

In the case of the `DrawMaze`, we only want to deal with clicks. If the user clicks with the normal click button, we will toggle the wall. If they use the secondary button, normally the right-button on a right handed mouse, it will set the location of the end of the maze. If they click with the middle button, it will change the location where the maze starts. This is implemented with the following code.

```
@transient private var lastTransform: Transform = null

    ...

override def mouseEvent(me: MouseEvent): Unit = me.eventType match {
  case MouseEvent.MouseClicked =>
    if (lastTransform != null) {
      val pnt = new Point2D(me.x, me.y)
      val clickPnt = lastTransform.inverseTransform(pnt)
      val gridX = (clickPnt.x / gridWidth).toInt
      val gridY = (clickPnt.y / gridWidth).toInt
      if (gridY >= 0 && gridY < maze.length && gridX >= 0 && gridX <
          maze(gridY).length) {
        me.button match {
          case MouseButton.Primary => maze(gridY)(gridX) = -1 - maze(gridY)(gridX)
```

```
        case MouseButton.Secondary =>
          endX = gridX; endY = gridY
        case MouseButton.Middle =>
          startX = gridX; startY = gridY
        case _ =>
      }
      drawing.draw()
    }
  }
  case _ =>
}
```

The `lastTransform` member is used so that we can convert from coordinates on the `Canvas` to a location in the maze. To make this work, the `draw` method starts with the line `lastTransform = gc.getTransform`. Calling `inverseTransform` gives us a transform that does the opposite of what was happening in the drawing to the provided point. If that is in the bounds of the maze, we check what button was clicked and perform the proper reaction, then tell the drawing to refresh.

In practice, there could be multiple `Drawables` stacked on top of one another and our mouse interactions should only go to, at most, one of them. To make this work, the following modifications were made to the `makeDrawingTab` in `DrawingMain`.

```
var currentClickable: Option[Clickable] = None
drawingTree.selectionModel.value.selectedItem.onChange {
  val selected = drawingTree.selectionModel.value.selectedItem.value
  if (selected != null) {
    propertyPane.content = selected.getValue.propertiesPanel()
    selected.getValue match {
      case c: Clickable => currentClickable = Some(c)
      case _ => currentClickable = None
    }
  } else {
    currentClickable = None
    propertyPane.content = new Label("Nothing selected.")
  }
}


// Canvas Mouse Interactions
canvas.onMouseClicked = (me: MouseEvent) =>
    currentClickable.foreach(_.mouseEvent(me))
canvas.onMousePressed = (me: MouseEvent) =>
    currentClickable.foreach(_.mouseEvent(me))
canvas.onMouseDragged = (me: MouseEvent) =>
    currentClickable.foreach(_.mouseEvent(me))
canvas.onMouseReleased = (me: MouseEvent) =>
    currentClickable.foreach(_.mouseEvent(me))
canvas.onMouseMoved = (me: MouseEvent) =>
    currentClickable.foreach(_.mouseEvent(me))
```

We add a var for an `Option[Clickable]` that will keep track of where the mouse input should go based on selection in the tree. It will store the appropriate `Drawable` if there is a selected item that is `Clickable`. Otherwise, it will have the value of `None`. We then register handlers for clicks, pressed, dragged, and released that use a `foreach` on the `Option` to send the event through to the `Clickable` if there is one.

Having this in place, you might want to think of how other `Drawables` might interact with the mouse and make them `Clickable`. An example that you could implement is to make the `DrawTransform` be `Clickable`. This interaction would likely use pressed, dragged, and released events. What they do would vary based on which transformation type was selected. In the case of translate, you could make it so that `value1` and `value2` were updated by the user dragging the mouse in x and y respectively. This would have the visual effect where the user feels like they can drag around anything below that transform in the drawing.

At this point we present the full code for the `DrawMaze` so that you can go through and see how everything fits together. Note how it `extends Drawable(d) with Clickable`. You should also notice that the `draw` method on lines 39–60 has been extended to show different colors based on `drawVisited` and the values stored in the maze.

```scala
package recursion.drawing

import collection.mutable
import scalafx.Includes._
import scalafx.scene.layout.VBox
import scalafx.scene.canvas.GraphicsContext
import scalafx.scene.Node
import scalafx.scene.paint.Color
import scalafx.scene.control.Button
import scalafx.event.ActionEvent
import stackqueue.adt.ArrayQueue
import scalafx.application.Platform
import scala.concurrent.Future
import scala.concurrent.ExecutionContext.Implicits.global
import scalafx.scene.canvas.Canvas
import scalafx.scene.input.MouseEvent
import scalafx.scene.input.MouseButton
import scalafx.scene.transform.Transform
import scalafx.geometry.Point2D

class DrawMaze(
    d: Drawing,
    private val maze: Array[Array[Int]],
    private var startX: Int = 0,
    private var startY: Int = 0,
    private var endX: Int = 9,
    private var endY: Int = 9) extends Drawable(d) with Clickable {

  private val gridWidth = 20
  private val offsets = Seq((0, -1), (1, 0), (0, 1), (-1, 0))

  @transient private var drawVisited = Map[(Int, Int), Color]()

  @transient private var propPanel: Node = null
  @transient private var lastTransform: Transform = null

  override def toString = "Maze"

  def draw(gc: GraphicsContext): Unit = {
    lastTransform = gc.getTransform
    if (drawVisited == null) drawVisited = Map[(Int, Int), Color]()
    gc.fill = Color.Black
    gc.fillRect(-1, -1, 2 + maze(0).length * gridWidth, 2 + maze.length * gridWidth)
    val max = maze.map(_.max).max
```

```scala
45    for (j <- maze.indices; i <- maze(j).indices) {
46      if (maze(j)(i) == -1) {
47        gc.fill = Color.Black
48      } else {
49        gc.fill = if (drawVisited.contains(i -> j)) drawVisited(i -> j)
50          else if (maze(j)(i) < -1) Color.Red
51          else if (maze(j)(i) > 0) Color(0.0, 1.0, 1.0, 0.5 + 0.5 * maze(j)(i) / max)
52          else Color.White
53      }
54      gc.fillRect(i * gridWidth, j * gridWidth, gridWidth, gridWidth)
55    }
56    gc.fill = Color.Green
57    gc.fillOval(startX * gridWidth, startY * gridWidth, gridWidth, gridWidth)
58    gc.fill = Color.Red
59    gc.fillOval(endX * gridWidth, endY * gridWidth, gridWidth, gridWidth)
60  }
61
62  def propertiesPanel(): Node = {
63    if (propPanel == null) {
64      if (drawVisited == null) drawVisited = Map[(Int, Int), Color]()
65      val panel = new VBox
66      val bfs = new Button("Breadth First Shortest Path")
67      bfs.onAction = (ae: ActionEvent) => Future { breadthFirstShortestPath() }
68      val dfs = new Button("Depth First Shortest Path")
69      dfs.onAction = (ae: ActionEvent) => Future {
70        drawVisited = Map()
71        DrawingMain.showMessageDialog("The shortest path is "
            +depthFirstShortestPath(startX, startY)+" steps.")
72      }
73      val dfsfast = new Button("Fast Depth First Shortest Path")
74      dfsfast.onAction = (ae: ActionEvent) => Future {
75        drawVisited = Map()
76        DrawingMain.showMessageDialog("The shortest path is "
            +depthFirstShortestPathFast(startX, startY, 0)+" steps.")
77        for (j <- maze.indices; i <- maze(j).indices) if (maze(j)(i) > 0) {
78          maze(j)(i) = 0
79          drawVisited += (i -> j) -> Color.Green
80        }
81      }
82      val longest = new Button("Longest Path")
83      longest.onAction = (ae: ActionEvent) => Future {
84        drawVisited = Map()
85        DrawingMain.showMessageDialog("The shortest path is "+longestPath(startX,
            startY)+" steps.")
86      }
87      panel.children = List(bfs, dfs, dfsfast, longest)
88      propPanel = panel
89    }
90    propPanel
91  }
92
93  def toXML: xml.Node = {
94    <drawable type="maze" startX={ startX.toString() } startY={ startY.toString() }
        endX={ endX.toString() } endY={ endY.toString() }>
95    { maze.map(r => <row>{ r.mkString(",") }</row>) }
```

```scala
 96        </drawable>
 97      }
 98
 99      override def mouseEvent(me: MouseEvent): Unit = me.eventType match {
100        case MouseEvent.MouseClicked =>
101          if (lastTransform != null) {
102            val pnt = new Point2D(me.x, me.y)
103            val clickPnt = lastTransform.inverseTransform(pnt)
104            val gridX = (clickPnt.x / gridWidth).toInt
105            val gridY = (clickPnt.y / gridWidth).toInt
106            if (gridY >= 0 && gridY < maze.length && gridX >= 0 && gridX <
                   maze(gridY).length) {
107              me.button match {
108                case MouseButton.Primary => maze(gridY)(gridX) = -1 - maze(gridY)(gridX)
109                case MouseButton.Secondary =>
110                  endX = gridX; endY = gridY
111                case MouseButton.Middle =>
112                  startX = gridX; startY = gridY
113                case _ =>
114              }
115              drawing.draw()
116            }
117          }
118        case _ =>
119      }
120
121      private def breadthFirstShortestPath(): Option[List[(Int, Int)]] = {
122        val queue = new ArrayQueue[List[(Int, Int)]]
123        queue.enqueue(List(startX -> startY))
124        var solution: Option[List[(Int, Int)]] = None
125        val visited = mutable.Set[(Int, Int)]()
126        while (!queue.isEmpty && solution.isEmpty) {
127          val steps @ ((x, y) :: _) = queue.dequeue
128          for ((dx, dy) <- offsets) {
129            val nx = x + dx
130            val ny = y + dy
131            if (nx >= 0 && nx < maze(0).length && ny >= 0 && ny < maze.length &&
                   maze(ny)(nx) == 0 && !visited(nx -> ny)) {
132              if (nx == endX && ny == endY) {
133                solution = Some((nx -> ny) :: steps)
134              }
135              visited += nx -> ny
136              queue.enqueue((nx -> ny) :: steps)
137
138              // Code for animation
139              drawVisited = visited.map(t => t -> Color.Red).toMap
140              drawVisited ++= ((nx -> ny) :: steps).map(t => t -> Color.Green)
141              Platform.runLater(drawing.draw())
142              Thread.sleep(100)
143            }
144          }
145        }
146        // Code for animation
147        solution.foreach(path => drawVisited = path.map(t => t -> Color.Red).toMap)
148        Platform.runLater(drawing.draw())
```

```scala
149    Thread.sleep(100)
150
151    solution
152  }
153
154  private def depthFirstShortestPath(x: Int, y: Int): Int = {
155    if (x == endX && y == endY) 0
156    else if (x < 0 || x >= maze.length || y < 0 || y >= maze(x).length ||
           maze(x)(y) < 0) {
157      1000000000
158    } else {
159      maze(x)(y) = -2
160
161      // Code for animation
162      Platform.runLater(drawing.draw())
163      Thread.sleep(100)
164
165      val ret = 1 + (depthFirstShortestPath(x + 1, y) min
166        depthFirstShortestPath(x - 1, y) min
167        depthFirstShortestPath(x, y + 1) min
168        depthFirstShortestPath(x, y - 1))
169      maze(x)(y) = 0
170      ret
171    }
172  }
173
174  private def depthFirstShortestPathFast(x: Int, y: Int, steps: Int): Int = {
175    if (x == endX && y == endY) 0
176    else if (x < 0 || x >= maze.length || y < 0 || y >= maze(x).length ||
           maze(x)(y) < 0) {
177      1000000000
178    } else if (maze(x)(y) > 0 && maze(x)(y) <= steps) {
179      1000000000
180    } else {
181      maze(x)(y) = steps
182
183      // Code for animation
184      Platform.runLater(drawing.draw())
185      Thread.sleep(100)
186
187      val ret = 1 + (depthFirstShortestPathFast(x + 1, y, steps + 1) min
188        depthFirstShortestPathFast(x - 1, y, steps + 1) min
189        depthFirstShortestPathFast(x, y + 1, steps + 1) min
190        depthFirstShortestPathFast(x, y - 1, steps + 1))
191      ret
192    }
193  }
194
195  private def longestPath(x: Int, y: Int): Int = {
196    if (x == endX && y == endY) 0
197    else if (x < 0 || x >= maze.length || y < 0 || y >= maze(x).length ||
           maze(x)(y) < 0) {
198      -1000000000
199    } else {
200      maze(x)(y) = -2
```

```scala
201
202        // Code for animation
203        Platform.runLater(drawing.draw())
204        Thread.sleep(100)
205
206        val ret = 1 + (longestPath(x + 1, y) max
207          longestPath(x - 1, y) max
208          longestPath(x, y + 1) max
209          longestPath(x, y - 1))
210        maze(x)(y) = 0
211        ret
212      }
213    }
214 }
215
216 object DrawMaze {
217   def apply(d: Drawing) = {
218     val maze = Array(
219       Array(0, -1, 0, 0, 0, 0, 0, 0, 0, 0),
220       Array(0, -1, 0, 0, -1, 0, -1, 0, -1, 0),
221       Array(0, -1, 0, -1, -1, 0, -1, 0, -1, 0),
222       Array(0, -1, 0, 0, -1, 0, -1, 0, -1, 0),
223       Array(0, -1, -1, 0, -1, 0, -1, 0, -1, 0),
224       Array(0, 0, 0, 0, 0, 0, -1, 0, -1, 0),
225       Array(0, -1, -1, -1, -1, -1, -1, 0, -1, -1),
226       Array(0, -1, 0, 0, 0, 0, 0, 0, 0, 0),
227       Array(0, -1, 0, -1, -1, -1, -1, 0, -1, 0),
228       Array(0, 0, 0, -1, 0, 0, 0, 0, -1, 0))
229     new DrawMaze(d, maze)
230   }
231
232   def apply(n: xml.Node, d: Drawing) = {
233     val maze = (n \ "row").map(_.text.split(",").map(_.toInt)).toArray
234     val startX = (n \ "@startX").text.toInt
235     val startY = (n \ "@startY").text.toInt
236     val endX = (n \ "@endX").text.toInt
237     val endY = (n \ "@endY").text.toInt
238     new DrawMaze(d, maze, startX, startY, endX, endY)
239   }
240 }
```

Each of the different path algorithms has some lines that include a call to `Platform.runLater(drawing.draw())` and a `sleep` to allow the drawing of the animation. You should enter this code or pull it down from GitHub to see this in action. Note that there are two path algorithms that are included in the code that we have not discussed yet.

One other thing to note about this code is that it is not thread safe. The different path algorithms are started in separate `Futures`, but we do nothing to prevent multiple path algorithms from being run at the same time, or to prevent the user from editing the path while an algorithm is running. This would be a good use of a lock from `java.util.concurrent.locks`. It is left as an exercise for the reader to make this addition so that the code is thread safe.

15.2.3 Longest Path

We have already said that the recursive shortest path is not as good as the breadth-first shortest path that we wrote previously. This is because our recursive approach is written to take every possible path through the maze and only return the length of the shortest. We will optimize this some in the next section, but first we should look at why this approach is of value even if it is not the ideal for the shortest path.

The advantage of the recursive method is flexibility. The breadth-first approach only works for shortest path. To illustrate this, here is code that finds the longest path using this same approach.

```scala
private def longestPath(x: Int, y: Int): Int = {
  if (x == endX && y == endY) 0
  else if (x < 0 || x >= maze.length || y < 0 || y >= maze(x).length ||
      maze(x)(y) < 0) {
    -1000000000
  } else {
    maze(x)(y) = -2
    val ret = 1 + (longestPath(x + 1, y) max
      longestPath(x - 1, y) max
      longestPath(x, y + 1) max
      longestPath(x, y - 1))
    maze(x)(y) = 0
    ret
  }
}
```

What you should immediately notice is that this code is remarkably similar to that for shortest path. Indeed, there are only two significant changes. First, instead of taking the min of the four recursive calls, this code takes the max. The reason for this change should be fairly obvious. Second, instead of returning 1000000000 for invalid locations, this code returns −1000000000. As before, we know that this cannot possibly be the answer. In addition, we can feel confident that it will not be the value selected by max. Other than those changes, the code is identical.

Longest path is not the only thing that we can calculate using recursion that we could not do with the breadth-first search. Another version of this code you could write is a countPaths. This counts how many distinct paths there are from the start to the end. We leave it as an exercise for the reader to implement this code. The hint is that the changes are similar to those going from shortest to longest path. You have to decide how you combine the results of the four recursive calls and what values should be returned in the base cases.

15.2.4 Optimizing the Maze

If you played around with the maze some, you might have noticed something. If you keep taking out wall after wall and run the shortest path function, as you remove walls, the function will start to run more and more slowly.[6] At a certain point, you will make the maze empty enough that the function will take longer than you have the patience to let it run. This happens because the basic algorithm tests all possible paths to find the shortest. The number of possible paths is extremely large for an empty maze. Our rather small 10

---

[6]If you are going to play with this much, you should probably reduce the length of the sleep call from 100 milliseconds to 10 milliseconds or less.

by 10 maze is sufficient to take many times longer than a human life if left empty; going to 20 by 20 or larger makes things far worse.

The breadth-first approach does not suffer from this issue because it can only visit each location once. Even using recursion, we have do things more intelligently to reduce the total number of calls for shortest path.[7] There are ways around this. The way we will discuss here makes the breadcrumbs smarter so that they can provide additional information so that paths which have no chance of being ideal are eliminated from consideration before they have been fully explored. The code for doing this is shown here.

```scala
private def depthFirstShortestPathFast(x: Int, y: Int, steps: Int): Int = {
  if (x == endX && y == endY) 0
  else if (x < 0 || x >= maze.length || y < 0 || y >= maze(x).length ||
      maze(x)(y) < 0) {
    1000000000
  } else if (maze(x)(y) > 0 && maze(x)(y) <= steps) {
    1000000000
  } else {
    maze(x)(y) = steps
    val ret = 1 + (depthFirstShortestPathFast(x + 1, y, steps + 1) min
      depthFirstShortestPathFast(x - 1, y, steps + 1) min
      depthFirstShortestPathFast(x, y + 1, steps + 1) min
      depthFirstShortestPathFast(x, y - 1, steps + 1))
    ret
  }
}
```

This looks very similar to the earlier version, but there is an extra argument to the function, one extra base case, and the breadcrumbs are not picked up. The extra argument keeps track of how many steps have been taken along the current path. The breadcrumbs, instead of just marking that you have been to a square, keep track of how many steps it took to get there on the shortest path so far. The extra base case is for a location that is reached taking more steps than the current step count. In that situation, the path cannot possibly be the shortest and the recursion is terminated. If you make the maze big enough, even this approach will be insufficient. If you test an empty 30 by 30 maze, you will discover this for yourself.

How Many Paths Are There?

You can empirically check to see that an empty maze takes a long time to solve. With a little work, we can estimate more accurately how long it would take. Imagine a 10 by 10 empty maze. In the recursive case, the function calls itself four times. Some of those will lead to base cases and immediately return. For our estimate we will say that two of the calls are not base cases. So the initial call splits to two, which split to four, and so on. After n steps in the recursion, there are roughly 2^n paths being worked on. The longest path in a 10 by 10 maze is 100 steps. To keep things simple we will stick with that number for how deep the recursion goes. It is an overestimate, but it is somewhat offset by assuming only two branches. This implies on the order of 2^{100} paths.

Hopefully, it is clear that 2^{100} is a very big number. Such big numbers can be hard for

---

[7]The optimization presented in the section does not work for longest path or counting paths. Those calculations require trying out all possible paths and cannot be further optimized.

people to really comprehend. After all, computers are really fast. Modern computers run at clock frequencies of several gigahertz. To help us put this number into perspective, assume that your computer can do 10^9 recursive calls each second.[8] We then use two other rough estimates: $2^{10} \sim 10^3$ and $1yr \sim 3 \times 10^7 sec$. This gives us the following:

$$2^{100} ops \sim 10^{30} ops \sim 10^{21} sec \sim 3 \times 10^{13} yr.$$

That is a few times longer than the best estimates for the current age of the Universe.

This type of exponential growth is common for recursive algorithms. If each non-base case calls the function m times and this goes n calls deep, then there will be m^n total calls. For some problems, this seems to be unavoidable. A lot of effort goes into trying to find ways of getting around this. You should learn about approaches to dealing with this later in your studies.

15.3   Graph Traversals

A maze is a special case of a very general data structure called a GRAPH. If you draw a bunch of dots and put lines between them, you have made a graph. If you are following the end-of-chapter projects, you might well have a form of a graph in the code that you are writing. For example, the multi-user text game has rooms that are connected by exits. That is an example of a graph. For a graphical game you might have different rooms/levels that have doors or other connections between them. This is also a graph.

Using the foundation of code that we created for the maze, it is not to hard to make a `DrawGraph` type which the user can interact with by clicking on the drawing. The full code is shown here. It is fairly long in large part because of the ability to interactively edit the graph. There is a text field that can be used to enter an integer weight for edges that are added to the graph.

```scala
package recursion.drawing

import scala.collection.mutable
import scala.concurrent.Future
import scala.concurrent.ExecutionContext.Implicits.global

import scalafx.Includes._
import scalafx.geometry.Point2D
import scalafx.scene.transform.Transform
import scalafx.scene.canvas.GraphicsContext
import scalafx.scene.paint.Color
import scalafx.scene.layout.BorderPane
import scalafx.scene.control.RadioButton
import scalafx.scene.Node
import scalafx.scene.input.MouseEvent
import scalafx.event.ActionEvent
import scalafx.scene.control.ToggleGroup
import scalafx.scene.layout.VBox
import java.awt.geom.Line2D
import scalafx.application.Platform
```

```scala
21
22   class DrawGraph(d: Drawing, nloc: Seq[Point2D], edges: Seq[(Int, Int, Int)], en:
         Int) extends Drawable(d) with Clickable {
23     import DrawGraph.{ GNode, GEdge }
24
25     private val nodes = mutable.Buffer[GNode]()
26     for (p <- nloc) nodes += new GNode(p.getX(), p.getY())
27     for ((f, t, w) <- edges) nodes(f).edges ::= new GEdge(nodes(f), nodes(t), w)
28
29     private var endNode = if (en < 0) {
30       if (nodes.nonEmpty) nodes(0) else null
31     } else nodes(en)
32
33     @transient private var propPanel: BorderPane = null
34     @transient private var lastTransform: Transform = null
35     @transient private var clickAction: MouseEvent => Unit = null
36     @transient private var hoverNode: GNode = null
37     @transient private var hoverEdge: GEdge = null
38     @transient private var startNode: GNode = null
39     @transient private var (dx, dy) = (0.0, 0.0)
40     @transient private var pathSet: Set[GNode] = null
41     private var currentWeight = 1
42
43     def draw(gc: GraphicsContext): Unit = {
44       def drawNode(n: GNode): Unit = {
45         gc.fill = if (n == endNode) Color.Green else Color.Black
46         gc.fillOval(n.x - 5, n.y - 5, 10, 10)
47         if (n == hoverNode) {
48           gc.fill = Color.Red
49           gc.fillOval(n.x - 4, n.y - 4, 8, 8)
50         }
51       }
52
53       def drawEdge(e: GEdge): Unit = {
54         gc.stroke = if (e == hoverEdge) Color.Red else Color.Black
55         gc.strokeLine(e.from.x, e.from.y, e.to.x, e.to.y)
56         gc.strokeText(e.weight.toString, 0.5f * (e.from.x + e.to.x).toFloat, 0.5f *
               (e.from.y + e.to.y).toFloat)
57       }
58
59       lastTransform = gc.getTransform
60       for (n <- nodes) {
61         drawNode(n)
62         for (e <- n.edges; if (nodes.indexOf(e.from) < nodes.indexOf(e.to))) {
63           drawEdge(e)
64         }
65         gc.stroke = Color.Black
66         if (startNode != null) gc.strokeLine(startNode.x, startNode.y, dx, dy)
67       }
68       if (pathSet != null) {
69         gc.fill = Color.Blue
70         for (n <- pathSet) {
71           gc.fillOval(n.x - 2, n.y - 2, 4, 4)
72         }
73       }
```

```scala
74   }
75
76   def propertiesPanel(): Node = {
77     if (propPanel == null) {
78       import DrawingMain.showMessageDialog
79       propPanel = new BorderPane
80       val weight = DrawingMain.labeledTextField("Weight", currentWeight.toString, s
                => currentWeight = s.toInt)
81       val addNodeButton = new RadioButton("Add Node")
82       addNodeButton.onAction = (ae: ActionEvent) => clickAction = e => e.eventType
              match {
83         case MouseEvent.MouseClicked =>
84           val p = lastTransform.inverseTransform(new Point2D(e.x, e.y))
85           nodes += new GNode(p.x, p.y)
86           if (nodes.length == 1) endNode = nodes(0)
87         case _ =>
88       }
89       val addEdgeButton = new RadioButton("Add Edge")
90       addEdgeButton.onAction = (ae: ActionEvent) => clickAction = e => e.eventType
              match {
91         case MouseEvent.MousePressed =>
92           if (hoverNode != null) {
93             startNode = hoverNode
94             dx = startNode.x
95             dy = startNode.y
96           }
97         case MouseEvent.MouseDragged =>
98           if (startNode != null) {
99             val p = lastTransform.inverseTransform(new Point2D(e.x, e.y))
100            dx = p.x
101            dy = p.y
102          }
103        case MouseEvent.MouseReleased =>
104          if (startNode != null && hoverNode != null) {
105            startNode.edges ::= new GEdge(startNode, hoverNode, currentWeight)
106            hoverNode.edges ::= new GEdge(hoverNode, startNode, currentWeight)
107          }
108          startNode = null
109        case _ =>
110      }
111      val moveNodeButton = new RadioButton("Move Node")
112      moveNodeButton.onAction = (ae: ActionEvent) => clickAction = e => e.eventType
              match {
113        case MouseEvent.MousePressed =>
114          if (hoverNode != null) {
115            startNode = hoverNode
116            dx = startNode.x
117            dy = startNode.y
118          }
119        case MouseEvent.MouseDragged =>
120          if (startNode != null) {
121            val p = lastTransform.inverseTransform(new Point2D(e.x, e.y))
122            startNode.x = p.x
123            startNode.y = p.y
124            dx = p.x
```

```scala
125            dy = p.y
126          }
127        case MouseEvent.MouseReleased =>
128          if (startNode != null && hoverNode != null) {
129            val p = lastTransform.inverseTransform(new Point2D(e.x, e.y))
130            startNode.x = p.x
131            startNode.y = p.y
132          }
133          startNode = null
134        case _ =>
135      }
136      val removeButton = new RadioButton("Remove")
137      removeButton.onAction = (ae: ActionEvent) => clickAction = e => e.eventType
                match {
138        case MouseEvent.MouseClicked =>
139          if (hoverNode != null) {
140            nodes -= hoverNode
141            for (n <- nodes) n.edges = n.edges.filter(_.to != hoverNode)
142            if (hoverNode == endNode && nodes.nonEmpty) endNode = nodes(0)
143            pathSet = null
144          } else if (hoverEdge != null) {
145            hoverEdge.from.edges = hoverEdge.from.edges.filterNot(_ sameAs hoverEdge)
146            hoverEdge.to.edges = hoverEdge.to.edges.filterNot(_ sameAs hoverEdge)
147          }
148        case _ =>
149      }
150      val setEndButton = new RadioButton("Set End")
151      setEndButton.onAction = (ae: ActionEvent) => clickAction = e => e.eventType
                match {
152        case MouseEvent.MouseClicked =>
153          if (hoverNode != null) {
154            endNode = hoverNode
155          }
156        case _ =>
157      }
158      val reachableButton = new RadioButton("Reachable")
159      reachableButton.onAction = (ae: ActionEvent) => clickAction = e =>
                e.eventType match {
160        case MouseEvent.MouseClicked =>
161          if (hoverNode != null) Future {
162            if (endNode != null) {
163              val reachable = canReach(hoverNode, mutable.Set())
164              Platform.runLater { showMessageDialog("The end node is"+(if
                    (reachable) "" else " not")+" reachable.") }
165            } else Platform.runLater { showMessageDialog("There must be an end
                    node.") }
166          }
167        case _ =>
168      }
169      val shortestPathButton = new RadioButton("Shortest Path")
170      shortestPathButton.onAction = (ae: ActionEvent) => clickAction = e =>
                e.eventType match {
171        case MouseEvent.MouseClicked =>
172          if (hoverNode != null) Future {
173            if (endNode != null) {
```

```scala
        shortestPath(hoverNode, Set()) match {
          case None =>
            Platform.runLater { showMessageDialog("There is no path.") }
          case Some((len, ps)) =>
            Platform.runLater {
              showMessageDialog("There is a path of length "+len+".")
              pathSet = ps
              drawing.draw()
            }
        }
      } else Platform.runLater { showMessageDialog("There must be an end
          node.") }
    }
    case _ =>
  }
  val group = new ToggleGroup
  group.toggles = List(addNodeButton, addEdgeButton, moveNodeButton,
      removeButton, setEndButton, reachableButton, shortestPathButton)
  val vbox = new VBox
  vbox.children = List(addNodeButton, addEdgeButton, moveNodeButton,
      removeButton, setEndButton, reachableButton, shortestPathButton, weight)
  propPanel.top = vbox
  }
  propPanel
}

override def toString() = "Graph"

def toXML: xml.Node = {
  <drawable type="graph" en={ nodes.indexOf(endNode).toString }>
  { nodes.map(n => <node x={ n.x.toString } y={ n.y.toString }/>) }
  {
    for (n <- nodes; e <- n.edges) yield <edge from={
        nodes.indexOf(e.from).toString } to={ nodes.indexOf(e.to).toString }
        weight={ e.weight.toString }/>
  }
  </drawable>
}

def mouseEvent(me: MouseEvent): Unit = {
  hoverNode = null
  hoverEdge = null
  var lastDist = 1e100
  for (n <- nodes) {
    val p = lastTransform.inverseTransform(new Point2D(me.x, me.y))
    val dx = p.x - n.x
    val dy = p.y - n.y
    val dist = math.sqrt(dx * dx + dy * dy)
    if (dist < 10 && dist < lastDist) {
      hoverNode = n
      lastDist = dist
    }
    if (lastDist > 3) for (e <- n.edges; if (nodes.indexOf(e.from) <
        nodes.indexOf(e.to))) {
      val line = new Line2D.Double(e.from.x, e.from.y, e.to.x, e.to.y)
```

```scala
223        val edist = line.ptSegDist(p.x, p.y).abs
224        if (edist < 3 && edist < lastDist) {
225          hoverEdge = e
226        }
227      }
228    }
229    if (hoverNode != null) hoverEdge = null
230    if (clickAction != null) clickAction(me)
231    drawing.draw()
232  }
233
234  private def canReach(n: GNode, visited: mutable.Set[GNode]): Boolean = {
235    if (n == endNode) true
236    else if (visited(n)) false
237    else {
238      visited += n
239      n.edges.exists(e => canReach(e.to, visited))
240    }
241  }
242
243  private def shortestPath(n: GNode, visited: Set[GNode]): Option[(Int,
         Set[GNode])] = {
244    if (n == endNode) Some(0 -> visited)
245    else if (visited(n)) None
246    else {
247      val newVisited = visited + n
248      n.edges.foldLeft(None: Option[(Int, Set[GNode])])((last, e) => {
249        (last, shortestPath(e.to, newVisited)) match {
250          case (None, Some((len, v))) => Some((len + e.weight, v))
251          case (_, None) => last
252          case (Some((len1, _)), Some((len2, v))) => if (len1 <= len2 + e.weight)
               last else Some(len2 + e.weight, v)
253        }
254      })
255    }
256  }
257 }
258
259 object DrawGraph {
260   def apply(d: Drawing) = new DrawGraph(d, List(new Point2D(100, 100)), List(), -1)
261
262   def apply(n: xml.Node, d: Drawing) = {
263     val end = (n \ "@en").text.toInt
264     val pnts = (n \ "node").map(pn => {
265       val x = (pn \ "@x").text.toDouble
266       val y = (pn \ "@y").text.toDouble
267       new Point2D(x, y)
268     })
269     val edges = (n \ "edge").map(en => {
270       val from = (en \ "@from").text.toInt
271       val to = (en \ "@to").text.toInt
272       val weight = (en \ "@weight").text.toInt
273       (from, to, weight)
274     })
275     new DrawGraph(d, pnts, edges, end)
```

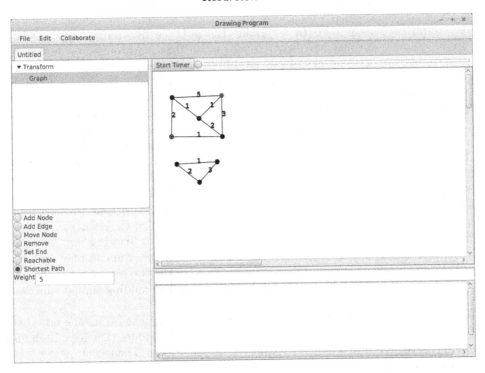

FIGURE 15.1: This is the drawing program with a sample graph created using `DrawGraph`.

```
276      }
277
278      private class GNode(var x: Double, var y: Double) extends Serializable {
279        var edges = List[GEdge]()
280      }
281
282      private class GEdge(val from: GNode, val to: GNode, val weight: Int) extends
             Serializable {
283        def sameAs(e: GEdge): Boolean = {
284          weight == e.weight && ((from == e.from && to == e.to) || (from == e.to && to
               == e.from))
285        }
286      }
287    }
```

To get you an idea of what you might build as a graph using this code, see figure 15.1.

While we do not care too much about the details of drawing the graphs, there are a few points that are worth discussing. Some deal with the inner workings of how the graph editing works, and others are abstraction elements that were included to keep the code shorter. Lines 35–39 declare some **transient** variables that are used for editing the graph. The `clickAction` is a function that takes a `MouseEvent` and does something. It starts off as **null** and is given a value in the `propertiesPanel` when one of the `RadioButtons` is selected. Those `RadioButtons` include options to make new nodes and new edges, move nodes, remove elements, and set what node is the end location.

The `hoverNode` and `hoverEdge` members are set in `mouseEvent` on lines 208–232. This method goes through and checks the mouse location relative to the various nodes and edges. Note that lines 222 and 223 use an element from the `java.awt.geom` package in the Java libraries. This is used to determine the distance between a line segment and the mouse pointer. We could have written code to do that math ourselves, but using the library is simpler. At the end of `mouseEvent`, both `hoverNode` and `hoverEdge` should be set appropriately based on what the mouse is close to. It then calls `clickAction`, which does the behavior specific to the selected edit option.

The actual representation of the graph in this code is at the very bottom on lines 278–280 which define `classes` called `GNode` and `GEdge`. These are declared `private` because they are implementation details that nothing outside of this file should care about.

The element that is significant for recursion is the last two radio button options which run algorithms on the graph. In general, the edges in graphs can be directed so that they only go from one node to another, and not back. The way we store the nodes in the example would allow that flexibility, but the code maintains two-way linkages so that the user interface does not have to be able to distinguish two connections between the same two nodes.

If you put this into your project, add appropriate code to `Drawing` and `Drawable` to handle adding this type and handle loading it from XML, then you can play with it for a while. The main aspect that we want to focus on here are the two methods that recursively run through the graph as some of the project options include adding similar functionality to your code.

The first method simply checks if the end node can be reached from the one that you click on. This method uses a mutable set to keep track of the nodes that have been visited. This set serves the purpose of "breadcrumbs" in this algorithm. The use of a `Set` type is for the purposes of efficiency. Having the set be mutable gives us a behavior like an algorithm that does not pick the breadcrumbs back up the way that the recursive shortest path did. This is fine here because we only want to know if a node can be reached. We are completely unconcerned with the path that is taken. Here is just that method.

```scala
private def canReach(n: GNode, visited: mutable.Set[GNode]): Boolean = {
  if (n == endNode) true
  else if (visited(n)) false
  else {
    visited += n
    n.edges.exists(e => canReach(e.to, visited))
  }
}
```

The base cases are reaching the end node and hitting a node that has already been visited. Each time a node is visited, that node is added into the `Set`. The recursion occurs in a call to `exists` on the edges coming out of the current node. The `exists` method is ideal for us because it will stop execution if any of the elements that are run through produce true. So, if the first edge from a node manages to reach the end, none of the other edges need to be taken.

The second method is another shortest-path algorithm, but this time on a weighted graph instead of a maze. Here again, a set is passed in to keep track of the visited elements. In this case, it is immutable because the path is significant and we need to pick up the breadcrumbs. This method also has a significantly more complex return type. It is `Option[(Int,Set[Node])]`. The `Option` part indicates that it is possible that there is no solution. This would be the case if the end were not reachable from the selected start node. This approach is generally preferred over using `null` to indicate that nothing worked. The `Int` is the length of the path, and the `Set[Node]` is all the nodes that were visited on the path that was found. This technically is not the same as returning a path, but it provides enough information that the user can easily figure out the path taken in most situations.

It was also easy to build as we are keeping track of the visited nodes using immutable sets anyway.

```
private def shortestPath(n: GNode, visited: Set[GNode]): Option[(Int,
    Set[GNode])] = {
  if (n == endNode) Some(0 -> visited)
  else if (visited(n)) None
  else {
    val newVisited = visited + n
    n.edges.foldLeft(None: Option[(Int, Set[GNode])])((last, e) => {
      (last, shortestPath(e.to, newVisited)) match {
        case (None, Some((len, v))) => Some((len + e.weight, v))
        case (_, None) => last
        case (Some((len1, _)), Some((len2, v))) => if (len1 <= len2 + e.weight)
            last else Some(len2 + e.weight, v)
      }
    })
  }
}
```

The base cases here are the same as for the reachable method. They simply return slightly different values. The recursive case is also a bit longer to deal with the different options. It uses a fold to compile an answer, running through the different edges out of the current node. The fold starts with a value of None specifying the appropriate type. The function in the fold does a match on the incoming value and the recursive call of the current edge. When there is a None, the other value is used. If both of the values are a Some, a pattern is used to pull out the length so the shorter one can be returned. The patterns for the match here display the use of _ to avoid giving names to values that we are not going to use.

This shortest path solution for the graph is just like the original solution to the maze and has the same pitfalls in that it can take exponentially long for large graphs. Here again, there are alternate approaches that can produce superior performance. In particular, you can do a version of breadth-first, but it has to be modified to take into account that the edges can have different weights. Implementing this is a good exercise for the reader.[9] Graphs can be used to represent many different systems, and algorithms on them are a significant concept that you should see much more of during your computing career.

15.4   Divide and Conquer

One standard problem solving technique that uses recursion is the divide and conquer approach. The idea is to take a big problem, and cut it into two or more smaller pieces, solve those pieces, then combine the results into a complete solution. This is similar to the idea behind problem decomposition. The standard approach to solving hard problems is that they should be broken down into easier problems. It is an essential aspect of programming, as programs can be arbitrarily complex. A similar approach can be used in programs to solve specific problems.

Divide and conquer is recursive, because the function that solves the whole problem is used to solve the pieces as well. The base case is whenever you get down to a size that you

---

[9]If you undertake this challenge, consider a priority queue instead of a regular queue to deal with the different weights. Doing that in the proper way produces an approach called Dijkstra's algorithm.

can solve without breaking it down. Often this has you go all the way down to the point of being trivial. The recursion moves you toward the base case as each recursive call is on a smaller problem than you started with. We will explore several different examples of divide and conquer algorithms to see how this approach works.

The first examples of divide and conquer that we will consider are sorting algorithms, merge sort and quicksort. Both of these sorts work by taking a sequence of numbers and breaking it into two pieces that are handled recursively. The default base case is a single element, as any sequence with one element is always properly sorted. However, it can be more efficient to stop the recursion at some larger number of elements and switch to some other, simpler sort to finish off the smaller segments.

15.4.1 Merge Sort

Merge sort works by recursively dividing the array in two equal parts[10] and recursively calls itself on those two parts. The real work in a merge sort happens as the code pops back up the call stack and the sorted subsections are merged together. One of the key advantages of a merge sort is that it is remarkably stable in how much work it does. It always divides down the middle so the recursion always goes $\log_2(n)$ levels deep. The merge operation for two lists of $n/2$ elements requires $n - 1$ comparisons and n memory moves. Given this, a merge sort will always scale as $O(n \log_2(n))$, regardless of the nature of the input.

The primary disadvantage of a merge sort is that it cannot be done in-place. This is because you cannot merge an array with two halves that are independently sorted into a single sorted array without using extra space. This makes a merge sort rather well suited for lists, which merge easily, but are immutable so the idea of in-place operations does not make sense. Writing a merge sort that works with arrays and does not consume more than $O(n)$ memory is a bit trickier.

We begin with the simplest form of merge sort. This is a sort of a List that uses a recursive merge. The code for that function is shown here.

```scala
def mergeSort[A](lst: List[A])(comp: (A, A) => Int): List[A] = {
  def merge(l1: List[A], l2: List[A]): List[A] = (l1, l2) match {
    case (_, Nil) => l1
    case (Nil, _) => l2
    case (_, _) =>
      if (comp(l1.head, l2.head) <= 0) l1.head :: merge(l1.tail, l2)
      else l2.head :: merge(l1, l2.tail)
  }

  val len = lst.length
  if (len < 2) lst
  else {
    val (front, back) = lst.splitAt(len / 2)
    merge(mergeSort(front)(comp), mergeSort(back)(comp))
  }
}
```

Note that the merge function is nested inside of mergeSort. This works well, as it prevents it from being accessed by outside code and gives it implicit access to the comparison function. The body of the primary method appears at the bottom, which has a base case for lists with a length of less than two. Otherwise, it splits the list at the midpoint, recursively calls sort

---

[10]Or within one of being equal in the case of odd numbers of elements.

on the halves, and merges the result. The length of the list is stored in a variable because it is an $O(n)$ call that we do not want to repeat more than is required.

The merge here is recursive with three cases. If either list is empty, the result is the other list. When neither is empty, a comparison of the two heads is performed and the lesser is consed on the front of the merge of what remains. This is fairly simple code for a sort that promises $O(n \log(n))$ performance, but unfortunately, it will not work well for long lists. The recursive merge function is not tail recursive and, as a result, it can overflow the stack if the lists being merged are too long. This limitation can be removed by using a **merge** function that employs a **while** loop.[11]

```scala
def merge(l1: List[A], l2: List[A]): List[A] = {
  var (lst1, lst2, ret) = (l1, l2, List[A]())
  while (lst1.nonEmpty || lst2.nonEmpty) {
    if (lst2.isEmpty || (lst1.nonEmpty && comp(lst1.head, lst2.head) <= 0)) {
      ret ::= lst1.head
      lst1 = lst1.tail
    } else {
      ret ::= lst2.head
      lst2 = lst2.tail
    }
  }
  ret.reverse
}
```

This function builds the merged list in reverse order, then calls the **reverse** method to turn it around. This might seem odd, but remember that adding to the head of a **List** is $O(1)$ while adding to the tail is $O(n)$. For this reason, it is significantly more efficient to take the approach shown here than to append and not reverse.

This gives us a reasonably good merge sort for lists. The use of an immutable type for sorting forces us to create a separate storage for the result. When dealing with an array, however, the mutability means that many sorts can use no more memory than the provided array and one temporary space holder. As was mentioned above, that is not an option for the merge operation. However, using the style we used for the list when dealing with an array would be extremely inefficient, requiring $O(n \log(n))$ memory. It turns out that we can fairly easily create an array-based version that uses $2n \in O(n)$ memory by making a second array and doing merges from the original array to the second array and back. The following code demonstrates that.

```scala
def mergeSort[A: ClassTag](a: Array[A])(comp: (A, A) => Int) {
  val data = Array(a, a.map(i => i))
  def mergeSortRecur[A](start: Int, end: Int, dest: Int) {
    val src = 1 - dest
    if (start == end - 1) {
      if (dest == 1) data(dest)(start) = data(src)(start)
    } else {
      val mid = (start + end) / 2 // Can fail for arrays over 2^30 in length
      mergeSortRecur(start, mid, src)
      mergeSortRecur(mid, end, src)
      var (p1, p2, pdest) = (start, mid, start)
      while (pdest < end) {
        if ((p2 >= end || comp(data(src)(p1), data(src)(p2)) <= 0) && p1 < mid) {
```

---

[11] It is also possible to write a tail recursive merge by passing in an additional argument that is the merged list. This prevents the code from having the :: waiting to happen after the recursive call has completed.

```
          data(dest)(pdest) = data(src)(p1)
          p1 += 1
        } else {
          data(dest)(pdest) = data(src)(p2)
          p2 += 1
        }
        pdest += 1
      }
    }
  }
  mergeSortRecur(0, a.length, 0)
}
```

Here the first line creates an `Array[Array[A]]` that holds the original array and a second array which is made as a copy using `map`. There is a nested recursive function declared that takes the start and end of the range we are sorting as well as the index of the destination in `data`. That destination alternates between 0 and 1 as we go down the call stack. The merge is done from `src` into `dest`. The initial call uses a destination of 0 so the final sorted version winds up being in the original array.

The comment on the calculation of `mid` is worth noting. For arrays with a bit more than a billion elements, the quantity `start+end` can overflow and `Int`. This will lead to a negative value for `mid`. You can get around this by using `mid = start+(end-start)/2` instead. This expression will work for all legal array sizes.

15.4.2 Quicksort

A second sorting example of divide and conquer is the quicksort. Unlike merge sort, the quicksort can be done in-place and, when written that way, it does all of its work going down the stack instead of on the way back up. Unlike merge sort, the quicksort is less stable and can degrade to $O(n^2)$ performance if things go poorly. This happens due to poor selection of a pivot. Quicksort works by selecting one element to be the "pivot" and then it moves elements around so that the pivot is in the correct location with elements that go before it located before it and those that go after it located after it. If pivots are selected which belong near the middle of the array, this produces $O(n \log(n))$ performance. However, if you consistently pick elements near the edges, the recursion has to go $O(n)$ levels deep and the resulting performance is truly horrible.

We can start with a version that works on `List`s for comparison with the original merge sort.

```
def quicksort[A](lst: List[A])(lt: (A, A) => Boolean):List[A] = lst match {
  case Nil => lst
  case h::Nil => lst
  case pivot::t =>
    val (before, after) = t.partition(a => lt(a, pivot))
    quicksort(before)(lt) ::: (pivot :: quicksort(after)(lt))
}
```

This code uses the first element of `lst` as the pivot. This is a really poor way to select the pivot, as sorted data will cause the pivot to always go at the beginning or end, making the algorithm really slow. Just selecting a pivot at random makes this type of behavior unlikely. With a bit more logic, we can make it almost impossible.

Of course, the advantage of quicksort over merge sort is that it can be done in place. This only works with a mutable collection like an `Array`. This next version is a general implementation of an in-place quicksort that can be easily modified in regard to picking a pivot. This first version always uses the first element, but it has that pulled off in a function so that it is easy to modify.

```scala
def quicksort[A](a:Array[A])(comp:(A,A)=>Int) {
  def pickPivot(start:Int,end:Int) = start

  def qsRecur(start:Int,end:Int) {
    if (start<end-1) {
      val pivot = pickPivot(start,end)
      val p = a(pivot)
      a(pivot) = a(start)
      a(start) = p
      var (low,high) = (start+1,end-1)
      while (low<=high) {
        if (comp(a(low),p)<=0) {
          low += 1
        } else {
          val tmp = a(low)
          a(low) = a(high)
          a(high) = tmp
          high -= 1
        }
      }
      a(start) = a(high)
      a(high) = p
      qsRecur(start,high)
      qsRecur(low,end)
    }
  }
  qsRecur(0,a.length)
}
```

This code retains that unfortunate characteristic that if it is used on a sorted array, it will have $O(n^2)$ performance. However, that can be corrected by making the following change.

```scala
def pickPivot(start: Int, end: Int) = start + util.Random.nextInt(end - start)
```

This modified `pickPivot` function picks a random value for the pivot. It can still have $O(n^2)$ behavior, but the odds are very low, especially for large arrays, as it requires randomly picking elements that are near the smallest or the largest in a range repeatedly.

The behavior of quicksort can be improved further by incorporating a bit more logic into the pivot selection and by using insertion sort when the segment we are sorting gets small enough. It might seem counterintuitive to rely on a sort that we know is $O(n^2)$ to help improve a sort that is supposed to be $O(n \log(n))$, however, this can be understood by remembering that order notation is really only relevant for large values of n. When n is small, the sorts that we normally think of as being inferior can actually demonstrate significantly better performance simply because they are simpler and have less overhead. Code that incorporates both of these improvements is shown here.

```scala
def quicksort[A](a: Array[A])(comp: (A, A) => Int) {
  def insertionSort(start: Int, end: Int) {
    for (i <- start + 1 until end) {
      val tmp = a(i)
      var j = i - 1
```

```scala
        while (j >= 0 && comp(a(j), tmp) > 0) {
          a(j + 1) = a(j)
          j -= 1
        }
        a(j + 1) = tmp
      }
    }

    def pickPivot(start: Int, end: Int) = {
      val mid = start + (end - start) / 2
      val sm = comp(a(start), a(mid))
      val se = comp(a(start), a(end - 1))
      if (sm <= 0 && se >= 0 || sm >= 0 && se <= 0) start
      else {
        val me = comp(a(mid), a(end - 1))
        if (sm <= 0 && me <= 0 || sm >= 0 && me >= 0) mid else end - 1
      }
    }

    def qsRecur(start: Int, end: Int) {
      if (start < end - 7) {
        val pivot = pickPivot(start, end)
        val p = a(pivot)
        a(pivot) = a(start)
        a(start) = p
        var (low, high) = (start + 1, end - 1)
        while (low <= high) {
          if (comp(a(low), p) <= 0) {
            low += 1
          } else {
            val tmp = a(low)
            a(low) = a(high)
            a(high) = tmp
            high -= 1
          }
        }
        a(start) = a(high)
        a(high) = p
        qsRecur(start, high)
        qsRecur(low, end)
      } else {
        insertionSort(start, end)
      }
    }
    qsRecur(0, a.length)
  }
```

The pivot selection works by finding the median of the first, middle, and last elements in the range. While technically this only eliminates selecting the smallest or largest element, it improves the probably of picking an element near the middle dramatically. It also provides ideal performance for arrays that were already sorted.

This method of picking a pivot also calls for using something other than quicksort when there are fewer than three elements. After all, the pivot selection process does not make sense if the first, middle, and last elements are not different. In this code, we have switched over to an insertion sort for any array with fewer than seven elements. This value was selected using some empirical testing to count the number of comparisons that were performed on random data sets.

15.4.3 Formula Parser

The last divide and conquer problem we will consider in this chapter is the problem of formula parsing. Consider an application where you want the user to be able to type in basic math formulas for evaluation. This would give you applications somewhat like graphing calculators, which can evaluate expressions the way you would write them on paper. So if the user were to enter "3+5*2" you want a function that can take that as a `String` and return the value 13.[12]

There are a number of different approaches to this. Probably the most well known is the "shunting-yard" algorithm developed by Edsger Dijkstra. An alternate approach uses divide and conquer. This approach divides the formula at the lowest-precedence operator, recurses on the two sides of that operator, and then applies the operator to the two return values to get an answer. If no operator is found, we have a base case for a number or an expression completely in parentheses, in which case the function should recurse on the contents of the parentheses.

An `object` with a method called `eval` for doing just this is shown here. It simply calls the recursive function, passing the formula with all spaces removed. The parse defines some variables, then has a loop that finds the lowest-precedence operator. It ends with a set of ifs to deal with different possibilities.

```scala
package recursion.util

object Formula {
  val ops = "+-*/".toSet

  def eval(form: String): Double = evalParse(form.filter(_ != ' '))

  private def evalParse(f: String): Double = {
    var opLoc = -1
    var parensCount = 0
    var i = f.length - 1
    while (i > 0) {
      if (f(i) == '(') parensCount += 1
      else if (f(i) == ')') parensCount -= 1
      else if (parensCount == 0 && (f(i) == '+' || f(i) == '-' && !ops(f(i - 1)))) {
        opLoc = i
        i = -1
      } else if (parensCount == 0 && opLoc == -1 && (f(i) == '*' || f(i) == '/')) {
        opLoc = i
      }
      i -= 1
    }
    if (opLoc < 0) {
      if (f(0) == '(') {
        evalParse(f.substring(1, f.length - 1))
      } else f.toDouble
    } else {
      f(opLoc) match {
```

---

[12]This example was chosen because it illustrates that we want to have proper order of operations so that multiplication is done before addition.

```
29      case '+' => evalParse(f.take(opLoc)) + evalParse(f.drop(opLoc + 1))
30      case '-' => evalParse(f.take(opLoc)) - evalParse(f.drop(opLoc + 1))
31      case '*' => evalParse(f.take(opLoc)) * evalParse(f.drop(opLoc + 1))
32      case '/' => evalParse(f.take(opLoc)) / evalParse(f.drop(opLoc + 1))
33    }
34   }
35  }
36 }
```

The first variable declared in the method on line 9 is `opLoc`, which should store the location of the operator. It is initialized to -1 and if no operator is found, it will still have that value at the end. After that is `parensCount`, which keeps track of how deeply nested we are in parentheses. The function is written so that no operator that is in parentheses can be considered to be the lowest-precedence operator. Last is a loop variable, `i`, which starts at the end of the string. The operators we are using are all left-associative, so the lowest-precedence operator will be the farthest to the right if there are multiple operators with the same level of precedence. To make this clear, $5 - 3 - 2$ is evaluated as $(5 - 3) - 2$, not $5 - (3 - 2)$, so the right minus sign has a lower precedence.

The `while` loop starts off with checks to see if there are parentheses. After that are checks to see if we have found different operators. There are some details to these that are worth noting. Both of the operator checks require that the parentheses count be zero. The check for a minus sign also requires that the character in front of it not be an operator. This is to make sure that formulas like "5+-3" work. In this situation, the '-' is negation, not a binary operator. When either '+' or '-' are found, the location is stored in `opLoc`, and the value of `i` is set to be negative so the loop will stop. This is because the first addition or subtraction that is found will always be the lowest-precedence operator.

For multiplication and division, the behavior is a bit different. These characters only matter if `opLoc` is -1 and they do not change the value of `i`. This is because finding a '*' or '/' does not automatically mean we have found the lowest-precedence operator. There could still be a '+' or '-' farther left that would have lower precedence. For that reason, the loop must continue working. However, if another '*' or '/' is found to the left of the first one seen, it should be skipped over because of the associativity.

We will continue playing with the formula parser in later chapters. In the next chapter we will add the ability to handle variables. That will let us include formulas in our project. We do not want to do it with this code, because string parsing is slow and we do not want to use code that parses the string every time a formula is evaluated.

15.5   End of Chapter Material

15.5.1   Summary of Concepts

This chapter was intended to give you a more advanced look at recursion and to explore some algorithms that use it in a bit of detail.

- Algorithms for optimal path finding through a maze.

  - The standard recursive approach is a brute-force method that checks all paths and picks the shortest. There can be a huge number of paths making this impractical for some situations.

- Using "smart breadcrumbs" allows for certain paths to be ignored as soon as it is found that they cannot be minimal.

- A drawable can be made to respond to mouse clicks on the drawing.

    - The drawable needs to inherit from a specific type so that the drawing knows if it should be listening for things.

    - Transforms can be considered by remembering the last ones used and using `inverseTransform`.

- We took our first look at graphs. These combine nodes and edges and can be used to represent many different types of problems.

    - Code for a reachability algorithm recursively follows edges and passes through a mutable set of visited nodes as nodes never have to be revisited.

    - Code for a shortest-path algorithm recursively follows edges and passes in an immutable set of visited nodes so that nodes are revisited on different paths.

- Divide and conquer algorithms. These take a large problem and break it into pieces, solve the pieces, and combine the solutions to get a final answer. The base case is typically when you get down to a size for which the solution is trivial.

    - We looked at merge sort and quicksort and explored a number of implementations of them.

    - We wrote a divide and conquer solution to formula parsing that allows basic operators and parentheses.

15.5.2 Exercises

1. Earlier in this chapter we saw the code for a longest-path algorithm. You can also construct a `countPaths` method with simple modifications to this approach. This counts how many distinct paths there are from the start to the end. Write an implementation of this. Hint: The changes are similar to those going from shortest to longest path. You have to decide how you combine the results of the four recursive calls and what values should be returned in the base cases.

2. Add a path count function and GUI option to the `DrawMaze`.

3. Make the animations thread safe by preventing other options from being activated while a maze solution is being calculated.

4. Compare the speed of the different merge sort and quicksort versions.

5. Extend the formula parser so that it can also do square roots.

6. Extend the formula parser so that is can also do the three basic trigonometric functions.

15.5.3   Projects

1. It is now time to implement a number of additional commands in your MUD. Hopefully you have been adding things to make it playable and working on having a real map. In a real MUD, map editing is something that can be done by certain users using standard commands from inside the MUD. You should not have to go out and edit XML by hand or anything like that. If you have not yet added this functionality to your MUD, you should now.

   There are some other commands that will be very handy for anyone trying to build a map, mainly to check that the result looks the way you want. In particular, you need to be able to jump to random rooms, check if you can get from one room to another, and see how many paths there are between rooms. These are not normal commands. They should only be available to users who have the rights for creating rooms. The last two can be written as recursive algorithms that run from room to room following the exits that connect them.

   If you want an extra challenge, write a "goto room" command where you specify the keyword of a room. The code should do a recursive search for a path. When it finds one (you can decide if you want the optimal one), it will schedule on the event queue that you wrote for priority queues, all of the moves with reasonable delays to take them there.

2. A good recursive problem for the web spider is to determine how many clicks it takes to get from one page to another. Closely related to that, you can measure distances, in clicks, between different data elements in a data set. Compare this to the version you wrote using a breadth-first search with a queue.

3. Recursion can be used for a number of different tasks in graphical games. Which you pick really depends on what your game is like. All are at least roughly related to the maze algorithm. If your game has multiple screens that are connected (picture classic Zelda) it could be useful to have a tool that tells you how different screens are connected.

   If your game has computer-controlled characters and obstacles that can block movement, then a recursive algorithm can be used to find a route from the character to the player. Note that in general, you do not want to allow an AI to follow the shortest route to the player. The reason being that it becomes impossible for the player to evade the enemy. Compare your code to what you did with the breadth-first search using a queue.

4. If you are doing the math worksheet project, you should add the ability for the user to define functions with names. Then make it so that the definition of one function can include a call to another function. This will use recursion in your code. You do not have to go all the way and include conditionals in the functions so that the functions themselves can be recursive. You are certainly allowed to try, but that is functionality that you will be able to add in later.

5. Every good paint program needs a flood fill option. Your Photoshop® project is no exception. The "Fill with color" fill option can be written in a very compact form using recursion. You should implement this and try it. What you will find if you have a large image is that you can easily overflow the stack by clicking in an open area.

   To avoid this, you have to switch from using recursion to using a loop with a manual stack or queue for keeping track of things. If you use a stack, the order of the fill will match what happens in recursion. If you use a queue, you get a breadth-first ordering instead. For this application, the difference will not be significant.

6. In chapter 13 you added a collisional simulation to the simulation workbench using a priority queue. That code has two different aspects that force every step to take $O(n^2)$ time, the comparison of particles to determine collision times, and the inserting of items into a sorted linked list priority queue. We will look at one way to improve on the first of these.

   Currently your code finds collisions by doing comparisons between every pair of particles in the simulation. In reality, collisions are localized. During the fairly short period of time represented by a time step in the simulation, it generally is not possible for a particle to collide with other particles on the opposite side of the simulation area. Technically, you can only collide with particles where $d < \Delta t |\vec{v_1} - \vec{v_2}|$, where d is the distance between the particles, Δt is the length of the time step, and $\vec{v_1}$ and $\vec{v_2}$ are the velocities of the two particles. For this reason, you really only need to compare to things that are close by. The problem is, you do not know in advance what is close by.

   One way to deal with this is to maintain, in each particle, knowledge of its nearest neighbors. Have each particle keep track of the closest m other particles. The value of m should probably be at least four for a 2-D simulation and six to eight for a 3-D simulation. Building those connections initially requires a full search and $O(n^2)$ checks. After it has been built, searches for particles to do collision checks against can be done with a recursive search running through the connections. The search can stop anytime it gets to a particle that is too far away.

   This same method can also be used to maintain the closest m particles. If the recursive search ever hits a particle that is closer than one of the ones currently being tracked, the new one can replace it. Every so often, another $O(n^2)$ search can be done to make sure the connections are all good, but this allows most of the steps to be done with each particle compared to a significantly smaller number of other particles.

7. Many users of the stock portfolio management system would like to have an optimal stock portfolio. Making an assumption that each stock would do as well next week as it did overall last week, write an optimization that makes a recommendation for each user as to what stocks and what amounts of each stock should be in each user's portfolio. Keep in mind that the minimum valuation of each portfolio should be \$10,000 and no individual stock's valuation should exceed one third of the total portfolio's value.

8. You would like to maximize user's satisfaction with your movie rental business. In order to do that, it is in your best interest to make sure that each user gets the highest ranked movie on their request list based on what you have available in inventory. Of course, you most likely do not have enough copies of every movie such that every user will get their first choice. Some users will end up with their second choice, and some with their third choice, etc. Obviously, you want each user to get a movie that is close to the top of their list as possible. Write an optimization that help you determine what group of movies you want to send out that would maximize total customer satisfaction.

9. You previously implemented the '[' and ']' characters for turtle graphics using a stack. You can also do this using recursion and the implicit call stack. Make your drawing method a recursive method. Any time you come across a '[' character, make a copy of your turtle and do a recursive call to the draw. When you hit a ']' character, you should return from the drawing function.

Chapter 16

Trees

"I think that I shall never see
A poem as lovely as a tree" - Joyce Kilmer

In this chapter you will learn about the concept of a tree as they are known in computer science. While we will not be writing poems about them, we will construct some lovely code to provide efficient implementations of some ADTs.

16.1 General Trees

Chapter 12 introduced the concept of the linked structure using its simplest form, a linked list. The most general expression of a linked structure is the graph that was briefly presented in chapter 15. A tree is a linked structure that falls somewhere between these two extremes. Like the other two, a tree is formed from vertices/nodes with edges between them. We consider these edges to have a direction, pointing from one vertex/node to another. Unlike a graph, where we can do pretty much anything we want with the edges between the nodes, to qualify as a tree, there are certain limitations. We will run through these and introduce the basic terminology, which is largely a mixture of metaphors from biology and genealogy.

A tree can include any number of nodes. One node is the ROOT of the tree. This node is special in that it has no incoming edges. Consider that you have a node C and another node P such that there is an edge from P to C. See figure 16.1 for an example. We call P the PARENT of C and C is the CHILD of P. To qualify as a tree, every node other than the root has one incoming edge. As a result, there is one and only one path from the root to any given node. Other nodes that have an edge going from P to them are the SIBLINGS of C. Going further with the genealogy metaphor, the other nodes that can be reached going down edges from a node are called the DESCENDANTS of that node. Similarly, the nodes that are on the path from a node to the root, including the root, are the ANCESTORS of the node.

Following the biology metaphor, any node that does not have outgoing edges is called a

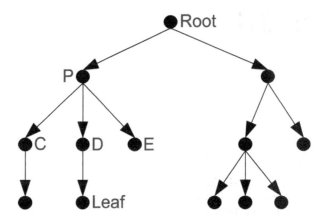

FIGURE 16.1: This figure shows a sample tree with some of the nodes labeled. Node P is the parent of nodes C, D, and E. D and E are siblings of C. Node P has a depth of 1 and a height of 2, while node C has a depth of 2 and a height of 1. The size of node P is six, while the size of node C is 2.

LEAF. The number of edges between the root and a node is called the DEPTH of the node. The number of edges between a node and the deepest leaf is called the HEIGHT of that node. The term SIZE refers to the number of descendants of a node, plus itself.

We can also define a tree in a recursive way. Using this definition, a tree is either an empty node or a node with edges to zero or more non-empty subtrees under it. This definition will be directly reflected in some of the implementations.

Note that for all these definitions of a tree, when you draw one out, there are no cycles of any kind. This is probably the key image to have in your mind when you think about trees—they do not allow cycles. You should also note from this description that a linked list is actually a perfectly happy tree. The root of the tree is equivalent to the head of the list, and each node has only one outgoing edge to a child. It might be a boring tree, but the standard, singly linked list definitely qualifies as a tree.

16.1.1 Implementations

A general tree allows each node to have an arbitrary number of children. This leads to a variety of different implementation strategies. One approach is to have the node store whatever data is needed along with a sequence of children. Our `Drawable` type defines a node in a tree and we used this approach. In that tree, the internal nodes are all of the `DrawTransform` type while the leaves are all the other types. The `DrawTransform` contains the following line of code.

```
private val _children: mutable.Buffer[Drawable],
```

This is the sequence of children. For this example, a mutable buffer was used, but other sequence implementations could work just as well.

Another example of a general tree is the file system on a computer. Depending on your operating system, you will have a drive or a root directory. Under that will be some combination of files and directories. The normal files are all leaves, while the directories can be internal nodes with children.

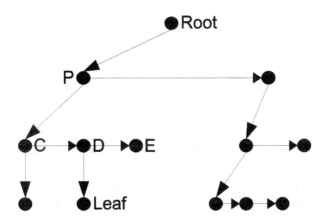

FIGURE 16.2: This figure shows the same tree that was displayed in figure 16.1, but draws arrows for the first-child, next-sibling approach to building the tree.

Instead of using a sequence in the node, a general tree can be built by explicitly utilizing a linked list. In this approach, each node keeps the data it needs along with references to the first child and the next sibling. The links for the next sibling provide a linked list of the children for a given node. A graphical representation of what this looks like in memory is breadth-first figure 16.2. Given this implementation, no node has to store more than two references to other nodes. This generally makes it a bit more efficient in memory, if somewhat more challenging to deal with in code.

16.1.2 Traversals

As with linked lists, trees are used as ways to store information. The tree simply has more structure to it. One of the most significant activities we want to perform on any collection of data is a traversal through the elements. For an array, we do this by counting indexes. For a linked list this was accomplished by having a reference, which we called **rover**, start at the head and follow links down the list. For a tree, things are a bit more complex. There is no single natural ordering for a tree. It makes sense for children to be handled from left to right, but there are questions about whether parents should come before or after children or whether children should be handled before siblings. This leads to the possibility for multiple different traversals of trees.

One way to do a traversal is breadth-first. This is the same term used for our shortest path maze algorithm because they operate in the same way. The breadth-first approach to the maze went to all locations that were n steps away before going to any that were $n + 1$ steps away. In the context of the tree, the breadth-first traversal considers all elements that are at a depth of n before considering any at a depth of $n + 1$. You can view this as going across each level in the tree from left to right. So in the tree in figure 16.3, a breadth-first traversal would visit the nodes in alphabetical order. The traversal of the maze can be viewed as a tree as well, where each node in the tree is a location in the maze and one node is a child of another if it was added to the queue as a neighbor of the parent.

You can also perform depth-first traversals of a tree. Again, the terminology is the same as used for the recursive searches of the maze. There are several variations on depth-first traversals. They all handle children before siblings. There are a family of options, because even with that rule, there is the question of when the parent is handled relative to children.

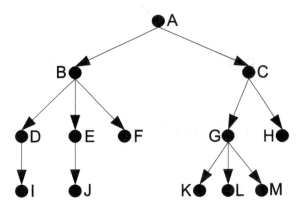

FIGURE 16.3: This figure shows the same tree that was displayed in figures 16.1 and 16.2. Here, all of the nodes are labeled with letters to aid the discussion of traversal.

If the parent is handled before the children, it is called a PRE-ORDER traversal. If the parent is handled after the children it is called a POST-ORDER traversal. When the parent is handled between children, it is an IN-ORDER traversal. We will go through these one at a time and consider the order of handling for the tree shown in figure 16.3. We will also consider the code we would write for such traversals.

All of the depth-first searches can be traced through the tree in the same way. You go from a node to its first child and everything below it, then pop back up to the node and go to the next child. Nodes are not handled/visited as soon as they are hit or every time they are hit, but instead when the desired traversal specifies.

A PRE-ORDER traversal handles parents before children and children are handled from left to right. With our sample tree, this gives the following order for the nodes: A, B, D, I, E, J, F, C, G, K, L, M, H. In the pre-order, parents always come before their children. As a result, the root is handled first. The lowest child on the far right side will be handled last. Assume that our `Node` class looks something like this,

```scala
case class Node(data: A, children: Seq[Node])
```

where `A` is the type of the data being stored in the tree. Then the code for a pre-order traversal might look like the following.

```scala
def preorder(n: Node, handler: A => Unit): Unit = {
  if (n != null) {
    handler(n.data)
    n.children.foreach(n2 => preorder(n2, handler))
  }
}
```

The `handler` is a function that is passed in that should be called at the appropriate time. Note that for a pre-order traversal, it is called before the recursion handles all of the children. A full traversal can be initiated by calling this function on the `root` node of the tree.

It is also possible to handle the parent after the children. This gives a POST-ORDER traversal. The post-order traversal for our sample tree has the following order for handling nodes: I, D, J, E, F, B, K, L, M, G, H, C, A. In this ordering, each node comes after all of its descendants. As a result, the root has to come last. The lowest element on the left side comes first. Using the `Node` class defined above, a post-order traversal could be written this way.

```
\begin{center}
def postorder(n: Node,handler: A => Unit): Unit = {
  if (n != null) {
    n.children.foreach(n2 => postorder(n2, handler))
    handler(n.data)
  }
}
\end{center}
```

Note that the only change is that the call to `handler` has moved below the processing of the children.

It is also possible to put the handling of the current node between some of the children. This produces an IN-ORDER traversal. For a general tree where each node can have an arbitrary number of children, this traversal is not particularly well defined. We will revisit this traversal later on when we have trees with more restricted structures.

The depth-first traversals, where children are handled before siblings, work well with recursion because the memory of the stack keeps track of what is happening for nodes where only some of the children have been processed. It turns out that we could write a depth-first traversal without recursion using the `Stack` types we wrote back in chapters 7 and 12. This code would push the root node, then loop while the stack is not empty, popping elements off, handling them, and pushing the children. The code for this looks like the following.

```
def loopPreorder(r: Node, handler: A => Unit):Unit = {
  val stack = new ListStack[Node]
  stack.push(r)
  while (!stack.isEmpty) {
    val n = stack.pop
    handler(n.data)
    n.children.reverse.foreach(stack.push)
  }
}
```

The call to `reverse` on the children insures that children are handled from left to right as the last one pushed to the stack will be the first one processed.

To make it clear, it is worth considering simple code for doing a breadth-first traversal of a tree. This type of traversal does not work with recursion. Instead of repeatedly trying different paths down from a node, which recursion does well, this simply runs across nodes of the same depth. This approach does not have the same branching type of behavior that characterizes recursion. Indeed, to implement this type of traversal, one uses a queue instead of a stack. Here is an example implementation.

```
def breadthFirst(r: Node,handler: A => Unit):Unit = {
  val queue = new ListQueue[Node]
  queue.enqueue(r)
  while (!queue.isEmpty) {
    val n = queue.dequeue()
    handler(n.data)
    n.children.foreach(queue.enqueue)
  }
}
```

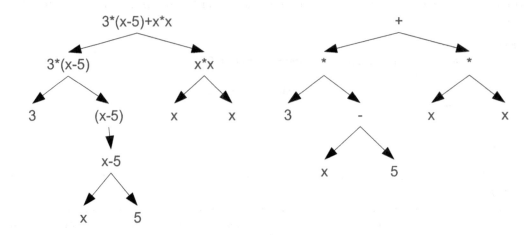

FIGURE 16.4: This figure shows the recursion that happens when we parse the formula 3*(x-5)+x*x on the left side. On the right is the tree that we would want to build from this formula.

Note how this looks nearly the same as the loop-based pre-order traversal except with a queue instead of a stack. If you run this on our sample tree, the order of handling is A, B, C, D, E, F, G, H, I, J, K, L, M.

16.2   Project Integration: Formula Parsing

As was already noted, there is a tree in the current project code that keeps track of, and gives order to the **Drawables**. In this section we will add a somewhat different type of tree to our formula parsing code from the last chapter that will make formula evaluation faster so that we can introduce variables and evaluate the same formula many times using different values for the variables. We can then add code to our project that utilizes the **Slider** that we added to the GUI at the beginning. The **Slider** will adjust a time variable that can be used in formulas in the different elements of the GUI so that our drawings can have animation.

Recall that the formula parser found the lowest-priority operator, and split on that. It recursively evaluated the two sides and combined their answers using the operator that was found. This involves a lot of string manipulation and running through the string to find things. However, this same process can generate a nice tree structure that is fast to evaluate.

Figure 16.4 shows two different trees. On the left is a tree showing the recursion, assuming the function were augmented to handle variables. Each node is a function call with the string that would be passed into that call. On the right is a tree with the significant operator, number, or variable from each call. This tree effectively represents the original formula and can be used for fast evaluation.

To do this, we make a **trait** that represents a node in the tree, and give it a method so that it can be evaluated. We then make **classes** that inherit from that **trait**, which define all the different types of nodes that we want to have. The **evalParse** function that we wrote previously can be changed to just do a parse and return a **Node** instead of a **Double**. The following code shows an implementation of that. In addition to parsing to a tree, this code adds support for some trigonometric functions and a square root function. It also uses the `apply` method and inherits from a function type so that the resulting formulas can be treated like functions.

```scala
1   package trees.util
2
3   import scala.collection.Map
4
5   class Formula(val formula: String) extends (Map[String, Double] => Double) with
        Serializable {
6     private val root = Formula.simplify(Formula.parse(formula))
7
8     def apply(vars: Map[String, Double]): Double = root(vars)
9
10    override def toString = formula
11  }
12
13  object Formula {
14    val ops = "+-*/".toSet
15
16    def apply(form: String, vars: Map[String, Double] = null): Double = {
17      val root = parse(form.filter(_ != ' '))
18      root(vars)
19    }
20
21    private def parse(f: String): Node = {
22      var opLoc = -1
23      var parensCount = 0
24      var i = f.length - 1
25      while (i > 0) {
26        if (f(i) == '(') parensCount += 1
27        else if (f(i) == ')') parensCount -= 1
28        else if (parensCount == 0 && (f(i) == '+' || f(i) == '-' && !ops.contains(f(i
              - 1)))) {
29          opLoc = i
30          i = -1
31        } else if (parensCount == 0 && opLoc == -1 && (f(i) == '*' || f(i) == '/')) {
32          opLoc = i
33        }
34        i -= 1
35      }
36      if (opLoc < 0) {
37        if (f(0) == '(') {
38          parse(f.substring(1, f.length - 1))
39        } else if (f.startsWith("sin(")) {
40          new SingleOpNode(parse(f.substring(4, f.length - 1)), math.sin)
41        } else if (f.startsWith("cos(")) {
42          new SingleOpNode(parse(f.substring(4, f.length - 1)), math.cos)
43        } else if (f.startsWith("tan(")) {
44          new SingleOpNode(parse(f.substring(4, f.length - 1)), math.tan)
45        } else if (f.startsWith("sqrt(")) {
46          new SingleOpNode(parse(f.substring(5, f.length - 1)), math.sqrt)
47        } else try {
48          new NumberNode(f.toDouble)
49        } catch {
```

```scala
50          case ex: NumberFormatException => new VariableNode(f)
51        }
52      } else {
53        f(opLoc) match {
54          case '+' => new BinaryOpNode(parse(f.take(opLoc)), parse(f.drop(opLoc +
                1)), _ + _)
55          case '-' => new BinaryOpNode(parse(f.take(opLoc)), parse(f.drop(opLoc +
                1)), _ - _)
56          case '*' => new BinaryOpNode(parse(f.take(opLoc)), parse(f.drop(opLoc +
                1)), _ * _)
57          case '/' => new BinaryOpNode(parse(f.take(opLoc)), parse(f.drop(opLoc +
                1)), _ / _)
58        }
59      }
60    }
61
62    private def simplify(n: Node): Node = n match {
63      case BinaryOpNode(l, r, op) =>
64        val left = simplify(l)
65        val right = simplify(r)
66        (left, right) match {
67          case (NumberNode(n1), NumberNode(n2)) => NumberNode(op(n1, n2))
68          case _ => BinaryOpNode(left, right, op)
69        }
70      case SingleOpNode(a, op) =>
71        val arg = simplify(a)
72        arg match {
73          case NumberNode(n) => NumberNode(op(n))
74          case _ => SingleOpNode(arg, op)
75        }
76      case _ => n
77    }
78
79    private sealed trait Node extends Serializable {
80      def apply(vars: Map[String, Double]): Double
81    }
82
83    private case class NumberNode(num: Double) extends Node {
84      def apply(vars: Map[String, Double]): Double = num
85    }
86
87    private case class VariableNode(name: String) extends Node {
88      def apply(vars: Map[String, Double]): Double = vars get name getOrElse 0.0
89    }
90
91    private case class BinaryOpNode(left: Node, right: Node, op: (Double, Double) =>
          Double) extends Node {
92      def apply(vars: Map[String, Double]): Double = op(left(vars), right(vars))
93    }
94
95    private case class SingleOpNode(arg: Node, op: Double => Double) extends Node {
96      def apply(vars: Map[String, Double]): Double = op(arg(vars))
97    }
98  }
```

This code includes a **class** called **Formula**, in addition to the object, so that you can instantiate an instance of the **Formula** type and that will parse the string once, then remember the tree for later evaluations. Calling **Formula.apply** on the companion object will be even slower than what was done before, but it has been left in for simple "one-off" calls. The import at the top makes it so that the **Map** type we are using is general and can be mutable or immutable.

There is another method in the companion **object** on lines 62–77 called **simplify**. This method is included for two reasons. It helps to optimize the evaluation of formulas, and in doing so, it can show you something that the tree approach can do that the text-based version could not have. It also gives you an example of a recursive method that operates on the formula tree so that you can see what such a thing might look like. The goal of this method is to make it so that the tree has as few operations as possible. It does this by finding either **BinaryOpNodes** or **SingleOpNodes** that operate on plain numbers and simplifies them to just be numbers. It does this recursively from the bottom of the tree. For example, $5 + 3 * 2$ should simplify to a single **NumberNode** with the value 11 while $x + 3 * 2$ should simplify down to $x + 6$ with just one binary operation.

With the **Formula** code written, we can add significant functionality to our drawing. First, we need to make the **Slider** at the top of the GUI update a value in the **vars** map of the **Drawing** that is associated with the name **"t"** for use in formulas.

```
slider.value.onChange {
  drawing.vars("t") = slider.value.value
  drawing.draw()
}
if(!drawing.vars.contains("t")) drawing.vars("t") = slider.value.value
  else slider.value = drawing.vars("t")
```

The **vars** map was added to the code when we added the RPN calculator to the commands. It can be used again here in connection with the **Formula** type. This is not just reuse, it adds power. Now the commands can be used to set variables that are referenced in formulas. All we need now is to make something that we draw utilize the formulas as well.

This is not hard to do either. For example, the x, y, width, and height of the **DrawRectangle** type can be changed to use formulas with the following alterations.

```
class DrawRectangle(
    d: Drawing,
    private var _x: Formula,
    private var _y: Formula,
    private var _width: Formula,
    private var _height: Formula,
    @transient private var _color: Color) extends Drawable(d) {

  ...

  def draw(gc: GraphicsContext): Unit = {
    gc.fill = _color
    gc.fillRect(_x(drawing.vars), _y(drawing.vars), _width(drawing.vars),
        _height(drawing.vars))
  }

  def propertiesPanel(): Node = {
    ...
      val xField = DrawingMain.labeledTextField("x", _x.toString, s => { _x = new
        Formula(s); drawing.draw() })
```

```scala
    val yField = DrawingMain.labeledTextField("y", _y.toString, s => { _y = new
        Formula(s); drawing.draw() })
    val widthField = DrawingMain.labeledTextField("width", _width.toString, s =>
        { _width = new Formula(s); drawing.draw() })
    val heightField = DrawingMain.labeledTextField("height", _height.toString, s
        => { _height = new Formula(s); drawing.draw() })
    ...
  }

  ...
}


object DrawRectangle {
  def apply(d: Drawing, x: String, y: String, width: String, height: String, color:
      Color) = {
    new DrawRectangle(d, new Formula(x), new Formula(y), new Formula(width), new
        Formula(height), color)
  }

  def apply(n: xml.Node, d: Drawing) = {
    val x = (n \ "@x").text
    val y = (n \ "@y").text
    val width = (n \ "@width").text
    val height = (n \ "@height").text
    val color = Drawing.xmlToColor(n \ "color")
    new DrawRectangle(d, new Formula(x), new Formula(y), new Formula(width), new
        Formula(height), color)
  }
}
```

The changes can be summarized as making x, y, width, and height be of type Formula
instead of Double, then altering the few other parts of the code that are required to make
those changes compile. You can make similar changes to DrawTransform by changing value1
and value2 from Double to Formula and then making minor alterations to fix the syntax
errors that creates. This same procedure of changing Double to Formula and fixing syntax
errors can be applied to any other Drawables that you want. The benefit of doing this is
that you can then make it so that your drawing changes as you move the slider. This allows
simple animations or other effects to be produced.[1]

16.2.1   Formula Tree Traversals and In-Order Traversal

For the formula tree, the in-order traversal is well defined if we make the rule that for
binary operator nodes, the left child is handled first, followed by the current node, and
then the right child. It is interesting to consider the three different traversals, pre-order,
post-order, and in-order, on the tree shown in figure 16.4.

- Pre-order: + * 3 - x 5 * x x

- Post-order: 3 x 5 - * x x * +

---

[1]If you made any of these types extend Clickable in the last chapter, you might find that it is not
straightforward to have that interact with the change to using a Formula. There is a reasonably simple
solution to this. Add methods to Formula called something like add and multiply that produce new instances
of Formula that add Nodes that do the proper operations. If you do this, try to do it intelligently so that
the tree does not grow too large. Make sure that calling add twice or multiply twice does add more than
one level to the height of the tree.

- In-order: 3 * x - 5 + x * x

The in-order traversal is the original formula, and is the way we are used to seeing it, but without the parentheses. The parentheses are required for in-order expressions to get the proper order of operations. The pre-order version is similar to the syntax of languages like LISP and Scheme. In Scheme, this expression would be written as (+ (* 3 (- x 5)) (* x x)). Parentheses are again needed to specify order of operation, but it is easy to extend the syntax so that more than two operands can be specified for any operator. The post-order traversal is exactly what you would enter into a RPN calculator, and it needs no parentheses to specify order of operation. While you might not normally think of equations as trees, it turns out that this form contains their meaning very nicely and translates easily into different useful versions of the expression.

16.3 Binary Search Trees: Binary Trees as Maps

The example above is a fun use of a tree and an interesting illustration of the power of trees. However, the most common use of these structures is for storing data. In this section we will explore the details of the binary search tree, BST, as a means of implementing the Map ADT.

The nature of the Map ADT was described in chapter 6 when we looked at the ones that are built into the Scala collections libraries. A Map is simply a collection that allows you to look up values based on keys of arbitrary types. You can implement a Map in many ways, including just an array or linked list of tuples, where one value in the tuple is the key and the other is the value. Searching for values in this implementation requires running through all the elements until a matching key is found, making it an $O(n)$ operation.[2] The binary search tree allows us to implement a Map where all the standard operations have an expected performance of $O(logn)$.

A binary search tree is first a binary tree. This is a tree that allows 0, 1, or 2 children per node. In the case of the BST, the children are typically called left and right, and it is possible to have a right child without having a left as well as a left without a right. At least part of the data stored in nodes includes a key that has a complete ordering. That means that for any two keys you can ask whether one is less than, equal to, or greater than the other. The primary criteria for a BST is that descendants to the left have lesser key values while descendants to the right have greater key values.

The name binary search tree should bring to mind the binary search on arrays. This is not by accident. A well-balanced BST will behave much like a binary search when the user looks for elements. To understand this, we will consider an example. Figure 16.5 shows an example BST that uses basic integers for keys to keep things simple. This tree contains the keys 0–9 in a proper BST organization.[3] The bold arrows indicate the path taken by a search for the value 2. The search starts at the root, and at each node compares the key value it is looking for to the one in the node. If it is equal, it has found it and can return the value in that node. If it is less, it moves to the left child of the node, otherwise, it moves to the right child of the node.

Adding new elements to a basic BST is fairly straightforward. If the tree is empty, there

---

[2]If the keys are sorted and an array is used, this can be reduced to $O(logn)$ using a binary search, but then adding to the Map becomes an $O(n)$ operation.

[3]As we will see shortly, there are many other ways that this could be drawn. The only requirement is that lesser values are to the left and greater values are to the right.

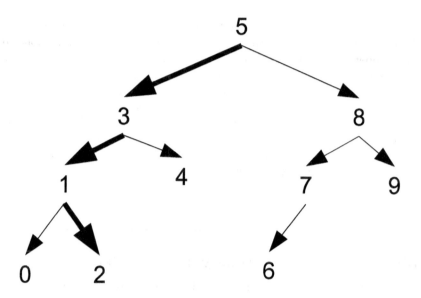

FIGURE 16.5: This is an example of a binary search tree using integer value keys. Searching for an element in the tree follows steps very much like those a binary search would perform. The bold arrows show the path that would be taken in a search for the value 2.

is no root; then we make a node with the current element and make that the root. For any other situation, we start at the root and walk down in the same way we would for a search, going to the left when the new key is less than what is stored in the node and the right when it is greater. Eventually this process reaches a point where there is no child in the indicated direction. A new node is created to store the new value and it is placed at that location in the tree. Figure 16.6 goes through this process adding the values 5, 3, 1, 8, 7, 4, 2, 9, 0, 6 in that order. The result is the tree shown in figure 16.5. Note that the order of adding the numbers is significant. While some other orderings would produce the same tree, most will not.

The process of removing from an BST is a bit more complex than adding. Consider the situation where we want to remove the node with the value 5 from our sample tree. This happens to be the root node. We cannot simply take it away as that would leave us with two separate trees. We have to do something that modifies the tree in a way so that the result is still a valid BST. One approach to this would be to put the right subtree at the bottom right of the left subtree or vice versa. In this case, that would put either the 8 to the right of 4 or the 3 to the left of 6. While these technically work, they are far from ideal because they have a tendency to unbalance the tree. We will see shortly how that can make a big difference.

The other approach is to move the data from a lower node, which is easier to remove, up to replace the 5. Nodes are easy to remove if they have zero or one child. Of course, the data that is moved has to be able to take the place of the node being removed without breaking the proper ordering of the tree. There are only two nodes that can do that, the greatest element on the left or the smallest element on the right. In this case, those are the values 4 and 6. If 5 were replaced by either of those, the rest of the tree structure remains perfectly intact and retains proper ordering. While this procedure is fairly easy to describe, unfortunately it has many special cases that have to be dealt with.

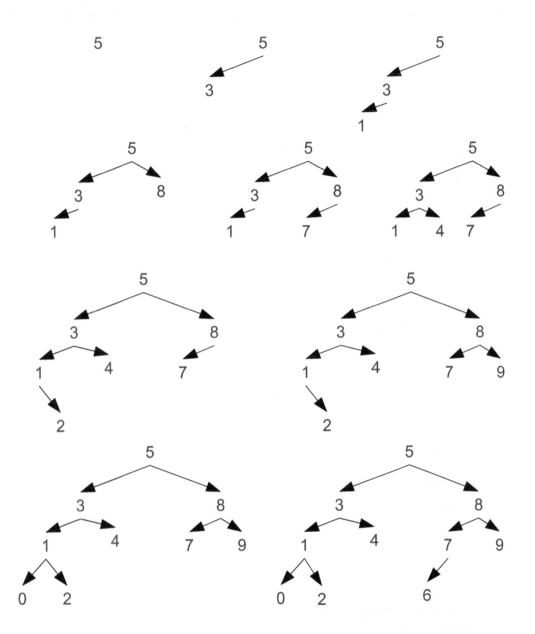

FIGURE 16.6: This figure shows the creation of the BST from figure 16.5. Each element is placed by running down the tree in the same manner as a search until we hit a location where there is not a child. A new node with the value is placed in that location.

The following is code that uses a BST to implement a mutable map from the Scala libraries. This has type parameters for both a key and value type, called K and V in this code. The `TreeMap` must be passed a comparison function at construction. This provides the complete ordering that is needed for the keys. The `mutable.Map` type has four abstract methods that must be implemented, `+=`, `-=`, `get`, and `iterator`. A complete implementation is shown here.

```scala
package trees.adt

import scala.annotation.tailrec
import scala.collection.mutable

class TreeMap[K, V](comp: (K, K) => Int) extends mutable.Map[K, V] {
  private class Node(var key: K, var data: V) {
    var left: Node = null
    var right: Node = null
  }

  private var root: Node = null

  def +=(kv: (K, V)) = {
    if (root == null) {
      root = new Node(kv._1, kv._2)
    } else {
      var rover = root
      var parent: Node = null
      var c = comp(kv._1, rover.key)
      while (c != 0 && rover != null) {
        parent = rover
        rover = if (c < 0) rover.left else rover.right
        if (rover != null) c = comp(kv._1, rover.key)
      }
      if (c == 0) {
        rover.key = kv._1
        rover.data = kv._2
      } else if (c < 0) {
        parent.left = new Node(kv._1, kv._2)
      } else {
        parent.right = new Node(kv._1, kv._2)
      }
    }
    this
  }

  def -=(key: K) = {
    def findVictim(n: Node): Node = {
      if (n == null) null
      else {
        val c = comp(key, n.key)
        if (c == 0) {
          if (n.left == null) n.right
          else if (n.right == null) n.left
          else {
            val (key, data, node) = deleteMaxChild(n.left)
            n.left = node
```

```scala
49          n.key = key
50          n.data = data
51          n
52        }
53      } else if (c < 0) {
54        n.left = findVictim(n.left)
55        n
56      } else {
57        n.right = findVictim(n.right)
58        n
59      }
60    }
61  }

63  def deleteMaxChild(n: Node): (K, V, Node) = {
64    if (n.right == null) {
65      (n.key, n.data, n.left)
66    } else {
67      val (key, data, node) = deleteMaxChild(n.right)
68      n.right = node
69      (key, data, n)
70    }
71  }

73  root = findVictim(root)
74  this
75  }

77  def get(key: K): Option[V] = {
78    var rover = root
79    var c = comp(key, rover.key)
80    while (rover != null && c != 0) {
81      rover = if (c < 0) rover.left else rover.right
82      if (rover != null) c = comp(key, rover.key)
83    }
84    if (rover == null) None else Some(rover.data)
85  }

87  def iterator = new Iterator[(K, V)] {
88    val stack = new linkedlist.adt.ListStack[Node]
89    pushRunLeft(root)
90    def hasNext: Boolean = !stack.isEmpty
91    def next: (K, V) = {
92      val n = stack.pop()
93      pushRunLeft(n.right)
94      n.key -> n.data
95    }
96    @tailrec def pushRunLeft(n: Node) {
97      if (n != null) {
98        stack.push(n)
99        pushRunLeft(n.left)
100     }
101   }
102 }
103 }
```

```
104
105    object TreeMap {
106      def apply[K, V](data: (K, V)*)(comp: (K, K) => Int): TreeMap[K, V] = {
107        val tm = new TreeMap[K, V](comp)
108        val d = data.sortWith((a, b) => comp(a._1, b._1) < 0).toIndexedSeq
109        def binaryAdd(start: Int, end: Int) {
110          if (start < end) {
111            val mid = (start + end) / 2
112            tm += d(mid)
113            binaryAdd(start, mid)
114            binaryAdd(mid + 1, end)
115          }
116        }
117        binaryAdd(0, data.length)
118        tm
119      }
120    }
```

The simplest method in this code is the **get** method. It makes a rover and a variable to store comparison results. Then it has a **while** loop that will go until the right key is found or the rover falls off the bottom of the tree. The variable c is used to store comparison results because the **comp** function might be expensive and we need to check for equality and at least one form of inequality each time through the loop. Storing the result in a variable reduces the number of calls by two and does not significantly hurt code readability.

The **+=** method is the next simplest and is heavily based on the **get** method. It has a special case for adding to an empty tree, as that has to change the value of **root** instead of a **left** or **right** in a node. The other case contains a **while** loop that is much like the one in **get** with the exception that it keeps a reference called **parent** that follows the rover down the tree. This happens because in the case of a new key, the rover will fall off the bottom of the tree. At that point, we need to know the parent to be able to add a node. This is much like adding to a singly linked list, where we needed a reference to the node before the location at which we are adding. That could also be done with a trailing node equivalent to **parent**, but it was easier for the linked list to simply check the **next** reference in the node. In this case, there are both **left** and **right** references so it is easier to use the trailing reference approach. After the loop breaks, there are three cases for the key being less than, greater than, or equal to the last node checked.

The condition where the key passed to **+=** is the same as one currently in the tree brings up something that was avoided above: What happens when there are duplicate keys? The map ADT is designed to only hold one value for any given key. For that reason, the code here replaces what is stored in the node with what is passed in. If a BST is used in a situation where duplicate keys are allowed, such as a multi-map, a different approach must be taken. One option would be to have the node store a sequence of values instead of a single one, other options, such as having equal values go left or right, can cause problems for some applications.

The **iterator** method is the next simplest. Any traversal would work, but there is a certain logic to using an in-order traversal as that traversal on a BST handles the nodes in sorted order by their keys. While it is simple enough to write a recursive traversal of a BST, that does not work well for the iterator, which needs to be able to provide elements one at a time instead of handling all elements as part of a single call. For this reason, our iterator needs to keep a stack to remember elements that have been passed over and not yet visited.

To get an in-order traversal, we repeatedly push all elements running down the left branch of any subtree. This is done by the **pushRunLeft** method. It is called on the root initially to set up the stack. Then each time a node is popped, that method is called on the right child of the node before the data element is returned. As long as the stack contains items, the iterator is not empty.

That leaves us with the **-=** method. The implementation shown here is not the most efficient, but it is far easier to understand than more efficient implementations. This version breaks the problem down a bit further and includes two nested functions that actually do the work. There are many possible special cases for removing from a BST. The overhead of these in the code is reduced by having the recursion set up so that values are assigned at each level upon return from the functions.

The first recursive method is called **findVictim** and it is repeated here. It returns the node that the parent, or root, should reference. As the name implies, the primary purpose of this method is to find the **Node** with the key that needs to be removed. The easy base case is when the function is called with a **null** node. That implies that the walk fell off the tree without finding the element so nothing needs to be deleted.

```scala
def findVictim(n: Node): Node = {
  if (n == null) null
  else {
    val c = comp(key, n.key)
    if (c == 0) {
      if (n.left == null) n.right
      else if (n.right == null) n.left
      else {
        val (key, data, node) = deleteMaxChild(n.left)
        n.left = node
        n.key = key
        n.data = data
        n
      }
    } else if (c < 0) {
      n.left = findVictim(n.left)
      n
    } else {
      n.right = findVictim(n.right)
      n
    }
  }
}
```

The second base case is when a matching key is found. If either the left or the right is **null**, the other one is returned so that it becomes the new child and the matching key is unlinked from the tree. The more challenging case is when the matching node has two children. In that case, the other method is invoked to find replacement data and remove the node it is in. The other two cases simply recurse to the left or right.

The second helper method also recurses on nodes going down the tree. In addition to returning the **Node** that should be connected to the parent, it also returns the key and data for the maximum key in that subtree. Note that the **findVictim** method calls this method on the left child.

```scala
def deleteMaxChild(n: Node): (K, V, Node) = {
  if (n.right == null) {
    (n.key, n.data, n.left)
  } else {
```

```scala
      val (key, data, node) = deleteMaxChild(n.right)
      n.right = node
      (key, data, n)
    }
  }
```

The recursion here is much simpler because it simply runs down until it finds a node with no right child. The base case returns the key and data of that node as well as the left child, which the parent of the node being removed should link to. As the stack pops back up, links are put back into place. The approach of making links while popping back up the stack requires extra reference assignments, as often the returned node was already the child. Without this though, the times when the child should be different become special cases and there are other challenges to building that code as well. Developing and testing such code is left as an exercise.

Both `get` and `+=` can be written using a recursive approach like what we used in `-=`. In the case of `get`, the code is of similar complexity and length. One key feature to note though is that the original version of `get` had to check if **rover** was **null** three times. This version only does it once.

```scala
  def get(key: K): Option[V] = {
    @tailrec def getter(rover: Node): Option[Node] = {
      if(rover==null) None
      else {
        var c = comp(key, rover.key)
        if(c==0) Some(rover)
        else if(c<0) getter(rover.left)
        else getter(rover.right)
      }
    }
    getter(root).map(_.data)
  }
```

Efficiency-wise, the two versions of `get` should be roughly identical, especially given that the recursive method is tail recursive.

Switching to a recursive approach for `+=` actually produces somewhat simpler code. As with `get`, the number of times we have to check for **null** gets cut down from three times to one time.

```scala
  def +=(kv: (K, V)) = {
    def add(rover: Node): Node = {
      if (rover == null) {
        new Node(kv._1, kv._2)
      } else {
        var c = comp(kv._1, rover.key)
        if (c < 0) rover.left = add(rover.left)
        else if (c > 0) rover.right = add(rover.right)
        else {
          rover.key = kv._1
          rover.data = kv._2
        }
        rover
      }
    }
    root = add(root)
    this
```

}

As with `-=`, this is doing extra reference assignments that could have an impact on efficiency, but you cannot be certain without doing empirical tests. Even if there is a performance impact, if you find that the recursive versions seem cleaner or more straightforward, you should consider using that approach in your own code.

16.3.1 Order Analysis

So what is the order of the methods for searching, adding to, or removing from a BST? It turns out, that depends a lot on the structure of the tree. To understand this, consider what happens if you add the values 0, 1, 2, 3, 4, 5, 6, 7, 8, and 9 to a BST in that order. These are the same values that were used in the previous example, only now they are sorted. It should be clear that doing this produces a tree with 0 and the root and the other values running down the right. This is a tree that has the structure of a singly linked list and the operations on it have the same performance as a singly linked list. That is to say, the three main operations are all $O(n)$ for this tree.

The structure of a tree is often described as being balanced or unbalanced based on how similar the heights of left and right subtrees are. The original example tree shown in figure 16.5 is balanced. There is no configuration of nodes for those key values that has a lower total height, and if you pick any node in the tree, the difference in height between the left and right subtrees is never more than one. On the other hand, the tree you get if those same key values are added in sorted order is completely unbalanced.

A balanced BST has $O(\log(n))$ performance for searching, insertion, and removal. This is a result of the fact that all of these operations run from the root down a single path through the tree. In a balanced tree, the number of nodes at a depth, d, scales as $O(2^d)$, so the total number depth for n nodes goes as $O(\log_2(n))$. This should not be surprising given the similarity to the binary search that was discussed earlier.

So an ideal tree gives $O(\log(n))$ behavior while the worst possible tree configuration is $O(n)$. These are called the BEST-CASE and WORST-CASE orders. Often we care more about the AVERAGE-CASE behavior of a system. If keys are added to a BST in random order, the resulting tree will be fairly well balanced, giving an average-case behavior of $O(\log(n))$. The average-case has a higher coefficient than the best-case, but it is still the same order. Unfortunately, the scenarios that produce worst-case performance in a BST are not all that uncommon. We have already seen that sorted, or nearly sorted, data will lead to worst-case behavior. It is also possible to start with a fairly well-balanced tree and have it become unbalanced. For example, if the usage of a BST is such that old nodes are removed and new ones added, but new ones tend to be above or below the average of the old ones, the tree might start off balanced, but it will skew over time. A possible solution to this is described in chapter 21.

This discussion should make it clear why the file ends with a companion object that has an `apply` method to build a tree using a recursive decent of the sorted data. This insures that the tree that is produced is perfectly balanced. Of course, this comes with the overhead of having to sort the sequence and convert it to an indexed sequence. That conversion is needed to prevent the random access in the `binaryAdd` method from producing $O(n^2)$ behavior.

16.3.2 Immutable BSTs

Now that you have seen how a BST can be used to implement a mutable map, what about an immutable one? To do this, one can use an approach very similar to what was done for the immutable singly linked list. The type for the immutable map itself is abstract with redefinitions of methods that are required to make it so that the return is a more specific type. There are subtypes for a node as well as for an empty tree. The latter is the equivalent of `Nil` for the `List` type. Here is a possible implementation of this type.

```scala
package trees.adt

abstract sealed class ImmutableTreeMap[K <% Ordered[K], +V] extends Map[K, V] {
  def +[B >: V](kv: (K, B)): ImmutableTreeMap[K, B]
  def -(k: K): ImmutableTreeMap[K, V]
}

private class Node[K <% Ordered[K], +V](
    private val key: K,
    private val data: V,
    private val left: ImmutableTreeMap[K, V],
    private val right: ImmutableTreeMap[K, V]) extends ImmutableTreeMap[K, V] {

  def +[B >: V](kv: (K, B)) = {
    if (kv._1 == key) new Node(kv._1, kv._2, left, right)
    else if (kv._1 < key) new Node(key, data, left + kv, right)
    else new Node(key, data, left, right + kv)
  }

  def -(k: K) = {
    if (k == key) {
      left match {
        case _: Empty[K] => right
        case i: Node[K, V] => {
          right match {
            case _: Empty[K] => left
            case _ => {
              val (k, d, newLeft) = i.removeMax
              new Node(k, d, newLeft, right)
            }
          }
        }
      }
    } else if (k < key) {
      new Node(key, data, left - k, right)
    } else {
      new Node(key, data, left, right - k)
    }
  }

  def get(k: K): Option[V] = {
    if (k == key) Some(data)
    else if (k < key) left.get(k)
    else right.get(k)
  }

  def iterator = new Iterator[(K, V)] {
    val stack = new linkedlist.adt.ListStack[Node[K, V]]
    pushRunLeft(Node.this)
    def hasNext: Boolean = !stack.isEmpty
```

```scala
      def next: (K, V) = {
        val n = stack.pop()
        pushRunLeft(n.right)
        n.key -> n.data
      }
      def pushRunLeft(n: ImmutableTreeMap[K, V]) {
        n match {
          case e: Empty[K] =>
          case i: Node[K, V] =>
            stack.push(i)
            pushRunLeft(i.left)
        }
      }
    }

    private def removeMax: (K, V, ImmutableTreeMap[K, V]) = {
      right match {
        case e: Empty[K] => (key, data, left)
        case i: Node[K, V] =>
          val (k, d, r) = i.removeMax
          (k, d, new Node(key, data, left, r))
      }
    }
  }

  private class Empty[K <% Ordered[K]] extends ImmutableTreeMap[K, Nothing] {
    def +[B](kv: (K, B)) = {
      new Node(kv._1, kv._2, this, this)
    }

    def -(k: K) = {
      this
    }

    def get(k: K): Option[Nothing] = {
      None
    }

    def iterator = new Iterator[(K, Nothing)] {
      def hasNext = false
      def next = null
    }
  }

object ImmutableTreeMap {
  def apply[K <% Ordered[K], V](data: (K, V)*)(comp: (K, K) => Int):
      ImmutableTreeMap[K, V] = {
    val empty = new Empty[K]()
    val d = data.sortWith((a, b) => comp(a._1, b._1) < 0)
    def binaryAdd(start: Int, end: Int): ImmutableTreeMap[K, V] = {
      if (start < end) {
        val mid = (start + end) / 2
        new Node(d(mid)._1, d(mid)._2, binaryAdd(start, mid), binaryAdd(mid + 1,
            end))
      } else empty
```

```
104      }
105      binaryAdd(0, data.length)
106    }
107  }
```

Most of the code here is located in the `Node` class. This code is similar to what we had for the mutable version, but there are some differences worth noting. One big difference is that there are no loops in this code. Instead, everything is done with recursion. Also, instead of checks for `null`, there are matches that look for `Empty`. This tree does not contain `null` references. This use of `match` is the proper style in Scala for determining types. There are methods called `isInstanceOf[A]` and `asInstanceOf[A]`, but their use is strongly discouraged as it commonly leads to runtime errors.

The `Empty` type is remarkably simple, but we make it anyway because it greatly simplifies the recursion and streamlines the code to avoid having the tree branches be `null` terminated. Unlike the `List`, where `Nil` could be an object, `Empty` has to be a class. This is because `ImmutableTreeMap` is invariant on type parameter K. This is a concept that is covered in the on-line appendices,[4] but what matters here is that there is no type that we can pick that will work for all key types. This could lead to significant wasted memory if every leaf of every tree referred to a different instance of `Empty`. Indeed, in that situation, the empties would take nearly as much memory as the rest of the trees. However, as this code is written, every tree that is made with the `apply` method of the `ImmutableTreeMap` object makes a single `Empty` object and that is shared as the terminator for everything in that tree, even as more elements are added and the number of leaves grows. The fact that both `Node` and `Empty` are private means that the `apply` method is the only way to instantiate one of these trees.

Another difference between this tree and the last one is that this tree requires K <: `Ordered[K]` instead of passing in a comparator. This is because each node in an immutable tree is a fully functional and independent tree of its own. Using a comparator function would require that all nodes keep a reference to that function. That is an overhead that we would rather avoid if possible. This was not an issue for the mutable BST as many nodes existed in each tree and only the enclosing tree needed to keep track of the comparator.

The immutable tree does have somewhat higher overhead for modifying operations. However, the overall performance is still $O(\log(n))$. The difference is that any time nodes are added or changed, everything must be replaced on the path from that node to the root. That allows any earlier versions of the tree to remain intact. So the immutable BST not only does $O(\log(n))$ comparisons and memory reads for the + and - operations, it also creates $O(\log(n))$ new nodes for those operations. This is still far superior to the $O(n)$ nodes that have to be created for random modifications of immutable lists. Remember also that the memory overhead of immutability becomes far more reasonable if your application is multithreaded, in which case mutable data would likely lead to race conditions.

16.4   End of Chapter Material

16.4.1   Summary of Concepts

- A tree is a structure with nodes and edges that go from one node to another. All nodes have exactly one incoming edge except the root, which has none.

---

[4]The appendices are available at `http://www.programmingusingscala.net`.

- The terminology for describing trees is a combination of genealogical and biological.
  * A parent node has an edge that links to the child node.
  * Children of the same parent are called siblings.
  * Everything below a node is called descendants while those in line above it are called ancestors.
  * Nodes with no children are called leaves.
  * The number of edges down from the root is called the depth.
  * The number of edges up from the lowest leaf is called the height.
  * The number of descendants plus self is the size of a node.

- There are multiple ways to traverse a tree.
  * Depth-first traversals handle the children before they handle later siblings.
    · Works well with recursion.
    · Pre-order: Handles a node before the children.
    · Post-order: Handles a node after the children.
    · In-order: Handles a node between the children. This makes most sense for trees limited to two children per node.
  * Breadth-first traversals run across an entire generation before going to the next one. This needs to be implemented with a loop and a queue.

- A tree makes a convenient way of representing a parsed formula. The recursive parsing code written earlier can be easily modified to produce this structure. The tree structure allows much faster evaluation than reparsing the string.

- A binary search tree, BST, is an efficient structure for storing values when there is a complete ordering on some key type. We can use it to implement the Map ADT.

  - Each node in a BST can have a left child and/or a right child.
  - The key of the left child must be smaller than the key of the parent and the key of the right child must be greater than the key of the parent.
  - Getting, adding, and removing from a balanced tree can all be done in $O(\log(n))$ operations.
  - A standard BST can degrade to a linked list. Such highly unbalanced trees give $O(n)$ behavior.

16.4.2 Exercises

1. Test the relative speed of the recursive parsing evaluation method we wrote in the last chapter to that with the tree built in this chapter. Modify the code from the last chapter to include variables and call the function on a formula a sufficient number of times to see your machine pause. Repeat the same thing using the tree version from this chapter, but for this one, build one `Formula` object and do the evaluations with different variable values.

2. Enhance the simplify method to include the following simplifications:
   - Addition with zero simplifies to the other value.
   - Subtracting zero simplifies to the other value.

- Multiplication by one simplifies to the other value.
- Division by one simplifies to the other value.
- Multiplication by zero simplifies to zero.
- Division with a zero numerator simplifies to zero.

For a bit of extra challenge, you could try to identify other simplifications and implement them as well.

3. Add a method to the formula node that will return the derivative of that node. Given the operations we have, this should require little more than remembering the product rule and the quotient rule.

4. Combine exercises 2 and 3 so that you have a derivative method that simplifies the expression as far as possible.

5. Alter the `-=` method of the `TreeMap` so that it is not recursive and only reassigns references when a child node actually changes. Make sure to test your code to insure that you have dealt with all possible cases. Once you feel confident it works, compare the complexity of the code and the execution speed to the recursive solution.

16.4.3 Projects

1. If you are writing a MUD, you should use a BST-based Map of rooms instead of the library one.

2. If you are writing the web spider, make a site tree and display it graphically. Allow the user to specify what page to use as the root. Link in other pages so that depth is the minimum click count from the root. If a page can be reached in the same click count from the root, favor the one that occurred earlier in the page of lowest depth.

3. If you are writing a graphical game, use a BST-based map as a way for you to quickly look up or jump to different elements of your game.

4. If you are doing the math worksheet, this chapter provides you with the ability to parse your formulas to an evaluation tree, which can make processing a lot faster. You can also use the tree as a means of doing symbolic manipulation. You will want to have simplification for this to work, but you can easily write code to take the derivative of a tree and produce a new tree. With a bit more effort, you can write code to do anti-derivatives/indefinite integrals for some functions. Note that not all functions have anti-derivatives with a closed form so some things simply will not work.

    If you are looking for even more of a challenge, try to do symbolic definite integrals and simplify them in a way that makes reasonable sense.

5. The expression trees can also be used to good effect in a Photoshop® program. You can give the user the ability to make custom filters by specifying a function that is used to combine the colors in pixels around the current pixel. Given the number of pixels in any real image, this would have been unacceptably slow with our earlier `String` parsing evaluation. The tree-based evaluation makes it feasible. You can decide what syntax you will have the user use to specify different adjacent pixel RGB values.

6. Fast evaluation of user-entered formulas also provides a significant benefit for the simulation workbench project. You can use custom forces to specify force functions for particle interactions, or as part of random number generation for discrete event simulations.

7. For the stocks project, replace your library Map with a BST-Map that you have written.

8. For the movies project, replace your library Map with a BST-Map that you have written.

9. For the L-systems project, replace your library Map with a BST-Map that you have written.

Chapter 17

Regular Expressions and Context-Free Parsers

Parsing text is a task that comes up in a lot of different programs, whether it is pulling data out of text files, dealing with user commands, evaluating formulas, or understanding programming languages, some form of processing is typically needed when dealing with text values. Our approach to this so far has been fairly ad hoc. When faced with a particular format of input, we have written specialized code to handle it. The two exceptions to this were using the built-in parser to deal with files that were formatted in XML and the library parser for JSON, both of which are standard text formats. In this chapter we will learn about some of the formal approaches that computer scientists have developed to deal with parsing text. These will give you the ability to deal with more complex forms of text input in fairly easy, standard, and flexible ways.

17.1 Chomsky Grammars

Our discussion begins with the theoretical basics. The hierarchy of grammars that underpins what will be discussed in this chapter, as well as a fair bit of the theory of languages in computer science, appeared in an article written by the linguist Noam Chomsky and was co-developed by Marcel-Paul Schützenberger [4].

In a formal sense, a Chomsky grammar consists of four elements:

- a finite set of non-terminal symbols, typically denoted by capital letters;

- a finite set of terminal symbols, typically denoted by lowercase letters;

- a finite set of production rules, each of which has a left-hand side a right-hand side that are composed of those symbols; and

- a special non-terminal denoted as the start symbol, often the character S.

Each grammar defines a language that is all the strings of terminal symbols that can be generated by that grammar. The way a string is generated is that a section of the string matching the left-hand side of a production is converted to the right-hand side of that same production. At any time, any production can be selected for any symbol or sequence of symbols in the current string. This process continues until the string we are left with contains only terminal symbols.

To illustrate this, consider the grammar that has the set S, A, B as non-terminal symbols, with S as the start symbol, a, b as terminal symbols, and the following productions.

$$S \to AB$$
$$A \to aA$$
$$A \to a$$
$$B \to bB$$
$$B \to b$$

To see what language this grammar generates, we can run through a series of productions to generate one string.

S	apply $S \to AB$
AB	apply $A \to aA$
aAB	apply $A \to aA$
$aaAB$	apply $B \to bB$
$aaAbB$	apply $A \to a$
$aaabB$	apply $B \to b$
$aaabb$	

If you play with this grammar for a little while, you should be able to see that the language it generates is one or more a symbols followed by one or more b symbols.

For the sake of brevity, when the same left-hand side can produce multiple different values on the right-hand side, it is common to list them on a single line using a pipe, |, between them. This can shorten the above set of productions down to the following form.

$$S \to AB$$
$$A \to aA \mid a$$
$$B \to bB \mid b$$

While this is a very simple example, hopefully it gives you the idea of how these systems work. Anyone who completed the L-system exercise in chapter 15 will find the Chomsky grammars to be somewhat familiar. L-systems are a different type of grammar. The rules are slightly different, but the general ideas are very similar.

Perhaps the most interesting thing about these grammars is that by limiting the complexity of productions in certain ways, one can subdivide all possible grammars into a hierarchy where each level in the hierarchy is fundamentally more powerful than the one below it. There are four specific levels of capability in the Chomsky hierarchy. We will look briefly at each of these in the following four subsections starting with the least powerful set of grammars.

17.1.1 Regular Grammars

In a regular grammar, all productions must have the form of a single non-terminal going to a terminal $(A \rightarrow a)$ or a single non-terminal going to a terminal and a non-terminal $(A \rightarrow aB)$. With some slight modifications to the example grammar made above, we can produce a regular grammar that generates the same language.

$$S \rightarrow aS \mid B$$
$$B \rightarrow bB \mid b$$

Run through some examples with this grammar to prove to yourself that it does generate the language of one or more a symbols followed by one or more b symbols.

Given these restrictions on the rules, the evolution of a string under a regular grammar is very straightforward. At any given time, there is only one non-terminal, and it is the last symbol in the string. All productions either grow the length of the string by one, inserting a new terminal in front of a non-terminal, or they replace the one non-terminal by a terminal and finish the process.

Despite these limitations, regular grammars are still fairly powerful. The primary limitation on them is that they have no memory. For example, it is impossible to make a regular grammar that generates the language that has a certain number of a symbols followed by the same number of b symbols. We refer to this language as $a^n b^n$. This is due to the fact that the system has no memory of how many a symbols it has generated and cannot enforce that the number of b symbols matches. By employing multiple non-terminals, it is possible to make a grammar that can match a and b symbols up to a certain number, but no finite regular grammar can do this for an arbitrary value of n.

17.1.2 Context-Free Grammars

The second rung up the Chomsky hierarchy is the context-free grammars. The productions in a context-free grammar have a single non-terminal on the left-hand side and any string of terminals and/or non-terminals on the right-hand side $(A \rightarrow \gamma)$. The string on the right can potentially be the empty string, typically written as ϵ.

The fact that context-free grammars are fundamentally more powerful than regular grammars can be seen with the following example grammar. The non-terminal set includes only the start symbol, S. The terminal set is a, b. The productions are $S \rightarrow aSb \mid ab$. It should not take long to convince yourself that this grammar produces the language $a^n b^n$ for $n \geq 1$. Context-free grammars can do this because they effectively have a limited memory in the form of a stack. Due to the limitations of the stack, however, there is no context-free grammar that generates the language $a^n b^n c^n$. This is because the work of making sure that there are n b symbols consumes the contents of the stack and it is then impossible to produce the same number of c symbols.

The term context-free indicates the limitation of these grammars. While it is possible to have productions that produce whatever you want, the nature of the production cannot depend on the context of the non-terminal that appears on the left-hand side. Despite this limitation, context-free grammars are remarkably useful in the field of computer science. The syntax of programming languages are typically specified as context-free grammars. In that situation, the symbols are the tokens of the language. The vast majority of sentences in natural language can also be generated by context-free grammars.

17.1.3 Context-Sensitive Grammars

The next step up the hierarchy removes the limitation that choice of production cannot depend on what is around the non-terminal. The general form of the rule is $\alpha A \beta \to \alpha \gamma \beta$, where α, β, and γ are all arbitrary strings of terminals and non-terminals and are potentially empty. Note that the α and β are not really involved in the production. Instead, they simply limit when the production can be applied.

Context-sensitive grammars are significantly harder to work with than context-free grammars and the number of applications that require context sensitivity is fairly limited. For that reason, we will not go into any detail on their usage.

17.1.4 Recursively Enumerable Grammars

The most powerful class of grammars in the Chomsky hierarchy is the set of recursively enumerable grammars. There are no restrictions on the productions of these grammars. You can have anything on the left-hand side produce whatever you want on the right-hand side. The truly remarkable thing about this class of grammars is that it is a full model of computation. Anything that is computable, in the theoretical sense of the Turing machine or the lambda calculus, can be computed with a properly designed recursively enumerable grammar. The implications of this are somewhat profound. It means that if you were willing to take the time to do so, you could rewrite any program you have done for this book or elsewhere using a recursively enumerable grammar. Granted, it would not naturally do graphics or communicate across networks, though if you came up with standards for how to interpret parts of strings for those functions, you could do even those things.[1]

This brief introduction to Chomsky grammars and the Chomsky hierarchy serves as a foundation for the rest of this chapter. However, it is a much broader topic and interested readers should do more exploration of the topic. If nothing else, a proper course on the theory of computer science should cover these topics in significantly greater detail than has been presented here.

17.2 Regular Expressions

The theoretical underpinnings of computer science are extremely important for a variety of reasons, and they play a significant role in major developments in the field. However, the primary focus of this book is on the development of problem-solving and programming skills. For that reason, we now turn our focus to the more practical applications of grammars and languages, beginning with regular expressions.

As the name implies, regular expressions are related to regular grammars. They provide a simple mechanism for matching and parsing text. As with regular grammars, they have their limitations. They are still quite powerful and, for the problems they are good at, they can make life a whole lot easier.

Regular expressions are typically written as strings with a particular format where certain characters have special meaning. They are used by many different tools, including some that you have likely been using for a while now. Both `grep` and the search feature of `vi` make use of regular expressions. Search in Eclipse and most other IDEs also typically

---

[1]To illustrate the power of recursively enumerable grammars, we have posted a code "gist" of one with 14 productions that can add arbitrary unsigned binary numbers at `https://gist.github.com/MarkCLewis/bf83f4e4a8c0dade34ed`.

includes a regular expression option. The tight integration of regular expressions into the programming language Perl was a significant part of why many people used that language for small text processing tasks.[2] Even the `split` method we have been using on `Strings` uses regular expressions.

While most of the aspects of regular expressions are fairly uniform, there are some features that differ from one implementation to another. Regular expressions in Scala are supported by the `scala.util.matching.Regex` class and companion `object` and they are built on top of the Java implementation in `java.util.regex`. The following subsections will go through a number of the major features of regular expressions. For a more definitive description of the features that are available, one can look at the API documentation for `java.util.regex.Pattern`. To turn a `String` into a `Regex` in Scala, simply call the `r` method.

```scala
scala> val str = "This is a string."
str: java.lang.String = This is a string.

scala> val regex = "This is a regular expression.".r
regex: scala.util.matching.Regex = This is a regular expression.
```

17.2.1 Characters and Character Classes

The reason that you have been able to use regular expressions in programs without even knowing you were using regular expressions is that most characters in regular expressions act just like the normal character. So the regular expression "Fred" matches the string "Fred" and nothing else. However, imagine that you want to find all the method invocations in a Scala program that are called on objects ending with "ing" using `grep`. So we want to search for the "ing." as the dot is used to invoke methods. Here is some of that output from attempting this on the `Drawing.scala` file with the matching sections in grey.

```
> grep ing. src/recursion/drawing/DrawingMain.scala
object DrawingMain extends JFXApp {
private var drawings = List[(Drawing, TreeView[Drawable])]()
val system = ActorSystem("DrawingSystem")
private val creators = Map[String, Drawing => Drawable](
"Rectangle" -> (drawing => DrawRectangle(drawing, 0, 0, 300, 300, Color.Blue)),
"Transform" -> (drawing => DrawTransform(drawing)),
"Text" -> (drawing => DrawText(drawing, 100, 100, "Text", Color.Black)),
"Maze" -> (drawing => DrawMaze(drawing)),
"Mandelbrot" -> (drawing => DrawMandelbrot(drawing)),
"Julia Set" -> (drawing => DrawJulia(drawing)),
"Cell Simulation" -> (drawing => DrawCellSim(drawing, 300, 300, 200, 1.0, 2.0,
10)),
"Bouncing Balls" -> (drawing => DrawBouncingBalls(drawing)),
"Graph" -> (drawing => DrawGraph(drawing)))
title = "Drawing Program"
val acceptItem = new MenuItem("Accept Drawings")
val sendItem = new MenuItem("Send Drawing")
val shareItem = new MenuItem("Share Drawing")
```

---

[2]One could argue that the inclusion of good regular expression support in other languages is responsible for the decline in Perl.

Clearly this finds everything that we wanted, but it is matching a lot of other things as well, things that do not include any period. This is a result of the fact that a period is a wild card that can match any character[3] in regular expressions. So what we really asked for is any sequence of four characters that starts with "ing". In order to force the period to actually match only a period, we need to put a backslash in front of it.[4] For all the special characters in regular expressions, you can force them to be treated as normal strings by putting a backslash in front of them. The fact that normal string literals in Scala also treat the backslash as an escape character becomes rather a pain when dealing with regular expressions. For this reason, it is standard to use the triple-quotes, raw string literals for regular expressions.

The period is actually one of many CHARACTER CLASSes in regular expressions. A character class matches one character as long as it comes from the right set. In the case of a period, the set is anything. You can also make your own character classes by putting characters you want to match in square brackets. For example, [abc] will match an 'a', a 'b', or a 'c'. A slightly more complex example is the regular expression b[aei]d, which will match the strings "bad", "bed", or "bid". For consecutive characters in ASCII, you can use a hyphen. So [0-9] matches any digit and [a-zA-Z] matches any letter, upper- or lowercase. If you put a ^ as the first character in the square brackets, the character class matches anything except what is specified. So [^aeiou] will match anything that is not a lowercase vowel. Of course, this means that square brackets are special characters in regular expressions and must be escaped with a backslash if you want to use them for purposes other than defining a character class. The same is true for the caret in certain usages and for the minus sign/hyphen inside of a character class.

Some character classes are used so frequently that there are shortcut names for them. In addition to the period, the following are defined in the Java regular expression library.

- \d - a digit, same as [0-9]

- \D - not a digit, same as [^0-9]

- \s - a whitespace character

- \S - a non-whitespace character

- \w - a word character, same as [a-zA-Z0-9]

- \W - not a word character, same as [^a-zA-Z0-9]

If you spend much time working with regular expressions, these character classes will likely be committed to memory.

17.2.2   Logical Operators and Capturing Groups

You have already seen that characters or character groups that are adjacent to one another match consecutive characters. So the regular expression cat can be read as 'c' and 'a' and 't'. Character classes can be used in any place that would take an individual character as well. If adjacency is like "and", how does one say "or"? With a single pipe, |?[5] So the regular expression cat|dog will match either "cat" or "dog". As with logical expressions, the "and" operation has higher precedence than the "or" does.

---

[3]In some usages the period will not match a newline.

[4]Like regular string literals in Scala, the Linux command line treats a backslash specially. As such, you have to include two backslashes on the command line for this to work.

[5]Hopefully the reader appreciates the fact that there is some similarity in the symbols used for expressing similar ideas across different tools.

Sections of a regular expression can also be grouped using parentheses. This gives you a way to control the parts of the regular expression you are taking the "or" of, such as a(bc|de)f. It also lets you bundle sections together for the quantifiers discussed in the next subsection. Lastly, it allows you to capture sections of the match. When you put parentheses in a regular expression, they define a capturing group. The code keeps track of all of the sections of the string that wind up being in parentheses so that you can pull them back out. These groups are numbered starting at one and they are ordered by the location of the opening parentheses. So it does not matter how long groups are or how they are nested, it is only the location of the opening parenthesis that matters.

Consider the simple example of a phone number in the format "(xxx) yyy-zzzz" where you want to capture each of the groups denoted by different letters and capture the standard 7-digit part of the number independently of the area code. To do this, you could use the regular expression \(($\backslash$d$\backslash$d$\backslash$d)\) (($\backslash$d$\backslash$d$\backslash$d)-($\backslash$d$\backslash$d$\backslash$d$\backslash$d)). This regular expression has four capturing groups in it numbered 1 through 4. The first one is the area code. The second one is the 7-digit phone number. The third and fourth groups are the subparts of the 7-digit number. They are numbered by the opening parentheses. Here is that same regular expression, with subscripts to make that clear, \(($_1\backslash$d$\backslash$d$\backslash$d)\) ($_2$($_3\backslash$d$\backslash$d$\backslash$d)-($_4\backslash$d$\backslash$d$\backslash$d$\backslash$d)). This example regular expression also illustrates that because parentheses are special characters used for grouping, they have to be escaped if you want to match one, as was required for the parentheses that are expected around the area code.

17.2.3 Greedy Quantifiers

A lot of the power of regular expressions comes in with quantifiers. Quantifiers allow you to have some control over the number of times an expression will occur. The default quantifiers are called "greedy" quantifiers because they will match as many characters as possible while still allowing the whole pattern to be matched. There are six options for the greedy quantifiers.

- X? - Matches the pattern X or nothing. Think of this as being 0 or 1 times.

- $X*$ - Matches nothing or the pattern X repeated any number of times. Think of this as 0 or more times.

- $X+$ - Matches the pattern X 1 or more times.

- $X\{n\}$ - Matches the pattern X repeated exactly n times.

- $X\{n,\}$ - Matches the pattern X repeated n or more times.

- $X\{n,m\}$ - Matches the pattern X repeated at least n times, but not more than m times.

The regular expression for the phone number could have been expressed with quantifiers as \(($\backslash$d$\{3\}$)\) (($\backslash$d$\{3\}$)-($\backslash$d$\{4\}$)). These quantifiers also allow us to express the languages for the grammars described in subsection 17.1.1. For example, a+b+ matches one or more "a"s followed by one or more "b"s. Note that there is not a good way, using these quantifiers, to allow the number of characters of each type to match without specifying how many times they occur. You cannot say **anbn** in a regular expression. You have to put an actual number in the brackets.

This should now make it clear why the argument passed to split when we wanted to break a string on spaces was " +". The + here is a quantifier so that this matches one or more spaces. Just in case the user also used tabs, it might be safer to use "\s+".

There are also options for quantifiers that are not greedy. The interested reader is directed to the Java API and other sources for information on reluctant and possessive quantifiers.

17.2.4   Boundary Requirements

The last option of regular expressions that we will consider here is boundary requirements. In the general usage, regular expressions can be used to match any part of a string and can be used to do multiple matches inside of a single string. There are situations where you want to restrict the generality of the matches. For this you can put characters at the beginning and the end of the regular expression to specify where it needs to begin or end. The following are some of the allowed options.

- ^ - As the first character in a regular expression, it requires the match to start at the beginning of a line.

- $ - As the last character in a regular expression, it requires the match to end at the end of a line.

- \b - Can be placed at the beginning or end of a regular expression forcing the match to start or end on a word boundary.

- \B - Like \b, except it forces a non-word boundary.

- \A - As the first character in a regular expression, it requires the match to start at the beginning of the input.

- \G - As the first character in a regular expression, it requires the match to start at the end of the previous match.

- \z - As the last character in a regular expression, it requires the match to end at the end of the input.

So if you want to match words that have four characters in them, you could use the regular expression \b\w{4}\b.

17.2.5   Using Regular Expressions in Code

Now that you know the basics of regular expressions, we can see how they are used in Scala. We already saw that you can produce a `scala.util.matching.Regex` by calling the r method on a textttString and that triple-quote strings are particularly useful for defining regular expressions because they do not treat the backslash as an escape character. The next question concerns what you can do with one of these `Regex` objects.

The two most useful methods in the `Regex` class are `def findAllIn(source: CharSequence): MatchIterator` and `def replaceAllIn(target: CharSequence, replacer: (Match) => String): String`. The type `CharSequence` is a supertype of `String` that is more flexible. For our examples we will just be using normal strings. The `Match` and `MatchIterator` types are declared in the `Regex` companion object. A `Match` object defines basic information from a single match of the regular expression to a string. The API can give you full details, but one of the things you will want to do a lot is find the parts of the string that were in the different capturing groups. This can be done with the `def group(i: Int): String` method. The argument is the number of the group you want the text for. If nothing matched that group, then the method returns `null`.

The `MatchIterator` extends `Iterator[String]` with `MatchData`. It also has a method called `matchData` that returns an `Iterator[Match]`. Putting these things together, we can take a block of text that has several embedded phone numbers along with the regular expression we made above and pull out just the parts of the phone numbers.

```scala
scala> val phoneNumbers = """For help you can try to following numbers:
     | Main line: (123) 555-4567
     | Secondary line: (123) 555-7022
     | Fax: (123) 555-5847"""
phoneNumbers: String =
For help you can try to following numbers:
Main line: (123) 555-4567
Secondary line: (123) 555-7022
Fax: (123) 555-5847

scala> val phoneRegex = """\((\d{3})\) ((\d{3})-(\d{4}))""".r
phoneRegex: scala.util.matching.Regex = \((\d{3})\) ((\d{3})-(\d{4}))

scala> for (m <- phoneRegex.findAllIn(phoneNumbers).matchData) {
     | println(m.group(1)+" "+m.group(3)+" "+m.group(4))
     | }
123 555 4567
123 555 7022
123 555 5847
```

The `for` loop runs through the `Matches` produced by calling `MatchData` on the results of `findAllIn`. The groups with indexes 1, 3, and 4 are printed from each match.

Hopefully it is clear how the use of regular expressions made this mode much easier to write than would have been required without regular expressions. Finding the phone numbers embedded in that text would have been rather challenging using other techniques that we have covered. Phone numbers have a nice, well-defined format to them, but it is not a specific string sequence that we can search for.

The `replaceAllIn` method has a second argument that is a function which takes a `Match` and returns a `String` that should replace the matched text in the first argument. We could use this method to alter the phone numbers, replacing their last four digits with the character 'x' using the following line.

```scala
scala> phoneRegex.replaceAllIn(phoneNumbers,m => {
     | "("+m.group(1)+") "+m.group(3)+"-xxxx"
     | })
res1: String =
For help you can try to following numbers:
Main line: (123) 555-xxxx
Secondary line: (123) 555-xxxx
Fax: (123) 555-xxxx
```

Passing in the `Match` gives you the ability to use the details of the `Match` instance in determining the replacement string. In this case, we wanted to preserve the first six digits, so groups 1 and 3 were used in the result. Again, if you think for a while about what would have been required to do this transformation using other techniques, you should appreciate the power provided by regular expressions.

There is one other approach to using regular expressions in Scala: patterns. We have previously seen patterns used with tuples, `Arrays`, and `Lists`.[6] They work with regular

---

[6]Though we did not look at it, Scala also supports limited pattern matching for XML.

expressions as well. When a regular expression is used as a match on a string, it pulls out the groups and only works if there is a match for the whole string. The following code can be put into the `Commands.scala` file in place of more traditional string parsing code that we had before that split on the first space.

```scala
val commandSplit = """\s*(\w+)(\s+(.*))?\s*""".r

def apply(input:String, drawing:Drawing):Any = {
  val commandSplit(command, _, rest) = input
  if (commands.contains(command)) commands(command)(rest, drawing)
    else "Not a valid command."
}
```

This does what we did previously with a call to `indexOf`, but this code is a bit more robust in how it handles white space, particularly leading white space. With the old code, if the user put spaces in front of the command, it would fail had we not trimmed the input before passing it to the command processor. This fixes that problem by leading with `\s*`.[7]

The place where regular expression patterns are probably most useful is in `for` loops. Recall that what you type before `<-` in a `for` loop is matched as a `val` pattern. Anything that does not match is skipped. This is remarkably helpful for regular expressions if you are running through a collection of strings where some might not match what you are looking for.

To see this, consider having a file where some of the lines are numbered and others are not. You only care about the ones that are numbered and they all have, for the first non-space characters, digits followed by a period. After the period is text you care about. You want to build a `Map[Int,String]` that lets you quickly look up the text from a line based on the number that was in front of it. The following code will do this for you.

```scala
val NumberedLine = """\s*(\d+)\.(.+)""".r
val source = io.Source.fromFile(fileName)
val lines = source.getLines
val numberedLines = (for (NumberedLine(num, text) <- lines) yield {
  num -> text
}).toMap
source.close
```

The `for` loop will only execute with lines matching the pattern and skip anything that does not match. This makes for compact and fairly easy-to-maintain code.

17.2.6 Drawback of Regular Expressions

The primary drawback of regular expressions is that they can be cryptic and somewhat difficult to maintain. Indeed, languages that highlight the use of regular expressions, such as Perl,[8] are often derided for their lack of readability. Coming up with exactly the right regular expression for a problem can also be quite challenging. This is expressed in this oft-repeated quote.

---

[7] Earlier versions of Scala would require that `commandSplit` start with a capital letter so that it was treated as a value instead of a new variable name in the match. You can force Scala to treat a name that begins with a lowercase letter as a previously declared value by putting backticks around the name. Current versions of Scala recognize that if a name used in a pattern is followed by parentheses, it is not a new variable declaration.

[8] There are quite a few other reasons why Perl is considered unreadable beyond the strong use of regular expressions.

Some people, when confronted with a problem, think
"I know, I'll use regular expressions." Now they have two problems.

- Jamie Zawinski (1997 Usenet post)

There is inevitably some truth to this quote. However, hopefully the examples above have shown you that in situations where regular expressions work well, they can make life much easier. Just do not try to use them to solve every problem you encounter.

17.3 Context-Free Parsers

What options do you have if you need to parse text that is beyond the power of regular expressions? Scala includes the package `scala.util.parsing`, which has a number of different subpackages that provide different parsing capabilities. We will focus on the `scala.util.parsing.combinator` package, which gives us the ability to parse context-free grammars.[9]

The introduction to context-free, CF, grammars given in subsection 17.1.2 used a notation fitting with theoretical computer science. Like mathematicians, theoretical computer scientists are quite happy to name things using single letters or symbols. In practical application, it is typically helpful to use longer, more informative names. The symbology is also altered a bit when expressing CF grammars for things like programming languages. To illustrate this, we will go back to the example of formula parsing that we have utilized in the last two chapters. The CF grammar for a basic mathematical expression using only the four primary operators can be written this way.

$$form ::= term \ \{(``+" \mid ``-") \ term\}$$
$$term ::= factor \ \{(``*" \mid ``/") \ factor\}$$
$$factor ::= floatingPointNumber \mid ``(" \ form \ ``)"$$

In this notation, the curly braces indicate elements that can be repeated zero or more times. It is also possible to use square brackets to indicate things that are optional.

This grammar says that a *formula* is a *term* followed by zero or more + or − signs followed by other terms. The *term* is a *factor* followed by factors with * or / between them. Lastly, the *factor* can be a number or a full formula in parentheses. Play with this to convince yourself that it can represent the full range of possibilities that are desired. It is also worth noting that this grammar does proper order of operations because * and / are bound a level below the + and -.

Using the combinator parser library, this converts fairly directly to the following Scala code.[10]

```scala
object Formula extends JavaTokenParsers {
  def form: Parser[Any] = term ~ rep(("+" | "-") ~ term)
  def term: Parser[Any] = factor ~ rep(("*" | "/") ~ factor)
  def factor: Parser[Any] = floatingPointNumber | "(" ~ form ~ ")"
}
```

---

[9]Technically there are limitations on the grammars that can be parsed with these parsers, but it is not a limitation on power. Grammars just have to be written in a way that facilitates the parsing. The details of this are beyond the scope of this book.

[10]The combinator parsing library is another library that was split into a separate module in Scala 2.11. For that reason, you might have to add it to your project if it is not included by default.

Each production in the original grammar becomes a method that returns a `Parser`, a `class` found inside `scala.util.parser.combinator.Parsers`. Currently this is a `Parser[Any]` to be completely general, but we will see how to change that later. Consecutive symbols in the original grammar are joined using the ~ operator. The curly braces are replaced with a call to `rep`. Had there been square brackets, they would have been replaced with a call to `opt`. There is also a method called `repsep` that can be used for alternating productions. This includes things like comma-separated lists where the separator does not contain information that you need and can be thrown away.

Using these rules, you can take any CF grammar that you might be interested in using and build a parser for it with little effort. You can use any of these parsers by calling the `parseAll` method that `Formula` gets via `JavaTokenParsers` in this example.[11] The following line of code will do just that and print the value output produced by the parser.

```
println(parseAll(form, "3+5*2").get)
```

17.3.1 Default Output

If you put that line of code in an app and run it, you get the following output.

```
((3~List())~List((+~(5~List((*~2))))))
```

The string parses, but the default output is a bit challenging to understand. To do so, we need to know what the default output of different parser elements are.

- `String` or regular expression - Gives back the text that was matched as a `String`.

- `P~Q` - Gives back `~(p,q)`, where `p` is the match for `P` and `q` is the match for `Q`. Note that `~(p,q)` prints out as `p~q`.

- `rep(P)` - Gives back lists of the matches: `List(p1,p2,p3,...)`.

- `repsep(P,Q)` - Gives back lists of the matches: `List(p1,p2,p3,...)`. Note that while `Q` can be a full parser, the output is ignored. This is why it is best used for things like comma-separated lists where you do not care about the separators.

- `opt(P)` - Gives back an `Option` on the match. If that optional part was not used, you get `None`. Otherwise you get `Some(p)`

Using this information and the original grammar, you can figure out how this parser produced the output that it did.

17.3.2 Specified Output

While the default output provides the full information from the parse, and you could run through it using match statements to extract the desired information, it is often more useful to specify your own output from the parser. This is done using the `^^` method. The signature of `^^` in `Parser[+T]` is `def ^^[U](f: (T) => U): Parser[U]`. It takes a function that converts from the normal output of this parser to some other type, `U`, and gives back a `Parser[U]`.

We can utilize this in the parser we wrote above to make it so that it outputs a numeric solution instead of the default parse output we had above. The first step in doing this is to

---

[11]This example uses `JavaTokenParsers` because that is where `floatingPointNumber` is defined. It also defines `ident`, which we will use later for identifier names.

change the return types of the various productions to `Parser[Double]`. That will introduce errors that we can fix by adding the appropriate transformations. The resulting code looks like this.

```
object Formula extends JavaTokenParsers with App {
  def form: Parser[Double] = term ~ rep(("+" | "-") ~ term) ^^ {
    case d ~ lst => lst.foldLeft(d)((n,t) => if (t._1=="+") n+t._2 else n-t._2)
  }
  def term: Parser[Double] = factor ~ rep(("*" | "/") ~ factor) ^^ {
    case d ~ lst => lst.foldLeft(d)((n,t) => if (t._1=="*") n*t._2 else n/t._2)
  }
  def factor: Parser[Double] = floatingPointNumber ^^ (_.toDouble) | "("~> form
      <~")"

  println(parseAll(form,"3+5*2").get)
}
```

The simplest application of `^^` in this example is after `floatingPointNumber`. That is a `Parser[String]` so the normal output is just a `String` that we can convert to a `Double` with a call to `toDouble`. The other two productions follow the `^^` with partial functions that get converted to full functions for us. Each of these has a `case` that matches a number followed by a tilde and a `List`. The number will be the result of the first `term` or `factor`. The `List` will be the result of `rep`, which is a `List[ [String,Double]]`. To get an answer, we need to run through the `List` accumulating a value. At each step, the proper operation should be applied to the current value and the parse of what follows the operator. This can be done in short form using `foldLeft`. Note that the `~` type works with infix pattern matching and has _1 and _2 methods like a tuple for getting to the first and second elements.

The parser for `factor` throws in something else to keep this code compact. It uses operators where a `<` or `>` is put before or after the `~`. These operators give you a shorthand way of ignoring parts of the match in the output. The direction the arrow points is toward the thing you actually care about. Without these, we would have had to follow that last case with a `^^` operator and a function that pulls off the parentheses and discards them. This does not have to be long. It could have taken the form `^^ case _ ~ f ~ _ => f`. The code is simpler without it though.

With these changes, you can run this program and see that now the output is `13.0`. Compare this to our original recursive parser that worked with these same operations and you will find that this is significantly shorter. However, this version is also completely opaque to anyone who is not familiar with CF grammars or combinator parsers.

There are, of course, many other options and possibilities with combinator parsers. Looking through the API at classes such as `scala.util.parser.combinator.Parsers` and `scala.util.parser.combinator.Parsers.Parser` can give you a quick view of some of the possibilities. One that we saw previously is the JSON parser.

17.4 Project Integration

Now it is time to put everything together, including some concepts from the last chapter. The `Formula` class can be expanded to parse to a tree instead of a `Double`. This will allow the use of variables with good speed. To show the true power of the combinator parsers, this implementation will include not only the four basic math operations and a few extra functions, it will also include exponentiation and an `if` conditional. These two features present interesting challenges. Exponentiation is left associative while the other operators are right associative. In order to have an `if`, we have to have comparison operators that give back Booleans. Given that, it is nice to have the standard Boolean operators as well. All of this needs to be integrated into a system that preserves proper order of operation.

To start off, we need to develop a CF grammar that will parse this little language that we are trying to implement. Here is the one we will use in the code.[12]

$$cond ::= "if(" \ bform \ ")" \ cond \ "else" \ cond \ | \ form$$
$$form ::= term \ \{(" + " \ | \ " - ") \ term\}$$
$$term ::= exp \ \{(" * " \ | \ "/") \ exp\}$$
$$exp ::= func \ \{"\hat{}" \ func\}$$
$$func ::= "sin" \ | \ "cos" \ | \ "tan" \ | \ "sqrt" \ "(" cond")" \ | \ factor$$
$$factor ::= floatingPointNumber \ | \ "(" \ form \ ")"$$
$$bform ::= bterm \ \{"\|" \ bterm\}$$
$$bterm ::= bnot \ \{"\&\&" \ bnot\}$$
$$bnot ::= "!("bform")" \ | \ bcomp$$
$$bcomp ::= cond \ (" == " \ | \ "! = " \ | \ " <= " \ | \ " >= " \ | \ " < " \ | \ " > ") \ cond$$

We convert this grammar into code, add two `Node traits` with subclasses, then make the parsers output nodes. The result is the following code.

```scala
package regexparser.util

import scala.util.parsing.combinator._

class Formula(formula: String) {
  private val root = Formula.parseAll(Formula.cond, formula).get

  def apply(vars: collection.Map[String, Double]) = root(vars)

  override def toString: String = formula
}

object Formula extends JavaTokenParsers {
  def apply(f: String) = new Formula(f)

  def eval(f: String, vars: collection.Map[String, Double] = null): Double =
    new Formula(f)(vars)

  private def cond: Parser[DNode] = """if\s*\(""".r ~> bform ~ """\)\s*""".r ~
    cond ~ """\s*else\s*""".r ~ cond ^^ {
      case b ~ _ ~ e1 ~ _ ~ e2 => new IfNode(b, e1, e2)
    } | form

  private def form: Parser[DNode] = term ~ rep(("+" | "-") ~ term) ^^ {
    case d ~ lst => new LeftAssocBinaryOpDNode(d, lst)
  }

  private def term: Parser[DNode] = exp ~ rep(("*" | "/") ~ exp) ^^ {
    case d ~ lst => new LeftAssocBinaryOpDNode(d, lst)
```

---

[12]This grammar includes some of those elements that go beyond the scope of this book. For example, if you change the first production to *cond ::= form | "if(" bform ")" cond "else" cond* it runs into problems because the string "if" is a valid identifier which matches *form*.

```scala
30    }
31
32    private def exp: Parser[DNode] = func ~ rep("^" ~> func) ^^ {
33      case d ~ lst => new PowBinaryOpDNode(d, lst)
34    }
35
36    private def func: Parser[DNode] = """(sin|cos|tan|sqrt)\("""".r ~ cond <~ ")" ^^ {
37      case f ~ n => new FunctionDNode(f, n)
38    } | factor
39
40    private def factor: Parser[DNode] = floatingPointNumber ^^ (s => new
           NumNode(s.toDouble)) |
41      ident ^^ (s => new VarNode(s)) | "(" ~> cond <~ ")"
42
43    private def bform: Parser[BNode] = bterm ~ rep("||" ~> bterm) ^^ {
44      case b ~ lst => new LeftAssocBinaryOpBNode(b, lst, _ || _)
45    }
46
47    private def bterm: Parser[BNode] = bnot ~ rep("&&" ~> bnot) ^^ {
48      case b ~ lst => new LeftAssocBinaryOpBNode(b, lst, _ && _)
49    }
50
51    private def bnot: Parser[BNode] = "!(" ~> bform <~ ")" ^^ (b => new BNotNode(b))
52        | bcomp
53    private def bcomp: Parser[BNode] = cond ~ ("""[=!><]=|<|>""".r) ~ cond ^^ {
54      case c1 ~ op ~ c2 => new CompNode(c1, op, c2)
55    }
56
57    private trait DNode {
58      def apply(vars: collection.Map[String, Double]): Double
59    }
60
61    private trait BNode {
62      def apply(vars: collection.Map[String, Double]): Boolean
63    }
64
65    private class LeftAssocBinaryOpDNode(first: DNode, restStr: List[~[String,
           DNode]]) extends DNode {
66      val rest = for (~(op, n) <- restStr) yield (op match {
67        case "+" => (_: Double) + (_: Double)
68        case "-" => (_: Double) - (_: Double)
69        case "*" => (_: Double) * (_: Double)
70        case "/" => (_: Double) / (_: Double)
71      }, n)
72      def apply(vars: collection.Map[String, Double]): Double =
73        rest.foldLeft(first(vars))((d, t) => {
74          t._1(d, t._2(vars))
75        })
76    }
77
78    private class PowBinaryOpDNode(first: DNode, rest: List[DNode]) extends DNode {
79      def apply(vars: collection.Map[String, Double]): Double =
80        math.pow(first(vars), rest.foldRight(1.0)((n, d) => math.pow(n(vars), d)))
81    }
```

```scala
82
83    private class NumNode(num: Double) extends DNode {
84      def apply(vars: collection.Map[String, Double]): Double = num
85    }
86
87    private class VarNode(name: String) extends DNode {
88      def apply(vars: collection.Map[String, Double]): Double = vars(name)
89    }
90
91    private class FunctionDNode(name: String, arg: DNode) extends DNode {
92      val f: Double => Double = name match {
93        case "sin(" => math.sin
94        case "cos(" => math.cos
95        case "tan(" => math.tan
96        case "sqrt(" => math.sqrt
97      }
98      def apply(vars: collection.Map[String, Double]): Double = f(arg(vars))
99    }
100
101    private class IfNode(cond: BNode, e1: DNode, e2: DNode) extends DNode {
102      def apply(vars: collection.Map[String, Double]): Double =
103        if (cond(vars)) e1(vars) else e2(vars)
104    }
105
106    private class LeftAssocBinaryOpBNode(first: BNode, rest: List[BNode],
107                                op: (Boolean, Boolean) => Boolean) extends
                                   BNode {
108      def apply(vars: collection.Map[String, Double]): Boolean =
109        rest.foldLeft(first(vars))((tf, b) => op(tf, b(vars)))
110    }
111
112    private class BNotNode(arg: BNode) extends BNode {
113      def apply(vars: collection.Map[String, Double]): Boolean = !(arg(vars))
114    }
115
116    private class CompNode(left: DNode, compStr: String, right: DNode) extends BNode {
117      val comp: (Double, Double) => Boolean = compStr match {
118        case "<" => _ < _
119        case ">" => _ > _
120        case "<=" => _ <= _
121        case ">=" => _ >= _
122        case "==" => _ == _
123        case "!=" => _ != _
124      }
125      def apply(vars: collection.Map[String, Double]): Boolean =
126        comp(left(vars), right(vars))
127    }
128  }
```

One of the first things to note about this code is that regular expressions are used in the parsers to keep things short and flexible. The other thing to note is that there are two top-level node traits called DNode and BNode. These both have an **apply** method, but they differ in that they return a **Double** and a **Boolean** respectively.

The binary operator nodes are a bit more complex than their counterparts from chapter 16. This is because these have the ability to take many operators that are at the same precedence level. Despite some parts of it being complex, the entire file and code for this is under 130 lines of code and it contains a significant amount of functionality. What is more, it is fairly easy to extend, particularly for adding single operator functions. This `Formula` class can now be substituted for the `Formula` class developed in the last chapter and used in the drawing project.

17.5 End of Chapter Material

17.5.1 Summary of Concepts

- Chomsky Grammars

  - Formal approach to generating languages and parsing languages. These form a hierarchy with four different levels of complexity. A grammar is defined as a set of terminals, a set of non-terminals, a set of productions, and a start symbol.

  - Regular grammars are the lowest level. Productions restricted to the form $A \rightarrow a \mid aB$.

  - Context-free grammars are next up in the level of complexity. Their productions have to be of the form $A \rightarrow \gamma$.

  - Context-sensitive grammars can take the symbols around a non-terminal into account when determining if a production is allowed. The productions must have the form $\alpha A \beta \rightarrow \alpha \gamma \beta$.

  - Recursively enumerable grammars have no restrictions on their productions. They are a complete model of computation.

- Regular expressions are a syntax for string parsing that have a power roughly equal to regular grammars.

  - Character classes can represent a set of characters that can be matched. You can build character classes with square brackets. There are also a number of built-in character classes for commonly used sets of characters.

  - Quantifiers specify that a character of grouping can be present certain numbers of times. Quantifiers include ?, *, and +.

  - Strings can be turned into `Regex` using the `r` method. The `Regex` object has methods called `findAllIn` and `replaceAllIn`. They can also be used as patterns.

- Context-Free Parsers

  - The combinator parser library in Scala makes it easy to build a parser from a CF grammar.

  - Default output of parsers includes a mix of strings, ~ objects, and lists. This can be challenging to work with.

  - Using the ^^ operator you can force parsers to output specific types that you want.

17.5.2   Exercises

1. Write a simple program that will take a Chomsky grammar and allow the user to generate strings from the language.

2. For each of the following grammars, try to figure out what languages they generate. For those that are regular, create an appropriate regular expression. For those that are context-free, write appropriate parsing code.

    - $S \to A$
      $A \to aA|B$
      $B \to bB|b$

    - $S \to aA$
      $A \to aA|bC$
      $C \to cC|c$

    - $S \to ABC$
      $A \to aA|a$
      $B \to bB|b$
      $c \to cC|c$

    - $S \to AB$
      $A \to aA|a$
      $B \to AbB|b$

    - $S \to eLAe$
      $eLA \to eRA$
      $ALA \to AAL$
      $ARe \to ALAe$
      $ARA \to RAA$

3. Try to come up with, or look up, a simple context-free grammar for English. Write a parser for it and include a simple dictionary.

4. Write a grammar for roman numerals. Have it parse to a tree such that the output has an **eval** method that will return the numeric value as an **Int**.

17.5.3   Projects

1. If you have been working on the MUD project, this chapter gives you the ability to use commands with significantly greater complexity. This can be done with either regular expressions or a parser. You can decide which you want to use, but the suggestion would be to write a simple grammar for more complex commands that you could put into your MUD and then write a parser that will take those and build usable expression trees from them.

2. Regular expressions and parsers can be extremely useful in pulling data out of text files for the web spider. You likely wrote up a fair bit of ad hoc code for getting the data you want out of HTML files. Convert that code to using a parser with regular expressions.

3. Both the networked and non-networked graphical game programs avoid having significant text input. However, in both cases, there are often reasons when it is nice to be

able to give commands to the program that go beyond what a normal user should be able to do in order to help with testing and debugging.

For this project, you should add a small command language to your game. If there is a good reason to do this for normal users, feel free to do so. Networked games could integrate this with chatting features. Non-networked games could put it into the GUI, or you could use the logging socket that was added back in chapter 11.

4. The material from this chapter can be particularly beneficial to the math worksheet project. Using combinatorial parsers, you can create a more complex, powerful formula parser, introduce additional commands with relative ease, and even consider putting in a simple programming language. You can also include conditionals expressions in your parser so that your mathematical functions can use recursive definitions.

    For all of these additions, you should have the parser build an expression tree for faster evaluation. Add whatever set of features you feel takes the project in the direction you want to go.

5. Photoshop® is another project that included mathematical formulas, and which can benefit from having commands in a simple language. Given what you learned in this chapter, you should be able to add much more powerful parsing to your formulas that are used with filters.

6. The simulation workbench project will also benefit from having real formulas. All of the applications of formulas from chapter 16 apply when using a combinatorial parser as well. In addition, the combinatorial parser makes it reasonable to extend formulas into more of a real language.

7. We do not have a clear use for regular expressions and parsing for the stock portfolio management project. Perhaps you can think of something interesting to do with it. You could create a tool that allows you to write a regular expression that searches for a word in a company name. As spoken of in the chapter, this is a common programming tool used by programmers, thus it is likely that only you, as the administrator of the stock portfolio system, would need to have access to this tool. For you to test this, it would be best if you had several stocks that were owned by the same company, thus it is more likely that the company name will be part of more than one stock.

8. Create a tool that allows you to write a regular expression that searches for a word in a movie title. As spoken of in the chapter, this is a common programming tool used by programmers, thus it is likely that only you, as the administrator of the movie system, would need to have access to this tool.

9. If you have been doing the L-systems project, you have been playing with grammars a long time. The L-system grammars are fundamentally distinct from Chomsky grammars. For that reason, the material in this chapter does not integrate well with your project, unless you want to allow users to specify formulas for probabilities on probabilistic productions. You should also find it interesting to look into just how L-system grammars differ from Chomsky grammars and what impact that has on the processing power of the two approaches.

Chapter 18

Binary Heaps

Back in chapter 13 we looked at the priority queue ADT. In that chapter, the implementation was written using a sorted linked list. While it was easy to write, this implementation has the downside of an $O(n)$ `enqueue` method. This gives overall performance that is $O(n^2)$ for the size of the queue. In this chapter we will look at a different implementation that provides $O(\log(n))$ performance for both `enqueue` and `dequeue`, the binary heap.

18.1 Binary Heaps

There are quite a few different styles of heaps. A common theme is that they have a tree structure. The simplest, and most broadly used, is the binary heap. As the name implies, it is based on a binary tree type of structure, where each node can potentially have left and right children. For a binary tree to be a binary heap it has to have two properties:

- Complete - The tree fills in each level from left to right. There are never gaps in a proper binary heap.

- Heap ordering - Priority queues require that elements have a complete ordering. The rule for a binary heap is that parents always have a priority that is greater than or equal to their children.

The heap ordering means that the highest-priority element is always at the root of the tree. An example is shown in figure 18.1.

18.1.1 Binary Heaps as Arrays

While we conceptually think of the heap as a binary tree, we do not implement it that way. It is far more efficient to implement it as an array. This only works because the heap is complete. Figure 18.2 shows the same heap as in figure 18.1 with subscripts that number the nodes in breadth-first order. Looking closely at the subscripts, you should see that there is a nice mathematical relationship between parents and children. Specifically, if you divide an index by two, using integer division, the result is the index of the parent. The reverse

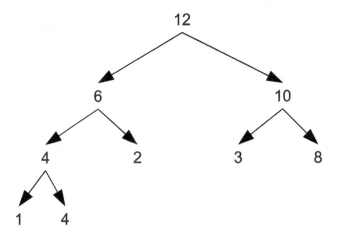

FIGURE 18.1: This figure shows an example of a binary heap. The numbers are priorities with higher values being higher priority. Note that there is no problem with duplicate priorities, unlike keys in a BST. There is also no order requirement between siblings.

relationship is that the children of a node have indices that are twice the node's index and twice the node's index plus one.

The figure shows the array leaving the 0 index spot empty. It is possible to implement a heap with all the elements shifted down one spot so that there is not a blank. However, this requires putting +1 and -1 in the code at many different points. Having the root at 0 rather than at 1 costs you an extra add to find the left child, and an extra subtraction to find the parent. This complicates the code, possibly making it harder to understand and maintain. While it removes a small, $O(1)$ waste in memory, it introduces an overhead in every call on the priority queue. For that reason, and to keep the code simpler, the implementation shown here will leave the first element of the array as a default value.

18.2　Heaps as Priority Queues

To use the binary heap as a priority queue, we have to be able to implement **enqueue**, **dequeue**, **peek**, and **isEmpty**. The last two are simple; the highest-priority element is always the root, so **peek** simply returns it. The **isEmpty** method just checks if there is anything present on the heap by checking an index value that has to be stored to indicate the first unused position in the array. The **enqueue** and **dequeue** are a bit more complex because they have to change the structure of the heap.

We start with **enqueue**, where an element is added to the queue. Adding to a BST was just a task of finding the proper place to put the new element. For a heap, the job is a bit different. The heap has to remain complete so when a new element is added, there is no question about what new location in the tree has to be occupied. The next open space in the row that is currently filling in or the first in the next row when the lowest row is complete has to be filled. However, the new element might not belong there. The way we deal with this is to put a "bubble" in the position that has to be filled and let it move up

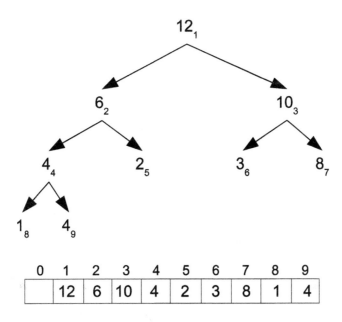

FIGURE 18.2: This figure shows the same heap as in figure 18.1 with subscripts that number the nodes in breadth-first order. Below that is an array with the elements at those locations.

until it reaches the location that the new element belongs. As long as the new element has a higher priority than the parent of the bubble, the parent is copied down and the bubble moves up. An example of adding a new element with a priority of 9 to the example heap is shown in figure 18.3.

A call to **dequeue** should remove the element from the root of the tree, as that will always have the highest-priority. However, to keep the tree complete, the location that needs to be vacated is at the end of the last row. To deal with this, we pull the last element out of the heap and put it in a temporary variable, then put a placeholder we will call a "stone" at the root. This stone then sinks through the heap until it gets to a point where the temporary can stop without breaking heap order. When moving down through the heap there is a choice of which child the stone should sink to. In order to preserve the heap order, it has to move to the higher-priority child as we are not allowed to move the lower-priority child into the parent position above its higher-priority sibling. This process is illustrated in figure 18.4.

Code to implement these operations using an array-based binary heap is shown below. This extends the **PriorityQueue[A]** type that was created in chapter 13. Like the sorted linked list implementation, a comparison function is passed in to provide the complete ordering of elements for instances of this **class**. The only data needed by this **class** is an **Array** we call **heap** and an integer called **end** that keeps track of where the next element should be added. These are created with 10 elements to begin with, one of which will never be used as index zero is left empty, and **end** is set to be 1.

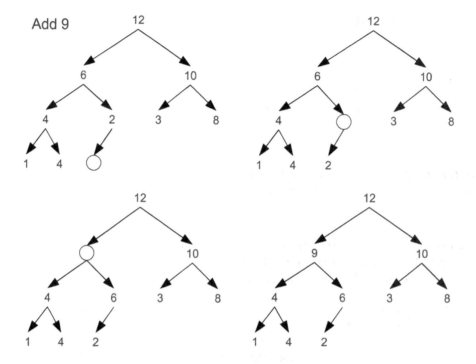

FIGURE 18.3: Adding a new element with a priority of 9 to the heap shown in figure 18.1 goes through the following steps. A bubble is added at the next location in a complete tree. It moves up until it reaches a point where the new value can be added without breaking the heap order.

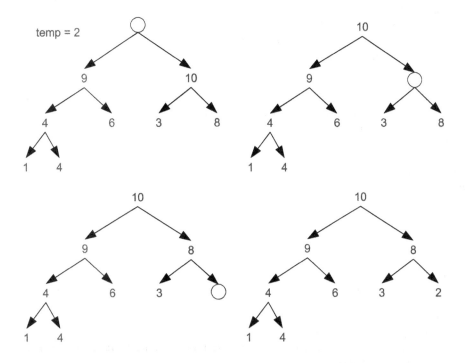

FIGURE 18.4: Calling `dequeue` on the heap at the end of figure 18.3 forces the tree to go through the following steps. The last element is moved to a temporary variable, and a stone is placed at the top of the heap. This stone sinks down until it reaches a point where the temporary can stop without breaking heap order. The stone always sinks in the direction of the higher-priority child. If the children have equal priority, it does not matter which side the stone goes to.

```scala
package heaps.adt

import scala.reflect.ClassTag

import priorityqueues.adt.PriorityQueue

class BinaryHeapPriorityQueue[A: ClassTag](comp: (A, A) => Int) extends
    PriorityQueue[A] {
  private var heap = new Array[A](10)
  private var end = 1

  def enqueue(obj: A): Unit = {
    if (end >= heap.length) {
      val tmp = new Array[A](heap.length * 2)
      Array.copy(heap, 0, tmp, 0, heap.length)
      heap = tmp
    }
    var bubble = end
    while (bubble > 1 && comp(obj, heap(bubble / 2)) > 0) {
      heap(bubble) = heap(bubble / 2)
      bubble /= 2
    }
    heap(bubble) = obj
    end += 1
  }

  def dequeue(): A = {
    val ret = heap(1)
    end -= 1
    val temp = heap(end)
    heap(end) = heap(0) // Clear reference to temp
    var stone = 1
    var flag = true
    while (flag && stone * 2 < end) {
      var greaterChild = if (stone * 2 + 1 < end && comp(heap(stone * 2 + 1),
          heap(stone * 2)) > 0)
        stone * 2 + 1 else stone * 2
      if (comp(heap(greaterChild), temp) > 0) {
        heap(stone) = heap(greaterChild)
        stone = greaterChild
      } else {
        flag = false
      }
    }
    heap(stone) = temp
    ret
  }

  def peek: A = heap(1)

  def isEmpty: Boolean = end == 1
}
```

The enqueue method begins with a check to see if there is space in the array for a new element. If not, a bigger aArray is created. As with other similar pieces of code written in this book, the size of the `Array` is doubled so that the amortized cost of copies will be $O(1)$. The rest of the `enqueue` method makes an `Int` called `bubble` that will walk up the heap to find where the new element should go. It is initialized to `end`, the first unused location in the `Array`. The main work is done in a `while` loop that goes as long as the bubble is greater than an index of 1 or the parent of `bubble`, located at index `bubble/2`, has a higher priority than the new element. When the loop terminates, the new element is placed at the location of the `bubble` and `end` is incremented.

The `dequeue` method starts off by storing the current root in a variable that will be returned at the end of the method. It then decrements `end`, stores the last element in the heap in a temporary, and clears out that element with the default value stored at `heap(0)`. A variable called `stone` is created that starts at the root and will fall down through the heap. There is also a `flag` variable created to indicate when the `while` loop should terminate. This style is helpful when the determinant of whether to stop the loop is complex and would be difficult to put in a single line. In this case, part of the determination is easy; if the stone has gotten to a leaf, it must stop there. Otherwise, it requires checking the temporary against the child or children to see if the temporary has lower-priority than the higher-priority child. That would require a complex `Boolean` expression that would be hard to read. The use of a `flag` generally improves the readability and maintainability of the code.

Inside the loop, the first task is determining which child has a higher priority. The check in the `while` loop condition guarantees that there is at least one. If there is a second, and it has higher priority, it is the one we want, otherwise we want the first child. Then that child is compared to the `temp` to see if the stone should keep sinking. If it should not keep sinking, then the loop needs to stop, so `flag` is set to `false`. When the loops terminates, the temporary is moved to the location where the `stone` stopped and the original root value is returned.

Using the image of the tree we can argue that both `enqueue` and `dequeue` are always $O(\log(N))$. This can also be argued from the code. The variable `end` is effectively n. In the `enqueue` method, the `bubble` starts at `end` and is divided by 2 in every iteration through the loop. This can only happen $\log_2(n)$ times before it gets to 1. With `dequeue` the `stone` starts at 1 and in each iteration it changes to either `stone*2` or `stone*2+1`. This too can only happen $\log_2(n)$ times before `stone` gets to `end`.

Testing this code is a simple matter of copying the test for the sorted linked list priority queue and changing what type is instantiated. It can also be used to improve the performance of any other code that requires only the priority queue interface to operate. For example, the cell division example from chapter 13 could be altered and would gain a significant performance boost for large populations.

Priority Queue with Frequent Removes

Back in subsection 13.2.2 we introduced a way of handling collisions in animations. In that subsection, the collisions were between balls, but to speak in more general terms, we will be speaking about collisions between particles. There are areas of study where people do simulations of collisions between millions of particles. When a sorted linked list was used for an event-based simulation of collisions, we had to add a method called `removeMatches`. This method was needed because whenever a collision between two particles is processed, all future events involving either of those particles needs to be removed from the queue as the new velocity of the particle will make those events invalid. This method was not hard to add to the sorted linked list implementation and the fact that it was $O(n)$ was not a problem as the `enqueue` method was also $O(n)$.

In chapter 20 we will see how adding spatial data structures can improve the overall performance of a time step from $O(n^2)$ to $O(n\log(n))$ or even, in certain ideal cases, $O(n)$. However, the sorted linked list priority queue will still limit performance to $O(c^2)$, where c is the number of collisions in the time step. That number generally does increase with the number of particles, so using a sorted linked list is still a potential performance bottleneck.

We have just seen that a heap-based priority queue gives $O(\log(n))$ performance for the main operations, which would provide $O(n\log(n))$ overall performance for the number of elements added and removed. This could provide a significant performance improvement for collisions if we could implement `removeMatches` on the heap. Unfortunately, elements jump around in a binary heap, and locating them is $O(n)$ and cannot be improved significantly when done in bulk. Fortunately, collision handling has aspects that are predictable and can be used to make an alternate implementation.

What is important about collision handling is that from one time step to the next, the number of collisions is fairly consistent, and the distribution of times for collisions is fairly uniform. This points to an implementation that combines arrays and linked lists. Let us say that in the previous time step there were c collisions for n particles and each time step lasts Δt time units. We start with an array that spans time from 0 until Δt with c elements that each represent $\frac{\Delta t}{c}$ time. The array references nodes for linked lists. There is another array with n elements, one for each particle in the simulation, that also keeps references to nodes. Each node is part of three orthogonal doubly linked lists and stores information about an event. One doubly linked list going through a node is sorted and based on time. The others are not sorted, represent particles, and there is one for each particle involved. So particle-particle collisions are part of two of these and particle-wall collisions are part of two.

A graphical representation of this structure is shown in figure 18.5. The array across the top is for time. The first cell keeps the head of the list for any events between 0 and $\frac{\Delta t}{c}$. The second is for events between $\frac{\Delta t}{c}$ and $2\frac{\Delta t}{c}$, etc. The array on the right has one cell for each particle in the simulation and the list keeps track of any potential collisions that particle is involved in. The lists are doubly linked so that any element can be removed without knowing about the one before it in any given list. This figure is greatly simplified in many ways to make it understandable. In a real situation, the number of elements in each array should be at least in the thousands for this to be relevant. In addition, while the vertical lists are sorted and there will be a nice ordering for elements, the horizontal links will often double back around and produce a far more complex structure than what is shown here.

So what is the order of this data structure? Linked lists are inherently $O(n)$ for random access. This includes adding to sorted linked lists. On the other hand, adding to the head of a linked list is $O(1)$. In the worst case, all the events could go into a single, long linked list. That would result in $O(n^2)$ performance. Such a situation is extremely unlikely given the nature of collision events. We expect roughly c events that are roughly evenly distributed. That would result in sorted linked lists of an average length of 1 and it should be very rare to have a length of longer than a few. The `dequeue` operation has to remove all the other events for any particles involved, but that is required by any implementation and the advantage of having doubly linked lists by particles is that this can be done as fast as possible considering only the nodes involving those particles have to be visited. So the expected time for all operations is the fastest it could possibly be. For **dequeue** it is always $O(1)$, and the implementation simply keeps track of the element in the time array that has the first element. The **enqueue** could be $O(n)$, but we expect $O(1)$ performance because the time lists are short. The **removeParticle** method will perform $O(n)$ operations in the number of events for that particle, but this too should be a small number as we do not expect any single particle to be involved in collisions with more than a few other particles during a time step for a single velocity.

You might wonder why, if this data structure is so great, it is not used more generally. The reason is that it is not generally good. It requires knowing that there should be about c events, they will all lie in a certain finite time range, and they will be evenly distributed. Change any of those conditions and this structure can be worse than a standard sorted linked list. If you ever wonder why you should bother to learn how to create data structures, like linked lists, when there are perfectly good implementations in libraries, there is a reason why. The standard library implementations do not work well for all problems. Library priority queues, for instance, are typically implemented with a heap. For the general case, that is the ideal structure. Not all problems fit the general case though. In this situation, a heap is probably worse than a sorted linked list and the only way you can get significantly better performance is if you write your own data structure.

18.3 Heapsort

Any priority queue implementation can be turned into a sort. Simply add all the elements to the queue and remove them. They will come off in sorted order. Insertion sort is roughly what you get if you do this with a sorted array or linked list. The priority queue described in the above text box gives something like a bucket sort. If you build a sort from a heap-based priority queue, you get the heapsort. The fact that the ideal heap uses an array means that heapsort can be done completely in place. Simply build the heap on one side of the array and extract to the other.

Here is an implementation of a heap sort. This builds the heap using the value as the priority so higher values will dequeue earlier. They are then dequeued to fill in the array starting with the end.

```
def heapsort[A](a: Array[A])(comp: (A, A) => Int): Unit = {
  for (end <- 1 until a.length) {
    val tmp = a(end)
    var bubble = end
    while (bubble > 0 && comp(tmp, a((bubble + 1) / 2 - 1)) > 0) {
      a(bubble) = a((bubble + 1) / 2 - 1)
      bubble = (bubble + 1) / 2 - 1
    }
    a(bubble) = tmp
  }
  for (end <- a.length - 1 until 0 by -1) {
    val tmp = a(end)
    a(end) = a(0)
    var stone = 0
```

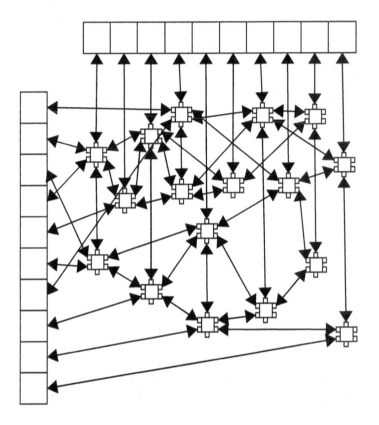

FIGURE 18.5: This shows a pictorial representation of the alternate data structure that can be used as an efficient priority queue for collision handling. The top array keeps the heads of lists for different time brackets during the time step. The left array is the heads of lists for different particles. All lists are doubly linked so that nodes can be removed without finding the one before it.

```
      var flag = true
      while (flag && (stone + 1) * 2 - 1 < end) {
        val greaterChild = if ((stone + 1) * 2 < end && comp(a((stone + 1) * 2),
            a((stone + 1) * 2 - 1)) > 0)
          (stone + 1) * 2 else (stone + 1) * 2 - 1
        if (comp(a(greaterChild), tmp) > 0) {
          a(stone) = a(greaterChild)
          stone = greaterChild
        } else {
          flag = false
        }
      }
      a(stone) = tmp
    }
  }
```

You should notice similarities between parts of this code and the heap implementation. The first **for** loop contains code that looks like **enqueue**. The second loop contains code that looks like **dequeue**. Some modifications had to be made because the array that is passed in does not include a blank element at the front.

This sort is $O(n \log(n))$, the same as for merge sort and quicksort. Unlike those two, the heapsort is not recursive. It is also stable, like the merge sort, and can be done in place, like the quicksort. These factors make it a nice sort for a number of different applications, though in general it is slower than a well-written quicksort.

18.4   End of Chapter Material

18.4.1   Summary of Concepts

- Binary heaps are one type of data structure that allows quick retrieval of a maximum priority element. This makes them efficient for implementing the priority queue ADT.

  - They are based on a binary tree structure.
  - They are always complete and have heap ordering.
  - Numbering tree elements beginning with one going in breadth-first order gives a simple mathematical relationship between indices of parents and children.
  - This leads to the most efficient implementation using an array because the heap is always complete.

- Binary heaps can be used to implement efficient priority queues. Both **enqueue** and **dequeue** are always $O(\log(n))$ because the heap is a complete tree.

- The heap can be used as the basis for a sort called the heapsort.

18.4.2   Exercises

1. Perform a speed test to compare the speed of the sorted list based priority queue from chapter 13 to the heap version presented in this chapter. Use exponentially spaced numbers of elements to get a broad range of n values.

2. Implement a binary heap that does not leave the 0 element empty. Test and debug it to make sure with works.

3. If you did the exercise above, you can now do a speed comparison of the heap code in the text to the code that starts the heap elements at the zero index.

4. The last text box in the chapter describes an alternate implementation for a priority queue that works well for the collision simulation where scheduled events have to be removed regularly. Write code for this type of queue and compare its speed to other priority queue implementations that you have.

18.4.3 Projects

For this chapter, you should convert whatever you did in chapter 13 from the implementation you used in that chapter to one of the implementations presented in this chapter. For the simulation workbench, if you have done a collisional simulation, consider doing the special priority queue that makes it easy to remove events.

Chapter 19

Direct Access Binary Files

Basic text files have the advantage of being simple. They are reasonably easy to program and the files themselves can be edited with a standard text editor. Unfortunately, they do not have any inherent meaning and they are very brittle to edit. Small mistakes can easily invalidate the whole file. These problems are corrected by using XML, which was introduced in chapter 10. Both flat text and XML have the downfall that they are slow and bloated in size. In these areas, XML is worse than flat text. To fix those problems we need to move in the opposite direction and go to binary files. Unfortunately, this move takes away the advantage that text and XML have that they can be read and edited easily with text editors. We saw how we could read and write in native binary in chapter 10 with the **DataInputStream** and **DataOutputStream** as well as how whole objects could be serialized to binary using **ObjectInputStream** and **ObjectOutputStream**.

In this chapter, we are going to take another step with binary files to make it so that they can be used like large arrays with the ability to access and read values at random locations. This makes many more things possible with files and is an essential part of applications that need to deal with extremely large data sets.

19.1 Random Access Files

Back in chapter 10 we focused on streams. The term stream matches their nature. Like water flowing by, data from a stream can be read as it goes by, but once it has gone by you cannot bring it back. You also cannot choose to just be at whatever position you want. For this capability we turn to a different **class** in the **java.io** package, **RandomAccessFile**. Looking at this class in the API will show you that it includes all the read methods from **DataInputStream** and the write methods from **DataOutputStream**. It is immediately clear that this one **class** gives you the ability to do both input and output in binary format. The real power of this **class** comes from the **seek(pos: Long)** method, which can jump to any location in a file. The partner of **seek** is **getFilePointer(): Long**, which returns the current location in the file. Putting this all together means that you can jump around in a file and read or write data in any way that you want.

To create a new **RandomAccessFile**, you pass arguments of the file to open, either as a

Type	Size in Bytes
Byte	1
Short	2
Int	4
Long	8
Float	4
Double	8
Boolean	1
Char	2

TABLE 19.1: This table shows how many bytes are written to disk for each of the basic value types using either `DataOutputStream` or `RandomAccessFile`.

`java.io.File` or a `String`, and a `mode: String`. The `mode` specifies how that file is going to be accessed. It is only allowed to have one of the following values.

- "r" - Open the file for reading only. If this is used and any of the write methods are invoked, an `IOException` will be thrown.

- "rw" - Open the file for reading and writing. This mode will try to create a file if it does not exist.

- "rws" - Open the file for reading and writing with the additional requirement that if a call modifies the file in any way, that call synchronously goes to the storage device. This will slow things down as it makes buffering of disk activity impossible, but is important for critical applications that need the file to be in a consistent configuration regardless of what happens to the application.

- "rwd" - Much like "rws", but only requires synchronization when the contents of the file are altered.[1]

If you use any other value for the `mode`, an exception will be thrown. We will typically use "rw" as the mode so it will look like this.

```
val raf = new RandomAccessFile("binaryFile.bin", "rw")
```

19.1.1 Fixed Record Length

The simplest usage of a binary file is with fixed-length records. This is where you store data in blocks of the same size. This is fairly easy when your data is composed of a fixed number of fixed size elements like `Int` and `Double`. In that situation you simple add up the sizes of all the elements. Table 19.1 shows the sizes of the basic types in Scala in bytes. So if the record to be written was a point in 3-D space that has an integer value associated with it, you would write out three `Doubles` and one `Int` using a total of $3 * 8 + 4 = 28$ bytes.

Many types of data will make the process of determining a length a bit more complex. For example, table 19.1 is missing the `String` type. That is because strings will take up different amounts of space based on how long they are. If you want to use a fixed-length record file with records that include strings, you have to determine a maximum length that you will allow your string to be, then write it out as a sequence of `Char` or `Byte`. Which you

---

[1] Files also have metadata associated with them. The "rws" mode will force synchronization when either metadata or contents are changed.

pick will depend on whether you expect to have characters outside of standard English in the text. Normal English characters are part of lower ASCII, which fits completely in a Byte. If the application is going to be internationalized or include special symbols from Unicode, you should probably stick with a Char. You can use the getBytes(): Array[Byte] method of String if you choose to use the Byte approach. In your code, you can decide how to deal with strings that are too long, whether you want to throw an exception or silently truncate them. Either way, you have to enforce a maximum size, make sure that you are able to read back in what you write out, and never write data beyond the record boundary.

Records that contain collections have to be treated much like Strings. You pick a maximum number of elements that are going to be stored to file. Calculate the size of an element and multiply by the maximum number you will store to determine the size contribution of the collection to the total record size.

To apply this idea, we will write a class that uses a fixed-length record random access file as the backing for a mutable.IndexedSeq so that data can be preserved from one run to the next. If you attach to the same file in separate runs, you will automatically get access to the same data. Using a file also allows you to have an extremely large collection without consuming that much memory. The downside is that accessing this collection will be significantly slower.

```scala
package binaryfiles.adt

import java.io.DataInput
import java.io.DataOutput
import java.io.File
import java.io.RandomAccessFile

import scala.collection.mutable

class FixedRecordSeq[A](
    file: File,
    recordLength: Int,
    reader: DataInput => A,
    writer: (DataOutput, A) => Unit) extends mutable.IndexedSeq[A] {

  private val raf = new RandomAccessFile(file, "rw")

  def apply(index: Int): A = {
    raf.seek(recordLength * index)
    reader(raf)
  }

  def update(index: Int, a: A): Unit = {
    raf.seek(recordLength * index)
    writer(raf, a)
  }

  def length: Int = (raf.length() / recordLength).toInt

  def close(): Unit = raf.close()
}
```

This class inherits from mutable.IndexedSeq, and it has fundamentally different behavior for the update method. Assigning out of bounds on this collection does not throw an exception. Instead, it grows the collection to the size of that index. This does not technically contradict the API documentation for the supertype, but it is something that users should be aware of.

The apply method seeks to the position in the file given by the index times the record length, then does a read. This simplicity is the real advantage of fixed record length files. Given an index, a simple mathematical operation gives the location of the data in the file. The update method is equally simple with a call to seek followed by a call of the write function.

To make it clear how this class would be used, we can look at some unit tests for it.

```scala
package test.binaryfiles.adt

import java.io.DataInput
import java.io.DataOutput
import java.io.File

import org.junit.Assert._
import org.junit.Before
import org.junit.Test

import binaryfiles.adt.FixedRecordSeq

class TestFixedRecordSeq {
  private var file: File = null
  private case class PointRec(x: Double, y: Double, z: Double, d: Int)
  private case class StudentRec(first: String, last: String, grades: List[Int])

  @Before def setFile: Unit = {
    file = new File("TextFile.bin")
    file.deleteOnExit()
    if (file.exists()) file.delete()
  }

  @Test def emptyStart: Unit = {
    val frs = new FixedRecordSeq[Int](file, 4, din => din.readInt(), (dout, i) =>
        dout.writeInt(i))
    assertTrue(frs.length == 0)
  }

  @Test def write4: Unit = {
    val frs = new FixedRecordSeq[Int](file, 4, din => din.readInt(), (dout, i) =>
        dout.writeInt(i))
    frs(0) = 3
    frs(1) = 8
    frs(2) = 2
    frs(3) = 9
    assertEquals(4, frs.length)
    assertEquals(3, frs(0))
    assertEquals(8, frs(1))
    assertEquals(2, frs(2))
    assertEquals(9, frs(3))
  }

  @Test def write100: Unit = {
    val frs = new FixedRecordSeq[Double](file, 8, din => din.readDouble(), (dout,
        d) => dout.writeDouble(d))
```

```scala
44    val nums = Array.fill(100)(math.random)
45    for (i <- nums.indices) frs(i) = nums(i)
46    assertEquals(nums.length, frs.length)
47    for (i <- nums.indices) assertEquals(nums(i), frs(i), 0)
48  }
49
50  private def readPoint(din: DataInput): PointRec = {
51    new PointRec(din.readDouble(), din.readDouble(), din.readDouble(),
         din.readInt())
52  }
53
54  private def writePoint(dout: DataOutput, p: PointRec): Unit = {
55    dout.writeDouble(p.x)
56    dout.writeDouble(p.y)
57    dout.writeDouble(p.z)
58    dout.writeInt(p.d)
59  }
60
61  @Test def writePoints: Unit = {
62    val frs = new FixedRecordSeq[PointRec](file, 3 * 8 + 4, readPoint, writePoint)
63    val pnts = Array.fill(10)(new PointRec(math.random, math.random, math.random,
         util.Random.nextInt))
64    for (i <- pnts.indices) frs(i) = pnts(i)
65    assertEquals(pnts.length, frs.length)
66    for (i <- pnts.indices) assertEquals(pnts(i), frs(i))
67  }
68
69  @Test def rewritePoints: Unit = {
70    val frs = new FixedRecordSeq[PointRec](file, 3 * 8 + 4, readPoint, writePoint)
71    val pnts = Array.fill(10)(new PointRec(math.random, math.random, math.random,
         util.Random.nextInt))
72    for (i <- pnts.indices) frs(i) = pnts(i)
73    assertEquals(pnts.length, frs.length)
74    for (i <- pnts.indices) assertEquals(pnts(i), frs(i))
75    for (i <- 0 until pnts.length / 2) {
76      val index = util.Random.nextInt(pnts.length)
77      pnts(index) = new PointRec(math.random, math.random, math.random,
           util.Random.nextInt)
78      frs(index) = pnts(index)
79    }
80    for (i <- pnts.indices) assertEquals(pnts(i), frs(i))
81  }
82
83  private def readStudent(din: DataInput): StudentRec = {
84    val buf = new Array[Byte](20)
85    din.readFully(buf)
86    val first = new String(buf.takeWhile(_ > 0))
87    din.readFully(buf)
88    val last = new String(buf.takeWhile(_ > 0))
89    val grades = List.fill(10)(din.readInt()).filter(_ > Int.MinValue)
90    new StudentRec(first, last, grades)
91  }
92
93  private def writeStudent(dout: DataOutput, s: StudentRec): Unit = {
94    dout.write(s.first.take(20).getBytes().padTo(20, 0.toByte))
```

```scala
95      dout.write(s.last.take(20).getBytes().padTo(20, 0.toByte))
96      s.grades.padTo(10, Int.MinValue).foreach(dout.writeInt)
97    }
98
99    @Test def writeStudents: Unit = {
100     val frs = new FixedRecordSeq[StudentRec](file, 20 + 20 + 4 * 10, readStudent,
            writeStudent)
101     val students = Array.fill(100) {
102       val first = Array.fill(util.Random.nextInt(15))((’a’ +
              util.Random.nextInt(26)).toChar).mkString
103       val last = Array.fill(util.Random.nextInt(15))((’a’ +
              util.Random.nextInt(26)).toChar).mkString
104       val grades = List.fill(util.Random.nextInt(5) + 6)(60 +
              util.Random.nextInt(40))
105       new StudentRec(first, last, grades)
106     }
107     for (i <- students.indices) frs(i) = students(i)
108     assertEquals(students.length, frs.length)
109     for (i <- students.indices) assertEquals(students(i), frs(i))
110   }
111
112   @Test def writeStudentsLong: Unit = {
113     val frs = new FixedRecordSeq[StudentRec](file, 20 + 20 + 4 * 10, readStudent,
            writeStudent)
114     val students = Array.fill(100) {
115       val first = Array.fill(10 + util.Random.nextInt(15))((’a’ +
              util.Random.nextInt(26)).toChar).mkString
116       val last = Array.fill(10 + util.Random.nextInt(15))((’a’ +
              util.Random.nextInt(26)).toChar).mkString
117       val grades = List.fill(util.Random.nextInt(5) + 6)(60 +
              util.Random.nextInt(40))
118       new StudentRec(first, last, grades)
119     }
120     for (i <- students.indices) frs(i) = students(i)
121     assertEquals(students.length, frs.length)
122     for (i <- students.indices) {
123       if (students(i).first.length > 20 || students(i).last.length > 20) {
124         assertEquals(students(i).copy(first = students(i).first.take(20), last =
              students(i).last.take(20)), frs(i))
125       } else {
126         assertEquals(students(i), frs(i))
127       }
128     }
129   }
130 }
```

The first tests work with simple Ints. The others work with two different case classes. The first case class, PointRec, has the structure described above with three Doubles and an Int. The task of reading and writing the Int type is so simple that function literals can be used there. For the case classes, helper functions were used. The ones for PointRec are both simple and straightforward.

The second case class, StudentRec, includes two Strings and a List[Int]. Having three elements that all have variable sizes is pretty much a worst-case scenario for a fixed record length file. The read and write methods determine how the records will be truncated. In this case, both strings are truncated to 20 characters and the list of grades is truncated to 10 grades. When writing anything shorter than that, the extra spots are padded with placeholders to complete the record. The last test includes strings that are longer than 20 characters to make sure that the truncation works properly.

19.1.2 Indexed Variable Record Length

A bit more work must be done if it is not possible to settle on a single fixed record length. There are a few reasons that you might run into this. One is that you cannot afford to lose any data and you cannot put any safe upper boundary on the size of some part of the record. A second is that you cannot lose data and the average case is much smaller than the maximum and you either cannot or are unwilling to waste the space that you would have to if you used a fixed record length.

In these situations you have to use a variable record length. The problem with this is that you cannot simply calculate the position in the file to go and read the nth element. One way to deal with this is to keep a separate index that has fixed-length records which gives locations for the variable-length record data. We will do this, and to keep things as simple as possible, we will put the index values in a separate file. Here is a sample implementation of such a sequence. It is built very much like the fixed record length version except that instead of a record size, there is a second file passed in.

```scala
package binaryfiles.adt

import collection.mutable
import java.io._

class VariableRecordSeq[A](
    index: File,
    data: File,
    reader: DataInput => A,
    writer: (DataOutput, A) => Unit) extends mutable.IndexedSeq[A] {

  private val indexFile = new RandomAccessFile(index, "rw")
  private val dataFile = new RandomAccessFile(data, "rw")

  def apply(index: Int): A = {
    indexFile.seek(12 * index)
    val pos = indexFile.readLong()
    dataFile.seek(pos)
    reader(dataFile)
  }

  def update(index: Int, a: A): Unit = {
    val baos = new ByteArrayOutputStream()
    val dos = new DataOutputStream(baos)
    writer(dos, a)
    val outData = baos.toByteArray()
    val (pos, len) = if (index < length) {
      indexFile.seek(12 * index)
      val p = indexFile.readLong()
      val l = indexFile.readInt()
      if (baos.size() <= l) (p, l) else (dataFile.length(), outData.length)
    } else (dataFile.length(), outData.length)
    dataFile.seek(pos)
    dataFile.write(outData)
```

```
35      indexFile.seek(12 * index)
36      indexFile.writeLong(pos)
37      indexFile.writeInt(len)
38    }
39
40    def length: Int = (indexFile.length() / 12).toInt
41
42    def close(): Unit = {
43      indexFile.close()
44      dataFile.close()
45    }
46  }
```

Using an index file makes the **apply** method a bit more complex. It starts with a **seek** in the index file and a read of the position where the data is in the main data file. That value is then passed to a **seek** on the data file and the value we want is read from that file. You can see from this code that the index file is basically a fixed record length file with a record length of 12. The value of 12 comes from the fact that we are storing both a **Long** for the position in the data file and an **Int** for the length of the data record in that file. Storing the length is not required, but it allows us to make the update a bit more efficient in how it stores things in the data file.

Most of the effort for using variable record length files goes into the **update** method. When a new value is written to file, if it fits in the space of the old value, it should be written in that same space. If it is a completely new record, or if the new value needs more space than the old one, the easiest way to deal with this is to put it at the end of the data file. These criteria are easy to describe, but they have one significant challenge, you have to know the length of what will be written before it is written to file. What is more, we really do not want to force the user of this class to have to pass in an additional function that calculates the size of a record. To get around this, the **update** method makes use of a **ByteArrayOutputStream**. This is an **OutputStream** that collects everything that is written to it in an **Array[Byte]**. This is wrapped in a **DataOutputStream** and passed to the **writer** function before anything else is done.

Doing this at the beginning of **update** allows us to get an array of bytes that we need to write to the data file without having an extra function for size or writing to the real data file. After this, a little logic is performed to see if the data can be written over an old value or if it needs to go at the end of the file. If this is a new record, or the length of **outData** is longer than the length of the space in the data file, it has to be put at the end of the data file and the length is the length of this output. Otherwise it goes at the old location and we use the old length because that is the amount of space we can safely use there.

The **update** method ends by jumping to the proper location in the data file and writing the array out to the file, then jumping to the correct position in the index file and writing the position and length there. This code will work with the same test code as the fixed record length version with only a few changes to **readStudent**, **writeStudent**, and the last test. There is no longer a reason to cut data down or pad things in this version. Here is the modified code with a stronger version of the last test to make certain the changing records works.

```
1    private def readStudent(din: DataInput): StudentRec = {
2      val first = din.readUTF()
3      val last = din.readUTF()
4      val num = din.readInt()
5      val grades = List.fill(num)(din.readInt())
6      new StudentRec(first, last, grades)
```

```scala
7    }
8
9    private def writeStudent(dout: DataOutput, s: StudentRec): Unit = {
10     dout.writeUTF(s.first)
11     dout.writeUTF(s.last)
12     dout.writeInt(s.grades.length)
13     s.grades.foreach(dout.writeInt)
14   }
15
16   @Test def rewriteStudents: Unit = {
17     val frs = new VariableRecordSeq[StudentRec](iFile, dFile, readStudent,
         writeStudent)
18     val students = Array.fill(100) {
19       val first = Array.fill(10 + util.Random.nextInt(15))(('a' +
           util.Random.nextInt(26)).toChar).mkString
20       val last = Array.fill(10 + util.Random.nextInt(15))(('a' +
           util.Random.nextInt(26)).toChar).mkString
21       val grades = List.fill(util.Random.nextInt(5) + 6)(60 +
           util.Random.nextInt(40))
22       new StudentRec(first, last, grades)
23     }
24     for (i <- students.indices) frs(i) = students(i)
25     assertEquals(students.length, frs.length)
26     for (i <- students.indices) assertEquals(students(i), frs(i))
27     for (i <- 0 until students.length / 2) {
28       val index = util.Random.nextInt(students.length)
29       students(index) = {
30         val first = Array.fill(10 + util.Random.nextInt(15))(('a' +
             util.Random.nextInt(26)).toChar).mkString
31         val last = Array.fill(10 + util.Random.nextInt(15))(('a' +
             util.Random.nextInt(26)).toChar).mkString
32         val grades = List.fill(util.Random.nextInt(5) + 6)(60 +
             util.Random.nextInt(40))
33         new StudentRec(first, last, grades)
34       }
35       frs(index) = students(index)
36     }
37     for (i <- students.indices) assertEquals(students(i), frs(i))
38   }
```

Notice that this code uses `readUTF` and `writeUTF`. We could not do this with the fixed record length file because the number of bytes it writes to file can vary by string contents, not just length. This freedom to use things that we do not know the exact size of can be quite useful. With a little modification, this class could be changed to work with serializable objects. We do not have strong control over the number of bytes written when an object is serialized, but using this approach, that is not a problem.

This code does have a weakness though. Every time you update a record with a longer record, the old position in the data file becomes wasted space. Nothing will ever be written there again. Over time, this wasted space can build up and become problematic. To get around that weakness, you can add another method that will build a new data file without the gaps and get rid of the old one. This is left as an exercise for the reader.

19.2   Linked Structures in Files

The two file-backed sequences that we just wrote behave like arrays. They have fast direct access, but trying to delete from or insert to any location other than the end will be very inefficient. This discussion should remind you of a similar one in chapter 12 that motivated our initial creation of a linked list. We now want to do that same thing, but in a file. That way we can have the benefits of data retention between runs and large data sets along with the ability for inserts and deletes.

When we built the linked list in memory, the links were created by references from one node to another. The nature of those references is not explicit in Scala. However, it is likely that they are stored as addresses in memory. Languages like C and C++ explicitly expose that detail in the form of constructs called pointers. A pointer stores an address in memory, and allows mathematical operations on it that move the pointer forward or backward through the memory. The languages that expose pointers are, in many ways, ideal for binary files and random access. This is because they allow direct copies of memory to and from disk. For most values, this is ideal and highly efficient. It does not work at all for pointers. Pointers are attached to memory and do not have meaning when they are separated from the memory of the execution of the program. The addresses of values will change from one run of a program to another and there is no simple way to go from an address in memory to another location in a file to recover the meaning of the link. For this reason, pointers have to be replaced with something that keeps a meaning in the file. The same is true of references.

Whether it is referred to as a pointer or a reference, the equivalent in a file is a position or something roughly equivalent to it. Either one would be stored as an integer value. If it were a true position in the file, it should probably be stored as a Long.

There is another detail when doing a linked list, or other linked structure, that we were able to ignore when working in memory, but which becomes fairly significant when we are responsible for dealing with things on our own in a file. The detail is what to do about space in the file that was part of a node that has been deleted. A very sloppy implementation could ignore this and always add new records to the end of the file. Such an implementation would require an extra function to compact the file that would need to be called fairly often. A much more efficient approach is to keep track of those places in the file using a separate linked list that we will call the "free list". The name comes from the fact that it is a linked list of nodes that are free to be used again.

Putting an array into a file is quite straightforward. That is why the code for the fixed record length sequence was so simple. Doing the same for a linked list requires a bit more planning as the linked list needs to keep track of certain information. Think back to our singly linked list implementation in chapter 12 and the values it stored. It included a head reference, a tail reference, and a length. Our file version will need those as well as the head of the free list. These values will be written at the front of the file. After that will be the nodes, each having a next position at the beginning. This format is drawn out in figure 19.1.

To take advantage of the functionality of a linked list, the code will implement the Buffer trait. To make things a bit simpler, it will assume a fixed record size. The class keeps a single RandomAccessFile as well as variables to store the current head, tail, length, and first free position. These variables are not directly accessed. Instead, there are methods for getting and setting each. The setters include code that seeks to the proper part of the beginning of the file and writes the new value to the file. This way the variables and the contents of the file should always agree. There is also a method to get the position of a new node. If there are no free nodes, it returns the end of the file. Otherwise it will return the

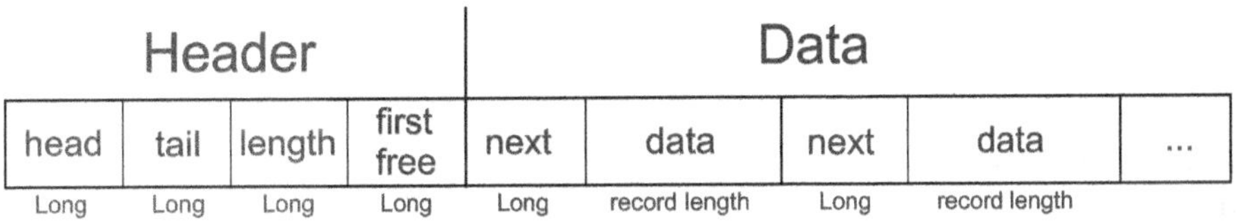

Header				Data				
head	tail	length	first free	next	data	next	data	...
Long	Long	Long	Long	Long	record length	Long	record length	

FIGURE 19.1: The binary file holding the linked list has the following format. The beginning of the file is a header with four values written as **Longs**. They are the position of the head node, the position of the tail node, the number of nodes, and the position of the first free node.

first free node, but before doing that it jumps to the node, reads its next reference, and stores that as the new **firstFree**.

```scala
package binaryfiles.adt

import java.io.DataInput
import java.io.DataOutput
import java.io.File
import java.io.RandomAccessFile

import scala.collection.mutable

class FixedRecordList[A](
    file: File,
    reader: DataInput => A,
    writer: (DataOutput, A) => Unit) extends mutable.Buffer[A] {

  private val raf = new RandomAccessFile(file, "rw")
  private var (localHead, localTail, localLen, localFirstFree) = {
    if (raf.length() >= 3 * 8) {
      raf.seek(0)
      val h = raf.readLong()
      val t = raf.readLong()
      val l = raf.readLong()
      val ff = raf.readLong()
      (h, t, l, ff)
    } else {
      raf.seek(0)
      raf.writeLong(-1)
      raf.writeLong(-1)
      raf.writeLong(0)
      raf.writeLong(-1)
      (-1L, -1L, 0L, -1L)
    }
  }

  private def lhead: Long = localHead

  private def lhead_=(h: Long): Unit = {
    raf.seek(0)
```

```scala
38      raf.writeLong(h)
39      localHead = h
40    }
41
42    private def length_=(len: Long): Unit = {
43      raf.seek(8 * 2)
44      raf.writeLong(len)
45      localLen = len
46    }
47
48    private def ltail: Long = localTail
49
50    private def ltail_=(t: Long): Unit = {
51      raf.seek(8)
52      raf.writeLong(t)
53      localTail = t
54    }
55
56    private def firstFree: Long = localFirstFree
57
58    private def firstFree_=(ff: Long): Unit = {
59      raf.seek(3 * 8)
60      raf.writeLong(ff)
61      localFirstFree = ff
62    }
63
64    private def newNodePosition: Long = if (firstFree == -1L) raf.length else {
65      val ff = firstFree
66      raf.seek(ff)
67      firstFree = raf.readLong()
68      ff
69    }
70
71    def +=(elem: A) = {
72      val npos = newNodePosition
73      raf.seek(npos)
74      raf.writeLong(-1L)
75      writer(raf, elem)
76      if (lhead == -1L) {
77        lhead = npos
78      } else {
79        raf.seek(ltail)
80        raf.writeLong(npos)
81      }
82      ltail = npos
83      length += 1
84      this
85    }
86
87    def +=:(elem: A) = {
88      val npos = newNodePosition
89      raf.seek(npos)
90      raf.writeLong(lhead)
91      writer(raf, elem)
92      lhead = npos
```

```scala
 93      if (ltail == -1L) ltail = npos
 94      length += 1
 95      this
 96    }
 97
 98    def apply(n: Int): A = {
 99      if (n >= length) throw new IllegalArgumentException("Requested index "+n+
100        " of "+length)
101      var i = 0
102      var pos = lhead
103      while (i <= n) {
104        raf.seek(pos)
105        pos = raf.readLong()
106        i += 1
107      }
108      reader(raf)
109    }
110
111    def clear(): Unit = {
112      raf.seek(ltail)
113      raf.writeLong(localFirstFree)
114      localFirstFree = lhead
115      lhead = -1
116      ltail = -1
117      length = 0
118    }
119
120    def insertAll(n: Int, elems: Traversable[A]): Unit = {
121      if (n > length) throw new IllegalArgumentException("Insert at index "+n+
122        " of "+length)
123      var i = 0
124      var (prev, next) = if (n == 0) (-1L, lhead) else {
125        var (pp, nn) = (lhead, -1L)
126        while (i < n) {
127          raf.seek(pp)
128          if (i < n - 1) pp = raf.readLong()
129          else nn = raf.readLong()
130          i += 1
131        }
132        (pp, nn)
133      }
134      if (prev != -1L) raf.seek(prev)
135      for (elem <- elems) {
136        val npos = newNodePosition
137        if (prev == -1L) {
138          lhead = npos
139          prev = npos
140        } else raf.writeLong(npos)
141        raf.seek(npos + 8)
142        writer(raf, elem)
143        raf.seek(npos)
144      }
145      if (next == -1L) ltail = raf.getFilePointer()
146      raf.writeLong(next)
147      length += elems.size
```

```
148      }
149
150      def iterator = new Iterator[A] {
151        var pos = lhead
152        def hasNext = lhead > -1L
153        def next = {
154          raf.seek(pos)
155          pos = raf.readLong()
156          reader(raf)
157        }
158      }
159
160      def length: Int = localLen.toInt
161
162      def remove(n: Int): A = {
163        if (n >= length) throw new IllegalArgumentException("Remove index "+n+
164          " of "+length)
165        var i = 0
166        var pos = lhead
167        var last, next = -1L
168        while (i <= n) {
169          raf.seek(pos)
170          if (i == n) {
171            next = raf.readLong()
172          } else {
173            last = pos
174            pos = raf.readLong()
175          }
176          i += 1
177        }
178        val ret = reader(raf)
179        if (last == -1L) {
180          lhead = next
181        } else {
182          raf.seek(last)
183          raf.writeLong(next)
184        }
185        if (pos == ltail) {
186          ltail = last
187        }
188        length -= 1
189        ret
190      }
191
192      def update(n: Int, elem: A): Unit = {
193        if (n >= length) throw new IllegalArgumentException("Updating index "+n+
194          " of "+length)
195        var i = 0
196        var pos = lhead
197        while (i <= n) {
198          raf.seek(pos)
199          pos = raf.readLong()
200          i += 1
201        }
202        writer(raf, elem)
```

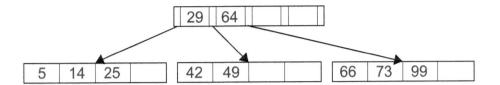

FIGURE 19.2: The nodes of a B-tree can each store a certain number of records. The number of records is selected to match an optimal size for disk reads and writes. Non-leaf nodes have references to children at the beginning, end, and between each record. The records in a node are sorted by a key. In addition, the nodes in children have to have key values that fall between the keys of the records they are between, or just below and above for the beginning and end children. B-trees are always balanced and require few disk accesses to find a record, but they do potentially waste some space.

```
203    }
204
205    def close(): Unit = raf.close()
206  }
```

Going through this code, you will notice that there are many places that have a call to `seek` followed by either a `readLong` or a `writeLong`. This is because the first eight bytes in each node are used to store the position of the next value.

As with a normal linked list, this code has quite a few `while` loops that run through the list counting. It also has the overhead of boundary cases that deal with the head and tail references. Those special cases, combined with the fact that file access is more verbose than accessing memory makes this code fairly long, though the length is largely made up of short, simple lines.

We stated above that this code uses a fixed record length, but there is no record length passed in. This code does not have to know the length, but it will fail if a fixed length is not used for all nodes. The assumption of a fixed length is made in the fact that this code assumes that any freed node can be used for any new data. This assumption breaks down if different pieces of data have different lengths. To make this code work with different length records requires three changes. First, each node must keep a length in bytes in addition to the next reference. Second, the `writer` must be called using a `ByteArrayOutputStream` as was done earlier. Third, the `newNodePosition` must be passed the size that is required and it will have to walk the free list until it finds a node that is big enough. If none are found, it will use the end of the file, even when nodes have been made free. Making these alternations is left as an exercise for the reader.

B-Trees

The file-based linked list works, but it is, in many ways, a horrible way to do things in a file. Disk reads and writes are extremely slow, especially done in small pieces. Disks work much better if you read a large chunk at a time. All the seeking in this code, and the fact that over time the list is likely to evolve such that consecutive nodes are very distant in the file, will cause very poor performance and likely make this unusable for most applications.

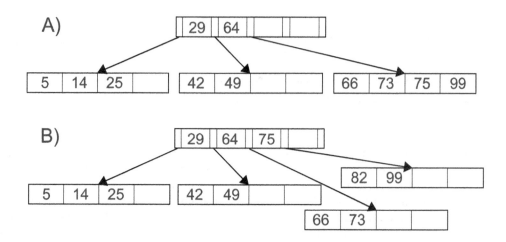

FIGURE 19.3: This figure illustrates how adds are done for B-trees. A) shows the tree from figure 19.2 with the addition of a record whose key value is 75. The 75 is simply placed in sorted order in a leaf because there is still space left in that node. For B), a record with a key value of 82 has been added. It would go right after the 75 record, but there is no more space in that node. Of the five records that would be in the node, the 75 is the median so it is moved into the parent and the other records are split into two different nodes. This process can propagate to higher levels in the tree if the parent is also full. A new level is created when the root node fills and has to split.

To deal with the special requirements of disk access, certain data structures have been developed that are specifically designed to access large chunks of data at a time and to do few such accesses. One of the easiest such structures is the B-tree, which was originally described in 1972 [3]. The B-tree is an extension on the idea of a BST. To make it more efficient for disk access, each node can hold multiple records and for non-leaves, there are children nodes not only for lower and higher keys, but also between each of the records. A sample B-tree is shown in figure 19.2.

Unlike a normal BST, a B-tree is always balanced. In addition, new nodes are added higher up the tree, instead of at the leaves. By default, a new record is added in a leaf. As long as that leaf still has spots left, that new record is inserted in the proper location and the add is done. This is illustrated in figure 19.3-A. If the leaf node where the new record belongs is already full, then a new node has to be made. This is done by splitting that one node into two, pushing the median record up into the parent, and putting links to the two children nodes on either side. This is shown in figure 19.3-B.

Real B-trees will typically have many more records per node than is drawn in these figures. The fact that their average splitting factor is very high means that they can hold an extremely large number of records in very few levels. To see this, consider a B-tree with 200 records per node that is roughly half filled. That means that each node has roughly 100 records in it and each internal node has roughly 100 children. So the top level would hold 100 records. The next would hold 10,000. The next would hold 1,000,000. This is still $O(\log(n))$ behavior, but it is technically $O(\log_{100}(n))$, which grows very slowly. A tree with a billion records will only barely crack five levels. This is significant, because each level winds up being a **seek** and a **read** when trying to find records.

19.3 End of Chapter Material

19.3.1 Summary of Concepts

- Direct access, or random access files allow you to jump from one position to another, reading and writing.

- This functionality allows them to be used as array-like data structures with extremely large capacities.

- It is easiest to deal with files where all records have the same length as that allows us to easily calculate where any given record is in the file.

- Variable record lengths can be used, but they require more bookkeeping. This is typically done by having a fixed record length index that tells us the position of elements in the data file.

- We can mimic references using file positions to create linked data structures in files.

  - Putting header information in a file allows us to keep track of things like the head or tail of a list, or the root of a tree.

  - Additional information can be stored with each record to include references to other nodes in the data structure.

  - While we can code a linked list in a direct access file, it is very inefficient because it requires a large number of disk accesses.

  - Other linked structures exist specifically to get around the problem of slow access times. The B-tree is an example of one.

19.3.2 Exercises

1. Write code to do speed comparisons of different file-backed sequences (array-like and linked list-like) to one another and compare them to memory-backed sequences.

2. Write a `compact` method for the `VariableRecordSeq`. This method should be called to eliminate blank space in the data file. It does this by creating a new data file and moving records from the old one to the new one, one at a time, updating the index for the new positions appropriately. Once this is done, the old data file can be deleted and the name of the new one can be changed to match the original name.

3. Make appropriate changes to the `FixedRecordList` to make a `VariableRecordList`.

4. Write a `linearize` method for the linked list that will create a new file where all the nodes have been reordered to the order they appear in the list and all free nodes have been removed.

5. Write a B-tree that only works in memory.

6. Convert your B-tree from the previous exercise so that it works using disk access.

19.3.3   Projects

All of the projects in this chapter have you writing direct access binary versions of some of your saved files for your project.

1. One of the things you have probably noticed if you are working on the MUD project is that saving individual players out to disk either requires separate files for each player, leading to a very large number of files, or it requires writing out all the players just to save one. A direct access binary file can fix this. You might have to make some sacrifices, such as specifying a maximum number of items that can be saved. Write code to store players in a fixed-length record and change the saving and loading code so that only single records are read or written as needed.

2. Depending on what data you are storing, there might be very different things that you put in a direct access binary file for the web spider. Things like box scores can be made much more efficient using a binary format instead of text. Images will be a bit more challenging. If you really want a challenge, try to put your data in a B-tree. The web spider is a perfect example of a project that uses enough data that having things stored on disk in a B-tree instead of in memory can be very beneficial.

3. If you have a graphical game that has different boards or screens, edit the storage so that it uses a direct access file to efficiently get to any screen in $O(1)$ time.

4. The math worksheet project can benefit from saving the worksheet in a binary file so that saving can be done on individual sections of the worksheet, instead of the whole thing.

5. If you are doing the Photoshop® project, you can put layers in variable-sized records of a binary file. Note that you cannot use default serialization for images, so you will likely have to write some form of custom serialization.

6. If you are doing the simulation workbench, you should make it so that large particle simulations write out using a custom binary format where the data for each particle is written out as the proper number of binary `Doubles`. This will allow you to write tools for reading in only specific particles.

7. There is no specific task associated with random access binary files for the stock project.

8. There is no specific task associated with random access binary files for the movies project.

9. There is no specific task associated with random access binary files for the L-systems project.

Chapter 20

Spatial Trees

A lot of the data that you deal with in certain fields of computing is spatial, meaning that it has ordering on multiple orthogonal axes. There are certain data structures that you can use to make the process of finding elements in certain parts of space more efficient. These spatial data structures can increase the speed of certain operations dramatically, especially when there are large numbers of elements involved. This chapter will consider a few of the many options for doing this.

20.1 Spatial Data and Grids

Examples of spatial data can be bodies in a physics simulation, characters spread out across a large world in a game, cell phone location data, or any number of other things. A standard question that has to be asked when dealing with this type of data is, which items are close to other items. To illustrate the nature of this problem, let us consider a simplified example. You are given N points in some M dimensional space. For visualization purposes, we will typically treat $M = 2$, but it could be higher. We want to find all pairs of points (p_1, p_2) such that the distance between them is less than a given threshold, $|p_2 - p_1| < T$.

The easiest way to write a solution to this problem is a brute-force approach that runs through all points in one loop, with a nested loop that runs through all the points after that one in the sequence and calculates and compares distances. If we model the points as anything that can be converted to `Int => Double`, we can write a `class` that serves as the interface for all the approaches to the basic neighbor visits that we will write.[1]

```scala
package spatialtrees.adt

abstract class NeighborVisitor[A <% Int => Double](val dim: Int) {
  def visitAllNeighbors(tDist: Double, visit: (A, A) => Unit): Unit
  def visitNeighbors(i: Int, tDist: Double, visit: (A, A) => Unit): Unit
```

---

[1]This is written as a `class` instead of a `trait` to keep the code simpler. The `<%` symbol gets expanded out to include an implicit argument. As traits cannot take arguments, we would have to get into the more complex details of this in order to use a `trait`. See the on-line appendices at http://www.programmingusingscala.net for details.

```
7    var distCalcs = 0
8    def dist[A <% Int => Double](p1: A, p2: A): Double = {
9      distCalcs += 1
10     math.sqrt((0 until dim).foldLeft(0.0)((d, i) => {
11       val di = p1(i) - p2(i)
12       d + di * di
13     }))
14   }
15 }
```

This **class** has a **dist** method as all of its subtypes will need to perform this calculation. It also takes an argument for the dimensionality of the space that the points should be considered in. There are two methods declared here. One of them does a complete search across all the pairs of points. The other does a search relative to just one point. The member named **distCalcs** is there simply for the purposes of benchmarking our different methods to see how many times they have to calculate the distance between points. Real working implementations would leave this out.

The brute-force implementation of this looks like the following.

```
1  package spatialtrees.adt
2
3  class BruteForceNeighborVisitor[A <% Int => Double](
4      d: Int,
5      val p: IndexedSeq[A]) extends NeighborVisitor[A](d) {
6
7    def visitAllNeighbors(tDist: Double, visit: (A, A) => Unit): Unit = {
8      for {
9        i <- 0 until p.length
10       pi = p(i)
11       j <- i + 1 until p.length
12       pj = p(j)
13       if dist(pi, pj) <= tDist
14     } visit(pi, pj)
15   }
16
17   def visitNeighbors(i: Int, tDist: Double, visit: (A, A) => Unit): Unit = {
18     val pi = p(i)
19     for {
20       j <- 0 until p.length
21       if j != i
22       pj = p(j)
23       if dist(pi, pj) <= tDist
24     } visit(pi, pj)
25   }
26 }
```

The method for finding all pairs has an outer loop using the variable i that goes through all the indices. Inside that it runs through all the points after the current point in the sequence. This way it does not deal with the same pair of points, in opposite order, twice. The creation of the variables pi and pj where they are in the code is intended to prevent unnecessary indexing into the **IndexedSeq**, p.

While this code is easy to write, one can see that the number of calls to **dist** scales as $O(N^2)$. This is acceptable for smaller values of N, but it makes the approach infeasible when

N gets large.[2] For that reason, we need to use alternate approaches that allow neighbors to be found with less effort than running through all of the other points.

The simplest approach to doing this is a regular spatial grid. Using this approach, you break the space that the points are in down into a grid of regularly spaced boxes of the appropriate dimension. For each box in the grid, we keep a list of the points that fall inside of it. Searching for neighbors then requires only searching against points that are on lists in cells that are sufficiently close. Here is code for an implementation of `NeighborVisitor` which uses this method.

```scala
package spatialtrees.adt

import collection.mutable

class RegularGridNeighborVisitor[A <% Int => Double](
    val p: IndexedSeq[A]) extends NeighborVisitor[A](2) {

  private val grid = mutable.ArrayBuffer.fill(1, 1)(mutable.Buffer[Int]())
  private var currentSpacing = 0.0
  private var min = (0 until dim).map(i => p.foldLeft(1e100)((d, p) => d min p(i)))
  private var max = (0 until dim).map(i => p.foldLeft(-1e100)((d, p) => d max p(i)))

  def visitAllNeighbors(tDist: Double, visit: (A, A) => Unit): Unit = {
    if (tDist < 0.5 * currentSpacing || tDist > 5 * currentSpacing)
        rebuildGrid(tDist)
    val mult = math.ceil(tDist / currentSpacing).toInt
    val offsets = Array.tabulate(2 * mult + 1, 2 * mult + 1)((i, j) => (i - mult, j
        - mult)).
      flatMap(i => i).filter(t => t._2 > 0 || t._2 == 0 && t._1 >= 0)
    for {
      cx <- grid.indices
      cy <- grid(cx).indices
      (dx, dy) <- offsets
      gx = cx + dx
      gy = cy + dy
      if gx >= 0 && gx < grid.length && gy >= 0 && gy < grid(gx).length
      i <- grid(cx)(cy).indices
      pi = p(grid(cx)(cy)(i))
    } {
      if (dx == 0 && dy == 0) {
        for {
          j <- i + 1 until grid(cx)(cy).length
          pj = p(grid(cx)(cy)(j))
          if dist(pi, pj) <= tDist
        } visit(pi, pj)
      } else {
        for {
          j <- grid(gx)(gy)
          pj = p(j)
          if dist(pi, pj) <= tDist
```

---

[2]One of the authors does large-scale simulations where N typically varies from 10^5 to more than 10^8. The author once needed to check if a bug had caused any particles to be duplicated. As a first pass, the author wrote a short piece of code like this to do the check. When it did not finish within an hour, the author took the time to do a little calculation of how long it would take. That particular simulation had a bit over 10^7 particles. The answer was measured in weeks. Needless to say, the author took the time to implement one of the methods from this chapter instead of the brute-force approach.

```scala
39          } visit(pi, pj)
40        }
41      }
42    }
43
44    def visitNeighbors(i: Int, tDist: Double, visit: (A, A) => Unit): Unit = {
45      if (tDist < 0.5 * currentSpacing || tDist > 5 * currentSpacing)
             rebuildGrid(tDist)
46      val mult = math.ceil(tDist / currentSpacing).toInt
47      val offsets = Array.tabulate(2 * mult + 1, 2 * mult + 1)((i, j) => (i - mult, j
             - mult)).
48        flatMap(i => i)
49      val cx = ((p(i)(0) - min(0)) / currentSpacing).toInt
50      val cy = ((p(i)(1) - min(1)) / currentSpacing).toInt
51      val pi = p(i)
52      for {
53        (dx, dy) <- offsets
54        gx = cx + dx
55        gy = cy + dy
56        if gx >= 0 && gx < grid.length && gy >= 0 && gy < grid(gx).length
57        j <- grid(gx)(gy)
58        if i != j
59        pj = p(j)
60        if dist(pi, pj) <= tDist
61      } visit(pi, pj)
62    }
63
64    /**
65     * Rebuild the grid to a size that is appropriate for the searches. Note that
66     * this method was not written for efficiency. A true implementation should
67     * include a rewrite of this method.
68     */
69    private def rebuildGrid(spacing: Double): Unit = {
70      min = (0 until dim).map(i => p.foldLeft(1e100)((d, p) => d min p(i)))
71      max = (0 until dim).map(i => p.foldLeft(-1e100)((d, p) => d max p(i)))
72      val cells = (0 until dim).map(i => ((max(i) - min(i)) / spacing).toInt + 1)
73      if (grid.size < cells(0)) grid.append(mutable.Buffer.fill(cells(0) -
             grid.size)(mutable.ArrayBuffer.fill(cells(1))(mutable.Buffer[Int]())): _*)
74      else if (grid.size > cells(0)) grid.trimEnd(grid.size - cells(0))
75      for (col <- grid) {
76        if (col.size < cells(1)) col.append(mutable.Buffer.fill(cells(1) -
               col.size)(mutable.Buffer[Int]()): _*)
77        else if (col.size > cells(1)) col.trimEnd(col.size - cells(1))
78        col.foreach(_.clear)
79      }
80      for (i <- p.indices) {
81        val cx = ((p(i)(0) - min(0)) / spacing).toInt
82        val cy = ((p(i)(1) - min(1)) / spacing).toInt
83        grid(cx)(cy) += i
84      }
85      currentSpacing = spacing
86    }
87  }
```

Clearly this code is longer and more complex than the brute-force approach. However, there are many situations for which this will produce $O(N)$ behavior for the method that visits all neighbor pairs and $O(1)$ behavior for only the neighbors of a single point.

This grid is hard coded to support only two dimensions. While not impossible, it is significantly more challenging to write code that uses arrays of arbitrary dimensions. Even if you are willing to put in that effort, regular grids do not work well in high dimensions. Imagine you have a n-dimensional cube and your search radius is about $1/100th$ the size of the cell so you want 100 divisions in each dimension. For a 2-D space this gives a modest $10,000 = 10^4$ cells in the grid. For a 3-D space you get a cube with $1,000,000 = 10^6$ cells. As the dimensions go up, this continues to grow as 10^{2n}. This causes significant problems above three dimensions, especially as most problems will need more than 100 bins per dimension. We could write a 3-D grid, but there really are not many applications where it helps to go above that.

The situations when a grid approach is ideal are those where the points are fairly uniformly distributed, the search radius is small compared to the whole area, and the area the points are spread through increases as N increases as opposed to the density of points increasing when N increases. In other words, if the number of neighbors for every point is fairly similar and when N is increased, the points are added in such a way that the number of neighbors stays the same, then the grid is pretty much an optimal structure. To give an idea of how good the grid is, a simple run was done with 1000 points spread randomly and uniformly across a square region, an ideal configuration for a grid. The brute-force approach always does $N * (N - 1)/2 = 499500$ distance calculations. The number for the grid varies with the random distribution of the points and the search size. However, using a search distance $1/100th$ the size of the point distribution was never seen to need more than 500 distance comparisons to find between 100 and 200 neighbor pairs.

It is also important to note that building the grid is an $O(N)$ operation. So assuming that the input for the program you are working on fits the conditions listed above, you can build a grid and find all the neighbors for all the points in $O(N)$ operations. That is a dramatic improvement over the $O(N^2)$ behavior of the brute-force approach. The brute-force approach starts to become intractable around 10^6 points and it is slow on most machines for $N = 10^5$. With the grid approach, speed is likely not the limiting factor, instead it is likely that memory will be. However many points you can deal with in the memory of the machine, you can probably process them as well using the grid.

20.2 Quadtrees and Octrees

What about situations for which the grid is not well suited? The most obvious example is when the points are not uniformly distributed. Consider the distribution of points shown in figure 20.1. Here there are two regions of very high-density surrounded by reduced density regions. The sample grid drawn on the region would work fairly well for the low-density regions as each cell has only a few points in it. However, the grid is not helping much in the high-density regions. The grid-based search would still be running through a very large number of point pairs. However, if the grid were drawn with a size appropriate for the high-density regions, the low density regions would be mostly empty cells and lots of time could be wasted processing them. It is even possible that the memory overhead of keeping all those empty cells could be a problem.

The way around this is to use a tree. Just like the BST is a more dynamic and adjustable structure than a sorted array, spatial trees have the ability to adjust resolution to the nature

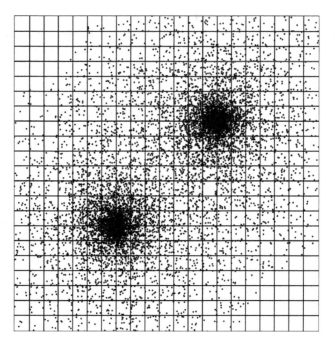

FIGURE 20.1: This figure shows an example set of points with a very non-uniform spatial distribution. A regular grid is laid over this to make it clear how the point density in the grid cells will vary dramatically. The grid size shown here works well for many of the low-density regions where there are only a handful of points per cell. A grid that only had a few points per cell in the high-density regions would be a very fine grid with an extremely large fraction of the cells being empty in the low-density areas.

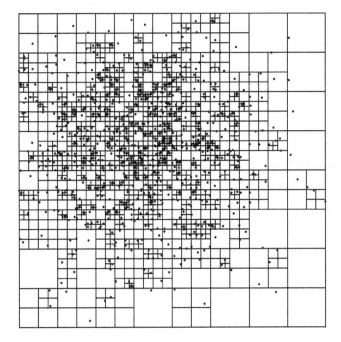

FIGURE 20.2: This is a quadtree with a number of non-uniformly distributed points to illustrate how the space is partitioned. For this example, each node is only allowed to hold a single point. The key advantage of the tree-based approach is that the resolution does not have to be uniform. Higher-density regions can divide the space much more finely.

of the data in different areas. In some ways, you could picture a BST tree as a spatial tree in a 1-D space. There is a complete ordering of values along a given line. Some thought has to be given to how this can be extended to deal with systems where there are multiple, orthogonal orderings. Many different approaches to this have been developed. We will consider two of them explicitly in this chapter.

The easiest spatial tree to understand is probably the quadtree. There are multiple types of quadtree. We will focus on a region-based quadtree. For this tree, each node represents a square region of space. We start with a bounding square that contains all of the points we are dealing with. That region is represented by the root of the tree. Each node can have four children. The children each represent the equally sized quadrants of their parent. With this tree the points all go into the leaves. How many points are allowed in each leaf varies with the application. Empirical testing can be used to find what works best. When there are more points in a region than is allowed, that region splits into the four children and the points are moved down.[3]

The results of building a quadtree with some non-uniform data is shown in figure 20.2. This tree has roughly 1000 data points in it. You can see that in regions where there are few points, the divisions are large. However, in the higher-density sections of the tree, the cells divide down to very small sizes to isolate the points.

The code for this tree is shown below. Like the grid, the quadtree is used for a 2-D spatial division and that is hard written into the code. The tree has a supertype called `Node` with two subtypes called `LNode` and `INode` for leaf node and internal node respectively. The

---

[3]It is worth noting again that this is something of a standard implementation of a region-based quadtree. Different aspects can be varied depending on the application. We will see an alternative approach later in this chapter for the closely related octree.

way this code is set up, only the leaf nodes hold points. For this reason, it makes sense to use the approach of having two subtypes as they store very different data other than the location information. The `INode` also has additional functionality for splitting points across the children and building the lower parts of the tree. The method `childNum` returns a number between 0 and 3, which is the index for a child based on location. The `groupBy` collection method is called as an easy way to get the different points in each quadrant.

```scala
1   package spatialtrees.adt
2
3   class QuadtreeNeighborVisitor[A <% Int => Double](
4       val p: IndexedSeq[A]) extends NeighborVisitor[A](2) {
5
6     private class Node(val cx: Double, val cy: Double, val size: Double)
7     private class LNode(x: Double, y: Double, s: Double, val pnts: IndexedSeq[Int])
8         extends Node(x, y, s)
8     private class INode(x: Double, y: Double, s: Double, pnts: IndexedSeq[Int])
          extends Node(x, y, s) {
9       val children = {
10        val groups = pnts.groupBy(pn => childNum(p(pn)(0), p(pn)(1)))
11        val hs = s * 0.5
12        val qs = s * 0.25
13        val ox = cx - qs
14        val oy = cy - qs
15        Array.tabulate(4)(i => makeChild(ox + hs * (i % 2), oy + hs * (i / 2), hs,
16          if (groups.contains(i)) groups(i) else IndexedSeq[Int]()))
17      }
18      def makeChild(x: Double, y: Double, s: Double, pnts: IndexedSeq[Int]): Node =
19        if (pnts.length > maxPoints) new INode(x, y, s, pnts)
20        else new LNode(x, y, s, pnts)
21      def childNum(x: Double, y: Double): Int = (if (x > cx) 1 else 0) + (if (y > cy)
          2 else 0)
22    }
23
24    private val maxPoints = 1
25    private val root = {
26      val minx = p.foldLeft(1e100)((mx, pnt) => pnt(0) min mx)
27      val maxx = p.foldLeft(-1e100)((mx, pnt) => pnt(0) max mx)
28      val miny = p.foldLeft(1e100)((my, pnt) => pnt(1) min my)
29      val maxy = p.foldLeft(-1e100)((my, pnt) => pnt(1) max my)
30      val cx = 0.5 * (minx + maxx)
31      val cy = 0.5 * (miny + maxy)
32      val s = (maxx - minx) max (maxy - miny)
33      if (p.length > maxPoints) new INode(cx, cy, s, p.indices)
34      else new LNode(cx, cy, s, p.indices)
35    }
36
37    def visitAllNeighbors(tDist: Double, visit: (A, A) => Unit): Unit = {
38      for (i <- 0 until p.length) {
39        val pi = p(i)
40        def recur(n: Node): Unit = n match {
41          case ln: LNode =>
42            ln.pnts.filter(j => j > i && dist(pi, p(j)) <= tDist).foreach(j =>
                visit(pi, p(j)))
43          case in: INode =>
44            val x = pi(0)
```

```
45      val y = pi(1)
46      if (x + tDist > in.cx) {
47        if (y + tDist > in.cy) recur(in.children(3))
48        if (y - tDist <= in.cy) recur(in.children(1))
49      }
50      if (x - tDist <= in.cx) {
51        if (y + tDist > in.cy) recur(in.children(2))
52        if (y - tDist <= in.cy) recur(in.children(0))
53      }
54    }
55    recur(root)
56  }
57 }
58
59 def visitNeighbors(i: Int, tDist: Double, visit: (A, A) => Unit): Unit = {
60   val pi = p(i)
61   def recur(n: Node): Unit = n match {
62     case ln: LNode =>
63       ln.pnts.filter(j => j != i && dist(pi, p(j)) <= tDist).foreach(j =>
             visit(pi, p(j)))
64     case in: INode =>
65       val x = pi(0)
66       val y = pi(1)
67       if (x + tDist > in.cx) {
68         if (y + tDist > in.cy) recur(in.children(3))
69         if (y - tDist <= in.cy) recur(in.children(1))
70       }
71       if (x - tDist <= in.cx) {
72         if (y + tDist > in.cy) recur(in.children(0))
73         if (y - tDist <= in.cy) recur(in.children(2))
74       }
75   }
76   recur(root)
77 }
78 }
```

The construction of the tree itself happens in the declaration of **root**. This type of quadtree must have bounds containing all of the points, so that is calculated first. After that is done, the proper node type is instantiated. That node serves as the root, and if it is an **INode**, it builds the rest of the tree below it. While this tree does not take advantage of it, one of the big benefits of the region-based quadtree is that it is fairly easy to write a mutable version that allows the user to add points one at a time as long as the user specifies an original bounding region.

The **visitAllNeighbors** and **visitNeighbors** methods include nested recursive functions that run through the tree looking for neighbors. The behavior on an **INode** is such that recursive calls are only made on children that the search area actually touches. When run on the same type of uniform random configuration used for the grid above, this tree did a bit fewer than twice as many distance calculations as the grid. This was still vastly better than the brute-force approach and remarkably close to the grid considering the configuration of points used was ideal for the grid.[4]

---

[4]The version of visitAllNeighbors shown here is very similar to putting visitNeighbors in a loop. This approach has one drawback; the recursion goes through the higher nodes many times, duplicating work. It is possible to write a function that recurses over two Node arguments that improves on this. It was not shown here because it is significantly more complex and difficult for the reader to understand.

As with the grid, the quadtree is inherently 2-D. A 3-D version that divides each node into octants is called an octree. We will build one of those in a later section. Also like the grid, this type of tree does not scale well to higher dimensions. This is due to the fact that the number of children contained in each child node scales as 2^n for an n-dimensional space. Given that the number of internal nodes scales as the number of points, having large internal nodes that store many children becomes inefficient. There are alternate ways of storing children, but if you really need a high dimensional space, the next tree we will discuss is your better bet.

Before leaving the topic of the quadtree, it is also worth noting that this region-based quadtree is inherently unbalanced. The depth of the tree is significantly larger in regions of high density. In practice, this does not introduce significant inefficiency, but it is worth noting as it is a real difference from the BST tree.

20.3   kD-Trees

What should you do if you have to deal with a high dimensional space, want a balanced spatial tree, and/or have dramatically different sizes in different dimensions that you want to resolve as needed? A possible solution to all of these is the use of a kD-tree. The "kD" in the name means k-dimensional, so the name itself points to the fact that this tree is intended to scale to higher dimensions.

The way that a kD-tree works is that internal nodes have two children split across a plane that is perpendicular to one axis. Put more simply, you pick a dimension and a value; anything less than or equal to the value in that dimension goes to the left and anything greater than it goes to the right. This behavior should be fairly easy to understand as it is remarkably similar to a BST, other than the fact that there are multiple possible dimensions to split in.

How you pick the dimension and the split value can vary greatly between applications. We will use a technique that tries to maximize the efficiency of each split and guarantees the tree is balanced. The split dimension will be the dimension with the largest spread. To pick a value to split on, we will use the median value in that dimension. That way, the number of points on the left and the right will be as close to equal as possible, resulting in a balanced tree. Figure 20.3 shows a 2-D tree generated using this approach allowing three points per leaf. The fact that there is no bounding box here is intentional. The kD-tree puts in divisions that span as far as needed. The ones at the ends are effectively infinite. The internal ones are cut off by other splits.

A sample code implementing `NeighborVisitor` using a kD-tree is shown here. This class is not hard coded to a specific dimensionality as there is nothing that ties it to a particular number of dimensions. There are many similarities between this code and the quadtree code, so we will focus on differences. The first is that the `IndexedSeq` that is passed in is converted to a private `Array`. This forces the additional requirement of a `ClassTag`, but it greatly improves the efficiency of building the tree.

```scala
package spatialtrees.adt

import scala.reflect.ClassTag

class KDTreeNeighborVisitor[A <% Int => Double: ClassTag](
    d: Int,
```

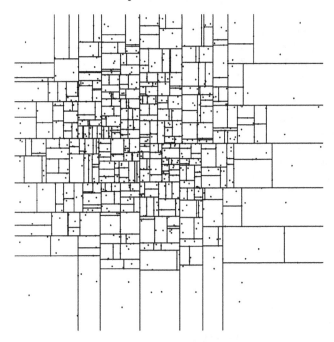

FIGURE 20.3: This is a kD-tree with a number of non-uniformly distributed points to illustrate how the space is partitioned. For this example, each node can hold up to three points. While this example only uses two dimensions, it is simple to scale the kD-tree higher without adding overhead.

```scala
     val pIn: IndexedSeq[A]) extends NeighborVisitor[A](d) {

   private val p = pIn.toArray
   private val maxPoints = 3

   class Node
   private class LNode(val pnts: Seq[Int]) extends Node
   private class INode(start: Int, end: Int) extends Node {
     val splitDim = {
       val min = (0 until dim).map(i => p.view.slice(start, end).
         foldLeft(1e100)((m, v) => m min v(i)))
       val max = (0 until dim).map(i => p.view.slice(start, end).
         foldLeft(-1e100)((m, v) => m max v(i)))
       (0 until dim).reduceLeft((i, j) => if (max(j) - min(j) > max(i) - min(i)) j
           else i)
     }
     val (splitVal, left, right) = {
       val mid = start + (end - start) / 2
       indexPartition(mid, start, end)
       (p(mid)(splitDim), makeChild(start, mid + 1), makeChild(mid + 1, end))
     }
     def indexPartition(index: Int, start: Int, end: Int): Unit = {
       if (end - start > 1) {
         val pivot = if (end - start < 3) start else {
           val mid = start + (end - start) / 2
           val ps = p(start)(splitDim)
```

```scala
32        val pm = p(mid)(splitDim)
33        val pe = p(end - 1)(splitDim)
34        if (ps <= pm && ps >= pe || ps >= pm && ps <= pe) start
35        else if (ps <= pm && pm <= pe || ps >= pm && pm >= pe) mid else end - 1
36      }
37      val ptmp = p(pivot)
38      p(pivot) = p(start)
39      p(start) = ptmp
40      var (low, high) = (start + 1, end - 1)
41      while (low <= high) {
42        if (p(low)(splitDim) <= ptmp(splitDim)) {
43          low += 1
44        } else {
45          val tmp = p(low)
46          p(low) = p(high)
47          p(high) = tmp
48          high -= 1
49        }
50      }
51      p(start) = p(high)
52      p(high) = ptmp
53      if (high < index) indexPartition(index, high + 1, end)
54      else if (high > index) indexPartition(index, start, high)
55    }
56  }
57  def makeChild(s: Int, e: Int): Node = {
58    if (e - s > maxPoints) new INode(s, e) else new LNode(s until e)
59  }
60  }

62  private val root = new INode(0, p.length)

64  def visitAllNeighbors(tDist: Double, visit: (A, A) => Unit): Unit = {
65    for (i <- 0 until p.length) {
66      val pi = p(i)
67      def recur(n: Node): Unit = n match {
68        case ln: LNode =>
69          ln.pnts.foreach(j => if (j > i && dist(p(j), pi) <= tDist) visit(pi,
                p(j)))
70        case in: INode =>
71          if (pi(in.splitDim) - tDist <= in.splitVal) recur(in.left)
72          if (pi(in.splitDim) + tDist > in.splitVal) recur(in.right)
73      }
74      recur(root)
75    }
76  }

78  def visitNeighbors(i: Int, tDist: Double, visit: (A, A) => Unit): Unit = {
79    val pi = p(i)
80    def recur(n: Node): Unit = n match {
81      case ln: LNode =>
82        ln.pnts.foreach(j => if (j != i && dist(p(j), pi) <= tDist) visit(pi, p(j)))
83      case in: INode =>
84        if (pi(in.splitDim) - tDist <= in.splitVal) recur(in.left)
85        if (pi(in.splitDim) + tDist > in.splitVal) recur(in.right)
```

```
86      }
87      recur(root)
88    }
89  }
```

The real meat of this code is in the creation of an **INode**. All that is passed in is a start and an end index into the **Array** to represent what points are supposed to be under this node. All manipulations are done in place reordering elements in the **Array, p**.

The first thing done upon construction of an **INode** is to find the dimension in which the specified points have the largest spread to make it the split dimension. After this is the declaration and assignment to **splitVal**, **left**, and **right**. Most of the work done here occurs in the method **indexPartition**. The purpose of this method is to arrange the elements of the **Array** along the split dimension such that the point which should be at **index** is located there and all the points below or above it along the appropriate dimension are above or below it in the array. In this case we happen to always be finding the median, but the code itself is more general than that. This function can find the element at any given index and do so in $O(N)$ time.

The way this is done is that the function works very much like a quicksort. The base case is a single element. If there are more elements, a pivot it selected and that pivot is moved into place just like it would be in a quicksort. Where this differs is that it only recurses to one side of the pivot. If the desired index is below the pivot, it recurses on the lower part. If it is above the pivot, the method recurses on the upper part. If the pivot happens to be at the desired index, the method returns. To see how this approach gives $O(N)$ performance, simply recall that $1 + \frac{1}{2} + \frac{1}{4} + \frac{1}{8} + \ldots = 2$. As long as the pivot is always close to the center, this is roughly the way the amount of work behaves with each call.

It is interesting to note that because the elements in the **p Array** are actually reordered, the **LNode** sequences of point indices are always contiguous ranges. The visit methods are fairly straightforward. They are actually a fair bit simpler than their counterparts in the quadtree because there are only two children and a single split in each node.

Running this code through the same 2-D tests that were done for the grid results in about four times as many distance calculations as we had for the quadtree. This is in part due to the fact that the leaves have three points in them. It is also somewhat due to the fact that nodes in the kD-tree can be elongated, even when the overall distribution of points is not. This is still far fewer distance calculations than the brute-force approach and you would have to do empirical tests to determine which option was truly best for your application.

Another issue that impacts both the quadtree and the kD-tree is the problem of identical points. The quadtree will run into problems if you have any cluster of points with more than **maxPoints** with a spacing that is extremely small. Such a cluster will lead to very deep leaves as the nodes have to split repeatedly to resolve those points. The kD-tree does not suffer from tight clusters in the same way, but both will crash if there are more than **maxPoints** that all have the same location. The fact that those points cannot be split, regardless of how far down the tree goes means that the code will try to create an infinite number of **INodes** to break them apart. The quadtree will run into this same problem if points are placed outside the bounds, specifically outside the corners.

20.4 Efficient Bouncing Balls

One example of an application where spatial data structures can be extremely beneficial is the bouncing balls problem that we made a `Drawable` for previously. For the 20 ball simulation that was performed previously, any spatial data structure is overkill. However, if the number of balls were increased by several orders of magnitude, a spatial data structure would become required.

If we were using a spatial data structure, the question of search radius arises. A single search radius that is safe for all balls is twice the maximum speed multiplied by the time step plus twice the maximum ball radius. As these values can be easily calculated in $O(N)$ time, it is not a problem to add them to the code. It would be slightly more efficient to use a unique search radius for each ball that uses the speed and size of that ball plus the maximum values to account for the worst possibility for the other ball.

There are few changes that need to be made in the code for using a spatial data structure. You can find the original code in chapter 13 on page 440. Anyplace there is a call to `findEventsFor`, there is a need for the spatial data structure. The `findEventsFor` method takes a `Seq[Int]` with the indices of the balls that are to be searched against. The current implementation is brute-force. It passes all indices after the current one when doing the complete search to initialize the queue and then all values other than the two involved in the collision for searching after a collision. The first one should use `visitAllNeighbors` with a visitor that calls the `collisionTime` method to see if the balls collide and adds them to the priority queue as needed. The call to `findEventsFor` is still needed to get wall collisions though it can be called with an empty list of other balls to check against as the call to `visitAllNeighbors` will have done that. The search for a single ball should use `visitNeighbors` to find the `against` list.

It is interesting to note the involvement of the time step in the search radius. The code for the bouncing balls uses the length of time since the last firing of the `AnimationTimer` as the time step. This occurs in `findEventsFor` where there is a cutoff on the time for events to be added to the queue. If this were shortened, the balls would move less between frames being drawn. For the old method, there is not a significant motivation to do this other than to have the gravity integration be more accurate. For the spatial data structures, it can dramatically speed up the processing of finding collision events as it reduces the area that has to be searched.

20.5 End of Chapter Material

20.5.1 Summary of Concepts

- Spatial data comes in many forms and is associated with many areas of computing.

- Brute-force techniques for doing things like finding nearby points in space tend to scale as $O(N^2)$.

- Spatial data structures can improve the scaling dramatically.

  - Grids are simple and for low-dimension, fairly uniform data, they are ideal.

  - Region-based quadtrees and octrees divide a region that encompasses all the data into successively smaller, uniformly sized parts. This provides variable resolution for the places where it is needed.

— The kD-tree scales to arbitrarily high dimensions. Each internal node splits the data at a particular value with lower values going to the left and higher values to the right.

20.5.2 Exercises

1. Write unit tests for the different spatial data structures that were presented in the chapter.

2. Write a quadtree that includes an `addPoint` method.

3. Parallelize any of the trees that were presented in this chapter so that searches happen across multiple threads. If you want a bit more challenge, make it so that construction the tree also occurs in parallel.

4. Pick a spatial tree of your choosing and do speed tests on it. One way you can do this is to put random points in the tree and then search for all pairs below a certain distance. For each tree that was described, you can vary the number of points that are allowed in a leaf node. Do speed testing with different numbers of points in the data set as well as different numbers of points allowed in leaves.

   It is also interesting to compare the relative speed of different trees of the same dimension.

20.5.3 Projects

1. Most readers probably will not have a use for spatial trees in their current MUD implementation. After all, the maps in a MUD are completely text based, not spatial. If you happen to have put "stats" on your characters though, you might be able to use this to do something interesting. Characters, including player controlled and non-player controlled, likely have stats that indicate how powerful they are in different ways. These stats can be used to place them in a higher dimensional space that you can map out with a kD-tree to enable functions like giving players the ability to find potential opponents or partners who are of similar capabilities.

2. Pretty much any data can be viewed as existing in a space of the appropriate dimensions. If you were working on the web spider, you should do exactly that to find things that are closely related in some way. What you do this on might depend on what data you are collecting. An obvious example would be if you were skimming box scores for a sport, you could make a space that has a different dimension for each significant statistical category. You could put all the entries you have in a kD-tree of that many dimensions, then do searches for everything within a certain distance of a particular point. This could show you things like players or teams that are comparable to one another in different ways.

   If the data you are storing is not numerical, it is still possible to assign spatial values based on text or other information. For example, you could select a set of words that are relevant for what you are looking at and use a function of the number of times each word appears as a value for a different axis. If you have 100 different words of interest, this would mean you have a 100-dimensional space that the pages would go into.[5] These can be put into the kD- tree and that can be used to find pages that are similar according to different metrics.

Even if you are cataloging images, you can associate a sequence of numeric values with images. The numeric values could relate to factors like size, color, brightness, saturation, contrast, or many other things. The image is made of a lot of RGB values, so there are many functions you could devise that produce something interesting.

3. Graphical games, both networked and not, have an automatic spatial nature to them. While your game might not really need a spatial data structure for locating things efficiently, you should put one in to give yourself experience with them. If you did create a game with a really big world, or you have a large number of entities in the world that require collision checking, these spatial data structures could provide significant benefits.

4. One interesting type of analysis that you could put into the math worksheet project is the ability to measure the fractal dimension of a set of data. There are many different formal definitions of fractal dimensions. We will use one called the correlation dimension. The correlation is determined using the correlation integral is formally defined as:

$$C(\varepsilon) = \lim_{N \to \infty} \frac{g(\varepsilon)}{N^2},$$

where $g(\varepsilon)$ is the number of pairs of points with distance less than epsilon, ε. The correlation dimension can be found from this using the relationship:

$$C(\varepsilon) \sim \varepsilon^{\nu},$$

where ν is the correlation dimension and is equal to the slope of the correlation integral on a log-log plot. Finding this value requires two steps. First, we need to calculate $C(\varepsilon)$. Once we have that, we need to take the log of both x and y values for all points on the curve and then do a linear fit to those points to find the slope.

Actually calculating $C(\varepsilon)$ is $O(N^2)$ in the number of points, making it infeasible for a large data set. As the quality of the answer improves as $N \to \infty$, it is better to have very large data sets and so a complete, brute-force calculation becomes infeasible. That is where the spatial tree comes in. The number of points in nodes of different sizes provides an approximation of the number of pairs of points with a separation that length or shorter. To make this work, you need to use a tree where all the nodes at a particular level are the same size. This applies to our quadtree and octree constructions from the chapter. It can also apply to a kD-tree if the tree is constructed in a different way where splits are done down the middle spatially instead of at the median point. Once you have built such a tree, you can use the approximation that

$$C(\varepsilon) \sim \sum_{\text{cells of size } \varepsilon} n_{cell}^2.$$

This says that the number of pairs separated by ε or less goes as the sum over all cells of size ε of the number of particles in each cell squared. This technically over-counts connections in a cell by a factor of two, but it does not count any connections between cells. In addition, any multiplicative error will be systematic across all levels, so the value of ν will not be altered.

---

[5]Using word counts for the value is typically not ideal because the length of a page can dramatically impact where it appears in your space. For that reason you often want to do something like take the log of the word count or the relative fraction of the number of instances of the words.

Once you have a set of points representing $C(\varepsilon)$, the last thing to do is to perform a linear fit to the points $(\log(\varepsilon), \log(C(\varepsilon)))$. A linear least-squares fit can be done by solving a linear equation using Gaussian elimination. The idea is that you want to find optimal coefficients, c_i, to minimize the sum of the squares of the following differences for your data points, $(x_j, y_j) = (\varepsilon_j, C(\varepsilon_j))$.

$$y_j - (c_1 f_1(x_j) + c_2 f_2(x_j) + c_3 f_3(x_j) \cdots).$$

This can be done by solving the equation $A^T A x = A^T y$, where

$$A = \begin{bmatrix} f_1(x_1) & f_2(x_1) & f_3(x_1) & \cdots \\ f_1(x_2) & f_2(x_2) & f_3(x_2) & \cdots \\ f_1(x_3) & f_2(x_3) & f_3(x_3) & \cdots \\ \vdots & \vdots & \vdots & \ddots \end{bmatrix}$$

and y is the column vector of your y_j values.

For a simple linear equation, there are only two terms, so $f_1(x) = x$ and $f_2(x) = 1$. That would reduce down to a 2×2 matrix that is fairly easy to solve. For this problem, you can typically reduce the complexity even further by making the assumption that $C(0) = 0$. This is true as long as no two points ever lie at exactly the same location. In that case, $A^T A = \sum_{j=1}^{N} x_j^2$ and $A^T y = \sum_{j=1}^{N} x_j y_j$. The value of the one coefficient can be found by simple division.

5. If you have been working on the Photoshop® project, you can use a 2D kD-tree to create an interesting filter for images. The kD-tree filter breaks the pixels in an image up into small regions and colors each leaf a single color. This can give something of a stained glass effect with rectangular regions. You can choose how to do the divisions to give the effect that you want. One recommendation is to make cuts in alternating dimensions such that you balance the brightness or the total intensity of a color channel on each side. So if the first cut is vertical, you find the location where the total amount of red to the left of the cut and the total to the right of the cut are as close as possible. Then you do horizontal cuts on each side of that following the same algorithm. After a specified number of cuts, you stop and average the total color of all pixels in each region. Use that average color to fill that region.

6. Spatial trees can be used to improve collision handling in the simulation workbench. Project 6 in chapter 15 described how collision finding is $O(n^2)$ by default and looked at a way to improve that for most time steps. Spatial data structures can also be used to reduce the number of pairs that have to be checked. The simplest approach just uses the tree to search for particles that are within a particular search radius of the particle in question. The search radius is given by $\Delta v \Delta t + 2R_{max}$, where Δv is the relative velocity of particles, Δt is the time step, and R_{max} is the radius of the largest particle. This should take the number of pairs checked down from $O(n^2)$ to $O(n \log(n))$.

7. There is no specific task associated with spatial trees for the stock project.

8. There is no specific task associated with spatial trees for the movies project.

9. One of the things that makes L-systems interesting is their ability to produce fractal geometry. If you are willing to get a bit mathematical with this project, you can do the analysis described in project 4 above to find the fractal dimension of the geometry produced by an L-system. This works with points instead of segments, so you should use the location of the turtle at the end of every 'F' in the string.

Chapter 21

Augmenting Trees

Chapter 16 covered the basics of trees and how to use a binary search tree (BST) to implement the map ADT. This implementation gave us the possibility of $O(\log(n))$ performance for all the operations on that ADT. Unfortunately, it also allowed the possibility of $O(n)$ performance if the tree were built in a way that made it unbalanced. In this chapter we will see how we can fix that flaw by adding a bit more data to nodes and using the data to tell us when things are out of balance so that operations can be performed to fix the situation. In addition, we will see how we can use a similar approach to let us use a tree as the storage mechanism behind a sequence ADT implementation that is $O(\log(n))$ for direct access, inserting, and removing.

21.1 Augmentation of Trees

The general idea of this chapter is to add some additional data to our tree nodes to give the tree additional capabilities. If you go back and look at the implementation of the BST, each node kept only a key, the data, the left child, and the right child. Augmenting data can deal with data values stored in subtrees or they can deal only with the structure of the tree itself.

There is one key rule for augmented data; it must be maintainable in $O(\log(n))$ operations for any of the standard methods. If this is not the case, then the augmentation will become the slowest part of the tree and it will become the limiting factor in overall performance. Fortunately, any value that you can calculate for a parent based solely on the values in the children will have this characteristic, assuming it does not change except when tree structure changes. You can be certain such a value is maintainable in $O(\log(n))$ operations because operations like **add** and **remove** from a tree only touch $O(\log(n))$ nodes as they go from the root down to a leaf. Maintaining the augmented values only requires updating nodes going back from the leaf up to the root.

Example values include tree properties like height and size. The height of a node is the maximum of the height of the two children plus one. The size is the sum of the sizes of the two children plus one. Any value where the value of the parent can be found using a binary operation like `min`, `max`, `+`, or `*` on the children will generally work. Even more complex values that require more than one piece of information can be calculated. Consider the example of an average of value in a subtree. An average cannot be calculated as the simple binary combination of two averages. However, the average is calculated as the quotient of two sums. The first sum is the sum of all the values you want to average. The second sum is how many there are. This is basically size, which we just described as being a sum over the children plus one. If you annotate nodes with both of those values, it is simple to produce the average for any subtree.

21.2   Balanced BSTs

The first example of any augmented tree that we will consider is a form of self-balancing BST called an AVL tree. Recall from chapter 16 that the primary shortcoming of our BST was that it could become unbalanced and then exhibit $O(n)$ performance. Whether the BST was balanced or not was completely dependent on the order in which keys were added and removed. This is something that is very hard to control. A self-balancing tree does a bit of extra work to ensure that the tree always remains properly balanced, regardless of the order of the operations that are performed on it. The AVL tree was the first such tree to be described and is fairly simple to understand [11].

The idea behind an AVL tree is that we want the height of the two children of any given node to be within one of each other. So if one child has a height of 5, the other child has to have a height of 4, 5, or 6 to be acceptable. If it does not meet this condition, the tree has to be adjusted to bring it back into compliance. This requirement of children being within one in height of each other keeps the tree close to perfectly balanced. However, it is rather restrictive and leads to a fair number of adjustments. For that reason, it is more common for libraries to use other self-balancing trees like a red-black tree.

The AVL tree works exactly like a normal BST for searches. It is only operations that add or remove nodes that necessitate heights to be recalculated and adjustments to be made. As long as the AVL condition is maintained, we can be certain that all the standard methods will complete in $O(\log(n))$ operations.

21.2.1   Rotations

The typical approach to re-balancing trees that have gotten unbalanced is through rotations. The idea of a rotation is to pull one of the children of a node up above the parent and swap one grandchild across to the other child in such a way that the proper BST ordering is preserved while pulling a grandchild that has grown too tall up to restore balance. The two basic rotations, to the right and the left, are shown in figure 21.1. A single rotation to the right can be used when the outer left grandchild has gotten too tall so that the height of the left child is two greater than the right child. Conversely, a single left rotation can fix the problem when the outer right grandchild has gotten too tall. These types of imbalance can occur when an add operation adds a new child to a subtree that increases its total height or a remove reduces the height of a subtree.

Note that the subtrees A, B, and C in figure 21.1 are shown without any detail other than one of them is taller than the other two. This is intentional. We do not care about the internal structure of those subtrees, only their overall heights.

In the situation where an internal grandchild, that is to say the right child of the left

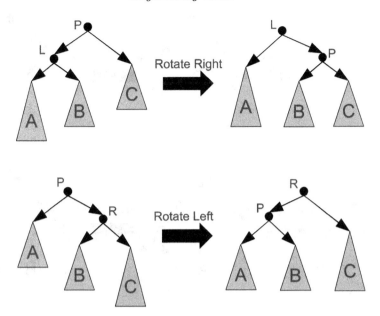

FIGURE 21.1: This figure shows single rotations to the left and the right. These rotations can be used when a node becomes unbalanced because the outer grandchild on the left or the right is too tall relative to the other subtree.

child or the left child of the right child, has grown to make the tree unbalanced, two rotations are required. The first rotation is applied to the child that has the greater height, moving it outward. The second rotation then moves the parent to the opposite side. Figure 21.2 shows how this works. Note that each step in these double rotations is one of the single rotations shown in figure 21.1. This is significant in code as well because after the single rotations have been implemented, it is a simple task to use them to build a double rotation.

21.2.2 Implementation

To implement an AVL tree, we will use the immutable BST written in chapter 16 as a base and edit it appropriately. Making a mutable AVL tree is left as an exercise for the reader. Like that tree, we have a public type that represents a tree with no functionality. We also have two private **classes** that implement the public type. The top-level type has two additions for functionality, **rotate** and **height**, as well as one for debugging purposes, **verifyBalance**. The **rotate** method is supposed to check the balance and perform appropriate rotations if needed. The **height** member stores the height of the node. It is an abstract **val** so the subtypes have to provide a value. Code for this is shown here.

```
package moretrees.adt

import linkedlist.adt.ListStack

abstract sealed class ImmutableAVLTreeMap[K <% Ordered[K], +V] extends Map[K, V] {
  def +[B >: V](kv: (K, B)): ImmutableAVLTreeMap[K, B]
  def -(k: K): ImmutableAVLTreeMap[K, V]
  def rotate: ImmutableAVLTreeMap[K, V]
  val height: Int
  def verifyBalance: Boolean
```

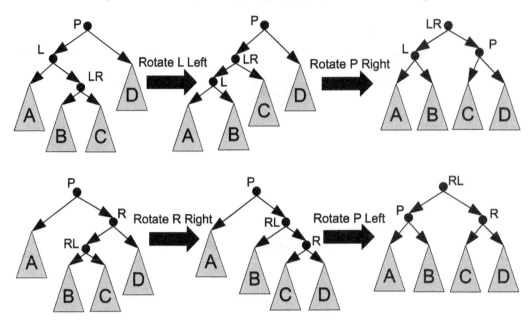

FIGURE 21.2: This figure shows double rotations to the left-right and the right-left. These rotations can be used when a node becomes unbalanced because the inner grandchild is too tall. Note that each of these is simply two applications of the rotations from figure 21.1

```scala
}

private final class AVLNode[K <% Ordered[K], +V](
    private val key: K,
    private val data: V, private val left: ImmutableAVLTreeMap[K, V],
    private val right: ImmutableAVLTreeMap[K, V]) extends ImmutableAVLTreeMap[K, V]
        {

  val height = (left.height max right.height) + 1

  def +[B >: V](kv: (K, B)) = {
    val ret = if (kv._1 == key) new AVLNode(kv._1, kv._2, left, right)
    else if (kv._1 < key) new AVLNode(key, data, left + kv, right)
    else new AVLNode(key, data, left, right + kv)
    ret.rotate
  }

  def -(k: K) = {
    val ret = if (k == key) {
      left match {
        case _: AVLEmpty[K] => right
        case i: AVLNode[K, V] => {
          right match {
            case _: AVLEmpty[K] => left
            case _ => {
              val (k, d, newLeft) = i.removeMax
              new AVLNode(k, d, newLeft, right)
            }
```

```scala
38              }
39            }
40          }
41        } else if (k < key) {
42          new AVLNode(key, data, left - k, right)
43        } else {
44          new AVLNode(key, data, left, right - k)
45        }
46        ret.rotate
47      }
48
49      def get(k: K): Option[V] = {
50        if (k == key) Some(data)
51        else if (k < key) left.get(k)
52        else right.get(k)
53      }
54
55      def iterator = new Iterator[(K, V)] {
56        val stack = new ListStack[AVLNode[K, V]]
57        pushRunLeft(AVLNode.this)
58        def hasNext: Boolean = !stack.isEmpty
59        def next: (K, V) = {
60          val n = stack.pop()
61          pushRunLeft(n.right)
62          n.key -> n.data
63        }
64        def pushRunLeft(n: ImmutableAVLTreeMap[K, V]): Unit = {
65          n match {
66            case e: AVLEmpty[K] =>
67            case i: AVLNode[K, V] =>
68              stack.push(i)
69              pushRunLeft(i.left)
70          }
71        }
72      }
73
74      def rotate: ImmutableAVLTreeMap[K, V] = {
75        if (left.height > right.height + 1) {
76          left match {
77            case lNode: AVLNode[K, V] => // Always works because of height.
78              if (lNode.left.height > lNode.right.height) {
79                rotateRight
80              } else {
81                new AVLNode(key, data, lNode.rotateLeft, right).rotateRight
82              }
83            case _ => null // Can't happen.
84          }
85        } else if (right.height > left.height + 1) {
86          right match {
87            case rNode: AVLNode[K, V] => // Always works because of height.
88              if (rNode.right.height > rNode.left.height) {
89                rotateLeft
90              } else {
91                new AVLNode(key, data, left, rNode.rotateRight).rotateLeft
92              }
```

```scala
 93          case _ => null // Can't happen.
 94        }
 95      } else this
 96    }
 97
 98    def verifyBalance: Boolean = {
 99      left.verifyBalance && right.verifyBalance && (left.height - right.height).abs <
            2
100    }
101
102    private def removeMax: (K, V, ImmutableAVLTreeMap[K, V]) = {
103      right match {
104        case e: AVLEmpty[K] => (key, data, left)
105        case i: AVLNode[K, V] =>
106          val (k, d, r) = i.removeMax
107          (k, d, new AVLNode(key, data, left, r).rotate)
108      }
109    }
110
111    private def rotateLeft: AVLNode[K, V] = right match {
112      case rNode: AVLNode[K, V] =>
113        new AVLNode(rNode.key, rNode.data, new AVLNode(key, data, left, rNode.left),
              rNode.right)
114      case _ => throw new IllegalArgumentException("Rotate left called on node with
            empty right.")
115    }
116
117    private def rotateRight: AVLNode[K, V] = left match {
118      case lNode: AVLNode[K, V] =>
119        new AVLNode(lNode.key, lNode.data, lNode.left, new AVLNode(key, data,
              lNode.right, right))
120      case _ => throw new IllegalArgumentException("Rotate right called on node with
            empty left.")
121    }
122  }
123
124  private final class AVLEmpty[K <% Ordered[K]] extends ImmutableAVLTreeMap[K,
        Nothing] {
125    def +[B](kv: (K, B)) = {
126      new AVLNode(kv._1, kv._2, this, this)
127    }
128
129    def -(k: K) = {
130      this
131    }
132
133    def get(k: K): Option[Nothing] = {
134      None
135    }
136
137    def iterator = new Iterator[(K, Nothing)] {
138      def hasNext = false
139      def next = null
140    }
141
```

```
142    def rotate: ImmutableAVLTreeMap[K, Nothing] = this

143

144    def verifyBalance: Boolean = true

145

146    val height: Int = 0
147  }

148

149  object ImmutableAVLTreeMap {
150    def apply[K <% Ordered[K], V](data: (K, V)*)(comp: (K, K) => Int):
           ImmutableAVLTreeMap[K, V] = {
151      val empty = new AVLEmpty[K]()
152      val d = data.sortWith((a, b) => comp(a._1, b._1) < 0)
153      def binaryAdd(start: Int, end: Int): ImmutableAVLTreeMap[K, V] = {
154        if (start < end) {
155          val mid = (start + end) / 2
156          new AVLNode(d(mid)._1, d(mid)._2, binaryAdd(start, mid), binaryAdd(mid + 1,
               end))
157        } else empty
158      }
159      binaryAdd(0, data.length)
160    }
161  }
```

The additions to the AVLEmpty type are simple. The height is always 0, the rotate returns this because it is always balanced, and the verifyBalance method always returns true.

The internal node type is slightly more complex. The height is set to the obvious value of (left.height max right.height)+1. The verifyBalance method checks the heights of the children and recursively calls itself on those children. The most significant part of the change is in the rotate method. This method checks the height of the children, and, if they are out of balance, it performs one or two rotations to return a node that has a balanced subtree. The rotations themselves are implemented in private methods called rotateLeft and rotateRight.

There are a number of matches done in the different rotate methods. This is due to the fact that they have to get hold of left and right grandchildren. This requires a match because the top-level type does not have left and right. This makes sense as they are not defined on AVLEmpty. The way the code is written, it should not be possible for those methods to be called with an AVLEmpty node. The rotation methods are written to throw exceptions should that happen.

The only other changes are the addition of calls to rotate in the + and - methods. These calls occur popping back up the stack so the entire tree is rebalanced every time. Testing for this class begins with test code from the original BST. To ensure that it is working, a tree must be built using + then checked with verifyBalance. The shortcut approach that is implemented in the companion object was created to always produce a perfectly balanced tree so using it does not actually test the rotation code.

21.3 Order-Statistic Trees: Trees as Sequences

The AVL tree used an augmentation of height to enable self-balancing. Augmenting by size can produce a tree that works as a sequence that is fairly efficient for both random

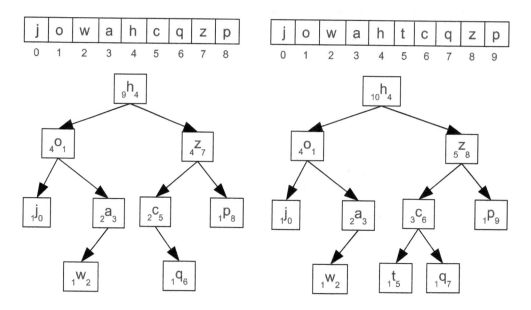

FIGURE 21.3: This figure shows two arrays and their possible tree representations. The second was arrived at by adding the letter 't' at index 5 of the first. In the tree, each node has the value, a letter, drawn large with a preceding number that is the size of the node and a following number that is the index.

access as well as random insertion and removal. By fairly efficient, we mean $O(\log(n))$, because it will keep the height augmentation as well and be a balanced tree using the AVL approach.

Both the BST and the AVL tree have been used to implement the Map ADT. The type we are describing here is a sequence. This is significant because the Map ADT needs a key type that has an ordering by which data can be accessed. In a sequence, things are accessed by an index, not a key. What is more, the index of an element can change if other items are inserted or removed before it. The key for a Map has to be immutable. Using mutable keys is a great way to produce errors with a Map. So how does this work, and what does the size augmentation have to do with it?

Assume a binary tree where the index is the key. This is much like a sequence where you look up values by their index. Figure 21.3 shows what this might look like. Letters have been used as the data here. On the left there is a sequence with nine characters in it and a balanced tree that has the same nine characters with numeric annotations. The number before the letter is the size of the node. The number after it is the index. On the right, the letter 't' has been inserted at index 5. Note how this changes the index values of all the letters that had indices greater than or equal to 5.

To use the tree as a sequence, we have to have an efficient way of finding nodes based on index. That would allow us to do direct access, as well as find nodes or positions for inserting and removing. The key to associating an index with nodes is in the size. The first indication of this is that the size of the node for 'o' is the same as the index of the root, 'h'. This is not coincidence. The size of 'o' is the number of nodes that are "before" the root in the sequence. Since indexes begin at 0, the index of the root should be the size of the left child. This logic holds throughout the tree, so having the size allows us to do a binary search through the index.

To see this, assume that we want to find the item at index 6. We start by looking at the size of the left child of the root. It is 4, which is less than 6. This tells us that what we are looking for is on the right side of the tree. This is the signature of a binary search, which by now you are well aware leads to $O(\log n)$ behavior. We do one comparison and it allows us to throw out half the data. So we go to the right child. We are not looking for index 6 on the right child though. Instead, we are looking for index 1. How did we determine that? The left child was the first 4 elements and the root itself was another. Because we have thrown out the first 5 elements, we subtract that number of elements from the index we are looking for. Thus, we can determine that we are looking for index 1.

After one comparison we are on the subtree with 'z' at the top looking for index 1. The size of the left child is 2. This is larger than the index we are looking for, so we go to the left. When we move to the left, the index stays the same because we are not jumping past any elements. After two comparisons we are on 'c' looking for index 1. The left child of 'c' is empty and therefore has a size of zero. That is smaller than the index we are looking for, so we go to the right and subtract 1 from the index we are looking for because we skipped the node with 'c'. We are now on the node with 'q' looking for index 0. The left child of 'q' is empty, giving it size zero, the same as the index we are looking for, and therefore 'q' is the value we are looking for.

This process illustrates how storing the size can let us find values at indexes in a number of operations proportional to the height of the tree. That is slower than for an array, but significantly faster than what a linked list would provide.

What would happen if we looked at index 6 again after adding the 't'? Again, we start at 'h' and see the left child has size 4. Then we move to 'z' on the right looking for index 1. The left child has size 3 so we move to the left, still looking for index 1. The left child of 'c' has a size of 1 meaning that 'c' is the node we are looking for.

This same type of searching could be done with an index augmentation. However, an index augmentation cannot be maintained efficiently. To see this, look closely at the differences between the numbers annotating the nodes in figure 21.3. The size values only changed for the nodes with 'c', 'z', and 'h'. That is the path from the new node up to the root, a path that will have length $O(\log(n))$ in a balanced tree. The index annotation has changed for 'c', 'q', 'z', and 'p', which is everything after the inserted value. In this case the difference is 3 to 4, but you can picture putting the new value at index 1 to see that the number of indexes that need to change scales as $O(n)$. This tells us that we want to augment our nodes with size, not index values.

The code below is an implementation of an order-statistic tree that extends the `mutable.Buffer` trait. This class uses internal `class`es for `Node` and `INode` as well as an internal `object` declaration for `Empty`. While this mirrors some of the style of the immutable tree in the last section, this tree is mutable and everything in `INode` is a `var`. While this uses a `class` structure similar to the immutable AVL tree, it has a very different implementation with more overloaded methods and a lot fewer uses of `match`.

The code for the `Buffer` methods themselves are generally short, as are the methods in `Empty`. The majority of the code is in `INode`, specifically in the methods for removing and rotating. The use of `Empty` with methods for `+`, `+:`, and `insert` keep the insertion code short. The recurring theme we see in this code is comparing an index to the size of the left child.

```scala
package moretrees.adt

import collection.mutable

class OrderStatTreeBuffer[A] extends mutable.Buffer[A] {
```

```scala
 6    abstract private class Node {
 7      def size: Int
 8      def height: Int
 9      def +(elem: A): Node
10      def +:(elem: A): Node
11      def apply(n: Int): A
12      def remove(n: Int): (A, Node)
13      def removeMin: (A, Node)
14      def update(n: Int, elem: A): Unit
15      def insert(n: Int, elem: A): Node
16      def rotate: Node
17    }
18
19    private class INode(var data: A, var left: Node, var right: Node) extends Node {
20      var size = 1 + left.size + right.size
21      var height = 1 + (left.height max right.height)
22
23      private def setSizeAndHeight: Unit = {
24        size = 1 + left.size + right.size
25        height = 1 + (left.height max right.height)
26      }
27
28      def +(elem: A): Node = {
29        right += elem
30        setSizeAndHeight
31        rotate
32      }
33
34      def +:(elem: A): Node = {
35        left += elem
36        setSizeAndHeight
37        rotate
38      }
39
40      def apply(n: Int): A = {
41        if (n == left.size) data
42        else if (n < left.size) left(n)
43        else right(n - left.size - 1)
44      }
45
46      def remove(n: Int): (A, Node) = {
47        val ret = if (left.size == n) {
48          val d = data
49          if (right == Empty) (data, left)
50          else if (left == Empty) (data, right)
51          else {
52            val (newData, r) = right.removeMin
53            right = r
54            data = newData
55            setSizeAndHeight
56            (d, this)
57          }
58        } else if (left.size > n) {
59          val (d, l) = left.remove(n)
60          left = l
```

```scala
61        setSizeAndHeight
62        (d, this)
63      } else {
64        val (d, r) = right.remove(n - left.size - 1)
65        right = r
66        setSizeAndHeight
67        (d, this)
68      }
69      (ret._1, ret._2.rotate)
70    }
71
72    def removeMin: (A, Node) = if (left == Empty) (data, right) else {
73      val (d, l) = left.removeMin
74      left = l
75      setSizeAndHeight
76      (d, rotate)
77    }
78
79    def update(n: Int, elem: A): Unit = {
80      if (n == left.size) data = elem
81      else if (n < left.size) left(n) = elem
82      else right(n - left.size - 1) = elem
83    }
84
85    def insert(n: Int, elem: A): Node = {
86      if (n <= left.size) left = left.insert(n, elem)
87      else right = right.insert(n - left.size - 1, elem)
88      setSizeAndHeight
89      rotate
90    }
91
92    def rotate: Node = {
93      if (left.height > right.height + 1) {
94        left match {
95          case lNode: INode => // Always works because of height.
96            if (lNode.left.height > lNode.right.height) {
97              rotateRight
98            } else {
99              left = lNode.rotateLeft
100             rotateRight
101           }
102         case _ => null // Can't happen.
103       }
104     } else if (right.height > left.height + 1) {
105       right match {
106         case rNode: INode => // Always works because of height.
107           if (rNode.right.height > rNode.left.height) {
108             rotateLeft
109           } else {
110             right = rNode.rotateRight
111             rotateLeft
112           }
113         case _ => null // Can't happen.
114       }
115     } else this
```

```scala
116        }
117
118        private def rotateLeft: INode = right match {
119          case rNode: INode =>
120            right = rNode.left
121            rNode.left = this
122            setSizeAndHeight
123            rNode.setSizeAndHeight
124            rNode
125          case _ => throw new IllegalArgumentException("Rotate left called on node with
                 empty right.")
126        }
127
128        private def rotateRight: INode = left match {
129          case lNode: INode =>
130            left = lNode.right
131            lNode.right = this
132            setSizeAndHeight
133            lNode.setSizeAndHeight
134            lNode
135          case _ => throw new IllegalArgumentException("Rotate right called on node
                 with empty left.")
136        }
137      }
138
139      private object Empty extends Node {
140        val size = 0
141        val height = 0
142        def +(elem: A): Node = new INode(elem, Empty, Empty)
143        def +:(elem: A): Node = new INode(elem, Empty, Empty)
144        def apply(n: Int): A = throw new IllegalArgumentException("Called apply on
                 Empty.")
145        def remove(n: Int): (A, Node) = throw new IllegalArgumentException("Called
                 remove on Empty.")
146        def removeMin: (A, Node) = throw new IllegalArgumentException("Called removeMin
                 on Empty.")
147        def update(n: Int, elem: A) = throw new IllegalArgumentException("Called update
                 on Empty.")
148        def insert(n: Int, elem: A): Node = new INode(elem, Empty, Empty)
149        def rotate: Node = this
150      }
151
152      private var root: Node = Empty
153
154      def +=(elem: A) = {
155        root = root + elem
156        this
157      }
158
159      def +=:(elem: A) = {
160        root = elem +: root
161        this
162      }
163
164      def apply(n: Int): A = {
```

```scala
165      if (n >= root.size) throw new IndexOutOfBoundsException("Requested index "+n+"
            of "+root.size)
166      root(n)
167    }
168
169    def clear(): Unit = {
170      root = Empty
171    }
172
173    def insertAll(n: Int, elems: Traversable[A]): Unit = {
174      var m = 0
175      for (e <- elems) {
176        root = root.insert(n + m, e)
177        m += 1
178      }
179    }
180
181    def iterator = new Iterator[A] {
182      var stack = mutable.Stack[INode]()
183      def pushRunLeft(n: Node): Unit = n match {
184        case in: INode =>
185          stack.push(in)
186          pushRunLeft(in.left)
187        case Empty =>
188      }
189      pushRunLeft(root)
190      def hasNext: Boolean = stack.nonEmpty
191      def next: A = {
192        val n = stack.pop
193        pushRunLeft(n.right)
194        n.data
195      }
196    }
197
198    def length: Int = root.size
199
200    def remove(n: Int): A = {
201      if (n >= root.size) throw new IndexOutOfBoundsException("Remove index "+n+" of
            "+root.size)
202      val (ret, node) = root.remove(n)
203      root = node
204      ret
205    }
206
207    def update(n: Int, newelem: A): Unit = {
208      if (n >= root.size) throw new IndexOutOfBoundsException("Update index "+n+" of
            "+root.size)
209      root(n) = newelem
210    }
211  }
```

Note that the rotations in this code look a bit different. This is because this code does mutating rotations where references are reset instead of building new nodes. That requires changing two references, them making sure that augmenting value are correct.

It is worth noting that the authors wrote this code first without the rotations. It was a non-AVL binary tree that implemented the `Buffer` trait. That code was tested to make sure that it worked before it was refactored to do the AVL balancing. A non-AVL version of this data structure is likely to be little better than a linked list unless you happen to be doing a lot of random additions. With the rotations, the code behavior looks identical from the outside, but we can be certain of $O(\log(n))$ performance for any single data element operation.[1] The tree and rotations do add overhead though, so unless your collection needs to be quite large and you really need a lot of random access as well as random addition and deletion, this probably is not the ideal structure.[2]

This data type is yet another example of why you need to know how to write your own data structures. The `scala.collection.mutable` package includes implementations of `Buffer` that are based on arrays and linked lists. It does not include an implementation based on an order-statistic tree. If your code needs the uniform $O(\log(n))$ operations, you really need to be capable of writing it yourself.

21.4 Augmented Spatial Trees

Binary trees are not the only trees that can be augmented. There are very good reasons why you might want to augment spatial trees as well. The data you augment a spatial tree with is typically a bit different from that in a normal tree. For example, you might store average positions of points in a subtree or the total mass of particles in a subtree. In other situations, you want to know the extreme values in a tree for different directions. In the case of the collisions that we looked at in chapter 20, the tree nodes could store information on maximum velocity and radius in a subtree so that searches could be done using a search radius that adjusts to the properties of the local population. This can make things much faster in situations where one particle happens to be moving unusually fast or is significantly larger than the others.

To illustrate an augmented spatial tree, the final example in this chapter is going to be a bit different. We will create an octree that is intended to provide fast calculations for things like collisions or ray intersections. This tree will be different from our previous spatial trees in a few ways. Obviously, it will be augmented with some other data. We will use the minimum and maximum values in x, y, and z for the augmentation. In addition, objects that are added to the tree will be able to go in internal nodes, not just leaves. The augmentation is intended to give us an estimate of "when" a cell becomes relevant.[3] The ability to store values high up in the tree is intended to let this structure work better for data sets where objects are of drastically different sizes. Something that is very large will have the ability to influence a lot more potential collisions or intersections than something very small. By placing objects at a height in the tree that relates to their size, we allow searches to "hit" them early and not have the bounds on lower levels of the tree altered by one large object. This is not the same as putting in a node that completely contains them. That would put any object, even small ones, that happen to fall near a boundary high, up in the tree.

---

[1]The `insertAll` method is $O(m \log(n))$ where m is the number of elements being added and n is the number of elements in the collection.

[2]It is worth noting that the `Vector` type in Scala, which is an immutable sequence, is implemented using a multiway balanced tree. It is more similar to a B-tree than the AVL tree, but the idea of using a tree to implement a sequence is similar.

[3]The quotes around "when" signify that we want to know a threshold value for a parameter, and that parameter might not be time.

The code for such a tree is shown below. The **class** is called an **IntersectionOctree**. There is a companion **object** that defines two **traits** that need to be extended by any code wanting to use this structure. One has methods needed by objects that are put in the tree and the other is a type that can be used to check for intersections based on different criteria. The type parameter on **IntersectionOctree** must inherit from the former. When an octree is built, the calling code needs to give it a sequence of objects, a center for the octree, a size for the octree, and a minimum size to split down to. This last parameter will prevent the tree from becoming extremely deep if a very small object is added.

```scala
 1  package moretrees.adt
 2
 3  import collection.mutable
 4
 5  class IntersectionOctree[A <: IntersectionOctree.IntersectionObject](
 6      objects: Seq[A], centerX: Double, centerY: Double, centerZ: Double,
 7      treeSize: Double, minSize: Double) {
 8
 9    private class Node(val cx: Double, val cy: Double, val cz: Double, val size:
          Double) {
10      val objs = mutable.Buffer[A]()
11      var children: Array[Node] = null
12      var min: Array[Double] = null
13      var max: Array[Double] = null
14
15      import IntersectionOctree.ParamCalc
16
17      def add(obj: A): Unit = {
18        if (obj.size > size * 0.5 || size > minSize) objs += obj
19        else {
20          if (children == null) {
21            val hsize = size * 0.5
22            val qsize = size * 0.25
23            children = Array.tabulate(8)(i =>
24              new Node(cx - qsize + hsize * (i % 2), cy - qsize + hsize * (i / 2 % 2),
                  cz - qsize + hsize * (i / 4), hsize))
25          }
26          children(whichChild(obj)).add(obj)
27        }
28      }
29
30      private def whichChild(obj: A): Int = {
31        (if (obj(0) > cx) 1 else 0) + (if (obj(1) > cy) 2 else 0) + (if (obj(2) > cz)
            4 else 0)
32      }
33
34      def finalizeNode: Unit = {
35        if (children != null) children.foreach(_.finalizeNode)
36        min = (0 to 2).map(i =>
37          objs.view.map(_.min(i)).min min
38            (if (children == null) 1e100 else
                children.view.map(_.min(i)).min)).toArray
39        max = (0 to 2).map(i =>
40          objs.view.map(_.max(i)).max max
41            (if (children == null) -1e100 else
                children.view.map(_.max(i)).max)).toArray
```

```scala
42        }
43
44        def findFirst(pc: ParamCalc[A]): Option[(A, Double)] = {
45          val o1 = firstObj(pc)
46          val o2 = firstChildObj(pc)
47          (o1, o2) match {
48            case (None, _) => o2
49            case (_, None) => o1
50            case (Some((_, p1)), Some((_, p2))) => if (p1 < p2) o1 else o2
51          }
52        }
53
54        private def firstObj(pc: ParamCalc[A]): Option[(A, Double)] = {
55          objs.foldLeft(None: Option[(A, Double)])((opt, obj) => {
56            val param = pc(obj)
57            if (opt.nonEmpty && opt.get._2 < param) opt else Some(obj, param)
58          })
59        }
60
61        private def firstChildObj(pc: ParamCalc[A]): Option[(A, Double)] = {
62          if (children == null) None else {
63            val cparams = for (c <- children; p = pc(min, max); if c !=
                  Double.PositiveInfinity) yield c -> p
64            for (i <- 1 until cparams.length) {
65              val tmp = cparams(i)
66              var j = i - 1
67              while (j > 0 && cparams(j)._2 > tmp._2) {
68                cparams(j + 1) = cparams(j)
69                j -= 1
70              }
71              cparams(j + 1) = tmp
72            }
73            var ret: Option[(A, Double)] = None
74            var i = 0
75            while (i < cparams.length && (ret.isEmpty || cparams(i)._2 > ret.get._2)) {
76              val opt = cparams(i)._1.findFirst(pc)
77              if (opt.nonEmpty && (ret.isEmpty || opt.get._2 < ret.get._2)) ret = opt
78              i += 1
79            }
80            ret
81          }
82        }
83      }
84
85      private val root = new Node(centerX, centerY, centerZ, treeSize)
86
87      objects.foreach(root.add)
88      root.finalizeNode
89
90      def findFirst(pc: IntersectionOctree.ParamCalc[A]): Option[(A, Double)] = {
91        root.findFirst(pc)
92      }
93    }
94
95    object IntersectionOctree {
```

```
 96    trait IntersectionObject extends (Int => Double) {
 97      def apply(dim: Int): Double
 98      def min(dim: Int): Double
 99      def max(dim: Int): Double
100      def size: Double
101    }
102
103    /**
104     * The A => Double should calculate the intersect parameter for an object.
105     * The (Array[Double],Array[Double) => Double should calculate it for a box
106     *   with min and max values
107     */
108    trait ParamCalc[A <: IntersectionOctree.IntersectionObject] extends (A => Double)
          with ((Array[Double], Array[Double]) => Double)
109  }
```

The way the tree is defined, it is mutable with an **add** method. However, before any tests can be done for intersections, the **finalizeNode** method has to be called on the tree. As it is written, all objects are added and the tree is finalized at the original construction.

The **finalizeNode** method has the task of setting the minimum and maximum bounds. This could be done as particles are added, but then higher-level nodes in the tree have their values checked and potentially adjusted for every particle that is added. Doing a finalize step at the end is more efficient.

The only method defined in **IntersectionOctree** is **findFirst**. This method is supposed to take a subtype of the **ParamCalc** type which can calculate intersection parameters for planes and objects in the tree. The work for this method is done in a recursive call on the root. The version in the node finds the first object hit as well as the first object in a child node hit. It gives back the first of all of these.

Where the tree comes into play is in **firstChildObj**. It starts by finding the intersection parameter for the bounds of each child node. These nodes can be overlapping. If any nodes are not intersected, they will not be considered at all. The nodes that are intersected are sorted by the intersection parameter using an insertion sort. This is one of those situations where an "inferior" sort can provide better performance. There are never more than eight items, so normal order analysis is not meaningful. After being sorted, the code runs through them in order, calling **findFirst** on each. This is done in a **while** loop so that it can stop if ever an intersection is found that comes before the intersection with a bounding region. The idea here is that if something is hit with a parameter lower than even getting into a bounding box, nothing in that bounding box could come before what has already been found. The way this works, the recursion is pruned first by boxes that are never hit and second by boxes that are hit after something in a closer box. For large data, sets this can reduce the total work load dramatically.

This tree requires the definition of two types to be useful. Those types would depend on the problem you want to solve. It is possible to use this for collision finding, but that might be better done with a method that, instead of finding the first intersection, finds all intersections up to a certain time. A problem that truly needs the first intersection is ray tracing. In computer graphics, ray tracing is a technique for rendering three-dimensional graphics. Here are implementations of a ray and a sphere to work with the tree. This is the code the data structure was tested with.

```
case class Ray(p0: Array[Double], d: Array[Double]) extends
    IntersectionOctree.ParamCalc[Sphere] {
  def apply(s: Sphere): Double = {
    val dist = (p0 zip s.c).map(t => t._1 - t._2)
```

```scala
    val a = d(0) * d(0) + d(1) * d(1) + d(2) * d(2)
    val b = 2 * (d(0) * dist(0) + d(1) * dist(1) + d(2) * dist(2))
    val c = (dist(0) * dist(0) + dist(1) * dist(1) + dist(2) * dist(2)) -
      s.rad * s.rad
    val root = b * b - 4 * a * c
    if (root < 0) Double.PositiveInfinity
    else (-b - math.sqrt(root)) / (2 * a)
  }

  def apply(min: Array[Double], max: Array[Double]): Double = {
    if ((p0, min, max).zipped.forall((p, l, h) => p >= l && p <= h)) 0.0 else {
      val minp = (0 to 2).map(i => (min(i) - p0(i)) / d(i))
      val maxp = (0 to 2).map(i => (max(i) - p0(i)) / d(i))
      var ret = Double.PositiveInfinity
      for (i <- 0 to 2) {
        val first = minp(i) min maxp(i)
        if (first >= 0 && first < ret) {
          val (j, k) = ((i + 1) % 3, (i + 2) % 3)
          if (first >= (minp(j) min maxp(j)) && first <= (minp(j) max maxp(j)) &&
              first >= (minp(k) min maxp(k)) && first <= (minp(k) max maxp(k)))
            ret = first
        }
      }
      ret
    }
  }
}

class Sphere(val c: Array[Double], val rad: Double) extends
    IntersectionOctree.IntersectionObject {
  def apply(dim: Int): Double = c(dim)
  def min(dim: Int): Double = c(dim) - rad
  def max(dim: Int): Double = c(dim) + rad
  def size: Double = rad * 2
}
```

The Ray type does the real work here. The first method takes a sphere and returns the parameter for when the ray intersects the sphere or positive infinity if it does not intersect. The second method takes arrays of min and max values for a bounding box and determines the parameter for the intersection of a ray and the box. If the ray starts in the box, it returns 0.0.

The Sphere class does little more than return basic information about a sphere. However, it would be fairly simple to make a whole hierarchy of geometric types and add appropriate code into the first apply method in Ray for intersecting that geometry to turn this into a useful data structure for a fully functional ray tracer.

21.5 End of Chapter Material

21.5.1 Summary of Concepts

- Augmenting trees is the act of storing additional data in the nodes that can be used to provide extra functionality. The data can relate to the tree structure or the information stored in the tree.

- Augmenting values must be maintainable in $O(\log(n))$ operations or better for standard operations in order to be useful.

- By augmenting with information related to the height of a node we can make self-balancing binary trees.

  - The AVL tree enforces a rule that the heights of siblings cannot differ by more than one.
  - When we find a parent with children that break the height restriction, the problem is corrected with the application of rotations.
  - An implementation was shown based on the immutable BST. It can also be done in a mutable manner.

- Augmenting a binary tree with size can be used to make an order-statistic tree which can be used as a sequence.

  - The index is like a key, but it is not stored so that inserts and removals can change the index.
  - Nodes can be found by index using a binary search where the desired index is compared to the size of the left child.

- Spatial trees can also be augmented.

  - This is often done using data stored in the tree.
  - An example was given where nodes store boundaries for objects inside of them to do more efficient searching for intersections.

21.5.2 Exercises

1. Write a mutable AVL tree.

2. Compare the speed of the AVL tree to the normal BST for random data as well as sequential or nearly sequential data.

3. Write a BST that is augmented with the number of leafs.

4. Compare the speed of different buffer implementations. Use the **ArrayBuffer** and **ListBuffer** that are part of the Scala libraries. Use the order-statistic tree written in this chapter as well. Compare methods of adding, removing, and indexing values for collections of different sizes.

5. Write **findUntil** for the octree. This is a method that will find all collisions up to a certain parameter value instead of just the first one. This is useful for finding collisions that can occur during a time step.

21.5.3   Projects

1. Use an order-statistic tree in place of a `Buffer` in your MUD. If you can think of another use for an augmented tree, implement that as well. You might consider the min-max tree described below in project 21.4.

2. If you have been working on the web spider, see if you can come up with an augmented tree structure that helps to partition and give additional meaning to your data. If you cannot think of anything, simply use the order-statistic tree in place of a `Buffer`.

3. If you are writing a graphical game where characters move around and there are obstacles, you could build an augmented spatial tree to help expedite the process of finding safe paths through the world. Such a tree would keep track of significant points in a map and store path information between those points. When the code needs to get a character from one location to another, it should do a recursive decent of the tree with each node holding augmentation data specifying which children are significant for moving in certain ways.

    The advantage of this is that exhaustive searches can be done in local spaces in the leaves of the tree. Those exhaustive searches are likely costly recursive algorithms which benefit greatly from only covering a small area. When moving over a larger area, paths are put together from one node to another. The recursive decent can prune out children that are known to be insignificant.

4. There is a commonly used construct in AI called the min-max tree, which is an augmented tree that compares moves from different players to help a given player find the optimal move. Such a tree could be implemented in any game where you want to give computer-controlled players reasonable intelligence. The idea of a min-max tree is that you assign values to an ending where each player wins. So if player A wins, that has a value of 1. If player B wins, that has a value of -1. In games that allow a tie, 0 represents a tie. In this tree, player A wants to pick paths that maximize the value and player B wants to pick paths that minimize the value. Since each player generally alternates moves, there are alternating levels that are maximizing and minimizing values. Hence the name min-max tree.

    If you can build a tree for all possible moves in a game, you can tell exactly who should win in a well-played game.[4] For most games of any significance, you cannot build this entire tree. Instead, you build it down to a certain level, and at that level you stop and apply a heuristic evaluation of the relative positions of the players. The heuristic should give a value between -1 and 1 based upon which player seems to be in the better position to win. No heuristic will be perfect, but using the min-max tree to evaluate many heuristics, you can generally get a reasonable prediction of what the ideal move is at any given time.

    How this works for games like checkers or chess is fairly obvious, but you are likely doing something different. In that case there are choices of where to move, what action to take, what to build, etc. Those can also be chosen using such a tree. Note that if the number of options is really large, the tree becomes huge. You will probably have to do a little work to keep the tree from blowing up by trimming out really bad choices early on.

5. If you have been working on the math worksheet, you can consider changing a `Buffer`

---

[4]Doing this for tic-tac-toe is an interesting exercise for the reader. The result is that a well-played game always ends in a tie.

implementation so that it uses an order-statistic tree. If you have added any features that would benefit from an augmented tree, you should write code for that.

6. Use the `IntersectionOctree` to add a ray-tracing element to your Photoshop® program.

7. The simulation workbench has two different augmentations that you can make to spatial trees to improve performance. For the collision tree, the search distance can vary with location. The earlier implementation, used a uniform search distance, which should be the maximum for any place in the tree. You can use a localized value by augmenting the tree.

   Recall that the search radius is given by $\Delta v \Delta t + 2R_{max}$, where Δv is the relative velocity of particles, Δt is the time step, and R_{max} is radius of the largest particle. Technically, these values vary for different subtrees. A more efficient implementation can be written that uses local values for those. Augment the tree with the values of Δv and R_{max} in each subtree.

   A tree can also be used to improve the efficiency of gravity calculations. The technique of gravity trees was originally developed by Barnes and Hut [2]. The original work used an octree, but many newer implementations use a kD-tree. The idea of a gravity tree code is that while the gravity from nearby masses needs to be handled exactly, the precise locations of individual bodies is not significant for things that are far away. For example, when calculating the gravity due to the Jupiter system on a mass, if the mass is near Earth, you do not need to do separate calculations for all the moons. Doing a single calculation for the total mass of the system at the center of mass of the system will be fine. However, if you are at the location of one of the outer satellites of Jupiter, the exact positions of some of the moons could be significant.

   To quickly approximate the force of gravity from distant collections of mass, the tree can be augmented with information about all the particles in a subtree. The simplest augmentation is that each node stores the total mass and center of mass location of that subtree. To calculate the gravity force on a particle, the code does a recursive decent of the tree. At any given node, a check is done comparing the size of the node to its distance from the particle currently under consideration. If the node meets the condition $d < \beta s$, where d is the distance between particle and node, s is the size of the node, and β is a parameter to adjust accuracy,[5] then the recursion must go down to the children. If that condition is not met, the force can be calculated against the center of mass of that node and the children can be ignored.

8. Use an order-statistic tree in place of a `Buffer` in your stock portfolio management system project. If you can think of another use for an augmented tree, implement that as well.

9. Use an order-statistic tree in place of a `Buffer` in your movies project. If you can think of another use for an augmented tree, implement that as well.

10. For the L-system project, you could consider representing the string of characters as a tree instead of as a normal string. Nodes could represent single characters or sequences of characters. Using a multiway tree, each generation you advance could add an extra layer to the bottom of the tree. Implement this and see how it compares to the approach of just using a normal `String`.

---

[5]For a monopole tree as described here, the value of β should be between 0.1 and 0.3. Smaller values made the calculations more accurate, but increase the amount of work that is done.

Chapter 22

Hash Tables

Chapter 16 showed how we could implement the Map ADT using a binary search tree. Chapter 6 had earlier introduced the **Map** types in the Scala standard libraries, including the **TreeMap** types, which are subtypes of their respective **SortedMap** types. You might recall that there were a number of other implementations of **Map** as well. In particular, there are both mutable and immutable versions of **HashMap**.

The **HashMap** uses a fundamentally different approach to implementing the Map ADT, so it serves as another example of separating interface from implementation. The fact that the **TreeMap** uses a BST for its implementation means that the basic operations for searching, adding, and removing are all $O(\log n)$. Our goal with hash maps is to get the search, insert, and remove methods all down to $O(1)$.

22.1 Basics of Hash Tables

A hash table is effectively a set of numbered bins with a function that maps keys to those bins. The bins are generally implemented as an array, because that provides $O(1)$ access time for finding a bin given its numeric index. Figure 22.1 shows a graphic representation of what happens with a hash table. The keys can be whatever type of values we want, and they do not have to have any type of natural ordering. We have a hash function that maps from the keys to numeric value in the range of valid indices in the array.

Hash functions cannot produce unique values. There are far more potential keys than there are positions in the array. For example, consider a **String** key with a number of possibilities limited by the maximum length of the **String**. Thus, there will inevitably be COLLISIONS where two values are mapped to the same location. We will talk later about the approaches that we use to deal with collisions.

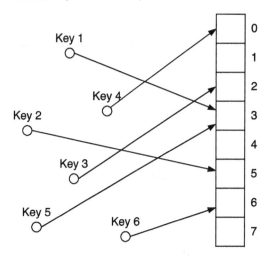

FIGURE 22.1: This figure shows a graphic representation of the idea behind a hash table. Keys can be any type of data and are not necessarily ordered. The keys are mapped, using some function that we call the hash function, to positions in the storage array. There are many more possible keys than there are locations in the array, so it is possible to have "collisions" where different keys map to the same location.

22.1.1   Hashing Functions

To make these functions that map from keys to indices generally involves two steps. The first step converts the key to an integer in the full range of integers between $2^{31} - 1$ and -2^{31}. Then we map that integer value into the range of valid indices. The first step in this process, generating a "hash code" for an object of an arbitrary type, is actually the hard part. Fortunately, Java and Scala have done a lot of that work for us. There is a method in the **Any** type called **hashCode**. So every value in the Scala language has the ability to calculate a hash code for you. The following code in a REPL demonstrates the hash codes that are generated for a number of different strings.

```
scala> words.map(w => w -> w.hashCode).foreach(println)
(This,2605758)
(is,3370)
(a,97)
(test,3556498)
(of,3543)
(the,114801)
(emergency,1629013393)
(broadcast,-1618876223)
(system.,-1737370657)
```

Unfortunately, this is not some magic solution that you can use blindly. The types in the library that you would consider using for keys will likely all have good implementations of **hashCode**. In addition, when you make a **case class**, one of the things that Scala adds into it is a valid implementation of **hashCode** and **equals** based on the elements in the **case class**. However, if you make your own **class** that is not a **case class**, the default implementations of **hashCode** and **equals** are unlikely to do what you want for keys, and you will probably need to override them. At this point, we would recommend that you generally stick with built-in types and **case class**es for your keys.

Overriding `hashCode` and `equals`

If you do decide to implement your own `hashCode`, there are some things that you should know. First, if two instances of your type say that they are equal to one another, they must produce the same `hashCode`. This is why you always need to override the `equals` method when you override the `hashCode` method. If you violate this rule, it will not only break the functionality of code in the chapter, but for the standard library `HashMaps` as well. Note that in Scala, when you say `==` in an expression, it is actually making a call to `equals` on the instance.

Second, a good `hashCode` returns different values for things that are not equal. Given the first rule, it is perfectly valid to write `override def hashCode = 42`. That satisfies the requirement that all equal values will have the same `hashCode`, but it will cause the `HashMaps` to have extremely poor performance because there are collisions between every value.

22.1.2 Hashing Methods

We now know how to accomplish the first step in the two-step process. You can use `hashCode` to go from your key to an integer value. To complete the hash function, we need to map those integers into the range of indices of our array. There are two approaches that we will consider for doing this. The simpler version is called the division method and it uses the modulo operator, which you should recall is what remained after division. We take the result of `hashCode` and modulo by the length of the array. There is one significant proviso to this, which shows up when we take this simple approach to the same strings from above assuming an array size of 10.

```scala
scala> words.map(w => w -> w.hashCode % 10).foreach(println)
(This,8)
(is,0)
(a,7)
(test,8)
(of,3)
(the,1)
(emergency,3)
(broadcast,-3)
(system.,-7)
```

The fact that the `Int` type in Scala is signed means that we can get negative values from `hashCode`. When you use `%` on a negative value, the result is negative. Given that the indices of our arrays are all non-negative, we will need to take the absolute value.

Unfortunately, not all values for length of arrays are good choices given this approach. For example, powers of two are particularly bad. The reason for this is that when you do modulo by a power of 2, say 16, you are really just pulling out the lower bits of the `hashCode` and throwing away the higher bits. Depending on how your `hashCode` method is written, this might be throwing away information about particular parts of the key, which can lead to more collisions in certain data sets. Really good values for the length of the array are primes that are not close to powers of two. Growing the array size as more elements are added in a manner that always uses primes that are far from powers of two is challenging.

So while we will change up our growth mechanism a bit, we will not be going to extreme lengths to keep them prime.

The other method of completing our hash function that we will discuss is called the multiplication method. Using this approach, we multiply the `hashCode` by a constant A in the range $0 < A < 1$. We then take the fractional part of that result, multiply it by the length of the array, and truncate it to an `Int`. While any value of A will work, some will behave better for some data sets than for others. Knuth [8] argued that $A = (\sqrt{5} - 1)2 \approx 0.618034$ works well for most data sets.[1]

For our code, we implement both of these approaches in an `object` that includes an enumeration so that users can easily specify which type of hashing method they want to use.

```scala
object HashingMethods extends Enumeration {
  val Division, Multiplication = Value

  private[adt] def buildHashFunction[K](method: Value): (K, Int) => Int = method
      match {
    case Division => (key, size) => key.hashCode().abs % size
    case Multiplication => (key, size) => {
      val x = key.hashCode() * A
      (size * (x - x.toInt).abs).toInt
    }
  }

  val A = (math.sqrt(5) - 1) / 2
}
```

There are other approaches to mapping our `hashCodes` into the range of indices for the array that can be useful in specific situations. However, these two are sufficient to illustrate how hash tables work, and they are the most generally used approaches because of their simplicity and speed.

22.2 Chaining

Now that you know how to map from your arbitrary keys to integer values of valid indices in our storage array, we need to decide how we are going to deal with the issue of collisions. The reality is that unless you know your data set in advance, it is impossible to completely avoid collisions, regardless of how good your hash function is. So when two values map to the same index in our table, we need to be able to deal with it. The simplest way of doing this, conceptually and in the memory of the computer, is to make it so that each entry in the table is actually a list of values, stored either as a linked list or a variable-length array like an `ArrayBuffer`. Using this approach in figure 22.1, when "Key 1" and "Key 5" both mapped to index 3, that entry would have a short list with two elements in it.

This is the approach that we use for our first full implementation of a hash map. The code below shows the `ChainingHashMap`, which extends the `mutable.Map` trait and implements the four abstract methods of that trait. It also implements its own versions of `isEmpty` and `size` to make them more efficient.

---

[1]That value happens to be the fractional part of the golden ratio.

```
 1   package hashtables.adt
 2
 3   import collection.mutable
 4
 5   class ChainingHashMap[K, V](method: HashingMethods.Value) extends mutable.Map[K,
         V] {
 6     private var numElems = 0
 7     private var table = Array.fill(11)(List[(K, V)]())
 8     private val hashFunc = HashingMethods.buildHashFunction(method)
 9
10     def +=(kv: (K, V)): ChainingHashMap.this.type = {
11       def appendOrUpdate(lst: List[(K, V)]): (Boolean, List[(K, V)]) = lst match {
12         case Nil => (true, List(kv))
13         case h :: t => if (h._1 == kv._1) (false, kv :: t) else {
14           val (added, newt) = appendOrUpdate(t)
15           (added, h :: newt)
16         }
17       }
18
19       if (numElems > ChainingHashMap.fillingFactor * table.length) growTable()
20       val index = hashFunc(kv._1, table.length)
21       val (added, lst) = appendOrUpdate(table(index))
22       table(index) = lst
23       if(added) numElems += 1
24       this
25     }
26
27     def -=(key: K): ChainingHashMap.this.type = {
28       def removeMatch(lst: List[(K, V)]): (Boolean, List[(K, V)]) = lst match {
29         case Nil => (false, Nil)
30         case h :: t => if (h._1 == key) (true, t) else {
31           val (found, newt) = removeMatch(t)
32           (found, h :: newt)
33         }
34       }
35
36       val index = hashFunc(key, table.length)
37       val (found, lst) = removeMatch(table(index))
38       table(index) = lst
39       if (found) numElems -= 1
40       this
41     }
42
43     def get(key: K): Option[V] = {
44       val index = hashFunc(key, table.length)
45       table(index).find(_._1 == key).map(_._2)
46     }
47
48     def iterator = new Iterator[(K, V)] {
49       private var i1 = table.iterator
50       private var i2 = nextBin()
51       private def nextBin(): Iterator[(K, V)] = {
52         if (i1.isEmpty) Iterator.empty
53         else {
54           var lst = i1.next
```

```
55        while (lst.isEmpty && i1.hasNext) {
56          lst = i1.next
57        }
58        lst.iterator
59      }
60    }
61
62    def hasNext: Boolean = i2.hasNext
63
64    def next: (K, V) = {
65      val ret = i2.next
66      if (i2.isEmpty) i2 = nextBin()
67      ret
68    }
69  }
70
71  override def isEmpty = numElems == 0
72
73  override def size = numElems
74
75  private def growTable(): Unit = {
76    val tmp = Array.fill(table.length * 2 + 1)(List[(K, V)]())
77    for (kv <- iterator) {
78      val index = hashFunc(kv._1, tmp.length)
79      tmp(index) ::= kv
80    }
81    table = tmp
82  }
83 }
84
85 object ChainingHashMap {
86   def apply[K, V](method: HashingMethods.Value, kvs: (K, V)*): ChainingHashMap[K,
          V] = {
87     val ret = new ChainingHashMap[K, V](method)
88     for (kv <- kvs) ret += kv
89     ret
90   }
91
92   val fillingFactor = 0.7
93 }
```

The class begins with declarations of numElems, table, and hashFunc. numElems is a
simple integer counter of the number of elements that are currently stored. The table is an
Array[List[(K, V)]] that begins with 11 elements that are all empty lists. The hashFunc
is our hash function using the code for HashingMethods shown earlier.

The simplest method in this class, other than isEmpty and size, is get on lines 43–46,
which should look up a key and return an Option of a value. The first line uses hashFunc
to calculate an index for the key. We then look up the List at that index and call find
to see if there is an element with a matching key. We map on the result of that to pull out
just the value. The fact that find returns an Option[(K, V)] causes this short code to do
exactly what we want.

The += method should add a new value to the Map. Unlike previous code, we actually
specify the return type here. Seeing this, you can probably understand why it was left off
previously. Lines 11–17 define a nested helper function called appendOrUpdate. When we

go to add a new key-value pair, there are two possibilities. The key could already be in the Map, in which case, we are supposed to update the value, or it might not be, in which case we are supposed to add it. This function handles that distinction. It walks the `List` looking for the key. If a match is found, it gives back a modified `List` with a new element in that location. If it gets to the end without finding a match, it returns a new `List` with the element appended to the end. It also returns a `Boolean` telling us if a new key-value pair was added.

Line 16 does a check to see if the table should get bigger. As we will discuss below, it is essential that each entry in the table has a short `List` in order to maintain efficiency. So once the number of elements gets above a certain size relative to the size of the table, we need to make a bigger table. The `growTable` method on lines 75–82 is similar to what we have seen in other collections with `Arrays` that had to be grown. It makes a bigger `Array`, using a factor of 2 plus one to help improve the behavior of the division method of hashing, then runs through the current contents, adding those elements to the new table. The add here can be simpler than the code in `+=` because we can be certain that there are no duplicate keys. As a result, each element is just consed onto its `List`. Once we have verified that the table is big enough, we calculate an index on line 20, then call `appendOrUpdate` on line 21 and store the new `List` back in the table on line 22. We finish by incrementing `numElems` if a new element was actually add and returning the current instance.

The `-=` method is very similar to `+=`. It too defines a helper function called `removeMatch` on lines 28–34 that goes through a list and removes a match if one is found. It returns both the new `List` and a `Boolean` so that we know if a match was found or not. Line 36 calculates the index in the table. Line 37 calls `removeMatch`. Line 38 updates the contents of the table with the new list and line 39 decrements the value of numElems if needed.

The most complex method in this code is the `iterator` method on lines 48–60. The reason for the complexity is that our table, as an `Array[List[(K, V)]]`, is a 2D data structure. What is more, some of the `Lists` in the table might be empty and should be skipped over completely. To deal with the 2D aspect, we declare two separate iterators on lines 49 and 50. The first one is an iterator through the `Array` and the second one is an iterator through a `List` that is in one element of the table. Lines 51–60 define a method called `nextBin` that are intended to advance the i1 iterator until it reaches a non-empty `List` and give back the iterator of that `List` to use as i2. With this in place, the implementations of `hasNext` and `next` for our implementation of `Iterator` are fairly straightforward. The only aspect that is at all challenging is that we have to remember to set i2 to the value returned by `nextBin` when it is empty.

One critical aspect to note about the `iterator` method is that it will run through the key-value pairs in what appears to be random order. Of course, it is not really random. It depends first on the hash function being used and then upon the order in which things are inserted when there are collisions. However, to outside code, it is basically impossible to determine the order in which things should be expected to be returned. This has an impact on testing. When testing a `TreeMap`, you actually want to verify that the iterator gives the keys back in sorted order, so you can write code that checks for that. With a `HashMap`, it is more challenging to verify if the `iterator` method is working. For our testing, we made use of the `toSet` method that `ChainingHashMap` inherits from `mutable.Map`. This method has to use `iterator` to get all the elements, and checks for equality on a Set are always order independent.

The companion `object` on lines 85–93 is fairly straightforward. It defines an `apply` method to make it easier to construct instances of the `class`. It also defines the constant `fillingFactor` that was used back on line 19 to determine if the table needed to grow.

Now that we have seen a full implementation of a hash map, we can explore the order of the various operations. In particular, we want to know how the amount of work done by get, +=, and -= scales with the number of elements in the Map. All three of these contain a call that runs through the contents of a `List` in one entry of the table. In `get` it is the `find` method that does this. In += and -= it is the helper methods that go through looking for matching keys. None of these methods do anything with more than one entry in the table.[2] So the order of these methods is really set by the length of the different `List`s in the table.

Unfortunately, we cannot say with absolute certainty how many elements will be in each table entry. That depends on the data and the hashing function. It is possible that every single key we added hashed to the same index, and every add was a collision. In that case, all of those methods would be $O(n)$. This is the worst-case behavior. Fortunately, as long as the `hashCode` method is well written, the odds of this happening on real data are extremely small. With a good `hashCode`, we expect the data to be uniformly distributed across the entries in the table. Given that our `fillingFactor` is less than one, this means that we expect the individual `List`s to be short, not more than a few elements. In that situation, all three methods will be $O(1)$, which is our expected behavior. To test this, we used this implementation with a `Double` as the key type and added one million elements using `math.random` to generate key values. Over several runs, the longest `List` in the table varied between 7 and 8. Hopefully this discussion makes it clear how important it is that we call `growTable` at appropriate intervals. Failure to do so would cause the `List`s in the table to grow in an $O(n)$ way, killing performance.

Immutability of Keys

It was mentioned in previous discussions of the Map ADT that it is important for key types to be immutable. It is worth taking a minute to consider this statement based on the implementation of a hash map. Think of what happens if you use a mutable key type. You put a key-value pair into the Map, then code edits the value of that key. (Remember that objects are passed around as references in Scala, so outside code can easily maintain a reference to a key instance that was added.) Changing the value of the key likely changed the value calculated by `hashCode`. That means that the key is now located in the wrong entry in the table. This makes it effectively impossible to find.

22.3   Open Addressing

An alternate approach to dealing with collisions is called "open addressing". Using this approach, the entries in the table are not some form of lists, they are simple values. When we go to put a new key-value pair in the table and find that it is occupied, we simply move to some other location. This moving to other positions is called PROBING, and there are a number of ways that it can be done.

---

[2]The exception is += when a call to `growTable` is made, but we have already addressed the fact that growing the table by multiples causes the amortized cost of this operation to be $O(1)$.

22.3.1   Probing Methods

The simplest form of probing is called LINEAR PROBING. Using this approach, if the current entry is full, we move to the next one in numeric order, wrapping back to the beginning if we go past the end. We can express this as $(i+j)\%size$ where i is the original index, j is the number of occupied locations we have found, and $size$ is the size of the table.

This approach is simple, and easy to implement, but it has a problem in that the location where the next add will place the value is not well distributed. Imagine the situation where we have 10 entries in our table and five values have been added that wound up all going in consecutive indices from 1–5. So index 0 is empty as are 6–9. If we are using linear probing, the next add actually has a 60% chance of going into index 6 assuming that the odds of the hash function producing an initial index are uniformly distributed. That uniform distribution means that there is a 10% chance the initial index will be 0, 10% for 1, 10% for 2, etc. However, due to the way that linear probing works, any initial index between 1 and 6 will wind up putting the new element in 6, so there are 60% odds it will go there. If that happens, there will be 70% odds the next one will go in 7.

This effect is called PRIMARY CLUSTERING and it is a big problem for linear probing. It tends to lead to long stretches of the table being filled which makes it more likely that when we go look for things, we will have to probe further. When doing open addressing, probing lengths are equivalent to the walking down lists with chaining. That is the thing that can change our Map from being $O(1)$ to something worse.

An alternative approach is called QUADRATIC PROBING. Using this approach, each probe location is given by $(i + c_1 j + c_2 j^2)\%size$, where once again, i is the initial index given by the has function, j is the count of how many times we have done probing, and $size$ is the size of the table. In addition, c_1 and c_2 are arbitrary integer constants. So instead of moving forward one entry each probe, it moves forward in jumps that get longer with each probe. This prevents primary clustering, but it is still subject to SECONDARY CLUSTERING because the probing sequence depends only on the one hash code. Fortunately, secondary clustering, which is where the probe sequence is the same for keys that hash to the same value, is much less problematic.

Here is an **object** that implements these two probing methods. Like **HashingMethods**, it is an enumeration of values that outside code is supposed to use. It also has a method used just in the library that generates functions for the probing. These functions take i, j, and $size$, and produce the appropriate index for the next probe.

```
object ProbingMethods extends Enumeration {
  val Linear, Quadratic = Value

  private[adt] def buildProbingFunction(method: Value): (Int, Int, Int) => Int =
      method match {
    case Linear => (i, j, size) => (i + j) % size
    case Quadratic => (i, j, size) => (i + 2 * j + 3 * j * j) % size
  }
}
```

Another approach to probing is called DOUBLE HASHING that alleviates both primary and secondary clustering. In this approach, we have two functions that generate different hash codes for a given key. If we call these h_1 and h_2, the probe location is given by $(h_1(key) + jh_2(key))\%size$. In our previous formulas, we have used i for what would be $h_1(key)$ here. By using a second hashing function, we make it so that keys that collide on the first probe are very unlikely to collide on the second probe, which effectively eliminates

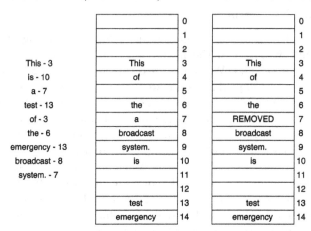

FIGURE 22.2: This figure illustrates the challenge that we run into when we have the option to remove a key-value pair when using open addressing. The left column shows our words and the value given by our hash function using the division method for an array with 15 elements. The middle column shows where all those keys would wind up in the table assuming linear probing. The third column shows it with "a" removed, which would cause problems if we go looking for "system."

clustering. We are not implementing this approach because we only have one `hashCode` method that we can call on our keys, and we do not want to implement a second one.

22.3.2 Removing with Open Addressing

The advantage of using open addressing is that we do not have to have a 2D data structure to store our values; everything goes in a single `Array`. This does cause a bit of a challenge when we implement a method to remove an element. To understand why this is, consider what is shown in figure 22.2. We have the words that we considered earlier on the left side with the indices that we would get using the division method with a table that has 15 entries. The middle column shows where those values would go if we added them in order to an open addressing hash map using linear probing. You can see that "of", "system.", and "emergency" all had to use probing and are at locations with higher indices than the hash function produced. In the third column we have taken out "a" and replaced it with "REMOVED". To understand why we do not just leave it blank, consider what happens when we go looking for "system.". The hash function says that "system." should be at index 7. If that spot were empty, we would think that "system." was not in the Map, which would be incorrect. On the other hand, we do not want the `get` and `-=` methods to have to run through the entire table if an entry is not there. Those methods should be able to stop at the first empty entry. For that to work, we have to mark the entries of removed elements in some way so that we know to keep searching beyond that location, but that it could be used for a subsequent add.

The way we will do that is to have our table store one of three types of values: `Empty`, `Removed`, or `Full`. That allows searches to skip past `Removed` entries without causing errors or forcing all searches to be $O(n)$ for keys that are not present.

22.3.3 Open Addressing Implementation

The code below shows an implementation of a `mutable.Map` using open addressing. As with the implementation using chaining, it starts with a series of declarations. This `class` has two extra declarations that we did not have before. In addition to `numElems`, there is a `numEntriesUsed` on line 10. The reason for this is that when we use open addressing, the decision of when to grow the table is not made based on how many elements are actually in the Map, but on how many entries have a state other than `Empty`. If we were to allow all the spots to be some combination of `Removed` and `Full`, even if very few are `Full`, then the behavior of searches will become $O(n)$.[3] There is also a second function called `probeFunc` declared that stores the probing function that we are using.

```scala
package hashtables.adt

import collection.mutable

class OpenAddressingHashMap[K, V](
    hashMethod: HashingMethods.Value,
    probeMethod: ProbingMethods.Value) extends mutable.Map[K, V] {
  import OpenAddressingHashMap._
  private var numElems = 0
  private var numEntriesUsed = 0
  private var table = Array.fill(11)(Empty: Entry[K, V])
  private val hashFunc = HashingMethods.buildHashFunction(hashMethod)
  private val probeFunc = ProbingMethods.buildProbingFunction(probeMethod)

  def +=(kv: (K, V)): OpenAddressingHashMap.this.type = {
    if (numEntriesUsed > OpenAddressingHashMap.fillingFactor * table.length)
        growTable()
    val index = hashFunc(kv._1, table.size)
    (0 until table.length).find { i =>
      val probeIndex = probeFunc(index, i, table.size)
      table(probeIndex) match {
        case Empty =>
          numElems += 1
          numEntriesUsed += 1
          table(probeIndex) = new Full(kv)
          true
        case Removed =>
          numElems += 1
          table(probeIndex) = new Full(kv)
          true
        case f: Full[K, V] if f.key == kv._1 =>
          table(probeIndex) = new Full(kv)
          true
        case _ => false
      }
    }
    this
  }

  def -=(key: K): OpenAddressingHashMap.this.type = {
```

---

[3]That assumes we are being careful. Searches that use a `while` loop could, in that situation, become infinite loops.

```scala
40      val index = hashFunc(key, table.size)
41      (0 until table.length).find { i =>
42        val probeIndex = probeFunc(index, i, table.size)
43        table(probeIndex).key == Some(key) || table(probeIndex) == Empty
44      }.foreach { i =>
45        val probeIndex = probeFunc(index, i, table.size)
46        if(table(probeIndex) != Empty) {
47          table(probeIndex) = Removed
48          numElems -= 1
49        }
50      }
51      this
52    }
53
54    def get(key: K): Option[V] = {
55      val index = hashFunc(key, table.size)
56      ((0 until table.length).find { i =>
57        val probeIndex = probeFunc(index, i, table.size)
58        table(probeIndex) == Empty || table(probeIndex).key == Some(key)
59      }).flatMap(i => table(probeFunc(index, i, table.size)).value)
60    }
61
62    def iterator = new Iterator[(K, V)] {
63      def advance() = while (index < table.length && table(index).key == None)
64        index += 1
65      var index = 0
66      advance()
67
68      def hasNext: Boolean = index < table.length
69      def next: (K, V) = {
70        val ret = table(index).kv.get
71        index += 1
72        advance()
73        ret
74      }
75    }
76
77    override def isEmpty = numElems == 0
78
79    override def size = numElems
80
81    private def growTable(): Unit = {
82      val tmp = Array.fill(table.length * 2 + 1)(Empty: Entry[K, V])
83      for (kv <- iterator) {
84        val index = hashFunc(kv._1, tmp.length)
85        var i = 0
86        var probeIndex = probeFunc(index, i, tmp.length)
87        while (tmp(probeIndex) != Empty) {
88          i += 1
89          probeIndex = probeFunc(index, i, tmp.length)
90        }
91        tmp(probeIndex) = new Full(kv)
92      }
93      table = tmp
94      numEntriesUsed = numElems
```

```
95      }
96    }
97
98    object OpenAddressingHashMap {
99      def apply[K, V](hashMethod: HashingMethods.Value, probeMethod:
            ProbingMethods.Value, kvs: (K, V)*): OpenAddressingHashMap[K, V] = {
100       val ret = new OpenAddressingHashMap[K, V](hashMethod, probeMethod)
101       for (kv <- kvs) ret += kv
102       ret
103     }
104
105     private val fillingFactor = 0.5
106
107     private sealed trait Entry[+K, +V] {
108       val key: Option[K]
109       val value: Option[V]
110       val kv: Option[(K, V)]
111     }
112     private object Empty extends Entry[Nothing, Nothing] {
113       val key = None
114       val value = None
115       val kv = None
116     }
117     private object Removed extends Entry[Nothing, Nothing] {
118       val key = None
119       val value = None
120       val kv = None
121     }
122     private class Full[K, V](val _kv: (K, V)) extends Entry[K, V] {
123       val key = Some(_kv._1)
124       val value = Some(_kv._2)
125       val kv = Some(_kv)
126     }
127   }
```

A key aspect to understanding the implementation of this **class** is found in the companion **object**. Lines 107–126 define a **sealed trait** called Entry along with two **objects** and a **class** that extend that **trait**. The implementations are called Empty, Removed, and Full, mirroring the names used above in our discussion of removing from an open addressing hash table. There are three abstract **val** declarations in Entry for the key, the value, and kv, which is a tuple with both. All three are Option types and both Empty and Removed simply give them the value None. The Full **class** gives them appropriate values for the key-value pair that it is initialized with.

As before, the simplest method is likely **get** on lines 54–60. This method does a **find** in the values from 0 until the length of the table. The **find** method stops as soon as it gets a true value, and there should never be a situation where we have to probe more times than the length of the table. The **find** will find the first element in the probe sequence that is either empty or has the key that is being looked for. It will keep searching on a Removed or a Full with the wrong key. It is the second equality check on line 56 that shows the value of putting abstract **vals** in the Entity type. Without those, we cannot use table(probeIndex).key and instead we have to check what subtype of Entity we are working with, which is a much more verbose operation.

Once the **find** method is done, it either gives us a None or a Some with the index that was found. Given that it stops on Empty and there should never be a situation where all the table entries are non-empty, it should be impossible to get back **None**. The code then calls **flatMap** on the resulting index and gives back the appropriate **Option** on a value. Here again, the ability to refer to **value** on an **Entry** greatly simplifies the code.

It is possible that you might feel uncomfortable with the use of **find** in the **get** method. It can certainly be written with a **while** loop as well. The following code could be put in place of the 4 lines from line 56–59 to use this approach.

```
var i = 0
var probeIndex = probeFunc(index, i, table.size)
while(table(probeIndex).key != Some(key) && table(probeIndex) != Empty) {
  i += 1
  probeIndex = probeFunc(index, i, table.size)
}
table(probeIndex).value
```

Note that there is one significant difference between these versions. If some other bug in your code caused all the entries in the table to be non-empty, this version would be an infinite loop. We could add an extra **i < table.length** to prevent that.

The **+=** method starts with a check to see if the table should be grown on line 16. Note that the value of **fillingFactor** used here is 0.5 instead of 0.7. Because the open addressing implementation does not have lists, it is critical that there be open slots. In addition, note that the check is done against **numEntriesUsed**, not **numElems**. Line 17 calculates the initial index, then lines 18–35 use **find** to run through the probing sequence again. In this case, we are not actually finding anything. We just need to run through the indices and stop at a certain point. This could also be done with **forall** or **exists**, but we felt that keeping the code consistent with **get** was a nice advantage of **find**. The function passed into **find** does a match on the entry at **probeIndex** and implements appropriate logic if a location is found where this key-value pair could be added.

The **-=** method on lines 39–52 uses a **find** to locate a matching key or an **Empty** entry. It then sends that index to a **foreach** that will replace the entry with **Removed** if a match was found.

The implementation of **iterator** is simpler with open addressing than with chaining because we do not have a 2D structure. We can make a single counter, we just have to have a bit of extra code that advances it to elements that store contents.

On the other hand, **growTable** is a bit more complex than before because we cannot simply insert into a **List**. Even when we grow the table, we still have to worry about probing as the new table is likely to still have collisions, even if there are no duplicate keys.

22.4   End of Chapter Material

22.4.1   Summary of Concepts

- Hash tables are an alternate implementation choice for the Map interface that can provide $O(1)$ performance for searching, adding, and removing.

- A hash table is an array of entries paired with a function, called the hash function, that can convert keys to integers in the range of valid indices for the array.

- The **Any** type includes a method called **hashCode** that will convert an instance of whatever type to an **Int**.

- The two main methods of going from an `Int` to one that is in the range of valid indices for the table are the division (which actually uses modulo) and multiplication methods.

- When two different keys map to the same index, it is called a collision. We discussed two different ways of dealing with collisions.

  - Chaining has each entry in the table store a list of the values that were mapped to that index.

  - Open addressing only stores one value per entry and uses probing to test other locations when entries are filled.

22.4.2 Exercises

1. You now potentially have many implementations of the Map ADT as we have discussed how it can be done with a sequence of tuples, a basic BST, an AVL-balanced BST, and now two different hash table implementations. Do performance testing for all of the implementations that you have to see which ones are fastest for different combinations of operations.

2. For the chaining implementation, there are valid arguments to use a `mutable.Buffer[(K, V)]` instead of a `List[(K, V)]` for the chaining. Modify the implementation from this chapter to do so, and compare the code for complexity as well as performance.

3. There is an art to making good implementation of `hashCode`. Do some research into how these work.

4. The chapter mentioned double hashing as a third approach to probing. You can implement this if you pass in a second function that will convert the key type to an `Int`. Try implementing this in the code.

22.4.3 Projects

Change one of your `Map` instances in your project to a `HashMap` that you have written.

Bibliography

[1] Gene M. Amdahl. Validity of the single processor approach to achieving large scale computing capabilities. In *Proceedings of the April 18-20, 1967, Spring Joint Computer Conference*, AFIPS '67 (Spring), pages 483–485, New York, NY, USA, 1967. ACM.

[2] J. Barnes and P. Hut. A hierarchical O(N log N) force-calculation algorithm. *Nature*, 324:446–449, December 1986.

[3] Rudolf Bayer and Edward M. McCreight. Organization and maintenance of large ordered indices. *Acta Inf.*, 1:173–189, 1972.

[4] Noam Chomsky. Three models for the description of language. *IRE Transactions on Information Theory*, 2:113–124, 1956.

[5] Martin Fowler. *Refactoring: Improving the Design of Existing Code*. Addison-Wesley, Boston, MA, USA, 1999.

[6] Erich Gamma, Richard Helm, Ralph E. Johnson, and John Vlissides. *Design Patterns: Elements of Reusable Object-Oriented Software*. Addison-Wesley, Reading, MA, 1995.

[7] D. Knuth. *Structured Programming with Go to Statements*, pages 257–321. Yourdon Press, Upper Saddle River, NJ, USA, 1979.

[8] Donald E. Knuth. *The Art of Computer Programming, Volume 3: (2Nd Ed.) Sorting and Searching*. Addison Wesley Longman Publishing Co., Inc., Redwood City, CA, USA, 1998.

[9] Mark C. Lewis and Lisa L. Lacher. *Introduction to Programming and Problem Solving Using Scala*. CRC Press, Taylor & Francis Group, 2016.

[10] Martin Odersky, Lex Spoon, and Bill Venners. *Programming in Scala: A Comprehensive Step-by-Step Guide, 3rd Edition*. Artima Incorporation, USA, 3rd edition, 2016.

[11] Robert Sedgewick. *Algorithms*. Addison-Wesley, Reading, Mass, 1983.

[12] Herb Sutter. The free lunch is over: A fundamental turn toward concurrency in software. *Dr. Dobbs Journal*, 30(3):202–210, 2005.

Index